Pure Mathematics
Volume one

D. Griffiths
Head of Upper School
The Bishop's Stortford High School

Harrap London

First published in Great Britain in 1984
by Harrap Limited
19–23 Ludgate Hill, London EC4M 7PD

© *David Griffiths* 1984

ISBN 0 245–54032–6

Typeset by Interprint Ltd. Malta
Printed & bound in Great Britain by
R.J. Acford, Chichester

Preface

The two volumes of this book together provide a course intended for students of single or double subject mathematics at A/S-level. I have aimed to cater for a wide range of syllabuses, and, although statistics is outside the scope of the book, three chapters on probability cover basic ideas (Chapter 22), probability distributions (Chapter 32) and Markov chains (Chapter 36).

The order of the chapters represents a suitable teaching order, although considerable variation is possible. Chapters 1 and 2 may seem particularly difficult if most of the ideas are unfamiliar, and these chapters could reasonably be postponed.

The use of informal numerical checks, applied on an *ad hoc* basis, is an essential part of any mathematician's technique, and in nearly every chapter it will be helpful to have a calculator within reach. By contrast, most of the various formal numerical methods are presented in a single chapter (Chapter 29); Simpson's rule, for example, seems much more at home here than in a chapter on integration.

The questions within the text are important, since they serve to consolidate, clarify or extend new ideas or techniques. Some questions are marked 'for discussion'; this indicates merely that there is no short answer which can be given at the back of the book; the reader does not need company in order to be able to attempt them.

The miscellaneous exercises at the ends of most chapters comprise past examination questions; they are all A-level except where marked 'O' or 'S'. I am grateful to the following examination boards for permission to reproduce these questions:

Associated Examining Board	AEB
University of Cambridge Local Examinations Syndicate	C
Joint Matriculation Board	JMB
University of London University Entrance and School Examinations Council	L
Oxford and Cambridge Schools Examination Board	O & C, MEI, SMP
Oxford Delegacy of Local Examinations	Ox
Southern Universities' Joint Board	SUJB
Welsh Joint Education Committee	W

Finally, I wish to express my thanks to my wife for undertaking the enormous task of reading the whole book, checking the exercises, and typing the final draft, and for making many valuable suggestions and corrections.

D.G.

Contents of Volume 1

Contents of Volume 2

Greek letters used in this book

	α	alpha
	β	beta
Γ	γ	gamma
Δ	δ	delta
	ε	epsilon
	θ	theta
	κ	kappa
Λ	λ	lambda
	μ	mu
	ν	nu
Π	π	pi
	ρ	rho
Σ	σ	sigma
Φ	ϕ	phi
	ψ	psi
Ω	ω	omega

1 Sets and the language of mathematics

1.1 Introduction

Mathematics is a precise subject and it requires a correspondingly precise language in which its ideas can be expressed and manipulated. In the first two chapters of this book, some of the fundamental concepts in mathematics are introduced, together with their associated terminology and notation. Mastery of these chapters at an early stage will certainly enhance the reader's understanding of the rest of the book; but, as is mentioned in the preface, this chapter and the next are harder than most of the rest of this volume, and they could reasonably be postponed.

1.2 Mathematical implication

Suppose p and q are statements which may be true or false. Then we write $p \Rightarrow q$ to mean *if* p *is true, then* q *must be true*. Note that
(1) $p \Rightarrow q$ is read as 'p implies q', but the word 'implies' here has a far stricter meaning than in normal usage, where it means little more than 'suggests'.
(2) $p \Rightarrow q$ tells us nothing about q if p is false. If $p \Rightarrow q$, and p is false, then q may be either true or false.
(3) $p \Leftarrow q$ is read as 'p is implied by q' and it means the same as $q \Rightarrow p$.
(4) $p \Leftrightarrow q$ means $p \Rightarrow q$ and $q \Rightarrow p$. It is read as 'p implies and is implied by q'. In this case p and q are either both true or both false.
(5) If we are required to prove a theorem of the form $p \Leftrightarrow q$, then there are *two* results to be proved: $p \Rightarrow q$ and $q \Rightarrow p$. The proof is therefore likely to be in two separate parts.

Qu.1 Decide whether the following are true or false:
(i) Peter is a sparrow $\Rightarrow$ Peter is a bird;
(ii) Paul is a bird $\Rightarrow$ Paul is a sparrow;
(iii) n is an even number $\Rightarrow$ n is divisible by 4;
(iv) n is an even number $\Leftarrow$ n is divisible by 4;
(v) n is a prime number $\Rightarrow$ n is odd;
(vi) n is odd $\Rightarrow$ n is prime;
(vii) n is a whole number $\Leftrightarrow$ n^2 is a whole number;
(viii) n is even $\Rightarrow$ $n(n+1)$ is even;
(ix) n is odd $\Rightarrow$ $n(n+1)$ is even;
(x) $n(n+1)$ is even $\Leftrightarrow$ n is odd;
(xi) a pentagon has equal sides $\Rightarrow$ it has equal angles;
(xii) a pentagon has equal angles $\Rightarrow$ it has equal sides;
(xiii) $xy=0 \Leftrightarrow x=0$ or $y=0$ or both;
(xiv) $xy \neq 0 \Leftrightarrow x \neq 0$ or $y \neq 0$ or both;
(xv) $xy=1 \Rightarrow x \neq 0$ and $y \neq 0$;

(xvi) $xy = x^2 \Rightarrow y = x$;

(xvii) $A \subseteq B \Leftrightarrow B' \subseteq A'$;

(xviii) $A \subseteq B \Leftrightarrow A \cap B' = \emptyset$;

(xix) $A \subseteq B \Leftrightarrow A \cap B = A \Leftrightarrow A \cup B = B$.

Other terminology is also used for mathematical implication. If $p \Rightarrow q$, we say that p is a *sufficient condition* for q, or that q is a *necessary condition* for p. (To avoid confusing these terms, remember that $p \Rightarrow q$ means 'p is sufficient to ensure that q is true' or that 'q necessarily follows from p'.) Then p *is necessary and sufficient for* q means that $q \Rightarrow p$ and $p \Rightarrow q$, i.e. that $p \Leftrightarrow q$. Another common expression is to say that p is true *if and only if* q is true; this also means $p \Leftrightarrow q$.

Qu.2 Decide whether the following are true or false.

(i) PQRS is a square is a sufficient condition for it to be a rectangle.

(ii) PQRS is a square is a necessary condition for it to be a rectangle.

(iii) A necessary and sufficient condition for n to be an odd number is that $\frac{1}{2}(n+1)$ is a whole number.

(iv) $x = y$ if and only if $x^2 = y^2$.

Definitions

(1) The *converse* of the statement '$p \Rightarrow q$' is the statement '$q \Rightarrow p$'. Clearly a statement may be true and its converse false. (See for example Qu.1 parts (iii) and (iv).)

(2) The *negation* of the statement p, denoted by p' or $\sim p$, is the exact opposite of p, and is true if and only if p is false. It is usually formed by inserting or removing the word 'not', although complicated sentences may need more subtle treatment. For example, the negation of '$x = 3$ and $y = 7$' is 'it is not the case that $x = 3$ and $y = 7$'. This negation might also be expressed as 'either $x \neq 3$ or $y \neq 7$ or both'. (See note below.)

Clearly the negation of a negation is the original statement; i.e. $\sim(\sim p) = p$.

Note: In mathematics, the word *or* is generally used inclusively, i.e. 'p or q' means 'p or q or both'; we do not need to add 'or both'.

Qu.3 Let P and Q be sets, and let p be the statement $x \in P$, and q the statement $x \in Q$. Verify (i.e. check that you agree) that 'p and q' corresponds to $x \in P \cap Q$.

(i) Membership of which set corresponds to

   (a) p or q;   (b) $\sim p$?

(ii) What can you say about P and Q if

   (a) $p \Rightarrow q$;   (b) $q \Rightarrow p$;   (c) $p \Leftrightarrow q$?

When giving a mathematical proof, care must be taken to make sure that the implication sign is not used the wrong way round, thereby giving the converse of the desired argument. This error often occurs when proving inequalities (which are dealt with fully in Chapter 16). For example, suppose we wish to prove that for all values of x, $x^2 + 1 \geqslant 2x$, and we argue thus:

$$x^2 + 1 \geqslant 2x \;\Rightarrow\; x^2 - 2x + 1 \geqslant 0$$
$$\Rightarrow\; (x - 1)^2 \geqslant 0;$$

but this is true for all values of x since no square is negative. Hence

$$x^2 + 1 \geqslant 2x \text{ for all } x.$$

This argument is completely invalid. We have started by *assuming* what we were trying to prove, and have ended up by showing something obvious! The argument can be corrected by reversing it thus:

No square is negative, so

$$(x - 1)^2 \geqslant 0$$
$$\Rightarrow\; x^2 - 2x + 1 \geqslant 0$$
$$\Rightarrow\; x^2 + 1 \geqslant 2x.$$

1.3 Two methods of proof

(1) *Counter-examples* Many mathematical statements or theorems include the words 'all', 'every' or 'any'. Even when these words do not appear, it may still be understood that the statement is a general statement. Thus, for example, when we say 'A triangle with two equal sides has two equal angles', we mean *every* such triangle, and are not referring to one particular triangle. Such statements, if true, may be difficult to prove. (Merely providing numerous examples proves nothing.) But if such a statement is *false*, we can prove that it is false by a single example known as a *counter-example*.

Example 1
Prove that the statement 'all odd numbers are prime' is false.
Solution
15 is an odd number, but it is not prime, being 3×5.

We saw above that $p \Rightarrow q$ means that if p is true, then q must be true. It follows that to disprove the statement $p \Rightarrow q$, we must find an example in which p is true and q is false. This is in effect what we have done in Example 1. The statement 'all odd numbers are prime' may be written 'n is odd $\Rightarrow n$ is prime'; our counter-example gives a value of n which is odd but not prime.

Qu.4 (For discussion.) Find counter-examples to disprove those parts of Qus. 1 and 2 that are false.

(2) *Contradiction* Proof by contradiction is a method of proving that a statement p is true. We start by assuming that p is *false* and show that this leads to either a contradiction or mere nonsense.

Example 2
Prove that there are infinitely many prime numbers.
Solution
Suppose on the contrary that there are only a finite number of primes.

Let k be the product of these primes, i.e.

$$k = 2 \times 3 \times 5 \times 7 \times \ldots \times l$$

where l is the last prime. Now the number $k+1$ is not divisible by any of these primes, since they are all factors of k. So either $k+1$ is itself a prime not in the list, or it has a prime factor not in the list. In either case this contradicts the assumption that 2, 3, 5, ..., l was a complete list of primes. Our original assumption (that the number of primes is finite) must therefore be false, and the required result is proved.

The method of proof by contradiction is also known as *reductio ad absurdum* (or simply R.A.A.).

Qu.5 Use the method of proof by contradiction to show that the quadratic equation $ax^2 + bx + c = 0$ cannot have a positive solution if $a > 0$, $b > 0$ and $c > 0$. (Make sure you begin by stating your assumption clearly.)

If $p \Rightarrow q$ then p cannot be true if q is false, i.e. $q' \Rightarrow p'$. Similarly if $q' \Rightarrow p'$, then if p is true (so that p' is false), q' is false (i.e. q is true), i.e. $p \Rightarrow q$. We have thus shown that

$(p \Rightarrow q)$ is equivalent to $(q' \Rightarrow p')$

or $(p \Rightarrow q) \Leftrightarrow (q' \Rightarrow p')$.

Definition
The statement $q' \Rightarrow p'$ is known as the *contrapositive* of the statement $p \Rightarrow q$.

Qu.6 Write in words the contrapositive of the statement 'If Peter is a sparrow, then he is a bird'. [*Hint*: first write the given statement using an implication sign.]

Proof by contradiction often takes the form of proving $q' \Rightarrow p'$ in order to show that $p \Rightarrow q$.

Example 3
If $y = 2 - 3x$, show that no two different values of x can give the same value for y.
Solution
We wish to show that if a and b are two different values of x then the corresponding y-values, $2 - 3a$ and $2 - 3b$, are different. We begin by assuming that for some a and b, the values of $2 - 3a$ and $2 - 3b$ are equal. (This is q' in the discussion above.)

$$2 - 3a = 2 - 3b \;\Rightarrow\; 3a = 3b$$
$$\Rightarrow\; a = b.$$

So (contrapositively)

$$a \neq b \;\Rightarrow\; 2 - 3a \neq 2 - 3b$$

as required.

1. In the following truth table (in which 1 means true and 0 false) the truth values of (p and q), and of $p \Rightarrow q$ are shown for each of the four possible combinations of true and false for p and q. (For example, $p \Rightarrow q$ is false only when p is true and q is false.) Complete the table.

p	q	p and q	$p \Rightarrow q$	p or q	q'	p and q'	p' or q
1	1	1	1				
1	0	0	0				
0	1	0	1				
0	0	0	1				

Check from the table that ($p \Rightarrow q$) is true if and only if (p' or q) is true.

2. A true statement about quadrilaterals is 'if it is a square, then its diagonals are perpendicular'.
   (i) Write this using the implication sign.
   (ii) State the converse in words.
   (iii) Give a counter-example to show that the converse is false.

3. Use the method of question 1 to show that
$$(p \Rightarrow q) \Leftrightarrow (q' \Rightarrow p').$$

4. (i) Show with the aid of a Venn Diagram that
$$P \subseteq Q \Leftrightarrow Q' \subseteq P'.$$
   (Compare this with the result in question 3.)
   (ii) Use Venn Diagrams to show *De Morgan's Laws:*
      (a) $(A \cap B)' = A' \cup B'$;
      (b) $(A \cup B)' = A' \cap B'$.
   (iii) Complete De Morgan's Laws for statements:
      (a) (p and q)$' = \ldots$;
      (b) (p or q)$' = \ldots$.

5. Let AB be the diameter of a circle. Use the well-known theorem
$$P \text{ lies on the circle} \Rightarrow \angle APB = 90°$$
and the method of contradiction to prove the converse of the theorem. [*Hint:* start by assuming that P does not lie on the circle, considering separately the cases in which P is inside and outside. Join BP and let it cut the circle at Q. Show clearly that $\triangle APQ$ has two right angles.]

6. Let p, q and r be the statements: n is a perfect square; n is even; and n is divisible by 4, respectively. Determine whether the following are true or false. Give counter-examples where appropriate.
   (i) $p \Rightarrow q$; (ii) $q \Rightarrow p$; (iii) (p and q) $\Rightarrow r$;
   (iv) $r \Rightarrow (p$ and $q)$; (v) $r \Rightarrow (p$ or $q)$.

7. Give a counter-example to show that $a > b \nRightarrow ac > bc$.

8. Disprove the claim that if n is a positive whole number, then $n^2 + n + 11$ is either prime or divisible by 11.

9. (i) Is this argument valid?
$$\begin{aligned} n = 1 &\Rightarrow n + 1 = 2 \\ &\Rightarrow (n+1)^2 = 4 \\ &\Rightarrow n^2 + 2n + 1 = 4 \end{aligned}$$

$$\Rightarrow n^2 + 2n - 3 = 0$$
$$\Rightarrow (n+3)(n-1) = 0$$
$$\Rightarrow n = 1 \text{ or } -3.$$

(ii) Is it true that $n = 1$ or $-3 \Rightarrow n = 1$? Which of the implication signs in (i) is not reversible?

(iii) Is it true that $n = -3 \Rightarrow (n+3)(n-1) = 0$?

10. Two sets P and Q are defined to be *equal* if and only if

$$x \in P \Leftrightarrow x \in Q.$$

(i) Write out this definition in non-mathematical language.

(ii) Are the sets $\{a, a, b, c\}$ and $\{a, b, c, c\}$ equal?

(iii) Give a similar formal definition of 'P is a subset of Q'.

11. The *Power Set* of a set S, denoted by $P(S)$, is the set of all subsets of S, including $\emptyset$ and S itself.

(i) List all members of $P(S)$ where $S = \{a, b, c\}$.

(ii) Find $n(P(S))$ when $n(S) =$
    (a) 2;   (b) 4;   (c) k.

12. Two sets A and B are said to be *disjoint* if $A \cap B = \emptyset$. Which of the following conditions are necessary and sufficient for A and B to be disjoint?

  (i) $x \in A \Rightarrow x \notin B$;   (ii) $A = B'$;

 (iii) $B' \subseteq A$;   (iv) $B \subseteq A'$;

  (v) $A \cap B' = A$;   (vi) $A' \cup B' = \mathscr{E}$.

13. Three or more sets $A, B, C, \ldots$ are said to be *pair-wise* disjoint if no two of them intersect. Draw a Venn Diagram to show that
$$A \cap B \cap C = \emptyset \nRightarrow A, B, C \text{ are pair-wise disjoint.}$$

1.4 Sets of numbers

Definitions

(1) An *integer* is a whole number (positive, negative or zero). The set of all integers is denoted by Z.

(2) The set of positive integers (or *natural numbers*) is denoted by Z^+ or N. Thus

$$Z^+ = \{1, 2, 3, 4, \ldots\}.$$

(3) A *rational number* is any number that can be expressed in the form $\frac{p}{q}$ where p and q are integers; i.e. it is any number that can be written as an ordinary fraction. In particular, any integer n is rational, since it may be written as $\frac{n}{1}$. The set of all rationals is denoted by Q, and the set of positive rationals by Q^+.

(4) A *real number* is any number at all. (Of course, this definition is not really adequate: it assumes that we already know what numbers exist. A precise definition of the set of real numbers is beyond the scope of this book. Note that 'non-real' numbers are introduced in Chapter 31, but they are not numbers in the sense in which the word is normally used.) The set of reals is denoted by R and the positive reals by R^+.

The set of real numbers can be represented by a single axis or number line, called the *real line* (Fig. 1.1). To every point of the line there corresponds a real number, and vice versa.

(5) A (real) number which is not rational is said to be *irrational*.

Figure 1.1

The question that immediately arises is that of whether irrational numbers actually exist. It would be quite reasonable to believe, on first thoughts, that every real number can be written as a fraction; but, as we now see, this is not the case.

Theorem 1.1

Some real numbers are irrational.

(Note: the theorem will be proved if we can find just one irrational number and show that it is irrational. Two proofs are given below; the second is renowned for its elegance; it is one of the best examples of *reductio ad absurdum*.)

Proof 1

We show first that when a rational number is written as a decimal, it either recurs or terminates. Consider $\frac{p}{q}$, where $p \in Z$ and $q \in Z^+$, and suppose that we turn it into a decimal by division. If the decimal does not terminate, then at each stage there is a 'remainder' which is a positive integer less than q. Once we are past the decimal point in the dividend, the quotient will begin to recur as soon as a 'remainder' occurs for the second time. (Try it with $\frac{2}{7}$, say.) Since there are at most $q-1$ remainders, the decimal must begin to recur on or before the qth decimal place.

It is now a simple matter to write down a decimal which neither terminates nor recurs, and which must therefore be irrational. An example would be

$$0{\cdot}101001000100001\ldots$$

(where the number of zeros between the ones is ever increasing).

Proof 2

We show, by the method of contradiction, that the number $\sqrt{2}$ is irrational. Suppose that $\sqrt{2}$ is rational, then we may assume that $\sqrt{2} = \frac{p}{q}$ where p and q have no common factor, i.e. the fraction will not cancel down any further. Then

$$2 = \frac{p^2}{q^2}$$

$\Rightarrow p^2 = 2q^2$ (1)
$\Rightarrow p^2$ is even
$\Rightarrow p$ is even
$\Rightarrow p = 2r$ for some $r \in Z$
$\Rightarrow (2r)^2 = 2q^2$, substituting into (1),
$\Rightarrow 2r^2 = q^2$
$\Rightarrow q^2$ is even
$\Rightarrow q$ is even.

We have thus shown that p and q are both even. But this contradicts the (reasonable) supposition that they have no common factor. It follows that our assumption that $\sqrt{2}$ is rational is false. So $\sqrt{2}$ is irrational.

Qu.7 Use the method of proof 2 to show that $\sqrt{3}$ is irrational. [*Hint*: you will not need to consider whether p is even, but whether it is divisible by 3.]

Qu.8 Show that the recurring decimal $0 \cdot 123123123 \ldots$ satisfies the equation $1000x - 123 = x$. Hence write down x as a vulgar fraction. (Do not bother to cancel.)

Qu.9 Use the method of Qu.8 to write the following as vulgar fractions.
  (i) $0 \cdot 616161 \ldots$;   (ii) $0 \cdot 34353435 \ldots$;
 (iii) $0 \cdot 100100100 \ldots$;
 (iv) $0 \cdot 351494949 \ldots$ (Consider $100\,000x - 1000x$).

We may summarize the relationships between the various sets of numbers that we have met thus:

$$Z^+ \subset Z \subset Q \subset R.$$

The following definition may be omitted at a first reading. (Polynomial equations are discussed in Chapter 10.)

Definition

An *algebraic number* is any number which is the root of some polynomial equation with integral coefficients. Certainly any rational $\frac{p}{q}$ is algebraic, since it is a root of $qx = p$; and many irrationals are algebraic. (For example, $\sqrt{2}$ is a root of $x^2 - 2 = 0$.) Denoting the set of algebraic numbers by A, we have

$$Q \subset A \subset R.$$

A number which is not algebraic is said to be *transcendental*. The numbers π and e (see Chapter 15) are transcendental, but it is not easy to prove that they are.

Notation

(1) When defining a set, a colon is used to mean 'such that'. (Sometimes a vertical line is used.) For example,

$$S = \{x \in Z^+ : x < 5\} \quad \text{reads}$$

'S is the set of all x belonging to Z^+ such that x is less than 5'. S is therefore the set $\{1, 2, 3, 4\}$.

(2) In other contexts, s.t. is a common abbreviation for 'such that'.

(3) A set of the form $\{x \in R : a \leqslant x \leqslant b\}$ is called a *closed* interval, and may be denoted by $[a, b]$. If, however, the end-points are not included in the set, then the set is an *open* interval and round brackets are used, so that

$$(a, b) = \{x \in R : a < x < b\}.$$

This notation can be extended so that, for example,

$$(a,b] = \{x \in R : a < x \leqslant b\}.$$

1.5 The symbols $\forall, \exists, \infty$

The symbol $\forall$ means 'for all' or 'for each', and $\exists$ means 'there exists'. Many sentences contain both symbols, and 'word order' can be vital. Consider, for example, the sentence

$$\forall x \in Z^+, \exists y \in R \text{ s.t. } y^2 = x. \tag{1}$$

This is perfectly true, since it expresses the fact that every positive integer x has a real square root y. (Actually there are two square roots.) But consider

$$\exists y \in R \text{ s.t. } \forall x \in Z^+, y^2 = x. \tag{2}$$

This looks similar to (1) above, but it states that there is some real number y which is a square root of every positive integer. Clearly no such number exists and (2) is false.

The following example shows how ordinary 'everyday' language can be ambiguous when we use it to deal with mathematical ideas.

Example 4
Ten people are standing in a bus queue. Every minute, a bus arrives with room for three people, and the front three get on; but at the same time, three more people join the back of the queue. Assuming that this process continues permanently, is it true to say 'Eventually, everyone gets on a bus'?

Solution
We cannot answer 'yes', because there are always ten people in the queue still waiting for a bus, so at no time has 'everyone' got on a bus; and we cannot answer 'no', because we cannot answer the challenge to name someone who does not get on a bus. The given statement has two meanings which are clarified below.

This problem may seem trivial or artificial, but it is analogous to many problems that arise in elementary analysis. In this particular case, the difficulty arises from the careless use of the words 'eventually' (which we take to mean 'after an infinite number of minutes') and 'everyone' (which here means 'an infinite number of people').

We may clarify the problem by calling the people in the queue P_1, P_2, P_3, ... (in order of queuing) and letting the number of minutes that have elapsed be t. We can then make the two true statements:

(1) Given $t \in Z^+$, $\exists n$ s.t. P_n has not got on a bus at time t.
(2) Given $n \in Z^+$, $\exists t$ s.t. P_n has got on a bus by time t.

Statement (1) says that *given a time*, we can then find a person who has not yet got on a bus; and statement (2) says that if a *person* is specified first, then there is a time by which he will have caught a bus.

We shall on various occasions in this book wish to use the word 'infinite' and its symbol ∞ (infinity). As we have seen in Example 4, great care is needed in dealing with problems involving 'infinity', and the reader is warned to use 'infinite' or ∞ only in the specific ways in which they are introduced in the text. The following definition gives one way in which the word may be used.

Definition

A set S is said to be an *infinite set* if and only if $\forall n \in Z^+$, S has at least n members.

Qu.10 (For discussion.) (i) Verify that the definition above agrees with your intuitive idea of an infinite set.
(ii) If you think that the definition sounds unnecessarily complicated, try to formulate a simpler definition.

Exercise 1b

1. Write out in full the sets
   (i) $\{x \in Z^+ : x < 6\}$;
   (ii) $\{x \in Z : x \leqslant 10\} \cap \{x \in Q : x > 3\}$.
2. Draw a Venn Diagram to show the relationship between the sets Z^+, Z, Q^+, Q, R^+ and R. (Take R as the universal set.) Put one representative number in each region of the diagram.
3. How does the proof that $\sqrt{2}$ is irrational break down for $\sqrt{4}$?
4. Show that
$$x \text{ is irrational} \Rightarrow \sqrt{x} \text{ is irrational.}$$
   [*Hint*: prove the contrapositive.]
5. Let u and v be rational and x and y be irrational. Determine whether the following must be rational or must be irrational or may be either. Give proofs or examples as appropriate.
   (i) $u + v$;   (ii) uv;   (iii) $u + x$;
   (iv) ux;   (v) $x + y$;   (vi) xy.
6. Are these statements true?
   (i) $\forall n \in Z^+$, $4n^2 + 1$ is prime.
   (ii) $\exists n \in Z^+$ s.t. $4n^2 + 1$ is prime.
   (iii) $\exists n \in Z^+$ s.t. $4n^2 + 1$ is not prime.
   Which two statements are mutual negations?
7. (i) Let $P(n)$ be a statement about the natural number n, and suppose that $P(1)$ is true, and that $\forall k \in N$, $P(k) \Rightarrow P(k+1)$. Show that $P(3)$ is true. Which is the smallest natural number s for which $P(s)$ is false?
   (ii) If, in part (i), $P(1)$ is true and $\forall k \in N$, $P(k) \Rightarrow P(k+2)$ what can be said about the set T where
$$T = \{n \in N : P(n) \text{ is true}\}?$$
8. Let $S = [0, 1)$. Explain the following statements briefly in your own words, saying whether or not they are true.
   (i) $\forall x \in S$, $\exists n \in Z^+$ s.t. $x^n < 10^{-6}$;
   (ii) $\forall n \in Z^+$, $\exists x \in S$ s.t. $x^n > 10^{-6}$.
9. Show that if, in question 8, $S = [0, 1]$, then one part of the question is false, and the other trivial.
10. Let $S = [0, 2)$.
    (i) What is the greatest member of S?
    (ii) Write down any real number u such that $\forall x \in S$, $u \geqslant x$. (Such a number is called an *upper bound* for S.)

(iii) Let U be the set of upper bounds for S. What is the least member of U?

(iv) Give an example of a subset of R which has no upper bound.

(v) Let T be a non-empty subset of R, and let V be the set of upper bounds for T. If $V \neq \emptyset$, can we be sure that V has a least member? Can this least member belong to T?

11. Prove that, given an irrational r, and a real $\varepsilon > 0$, then $\exists q \in Q$ with $0 < r - q < \varepsilon$. [*Hint*: the problem is to find a rational number q which is less than the given irrational r, but which is 'very close' to r. It may help to consider r in decimal form, and to remember that a terminating decimal is rational.]

12. (Hard) (i) Show that between any two rationals there is an irrational, and between any two irrationals there is a rational.

(ii) Explain why we cannot deduce from (i) that rationals and irrationals occur alternately on the real line.

13. One feature of integer arithmetic, as distinct from arithmetic with rational or real numbers, is that division can often not be done, i.e. a division may leave a remainder. The idea of divisibility leads on to the ideas of H.C.F. and L.C.M. The flow diagram below illustrates the *Euclidean Algorithm* for integers.

Figure 1.2

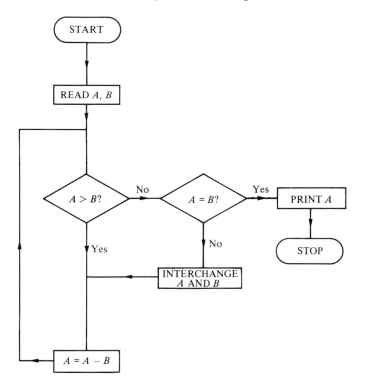

(i) Work through the flow diagram where the input numbers (A and B) are respectively

(a) 72 and 50;   (b) 83 and 16;   (c) 144 and 60.

(ii) What is the relationship of the output to the two input values?

(iii) (For discussion.) Explain how and why the algorithm works.

1.6 The Cartesian† product of sets

Definition

Let S and T be two sets. The *Cartesian product* of S and T, written $S \times T$ (and pronounced 'S cross T'), is the set of ordered pairs (s, t) where $s \in S$ and $t \in T$. (An ordered pair is a pair where the order matters, i.e. where (a, b) and (b, a) are to be regarded as different, unless $a = b$.)

Suppose, for example, that $S = \{a, b, c\}$ and $T = \{x, y\}$, then

$$S \times T = \{(a, x), (b, x), (c, x), (a, y), (b, y), (c, y)\}.$$

Qu.11 With S and T as above, list the members of (i) $T \times S$; (ii) $T \times T$.

The Cartesian product of three or more sets is defined similarly; thus $S \times T \times U$ is the set of all ordered triples (s, t, u) with $s \in S$, $t \in T$ and $u \in U$.

Of particular importance is the set $R \times R$, also denoted by R^2. If we take x- and y-axes in the usual way, then $R \times R$ may be represented by the set of all points in the plane. Indeed, the ordinary coordinates with which we are familiar are known as *rectangular Cartesian coordinates*—rectangular, because the axes are perpendicular. Similarly, the members of $R \times R \times R$ (or R^3) correspond to points in three-dimensional space whose coordinates may be referred to three mutually perpendicular axes. These axes are usually called the x-, y- and z-axes, and are conventionally orientated so that they form a *right-handed set*. This means that the directions of the positive x-, y- and z-axes are given by the directions of the thumb, index finger and middle finger *in that order* of a right hand, when those three fingers are pointed so as to be mutually perpendicular (Fig. 1.3). A further convention is that the z-axis points upwards. The x- and y-axes are then usually drawn as in Fig. 1.4a or as in Fig. 1.4b.

Figure 1.3

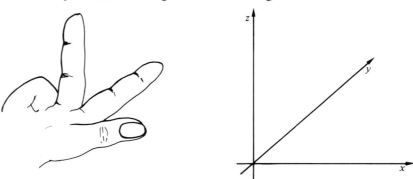

Qu.12 Check that Figs. 1.4a and 1.4b each form right-handed sets.

Qu.13 (For discussion.) (i) Verify that if three mutually perpendicular axes (x, y and z) do not form a right-handed set, then they form a left-handed set.

(ii) Verify that if the direction of one axis is exactly reversed, then so is the orientation (i.e. the left/right-handedness).

(iii) Verify that interchanging two axes changes the orientation.

†The adjective associated with the mathematician and philosopher Descartes.

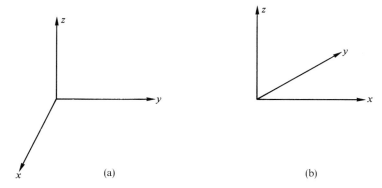

(a)   (b)

(iv) Verify that if the three axes are interchanged cyclicly (see Fig. 1.5) then the orientation is unchanged.

Figure 1.5

1.7 Relations

A relation from a set S to a set T is a rule connecting members of S with members of T. For example, if S is the set of women in a certain town, and T is the set of men, then possible relations would be: '... is the mother of ...', '... is taller than ...' and '... has different coloured eyes from ...'. If we let R stand for the first of these examples, then we can write sRt to mean 's is the mother of t', and $s\not Rt$ to mean 's is not the mother of t'. If R is a relation from S to T, then

$$\forall\ s \in S, t \in T, \text{ either } sRt \text{ or } s\not Rt$$

i.e. for every pair chosen from S and T, sRt is either true or false.

Qu.14   Let $S = \{1, 2, 6, 20\}$ and $T = \{1, 2, 4, 5, 10\}$, and let R be the relation 'is greater than the square of'. List the members, (s, t), of $S \times T$ which satisfy sRt.

If R is a relation from S to T, then the *inverse relation* from T to S is denoted by R^{-1} and defined by

$$tR^{-1}s \text{ if and only if } sRt.$$

Inverse relations may need careful wording. Suppose A is a set of people, and R is the relation 'is the mother of' defined from A to itself. We cannot express R^{-1} as 'is the son or daughter of' (why not?); we must choose instead between the equally inelegant 'has as mother' and 'is the son or daughter of the woman'.

1.8 Equivalence relations

Certain types of relation from a set to itself are of particular interest. We therefore make the following definition.

Definition
Let R be a relation from a set S to itself. Then R is said to be an *equivalence relation* on S if and only if $\forall\ a, b, c \in S$,

(i) aRa (the reflexive rule);
(ii) $aRb \Rightarrow bRa$ (the symmetric rule);
(iii) $(aRb$ and $bRc) \Rightarrow aRc$ (the transitive rule).

Consider, for example, the relation $\leqslant$ (is less than or equal to) defined on the set Z:
(i) It is reflexive, since any member of Z satisfies $a \leqslant a$.
(ii) It is *not* symmetric, since if $a \leqslant b$, we cannot deduce that $b \leqslant a$.
(iii) It is transitive, since if $a \leqslant b$ and $b \leqslant c$, we can deduce that $a \leqslant c$.
The failure of the symmetric rule means that $\leqslant$ is not an equivalence relation.

Qu.15   The following relations are defined either on Z or on A, the set of all people living in the U.K. Determine which are equivalence relations, and specify for the others which rule or rules are not satisfied. Give particular counter-examples where the failure of a rule is not obvious.
  (i) is taller than;
 (ii) was born in the same year as;
(iii) is less than;
(iv) has the same parity as (Two numbers have the same parity if they are both odd or both even.);
  (v) has the same mother as;
 (vi) is the parent of;
(vii) has a factor greater than 1 in common with;
(viii) has no factor except 1 in common with;
 (ix) is equal to.

If R is an equivalence relation on the set S, and $x \in S$, then the set of all elements of S which are 'equivalent to' or 'related to' x under R is denoted by $[x]$, i.e.

$$[x] = \{y \in S : yRx\}.$$

The set $[x]$ is known as the *equivalence class* of x. Clearly $x \in [x]$ by the reflexive rule.

Qu.16   Prove that any two members of $[x]$ are 'related'. (Write out your proof carefully, stating when any of the three rules is used.)

Theorem 1.2
Let R be an equivalence relation defined on S, and let $x, y \in S$. Then either $[x] = [y]$, or $[x]$ and $[y]$ are disjoint (i.e. $[x] \cap [y] = \emptyset$).
Proof
Suppose $[x]$ and $[y]$ are not disjoint; then $\exists z \in [x] \cap [y]$.

Then
$$zRx \text{ and } zRy,$$
$$\Rightarrow xRz \text{ and } zRy \quad \text{(symmetric rule)}$$
$$\Rightarrow xRy \quad \text{(transitive rule)}.$$

Now choose $a \in [x]$; then we have

$$aRx \text{ and } xRy$$

$$\Rightarrow aRy \quad \text{(transitive rule)}$$
$$\Rightarrow a\in[y].$$

Thus
$$a\in[x] \Rightarrow a\in[y]$$

i.e.
$$[x]\subseteq[y].$$

Similarly
$$[y]\subseteq[x],$$

and so
$$[x]=[y].$$

Therefore $[x]$ and $[y]$ are either disjoint or equal.

The proof above has a number of features which are worth noting:
(1) The main part of the proof is concerned with showing that two sets ($[x]$ and $[y]$) are equal. A common method of showing that two sets A and B are equal is to show that $A\subseteq B$ and $B\subseteq A$. This is done here.
(2) The usual method of proving that $A\subseteq B$ is to show that a typical member of A also belongs to B. This explains why, in the proof above, we say 'Now choose a $\in[x]$'.
(3) The word 'similarly' is a great time-saver when part of a proof would be almost an exact copy of a previous part.

Qu.17   Write out in full the part of the proof above which is covered by the word 'similarly'. Start with the line 'Now choose $b\in[y]$'.
Qu.18   Let $S=\{x\in Z^+ : x\leqslant 20\}$ and define a relation R on S by

$$aRb \iff \frac{a-b}{5}\in Z.$$

(i) Show that R is an equivalence relation.
(ii) List the members of $[2]$.
(iii) List the other equivalence classes, writing them as subsets of S as in (ii).
(iv) Check that the theorem above holds.

Definition
Let S be a set, and let $S_1, S_2, S_3 \ldots S_n$ be n subsets of S such that

(i) $i\neq j \Rightarrow S_i\cap S_j=\emptyset$

(which says that no two of the subsets overlap—unless, of course, they are the same subset);

and (ii) $\forall x\in S, \exists i$ s.t. $x\in S_i$

(which says that every element of S belongs to *some* subset; i.e. the subsets together cover S); then the set of subsets $S_1, S_2, \ldots S_n$ is called a *partition* of S.

The idea of a partition is a simple one and may be illustrated on a Venn Diagram.

In Fig. 1.6, the six sets $A, B \ldots F$ form a partition of S since they do not overlap, and together they cover S.

Figure 1.6

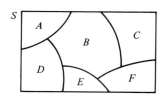

Qu.19 Let S be the set of pupils at a secondary school. Which of the following sets of subsets might not partition S. Give reasons for your answers.
  (i) Boys; girls.
  (ii) Those aged 11 or under; those aged 12; ...; those aged 17; those aged 18 or over.
  (iii) Those who play football; those who play cricket; those who play neither game.
  (iv) Those who passed at least one O-level; those who have failed all their O-levels.

The previous theorem showed that distinct equivalence classes do not overlap. In addition, we know that every element of S belongs to *some* equivalence class, since $\forall x \in S$, $x \in [x]$. It follows that any equivalence relation on S gives rise to a partition of S, namely the set of equivalence classes. The following question shows how a partition leads naturally to an equivalence relation.

Qu.20 Let $P = \{S_1, S_2, \dots S_n\}$ be a partition of a set S. Show that R is an equivalence relation on S, where R is defined by
  $aRb \Leftrightarrow a$ and b belong to the same member of P.
Verify that $S_1, S_2, \dots S_n$ are the equivalence classes.

1.9 Modulo arithmetic

When arithmetic is performed modulo 5, for example, each number is replaced by its remainder after division by 5. Thus 8 is replaced by 3, 11 by 1 and 35 by 0. This means that only the five numbers 0, 1, 2, 3, 4 ever appear; and it leads to strange-looking arithmetic such as

$$4 + 3 \equiv 2 \ (\text{mod } 5)$$

and
$$3 \times 3 \equiv 4 \ (\text{mod } 5).$$

The symbol $\equiv$ is used instead of $=$; it is read 'is congruent to'.

Qu.21 Evaluate the following in modulo 5 arithmetic:
(i) 3×4; (ii) $4 \times (3+4)$; (iii) $(2+3) \times (3+3)$.
Qu.22 Evaluate these mod 6:
(i) 5×4; (ii) $(5+3) \times 3$; (iii) 3^4; (iv) 2^5.

Counting downwards in modulo 5 gives the continuing sequence

$$4, 3, 2, 1, 0, 4, 3, 2, 1, 0, 4, \dots,$$

so we can replace negative numbers by one of these five digits; thus

SETS AND THE LANGUAGE OF MATHEMATICS

$$-1 \text{ is congruent to } 4$$
$$-2 \text{ is congruent to } 3$$
$$-3 \text{ is congruent to } 2$$
$$-4 \text{ is congruent to } 1$$
$$-5 \text{ is congruent to } 0$$
$$-6 \text{ is congruent to } 4$$
$$\text{etc.}$$

Qu.23 Evaluate
(i) 57^2 mod 59; (ii) 57^3 mod 59.
[*Hint*: $57 \equiv -2$ mod 59.]

Qu.24 (For discussion.) Re-examine Qu.18 with R applied to the whole of Z, and explain the relationship with modulo 5 arithmetic.

Qu.18 (extended as in Qu.24) suggests that when we use, say, the digit 2 in mod 5 arithmetic, it really represents the whole of the equivalence class [2]. We could equally well use 7 or 12 or -3 as a 'representative' of this class. A fundamental question now emerges: when adding or multiplying in mod 5 (or in any other modulo), does it matter which representatives of the classes we use? Fortunately, as we see in the theorem below and Qu.25, it does not matter; but if it did, then modulo arithmetic would lose all its elegance. We should have to make sure at each stage of any calculation that we were using the 'correct' representatives, and the two ways of evaluating even a simple problem such as $(2 \times 3) \times 4$ mod 5, as shown below, might lead to different answers.

$$
\begin{array}{ll}
(2 \times 3) \times 4 & \qquad (2 \times 3) \times 4 \\
\downarrow & \qquad \qquad \downarrow \\
6 \times 4 & \qquad \quad 6 \times 4 \\
\downarrow \text{ Reduce mod 5} & \qquad \quad \downarrow \\
1 \times 4 & \qquad \qquad 24 \\
\downarrow & \qquad \qquad \downarrow \text{ Reduce mod 5} \\
4 & \qquad \qquad 4
\end{array}
$$

Theorem 1.3

If $a \equiv c$ mod 5 and $b \equiv d$ mod 5, then $ab \equiv cd$ mod 5.
(This theorem says, in effect, that if alternative representatives—c and d—of two equivalence classes are used in place of a and b in a multiplication, then the answer cd will be in the same class as ab. Thus it does not matter which members of classes are used for multiplication.)

Proof

$$a \equiv c \text{ mod 5 and } b \equiv d \text{ mod 5}$$

$$\Rightarrow \exists \, m, \, n \in Z \text{ s.t. } a = c + 5m \text{ and } b = d + 5n.$$

So
$$ab = (c + 5m)(d + 5n)$$
$$= cd + 5(md + nc + 5mn).$$

But $\qquad md + nc + 5mn$ is an integer, k say.

So $\qquad ab = cd + 5k$ for $k \in Z$,

i.e. $\qquad ab \equiv cd$ mod 5.

Qu.25 Prove the corresponding result for *addition* mod 5.

Exercise 1c
1. For each of the following, draw a pair of axes (x and y) in the usual way, and illustrate the given subsets of $R \times R$. (For (v), shade the required region, and use the convention that a boundary which is not included is indicated by a dotted line.)
   (i) $Z \times Z$;   (ii) $R \times Z$;

   (iii) $Z \times R$;   (iv) $(Z^+ \times R) \cup (R \times Z^+)$;

   (v) $(1, 2] \times [0, 1]$;   (vi) $[2, 5] \times \{3\}$.

2. One of these two statements is not correct; decide which, and correct it.
   (i) If P and Q are finite sets, then
   $$n(P \times Q) = n(P).n(Q).$$
   (ii) If either (or both) of P and Q is infinite, then $P \times Q$ is infinite.

3. (A tetrahedron is a solid with four vertices and four plane faces.) Six metal rods, each of length 10 cm, are to be joined to form a tetrahedral framework $OABC$. Show, with the aid of suitable diagrams, that once the positions of O, A and B have been fixed in space, then there are two possible positions for C. Verify that the directions OA, OB and OC (in that order) form a non-perpendicular right- or left-handed set according to which of the two positions for C is chosen.

4. (i) In how many ways can a set of three elements be partitioned? (Include the case in which there is a single subset equal to the original set.)
   (ii) Repeat for four elements.

5. Give the inverses of the following relations which are defined from the given set to itself.
   (i) Z; is greater than.
   (ii) Z^+; is a factor of.
   (iii) All people; is the husband of.
   (iv) R; is the square of.

6. State whether each of the following relations is (a) reflexive; (b) symmetric; (c) transitive. The set on which each is defined is given.
   (i) The set of all statements; $\Rightarrow$.
   (ii) The set of all statements; $\Leftrightarrow$.
   (iii) R; $\neq$.
   (iv) R; differs by less than $1/10$ from.
   (v) Z^+; is a multiple of.
   (vi) $\{1, 2, 3, 4\}$; R defined by
   $$aRb \Leftrightarrow \max (a, b) \leqslant 2$$
   where max (a, b) means the larger of a and b, or their common value if they are equal.

7. Criticize the following argument which seems to suggest that any relation which is symmetric and transitive must also be reflexive.
   Let $x \in S$, and choose some $y \in S$ s.t. xRy.

   Then $xRy \Rightarrow yRx$ (symmetric rule).

   But xRy and $yRx \Rightarrow xRx$ (transitive rule).

Hence $\forall x \in S,\ xRx$.

8. In this question, r, s and t are short for reflexive, symmetric and transitive. Give examples of relations, stating the sets on which they are defined, which are
   (i) r and s but not t;   (ii) r and t but not s;
   (iii) s and t but not r;   (iv) r only;
   (v) s only;   (vi) t only;
   (vii) neither r nor s nor t.

9. Evaluate $1 \times 2 \times 3 \times 4 \times \ldots \times 12 \bmod 13$. (You should 'reduce mod 13' at each stage rather than at the end.)

10. Find the remainder when the following are divided by 7:
    (i) 6^{20};   (ii) 2^{50};
    (iii) $1000 \times 999 \times 998 \times \ldots 3 \times 2 \times 1$ where all multiples of 7 are omitted.

11. (i) If $a,\ b \in Z$, show that $a^2 + b^2$ cannot be congruent to 3 mod 4. (Consider the possible values of a^2 and b^2 mod 4.)
    (ii) By considering the possible values of cubes mod 7, show that $1\,000\,002$ cannot be expressed in the form $a^3 + b^3$ where $a,\ b \in Z$.

1.10 Binary operations

Given a set S, a *binary operation*, $*$, on S is a rule for combining two elements of S in a particular order to give another element (which may or may not belong to S). Simple examples of binary operations on sets of numbers are addition and multiplication. (These are, of course, usually denoted by $+$ and $\times$ rather than by $*$.)

It will be helpful, when considering properties of binary operations, to have some particular examples to refer to. We therefore give four examples, $A1$ to $A4$, below. We use the symbol $*$ in each case.

A1 The set Z under ordinary multiplication.

A2 The set $\{3, 6, 9, 12\}$ under multiplication mod 15.

A3 The set Z^+ under the operation 'double the first number and subtract the second', i.e. $a*b = 2a - b$.

A4 Consider the equilateral triangle XYZ shown in Fig. 1.7.

Figure 1.7

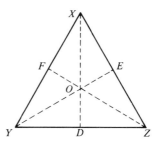

Let a be the operation 'rotate through $120°$ anticlockwise centre O'.
Let c be the operation 'rotate through $120°$ clockwise centre O'.
Let x be 'reflect in the line XD'.
Let y be 'reflect in the line YE'.
Let z be 'reflect in the line ZF'.
And let e be 'leave the triangle as it is'.
(To avoid any confusion, note that, for example, the operation x

is not defined as 'reflection in a vertical line', since the triangle may at some stage have been moved so that the line XD is not vertical.)

We now take S to be the set of these six operations, $\{a, c, x, y, z, e\}$, and define $*$ to mean 'followed by'. Then, for example,

$$a * x = z,$$

since the net effect of a followed by x is z, as shown in Fig. 1.8.

Figure 1.8

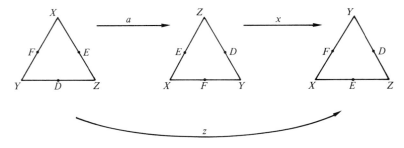

The answers to all possible 'products' under $*$ may be given in a *structure table*.

				2nd element			
$*$	a	c	x	y	z	e
a	c	e	z	x	y	a
c	e	a	y	z	x	c
x	z	y	e	a	c	x
y	x	z	c	e	a	y
z	y	x	a	c	e	z
e	a	c	x	y	z	e

1st element (labels the rows)

1.11 Properties of binary operations

Let $*$ be a binary operation on a set S. We now consider some important properties that may or may not be possessed by the pair $(S, *)$.

(1) *Closure.* We say that S is *closed* under $*$ if and only if

$$\forall\, a, b \in S,\ a * b \in S,$$

i.e. every possible product of elements in S is also in S. Examining our four examples above, it is clear that A4 is closed since its structure table contains only elements that were in the original set. A1 and A2 are also closed, but A3 is *not* closed, since, for example, $2 * 5 = -1$ which is not in Z^+.

(2) *Associativity.* We say that S is *associative* under $*$ if and only if

$$\forall\, a, b, c \in S,\ (a * b) * c = a * (b *).$$

The reader is left to verify that A1 and A2 are associative. A4 is *not* associative, since, for example, making use of the structure table, we have

SETS AND THE LANGUAGE OF MATHEMATICS

$$(x*y)*z = a*z = y$$
but
$$x*(y*z) = x*a = z.$$

Example A3 is interesting, since if we attempt to evaluate say $(3*8)*2$, we get $-2*2$, and this is not defined! (In A3, $*$ was defined for members of Z^+.) The lack of closure in this example means that it cannot be associative, since not all 'triple products' can be evaluated.

It is important that, when checking for associativity, the order of the elements is not inadvertantly changed. When substituting into $a*(b*c)$, care must be taken to put the value of a *in front of* the answer to $b*c$.

Once we know that an operation is associative, we may write $a*b*c$ without ambiguity, whereas for a non-associative operation, we should not know whether this meant $(a*b)*c$ or $a*(b*c)$. In the case of ordinary subtraction (or division) which is non-associative, we may write $a-b-c$ without ambiguity only because it is a convention that this means $(a-b)-c$ and not $a-(b-c)$.

(3) *Identity element.* S is said to have an *identity element* under $*$ if and only if $\exists e \in S$ s.t.

$$\forall a \in S, \ a*e = a \text{ and } e*a = a.$$

In other words, the identity element is a member of S which, when it is combined with any member of S, leaves that element unchanged. In example A1, the identity element is 1 (and for ordinary addition it is zero); in A2, the identity is 6, and in A4 it is e. The reader is left to show that A3 does not have an identity element. (It will help to read Example 5 first.) Note that the identity element must be a *fixed* member of S; it cannot be varied according to which element of S it is being multiplied by. An identity element is always unique. (See Exercise 1d, question 7.)

(4) *Inverses.* If S does possess an identity element e under $*$, then an element $a \in S$ is said to have an *inverse* (usually denoted by a^{-1}) if $\exists a^{-1} \in S$ s.t.

$$a*a^{-1} = e \text{ and } a^{-1}*a = e.$$

If S has no identity, then it is meaningless to ask about the existence of inverses.

In example A1, if we hunt for an inverse for 3, then we seek an element x such that

$$3 \times x = 1 \text{ and } x \times 3 = 1.$$

It is tempting to say that the inverse is $\frac{1}{3}$, but unfortunately, this is not in the given set Z. So 3 has no inverse. In fact, only two elements, 1 and -1, have inverses, each being *self-inverse* (i.e. each is its own inverse).

In A2, the inverse of 3 is 12 since

$$3*12 = 12*3 = 6, \text{ and 6 is the identity.}$$

The inverse of 12 is 3, and 6 and 9 are self-inverse.

In A4, e, x, y and z are self-inverse, $c^{-1} = a$ and $a^{-1} = c$.

If we say that a set 'has inverses' under an operation $*$, we mean that every element of the set has an inverse.

(5) *Commutativity.* S is said to be *commutative* under $*$ if

$$\forall\, a, b \in S,\ a * b = b * a.$$

A1 and A2 are commutative; A3 and A4 are not.

Qu.26 A binary operation $*$ on a set S of six elements is defined by the following structure table.

	2nd element					
$*$	p	q	r	s	t	u
p	p	q	r	s	t	u
q	q	r	p	u	s	t
r	r	p	q	t	u	s
s	s	t	u	p	q	r
t	t	u	s	r	p	q
u	u	s	t	q	r	p

(1st element is given by the left-hand column.)

Show that $(S, *)$ possesses the first four of the five properties discussed above. Specify the identity element and the inverse of each element. [*Note*: the proving of associativity presents a problem, since strictly we ought to check every possible arrangement of three elements from the set. For the time being, the reader should merely give a few examples that *suggest* associativity (but, of course, do not prove it). Methods of proving associativity are sometimes available; see Exercise 1d, question 1.]

Example 5
A binary operation $*$ is defined on the set of real numbers by $a * b = a + b - 5$. Determine
(i) whether there is an identity element;
(ii) whether there are inverses.
Solution
(i) We seek a number e such that, for every choice of a, $a * e = a$;

i.e.
$$a + e - 5 = a$$
$$\Rightarrow e = 5.$$

Thus $a * 5 = a$ for all $a \in R$.
We must also check that $5 * a = a$ for all a. This is indeed the case. So the identity element is 5.
(ii) If a^{-1} is the inverse of a, then

$$a^{-1} * a = 5$$
$$\Rightarrow a^{-1} + a - 5 = 5$$
$$\Rightarrow a^{-1} = 10 - a.$$

This suggests that $(10 - a)$ is the inverse of a. Testing, we have

$$(10-a)*a=(10-a)+a-5$$
$$=5(=e),$$
and
$$a*(10-a)=a+(10-a)-5$$
$$=5(=e).$$

So $(10-a)$ is the inverse of a for each a.

Definition

Given a non-empty set S and a binary operation $*$ on S, then the pair $(S, *)$ is said to form a *group* if and only if, under $*$,

G1 S is closed.
G2 S is associative.
G3 S has an identity element.
G4 Every element of S has an inverse.

If, in addition, S is commutative, then we say that S is a commutative group or an abelian† group under $*$.

Thus our examples A1, A3 and A4 above are not groups, but A2 is an abelian group.

The rules G1 to G4 in the definition above are called the *axioms* of a group. The axioms of a branch of mathematics are those laws which are taken as a starting point and from which the rest of the subject follows by *deduction*. Axioms therefore do not require proof: they are true *by definition*. (Some elementary Group Theory is included in Chapter 33.)

Suppose now that we have a set S on which *two* binary operations, $\triangle$ and $*$, are defined. We can, of course, ask whether the group axioms above are satisfied by S under each of the operations separately; but the following property is one way in which the two operations may be connected.

Definition

We say that $\triangle$ is *distributive* over $*$ if and only if, $\forall a, b, c \in S$,

$$a\triangle(b*c)=(a\triangle b)*(a\triangle c)$$
and
$$(b*c)\triangle a=(b\triangle a)*(c\triangle a).$$

For example, in ordinary arithmetic, multiplication is distributive over addition, since

$$a(b+c)=ab+ac \text{ and } (b+c)a=ba+ca.$$

But addition is not distributive over multiplication because, in general,

$$a+bc \neq (a+b)(a+c).$$

Qu.27 Show by drawing Venn Diagrams that each of the binary operations $\cap$ and $\cup$ (defined on the set of subsets of some universal set) is distributive over the other.

1.12 Division by zero Let G be a group under the operation $*$, and suppose that $a, b \in G$ and

†After the mathematician Abel.

that we wish to find $x \in G$ such that

$$a*x=b.$$

In a sense we wish to 'undo' the a on the left-hand side. (If $*$ were addition, we should subtract a from each side; if it were multiplication we should divide each side by a.) We have no inverse operation to $*$ itself, but we can proceed as follows.

'Pre-multiplying' each side by a^{-1},

$$a^{-1}*(a*x)=a^{-1}*b$$
$$\Rightarrow \quad (a^{-1}*a)*x=a^{-1}*b \quad \text{(associativity)}$$
$$\Rightarrow \qquad\qquad e*x=a^{-1}*b$$
$$\Rightarrow \qquad\qquad\quad x=a^{-1}*b.$$

We have in effect 'divided' by a by 'multiplying' by a^{-1}. (In the above solution, since we do not know that $*$ is commutative, we must take care that we either pre-multiply or post-multiply each side by a^{-1}; we must not mix the two. Post-multiplying by a^{-1} in this case would give $a*x*a^{-1}=b*a^{-1}$, and we cannot go on from here.)

Qu.28 Find x such that $x*a=b$.

If we now consider the real numbers under ordinary multiplication, we see that the process of division may be defined as 'multiplication by the inverse'. Every member of R has a multiplicative inverse except for the number zero, there being no number x satisfying $0 \times x = 1$. It follows that we may divide by any number except zero. *Division by zero is simply not defined and so cannot be done.*

Many simple fallacies in mathematics essentially involve division by zero, for example

$$x=1 \;\Rightarrow\; x+2=2x+1$$
$$\Rightarrow\; (x+2)^2=(2x+1)^2$$
$$\Rightarrow\; x^2+4x+4=4x^2+4x+1$$
$$\Rightarrow\; x^2+x-2=4x^2+x-5$$
$$\Rightarrow\; (x-1)(x+2)=(x-1)(4x+5)$$
$$\Rightarrow\; x+2=4x+5$$
$$\Rightarrow\; 3x=-3$$
$$\Rightarrow\; x=-1.$$

So
$$1=-1.$$

Qu.29 Show that the fallacy in this argument is caused by dividing by zero.

Qu.30 If the argument is reversed (starting with $x=-1$ and reversing all the implication signs to end with $x=1$), where is the fallacy in the reversed argument?

Qu.31 Is it true that $x^2=4x \Rightarrow x=4$?

1. Let $*$ be the binary operation defined on the set R by

$$a * b = a + b + 1.$$

Show that $(a * b) * c = a + b + c + 2$.
Expand $a * (b * c)$ similarly. What does this prove?

2. Determine whether each of the following (a) is closed; (b) is associative; (c) has an identity element; (d) has inverses; (e) is commutative.

Where your answers are not obvious, give proofs or counter-examples.

    (i) Z^+ under ordinary addition;
    (ii) Z under ordinary addition;
    (iii) Z under ordinary multiplication;
    (iv) $Z^+ \cup \{0\}$ under ordinary addition;
    (v) R under ordinary multiplication;
    (vi) $\{1, 2, 3, 4, 5\}$ under multiplication mod 6;
    (vii) $\{0, 1, 2, 3, 4, 5\}$ under addition mod 6;
   (viii) R under $*$ defined by $a * b = (a + b)/2$;
    (ix) Z^+ under 'to the power of';
    (x) $\{1, 2\}$ under multiplication mod 3;
    (xi) all 2×2 matrices under matrix addition;
    (xii) all 2×2 matrices under matrix multiplication;
   (xiii) $\{1\}$ under ordinary multiplication;
   (xiv) R under $*$ as in question 1;
    (xv) R under $\triangle$ defined by $a \triangle b = a + b + ab$;
   (xvi) the set of all subsets of some universal set under $\cup$ (union);
  (xvii) the set of all subsets of some universal set under $\cap$ (intersection);
 (xviii) the set $\{a, b, c, d\}$ under $*$ defined by the structure table:

$*$	a	b	c	d
a	a	b	c	d
b	b	a	d	c
c	c	d	a	b
d	d	c	b	a.

3. With $*$ and $\triangle$ defined as in question 2(xiv) and (xv),
(i) show that $a \triangle (b * c) = 2a + b + c + ab + ac + 1$;
(ii) expand $(a \triangle b) * (a \triangle c)$ and deduce that $\triangle$ is distributive over $*$.
(iii) Is $*$ distributive over $\triangle$? Give either a proof as in (i) and (ii) or a numerical counter-example.

4. Using the fact that multiplication is distributive over addition, show that

$$(a + b)(c + d) = ac + bc + ad + bd.$$

[*Hint*: start by considering $(a + b)$ as a single number and distributing it over the second bracket.]

5. Let $*$ be defined on $Z^+ \times Z^+$ by

$$(a, b) * (c, d) = (ad + bc, bd).$$

Show that the set is closed and associative under $*$. Is there an identity element? (Be careful!)

6. The set S under the operation $*$ forms a non-abelian group. Make x the subject of the equation $a * x * b = c$. Write out your solution showing only one step at a time, and stating which of the group properties is used at each stage.

7. Show that S can have at most one identity element under an operation $*$. [*Hint*: use R.A.A., starting by assuming that there are two, e and f, and consider $e * f$.]

8. (i) Show that if $(S, *)$ is closed and associative and has an identity element e, then no element can have more than one inverse. [*Hint*: suppose that a has inverses x and y, and consider $x * a * y$.]
(ii) Show that $(a^{-1})^{-1} = a$. [*Hint*: apply the result of part (i) to a suitably chosen element.]

9. (i) Give an example of a binary operation which is commutative but not associative.
(ii) Let $*$ be defined on Z by $a * b = a$ (i.e. it gives the first element of the two). Is $*$ (a) commutative;   (b) associative?

10. If $(S, *)$ is not associative, in how many ways may $a * b * c * d$ be evaluated? (Give a list with brackets inserted to leave no ambiguity.)

11. (i) Show that we may unambiguously write $a * a * a$ if $*$ is either associative or commutative.
(ii) Is the same true for $a * b * a$? (Give a proof or a counter-example.)

12. The algebra of matrices is similar to that of ordinary numbers, one important difference being that matrix multiplication is not commutative. Explain why this means that the expansion of $(\mathbf{A} + \mathbf{B})^2$ is not $\mathbf{A}^2 + 2\mathbf{AB} + \mathbf{B}^2$. What is the correct expansion?

13. Consider the equilateral triangle XYZ. Let e mean 'leave the triangle as it is', c mean 'rotate the triangle through 120° clockwise', and let a mean 'rotate the triangle through 120° anticlockwise'. Let u, v and w mean 'reflect the triangle in lines l, m, n respectively' (see Fig. 1.9) where these lines remain *fixed* as

Figure 1.9

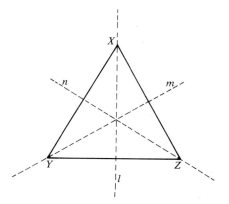

the triangle moves; i.e. they do not move with the triangle. Draw a structure table for the six operations e, c, a, u, v, w under the binary operation 'followed by', and show that they form a non-abelian group. (Contrast this with A4 in the text.)

14. Investigate the set of 8 operations on a square, corresponding to
(i) example A4 in the text;
(ii) question 13 above.

15. (i) Suppose that $(S, *)$ forms a (non-abelian) group. Evaluate $(a*b)*(b^{-1}*a^{-1})$ and deduce an expression for $(a*b)^{-1}$.
(ii) Obtain the same result by solving for x the equation $a*b*x=e$.
(iii) Write down the inverse of $(a*b*c)$ in terms of the inverses of a, b and c.

16. Let G be a finite group. Prove that, in a structure table for G, no element can appear twice in the same row or column.

17. This question concerns the set $\{1, 2, 3, 4\}$ under multiplication mod 5, and it is required to find an answer to $4 \div 3$.
(i) Explain why the answer is not $1\frac{1}{3}$.
(ii) Find the inverse of 3 under multiplication mod 5, and hence solve the problem by regarding it as 4×3^{-1}.
(iii) Rewrite the problem $4 \div 3 = x$ as $4 = x \times 3$, and find x by trying each member of the set.

Miscellaneous exercise 1

1. A binary operation $*$ is defined on the set S of ordered pairs (x, y) of real numbers, where $y \neq 0$, as

$$(x_1, y_1) * (x_2, y_2) = (x_1 + x_2, y_1 y_2).$$

Determine the identity element for this operation and the inverse of the element (x, y). Show further that $\{S, *\}$ is a commutative group.

[C (O)]

2. The non-empty subset S of Z (with normal arithmetic) is such that
$$m \in S, \ n \in S \Rightarrow m - n \in S.$$
Show that
(i) $0 \in S$ and
(ii) $m \in S \Rightarrow -m \in S$.
Hence prove that under addition S is an abelian group.

[SMP]

3. (a) A binary operation, $\circ$, is defined on the set R of real numbers by
$$a \circ b = \begin{cases} \text{maximum of } a \text{ and } b \text{ if } a \neq b \\ a \text{ if } a = b. \end{cases}$$

State whether $\circ$ is (i) closed, (ii) associative, giving your reasons.
Show that R does not contain an identity element for this operation. Suggest a set S on which this operation can be defined which does contain an identity. Does every element of your set S have an inverse in S?

[MEI]

4. Given the multiplication table

*	a	b
a	a	a
b	a	a

show that the binary operation $*$ on the set $S=\{a,b\}$ is both associative and commutative.

Construct a multiplication table for a binary operation $\circ$ on S such that $\circ$ is commutative but not associative. Show that neither $(S,*)$ nor $(S,\circ)$ forms a group.

[L]

5. The operations $+$ and $\times$ are such that for any given sets A and B, the sets $A+B$ and $A\times B$ are uniquely defined. The operations have the properties:

I $+$ and $\times$ are both commutative and associative,

II $\times$ is distributive over $+$,

III the following relationships hold for any set A:

(a) $\mathscr{E}+A=\mathscr{E}$
(b) $A+A'=\mathscr{E}$
(c) $A\times A=A$
(d) $A\times\mathscr{E}=A$

where A' and $\mathscr{E}$ have their normal meanings. The expression $A\times B$ may be abbreviated to AB. Prove, justifying each step in your argument by reference to the above properties:

(i) $ABC+A'BC=BC$,
(i) $A(A+B)=A+AB=A$.

[JMB]

6. Throughout this question you may assume the properties of the integers under the operations of addition, subtraction and multiplication, including the distributive laws. The set P is a subset of the integers with the properties:

I The set P is closed under addition.

II The set P is closed under multiplication.

III "$a>b$" $\Leftrightarrow$ "$a-b\in P$."

Without making any assumptions about the properties of inequalities, use the properties I–III above to prove that

(i) If $a>b$ and $b>c$ then $a>c$.
(ii) If $a>b$ and $c>0$ then $ac>bc$.
(iii) If $b>a>0$ and $d>c>0$ then $bd>ac$.

[JMB]

2 Functions and limits

2.1 Functions The word 'function' is used in mathematics and science in both a precise and an imprecise sense, so that many so-called functions are not strictly functions at all. We begin by explaining the general idea of a function, and then we shall examine what conditions need to be imposed to provide a rigorous definition.

If we say that *y is a function of* x, we mean that the value of y depends on the value of x. Thus

$$y = x^2, \quad y = \frac{1+x}{1+x^2}, \quad y = 10^x$$

are all examples of functions. The word is also used when the mathematical relationship between the variables is not stated (and possibly not known). Thus we may say that

(i) the temperature on a mountain is a function of the height above sea-level;

(ii) the pressure exerted by a certain quantity of gas is a function of its temperature and volume;

(iii) the price of carpets is a function of the cost of production and of the level of consumer demand for carpets (and, no doubt, of many other factors).

Functions are generally denoted by letters; if y is a function of x, we may call the function itself f, and write $y = f(x)$ (pronounced 'f of x'). Other notations are common, for example $x \xrightarrow{f} y$ and $y = (x)f$. If the value of y depends on more than one variable, v, w and x say, then we may write $y = f(v, w, x)$.

Suppose now that $y = f(x)$. We may wish to say 'the value of y when $x = 2$', this is conveniently denoted by $f(2)$.

Example 1
Let $f(x) = 3x^2 + 2x$.
   (i) Find $f(5)$.   (ii) Simplify $f(x+1)$.
(iii) Simplify $f(2p)$.
Solution
(i) $f(5) = (3 \times 5^2) + (2 \times 5) = 85$.
(ii) We must substitute $x + 1$ in place of x, so

$$\begin{aligned} f(x+1) &= 3(x+1)^2 + 2(x+1) \\ &= 3(x^2 + 2x + 1) + 2(x+1) \\ &= 3x^2 + 8x + 5. \end{aligned}$$

(iii) $f(2p) = 3(2p)^2 + 2(2p) = 12p^2 + 4p$.

Qu.1 Let $f(x)=2x+3$. Evaluate
(i) $f(7)$;   (ii) $f(7)+f(3)$;   (iii) $f(10)$.
Show that $\forall m, n \in R$
(iv) $f(m+n)=f(m)+f(n)-3$;
(v) $nf(m)-mf(n)=3(n-m)$.

The words *mapping* and *transformation* mean the same as function. 'Transformation' is usually used in geometrical contexts; 'mapping' is useful since it gives us the verb 'to map'.

The function in Example 1 may be regarded as a rule for operating on a number x to obtain some new number $f(x)$. When, in part (i), the 'input' was 5, the 'output' was found to be 85. In general, the input and output values of a function need not be numbers: they may be members of any sets. For example, if

$$A=\{q,r,s,t\} \text{ and } B=\{h,k,l,m,n\},$$

we may define a function f by specifying its effect on each member of A, thus:

$$f(q)=m, \ f(r)=m, \ f(s)=n, \ f(t)=l.$$

We may illustrate f on a diagram (Fig. 2.1).

Figure 2.1

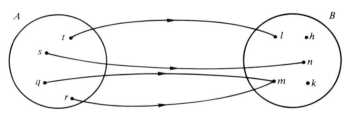

We say that s maps to n, or that n is the *image* of s. In this example, not every member of B is an image, and m is the image of two members of A.

The notation $f:A \rightarrow B$ means that f is a mapping from the set A to the set B. A is known as the *domain* of f, and B is the codomain of f. The set of image points (a subset of the codomain) is called the *range* of f, i.e.

$$\text{range of } f=\{y:y=f(x) \text{ for some } x \text{ in the domain}\}.$$

In Fig. 2.1 the range is $\{l,m,n\}$.

2.2 Well-defined functions

We are now in a position to give a precise, formal definition of a function.

Definition
A *function* is a rule which associates with every member of one set (the domain) one and only one member of another set (the codomain). The domain and codomain may, of course, be the same set.

The important feature of this definition is that it insists that a function should be *well-defined*, i.e. that every member of the domain has an image, and no member has two or more images. If some members of the domain have two or more images, the 'function' is said to be *many-valued*. Fig. 2.2 illustrates both ways in which a so-called function may fail to be well-defined.

Figure 2.2

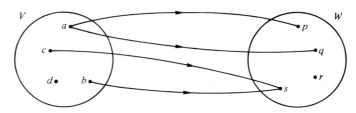

This is not a function since
(i) d has no image, and   (ii) a has two images.
(Note that it does not matter that r is not an image, nor that s is the image of two members of V.) If we wished to make this a genuine function, we could
(i) redefine the domain to be the set $U = \{a, b, c\}$, and
(ii) choose one of p and q to be the image of a (Fig. 2.3).

Figure 2.3

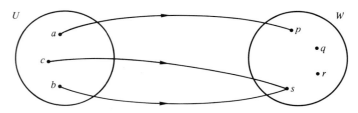

This is now an acceptable function.
    Examples 2 and 3 illustrate 'functions' which are not well-defined.

Example 2
Let f be a mapping from Q to Z defined by $f(p/q) = p + q$. This is not well-defined since, for example,

$$f(\tfrac{2}{3}) = 5 \text{ and } f(\tfrac{4}{6}) = 10.$$

So one fraction $(\tfrac{2}{3} = \tfrac{4}{6})$ is mapped to two different images. One way of overcoming this would be to redefine f by adding the condition 'where p/q has been cancelled as far as possible'.

Example 3
Let $g: R \to R$ be defined by $g(p) = \sqrt{p}$. (The sign $\sqrt{}$ means the *non-negative* square root, and so the fact that p may have two square roots is not the cause of the problem here.) Then g is not well-defined, since it is not defined at all for $p < 0$. The simplest solution is to write $g: R^+ \cup \{0\} \to R$, so that the domain is $R^+ \cup \{0\}$.

    When considering functions defined on R (i.e. with R as the domain), it is important to appreciate that the function is the *rule* which acts on the members of R. Thus when we write 'the function $y = x^2 + 2$' we really mean 'the function square-it-and-add-two'; the x acts merely as a dummy to illustrate the effect of the function.

Qu.2 Each of the following functions is from R to R. State whether or not they are well-defined. Where appropriate, justify your answers.

(i) $y = x^2$; (ii) $f(x) = 1/x$;

(iii) $g(x) = 1/(1 + x^2)$; (iv) $y = -\sqrt{x}$;

(v) $x \xrightarrow{f} \sqrt[3]{x}$; (vi) $f(x) =$ the nearest integer to x;

(vii) $h(x) =$ the smallest integer not less than x;

(viii) $\forall x \in R, f(x) = 0$.

2.3 One-to-one and onto functions

Definitions

(1) A function f is said to be *one-to-one* (or 1-1) if and only if no two distinct members of the domain have the same image; i.e.

$$f \text{ is } 1\text{-}1 \Leftrightarrow (f(a) = f(b) \Rightarrow a = b).$$

(The right-hand side says that if a and b have the same image, then they must really be the same element of the domain.)

(2) A function f is said to be *onto* if and only if every member of the codomain is the image of some element of the domain. The range and the codomain will then be equal. Thus

$$f \text{ maps } A \text{ onto } B \Leftrightarrow \forall y \in B, \exists x \in A \text{ s.t. } y = f(x).$$

The mapping in Fig. 2.4 is onto but not 1-1; the mapping in Fig. 2.5 is 1-1 but not onto.

Figure 2.4

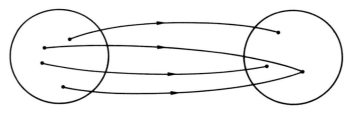

Figure 2.5

Example 4

Determine whether the function $f(x) = \dfrac{x+2}{3x+1}$ is 1-1 and onto R.

Solution

Writing $y = \dfrac{x+2}{3x+1}$, we wish to investigate the number of different values of x corresponding to each value of y. We have

$$y = \frac{x+2}{3x+1} \Rightarrow y(3x+1) = x+2$$

$$\Rightarrow 3xy - x = 2 - y$$

$$\Rightarrow \quad x = \frac{2-y}{3y-1} \tag{1}$$

Now if y takes any value other than $\frac{1}{3}$, then x has a unique value which can be calculated from (1). Examining the case $y = \frac{1}{3}$ separately, we have

$$\frac{1}{3} = \frac{x+2}{3x+1} \Rightarrow 3x+1 = 3x+6$$

which has no solution. Thus y cannot take the value $\frac{1}{3}$, but each other value of y may be obtained from a unique x. So f is 1–1 but not onto (since $\frac{1}{3}$ is not in the range).

Qu.3 Which real number is not in the domain of f?
Qu.4 Prove that f is 1-1 by showing that if $f(a)=f(b)$, then $a=b$.
Qu.5 Determine whether each of the following functions from R to R is (a) 1-1; (b) onto.

(i) $y=x^2$; (ii) $y=x^3$; (iii) $y = \begin{cases} x & \text{if } x \geqslant 0 \\ x+1 & \text{if } x < 0. \end{cases}$

2.4 Composite functions

Let $f:A \to B$ and $g:B \to C$ (Fig. 2.6) and let $x \in A$, so that $f(x) \in B$. If we now let g operate on $f(x)$, we get $g[f(x)] \in C$.

Figure 2.6

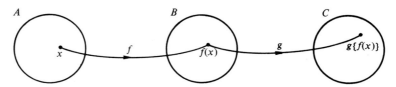

We may regard the mapping $x \to g[f(x)]$ as a single function (from A to C) and we denote this function by $g \circ f$, so that $g \circ f(x) = g[f(x)]$. When a function is made up of two or more functions performed successively, it is called a *composite function*. Note that $g \circ f$ means 'f followed by g' since, when we write $g \circ f(x)$, the f (being next to the element x) is done first.† Similarly, $h \circ g \circ f$ means 'f followed by g followed by h'. A composite function is sometimes referred to as a 'function of a function'.

Example 5
Let f and g be functions from R to R defined thus:

$$f(x) = x^2 \text{ and } g(x) = 2x+1.$$

Simplify (i) $g \circ f(x)$; (ii) $f \circ g(x)$.
Solution
(i) $g \circ f(x) = g(x^2) = 2x^2 + 1$.
(ii) $f \circ g(x) = f(2x+1)$
$\qquad = (2x+1)^2$
$\qquad = 4x^2 + 4x + 1$.

†This is the main problem with the convention of writing the function before the element rather than after it. Where the notation $(x)f$ is used, f followed by g becomes the more natural $f \circ g$. Matters are made worse by the fact that some functions *are* written after the element; thus for example we write x^2 and not ^2x. This inconsistency can be confusing when meeting, say, $\sin^2 x$ for the first time. (See Chapter 11).

Qu.6 With f and g as in Example 5, simplify
(i) $f \circ f(x)$;   (ii) $g \circ g(x)$.

2.5 Identity and inverse functions

Given a set A, the *identity function* on A is the mapping e from A to A which maps each element to itself, i.e.

$$\forall a \in A, \quad e(a) = a.$$

If f is any mapping from A to A, then

$$e \circ f = f \circ e = f.$$

Intuitively, the *inverse* of a function f (denoted by f^{-1}) is the function which 'reverses' the effect of f. Thus if f is the 'double-it' function, $f(x) = 2x$, then f^{-1} is the 'halve-it' function, $f^{-1}(x) = \frac{1}{2}x$. Similarly, squaring and square rooting are mutually inverse.

We cannot assume, however, that any function has an inverse. Consider, for example, the function illustrated in Fig. 2.3. Simply reversing the arrows, we get Fig. 2.7 which is not a well-defined function.

Figure 2.7

For f^{-1} to be well-defined, f must be 1-1 and onto. (This may be checked diagrammatically.) If $f: A \to B$ is 1-1 and onto, then $f^{-1}: B \to A$ is also 1-1 and onto, and $f^{-1} \circ f$ and $f \circ f^{-1}$ are the identity mappings on A and B respectively.

Qu.7 Determine whether the set of all mappings from R to R which are 1-1 and onto forms a group under the binary operation $\circ$.

Example 6
Let $f(x) = \dfrac{3x+1}{x^2-1}$. Find the inverse function of f.

Solution
(Note that $f(x)$ is not defined for $x = 1$ or -1. Also, at this stage, we do not know whether f is 1-1 and onto.)

Let
$$y = \frac{3x+1}{x^2-1}.$$

We rearrange this to express x in terms of y.

$$y(x^2 - 1) = 3x + 1$$
$$\Rightarrow yx^2 - 3x - (y+1) = 0. \tag{1}$$

Now two cases arise:
(i) If $y = 0$, (1) becomes $-3x - 1 = 0$.

So $x = -\frac{1}{3}$.

(ii) If $y \neq 0$, (1) is quadratic, so, by the formula†,

$$x = \frac{3 \pm \sqrt{\{9 + 4y(y+1)\}}}{2y}.$$

Combining the two cases, and writing $f^{-1}(y)$ in place of x, we have:

$$f^{-1}(y) = \begin{cases} -\dfrac{1}{3} & \text{if } y = 0 \\[2ex] \dfrac{3 \pm \sqrt{\{9 + 4y(y+1)\}}}{2y} & \text{if } y \neq 0. \end{cases}$$

(The plus-or-minus indicates that f^{-1} is not single-valued, i.e. that f was not 1-1.)

Exercise 2a

1. Let $f(x) = x^2 + 3x + 1$. Evaluate
   (i) $f(2)$;   (ii) $f(0)$;   (iii) $f(-3)$;
   and simplify
   (iv) $f(x-2)$;   (v) $f(2x)$.

2. Let $f(x) = x^2$, $g(x) = 2x + 1$, $h(x) = 3x$.
   (i) Is it true that

   $$\forall\, a,\, b \in R,\ f(a+b) = f(a) + f(b)?$$

   Give a proof or a counter-example.
   (ii) Repeat (i) for the functions g and h.

3. Determine whether the following functions are
   (a) 1-1;   (b) onto.
   (i) $f: R \to R$ defined by $f(x) = 2x$;
   (ii) $f: Z \to Z$ defined by $f(x) = 2x$;
   (iii) $f: R \to R$ defined by $f(x) = x^2$;
   (iv) $f: R^+ \to R^+$ defined by $f(x) = x^2$.

4. Let A and B be finite sets and let $f: A \to B$. What can be said about $n(A)$ and $n(B)$ if f is
   (i) 1-1;   (ii) onto;   (iii) 1-1 and onto?

5. (Compare this question with number 4.)
   (i) Let $f: Z^+ \to Z$ be defined by

   $$f(n) = \begin{cases} n/2 & \text{if } n \text{ is even} \\ (1-n)/2 & \text{if } n \text{ is odd}. \end{cases}$$

   Show that f is onto (even though $Z^+ \subset Z$).
   (ii) Let E be the set of even integers. Find a mapping from Z to E which is 1-1 (even though $Z \supset E$).

6. Let $f(x) = 2x + 3$, $g(x) = 5x + 2$. Simplify
   (i) $f \circ g(x)$;   (ii) $g \circ f(x)$.

7. Show that the mappings $f(x) = ax + b$ and $g(x) = cx + d$ commute under composition if and only if $d(a-1) = b(c-1)$.

8. Use the method of Example 6 to find the inverse functions of
   (i) $h(x) = 2x - 3$;   (ii) $u(x) = \dfrac{x-1}{2x+3}$.

†The formula for a quadratic equation is derived in Section 10.16.

9. If f and g are two mappings from A to B, suggest a formal definition for *equality* of these functions; i.e. complete the following
f and g are equal $\Leftrightarrow \dots$

10. (i) Find four different mappings from A to B, where $A = \{p, q\}$ and $B = \{s, t\}$.
(ii) How many different mappings are there from A to B if $n(A) = \lambda$ and $n(B) = \mu$.

11. Let $T = \{1, 2, 4, 8, \dots\}$ and let $S(T)$ be the set of all finite subsets of T. Find a mapping from $S(T)$ to $Z^+ \cup \{0\}$ which is 1-1 and onto. (There is a very simple one.)

12. A 1-1 and onto mapping from a set to itself is called a *permutation*. How many permutations are there of a set S if $n(S) = k$?

13. Describe the geometrical effects of the following transformations from R^2 to R^2. (The best way is to choose a few particular points, A, B, C etc., and plot them and their images, A', B', C' etc.. The vertices of the unit square, $(0, 0)$, $(1, 0)$, $(1, 1)$ and $(0, 1)$, are often used.)
   (i) $(x, y) \rightarrow (2x, y)$;
   (ii) $(x, y) \rightarrow (-x, y)$;
   (iii) $(x, y) \rightarrow (-y, x)$.

14. What is the geometrical effect of $f : R^2 \rightarrow R^2$ defined by $f((a, b)) = (a, 0)$? (This transformation is an example of a *projection*.)

15. Let T be the set of equivalence classes of a set S under some equivalence relation. Explain why $f : T \rightarrow S$ defined by $[x] \xrightarrow{f} x$ is not in general well-defined.

Exercise 2b (Harder questions)

1. Let α be the permutation of the numbers $\{1, 2, 3\}$ defined by $\alpha(1) = 1$, $\alpha(2) = 3$, $\alpha(3) = 2$. (We might write this as

$$\alpha \begin{pmatrix} 1 \\ 2 \\ 3 \end{pmatrix} = \begin{pmatrix} 1 \\ 3 \\ 2 \end{pmatrix}.)$$ Choose letters for the other five permutations,

specifying the effect of each. Write out a structure table for the six permutations under composition. Check that they form a group.

2. Let $f : A \rightarrow A$ such that $\forall a \in A$, $f \circ f(a) = a$. Show that f is 1-1 and onto.

3. (i) Let $f : S \rightarrow S$ where $n(S) > 2$, and suppose that $\forall T \subset S$ with $n(T) = 2$ we have $t \in T \Rightarrow f(t) \in T$. Show that f must be the identity mapping. Show how your proof breaks down if $n(S) = 2$.
(ii) Prove the corresponding result when $n(S) > 3$, and the subset property applies to subsets T of 3 elements. (You may need to consider separately the case when $n(S) = 4$.)

4. Find an example of a finite set S and a function $f : S \rightarrow S$ such that $\forall s \in S$, $f \circ f(s) = f(s)$ but f is not 1-1.

5. If $n(A) = 4$, how many distinct mappings f are there such that $\forall a \in A$, $f(a) \neq a$ and $f \circ f(a) = a$? Can you generalize to the case $n(A) = k$?

6. Let $n(A) = 4$, and let $P(A)$ be the power set (set of subsets) of A.

Let $f: A \to P(A)$.

  (i) Find an example of f such that
$$\forall\, a,\, b \in A,\ a \in f(b) \Rightarrow b \notin f(a).$$
  (ii) Prove that no f can exist such that
$$\forall\, a,\, b \in A,\ a \in f(b) \Leftrightarrow b \notin f(a).$$
  (iii) Prove that no function satisfying the condition in part (i) can also satisfy: $\forall\, a \in A,\ n(f(a)) \geqslant 2$. [*Hint*: it may help to consider the 4×4 rectangular array of the elements of $A \times A$.]

  (iv) If $n(A) = 5$, show that a function satisfying both conditions (see part (iii)) can be found.

7. Let $f: A \to B$. Show that R defined on A by
$$a_1\, R\, a_2 \Leftrightarrow f(a_1) = f(a_2)$$
is an equivalence relation, and that f induces a mapping from the set of equivalence classes to B which is 1-1.

8. Let f and g be mappings from R^2 to R defined by
$$f(x,y) = x + y,\ g(x,y) = y - x;$$
and let $\theta: R^2 \to R^2$ be defined by
$$\theta(x,y) = (f(x,y),\ g(x,y)) = (x+y,\ y-x).$$

  (i) Draw diagrams to illustrate the equivalence classes of R^2 induced by each of f and g in the manner of question 7.

  (ii) Is θ 1-1?

  (iii) Is θ onto?

  (iv) Find the inverse function of θ.

9. Let $f: A \to B$ and $g: B \to C$. Show that
$$g \circ f \text{ is 1-1} \not\Rightarrow g \text{ and } f \text{ are both 1-1},$$
and $g \circ f$ is onto $\not\Rightarrow g$ and f are both onto.

10. Find a 1-1 mapping from $Z^+ \times Z^+$ to Z^+.

2.6 The graph of a function

For the rest of this chapter, we shall restrict our attention to functions from R to R. Clearly such functions cannot be represented by the type of diagram used so far; instead, we represent the domain (or the *independent variable*) by a horizontal line (the x-axis), and the codomain (or the *dependent variable*) by a vertical line (the y-axis).

Figure 2.8

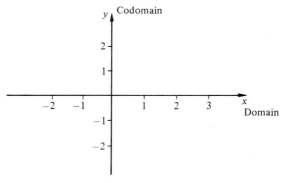

Given a function f, there corresponds to each $x \in R$ a point $(x, f(x))$. The set of such points is called the *graph* of f.

Qu.8 Illustrate the function $f(x) = x^2$ by plotting the set of points (x, x^2) for $-3 \leqslant x \leqslant 3$. (A rough diagram will do.)

One of the most important properties that a graph may, or may not, possess is that of *continuity*. A formal definition is given in Chapter 39; an intuitive definition is given below.

Definition
A function is said to be *continuous* if its graph has no breaks in it.

A simple example of a *discontinuous* function is $y=1/x$ (Fig. 2.9). It is · discontinuous (and, indeed, not defined) at $x=0$.

Figure 2.9

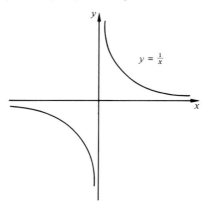

$y = \frac{1}{x}$

2.7 Periodic functions   When defining a function, it is sometimes necessary to split the domain into two or more parts and to define the function in different ways on different parts of the domain. An example would be the function

$$f(x)= \begin{cases} x & \text{for } x \leqslant 1 \\ 2-x^2 & \text{for } 1 < x < 2 \\ -2 & \text{for } x \geqslant 2. \end{cases}$$

The graph of f is in three distinct parts, and, in this case, is continuous (Fig. 2.10).

Figure 2.10

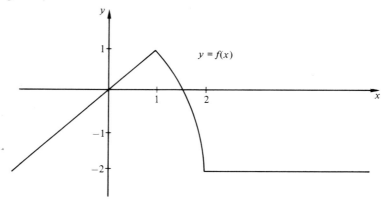

$y = f(x)$

Qu.9 Draw the graph of the following function from $x=-2$ to $x=6$. State the x-coordinate of the point at which it is discontinuous.

$$y= \begin{cases} 2 & \text{for } x < 0 \\ x+1 & \text{for } 0 \leqslant x < 2 \\ 5-x & \text{for } x \geqslant 2. \end{cases}$$

(Use a heavy dot on one of the 'loose ends' of the graph to indicate the value of the function at the discontinuity.)

When defining a function in different ways on different parts of the domain, care must be taken to ensure that it is well-defined. For example, if we write

$$g(x) = \begin{cases} x^2 & \text{for } x \leqslant 2 \\ x & \text{for } 2 \leqslant x < 3 \\ 3 & \text{for } x > 3 \end{cases}$$

then g is double-valued at $x = 2$, and not defined at all at $x = 3$.

Qu.10   Sketch the graph of the function

$$y = \begin{cases} x & \text{for } x \geqslant 0 \\ -x & \text{for } x < 0. \end{cases}$$

(This function is known as the *modulus function*; the modulus of x, denoted by $|x|$, is most simply thought of as the numerical value of x with a positive sign.)

A *periodic function* is one whose graph repeats itself at regular intervals, the interval being known as the *period*. For example, if

$$h(x) = \{(x-1)^2 \text{ for } 0 \leqslant x < 2$$
$$\text{and } h \text{ is periodic with period 2,}$$

then the graph of h is as illustrated in Fig. 2.11.

Figure 2.11

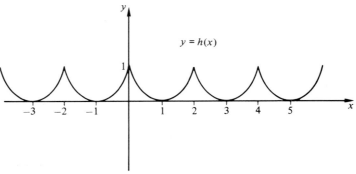

A periodic function may well be discontinuous at the end-points of each interval, and again care must be taken to ensure that it is well-defined.

Qu.11   Given that $y = x^2$ for $0 \leqslant x < 1$ and that y is periodic with period 1, sketch the graph of y against x.

We now make a formal definition.

Definition
A function f is said to be *periodic* if and only if $\exists \, a \in R$ s.t.

$$\forall \, x \in R, \ f(x+a) = f(x).$$

The number a is called the *period* of f.

2.8 Even and odd functions

Definitions

(1) A function f is said to be *even* if and only if

$$\forall\, x \in R,\ f(-x) = f(x).$$

(2) f is said to be *odd* if and only if

$$\forall\, x \in R,\ f(-x) = -f(x).$$

The first of these definitions says that if f is even, then the sign of x can be changed without altering the value of y. Thus if the point (a, b) lies on the graph of f, then so does the point $(-a, b)$. It follows that the graph of f has line-symmetry in the y-axis (Fig. 2.12). A simple example would be $y = x^2$.

Figure 2.12

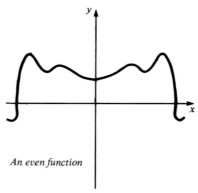

An even function

The second definition says that, for an odd function, changing the sign of x has the effect of changing the sign of y; i.e. if (a, b) lies on the graph, then so does $(-a, -b)$. The graph will then have $180°$ rotational symmetry (or S-symmetry) about the origin (Fig. 2.13). An example is $y = x^3$.

Figure 2.13

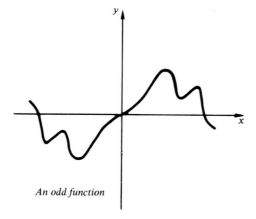

An odd function

2.9 Inverse functions and their graphs

Qu.12 (i) Let $f:R \to R$. Show that f is well-defined if and only if any vertical line meets the graph of f exactly once.

(ii) State the corresponding graphical properties if f is (a) 1-1; (b) onto.

FUNCTIONS AND LIMITS

Suppose we know what the graph of f looks like, and we wish to draw the graph of the inverse function f^{-1}. Let $y = f(x)$ so that $x = f^{-1}(y)$. In a sense, the graph of f is also a graph of f^{-1} if we regard the vertical axis as the domain, and the horizontal axis as the codomain. To preserve the usual convention, however, we must interchange the two axes, i.e. we reflect the graph in the line $y = x$, and then re-label the axes so that x remains the independent variable (Fig. 2.14).

Figure 2.14

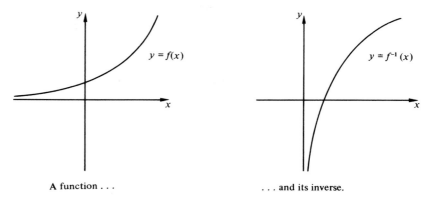

A function . . .   . . . and its inverse.

Qu. 13 Let

$$f(x) = \begin{cases} x & \text{for } -1 \leqslant x < 0 \\ 2x^3 & \text{for } 0 \leqslant x < 1 \end{cases}$$

and let f be periodic with period 2.
(i) Sketch the graph of f and state its range.
(ii) The domain of f is now restricted to the set $[-1, 1)$. Sketch the graph of the inverse function f^{-1}, and define it (by specifying its effect on its domain).

Exercise 2c

1. Sketch the graph of the function

$$f(x) = \begin{cases} 6+x & \text{for } x < -2 \\ x^2 & \text{for } -2 \leqslant x \leqslant 2 \\ 6-x & \text{for } x > 2. \end{cases}$$

State where it is discontinuous.

2. Sketch the graph of

$$f(x) = \begin{cases} 2 & \text{for } 0 \leqslant x \leqslant 4 \\ x-2 & \text{for all other } x. \end{cases}$$

State where the function is discontinuous.

3. (i) Sketch the graph of any function f with the property that

$$a > b \Rightarrow f(a) > f(b).$$

(Such a function is said to be an *increasing* function. *Decreasing* functions are defined similarly.)

(ii) Must such a function be onto?

(iii) Must it be 1-1?

4. (i) Verify graphically that any function which is continuous, 1-1 and onto is either increasing or decreasing.

(ii) Show that in part (i) the condition of continuity is essential, by sketching a graph which is 1-1 and onto, but neither decreasing nor increasing.

5. (i) Show from the definition of a periodic function that if f is periodic with period a, then it is also periodic with period 2a. Generalize this result. (Note that when we say that f is periodic with period p, we usually mean that p is the smallest period of the function.)

(ii) If f is periodic with period 3 and with period 5, what can be deduced?

(ii) Repeat (ii) with periods 6 and 9.

6. Let $f(x)=1/x$ for $0<x\leqslant 2$, and let f be periodic with period 2.

(i) State the value of $f(0)$ and $f(11)$.

(ii) On the same axes, sketch the graphs of f and of $y=x$. At how many points do the graphs intersect?

(iii) Suppose the graphs cross at the point where $x=k$ and k is 'very large'. Show that k is of the form $2n+h$ where $n\in Z^+$ and h is very small.

7. (i) Prove that if f is odd, then $f(0)$, if it exists, is zero.

(ii) Prove that there is only one (well-defined) function which is odd and even. What is it?

8. Show that $y=x^n$ is either odd or even if $n\in Z^+$. How can you tell whether it is odd or even? Is the same true if $n\in Z$?

9. Prove that if f and g are both odd, then $h(x)$ defined by $h(x)=f(x)+g(x)$ is also odd. Is the sum of two even functions even?

10. What can be said about the numbers a_0, a_1, a_2, a_3 if $f(x)=a_0+a_1 x+a_2 x^2+a_3 x^3$ is

(i) even; (ii) odd?

11. If f is even and g is odd, say whether h is odd or even or neither if

(i) $h(x)=xf(x)$;   (ii) $h(x)=xg(x)$;

(iii) $h(x)=x^2 g(x)$;   (iv) $h(x)=f(x)+g(x)$;

(v) $h(x)=f(x)g(x)$;   (vi) $h(x)=f(x)/g(x)$;

(vii) $h(x)=\{g(x)\}^2$.

12. (i) Let $f(x)=x$ and let $g(x)$ be defined on the set $S=\{x\in R:0\leqslant x\leqslant 1\}$ such that g is continuous, $g(0)\geqslant 0$ and $g(1)\leqslant 1$. Show by considering the graphs of f and g that

$$\exists s\in S \text{ s.t. } f(s)=g(s).$$

(ii) A rectangular table $ABCD$ is 1 m wide (see Fig. 2.15). A piece of string 1 m long is graduated from 0 to 1 and rests initially with the zero end at P and the 'one' end at Q. It is then picked up and dropped at random on the table. Show that at least one point on the string is the same distance from the edge AB as it was originally. [*Hint*: let the initial and final distances from AB of the

Figure 2.15

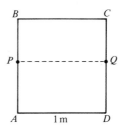

point marked x on the string be $f(x)$ and $g(x)$ and use part (i).]

13. (i) Write the statement $|x| < \frac{1}{2}$ without using the modulus notation.

(ii) Verify that $|a-b|$ is the difference between a and b.

(iii) Write the inequality $|x+2| < 1$ without the modulus notation, and hence find the possible values for x.

(iv) Sketch the graph of $y = |x-1|$.

14. Plot the graph of f from $x = -3$ to $x = 3$ where

$$f(x) = \begin{cases} \dfrac{|x|}{x} & \text{if } x \neq 0 \\ 0 & \text{if } x = 0. \end{cases}$$

15. Simplify $\{x \in R : |x+1| \leqslant 2\} \cap \{x \in R : |x - \frac{1}{2}| < 1\frac{1}{2}\}$.

16. (i) Sketch the graph of

$$f(x) = \begin{cases} 2x & \text{for } x < 0 \\ x^2 & \text{for } 0 \leqslant x \leqslant 2 \\ x+2 & \text{for } x > 2. \end{cases}$$

(ii) Sketch and define the inverse of f.

17. Find the inverse function of $f(x) = 1 + 2x - x^2$.

For what real numbers y is $f^{-1}(y)$

(i) not defined;

(ii) double-valued?

Illustrate your answers by giving a sketch of the graphs of f and of f^{-1}.

18. A function is defined on R by

$$f(x) = \begin{cases} 1 & \text{if } x \in Q \\ 0 & \text{if } x \notin Q. \end{cases}$$

Where is the function discontinuous?

2.10 Roots of equations

Consider an equation involving a single variable x. By taking every term to the L.H.S. (left-hand side), we can write the equation in the form $f(x) = 0$. (Thus, for example, the equation $\sqrt[3]{(1+x^2)} = 1 + x$ would be written $\sqrt[3]{(1+x^2)} - x - 1 = 0$, the L.H.S. of this now being called $f(x)$. It is not being suggested that this is necessarily a helpful step in solving the equation.)

A solution of an equation is known as a *root* of the equation; and when we talk of the *roots of a function* f, we mean the solutions of the equation $f(x) = 0$.

Graphically, the roots of f are the values of x at which the graph of f meets the x-axis. In Fig. 2.16, three roots of the function are shown, two negative and one positive.

Figure 2.16

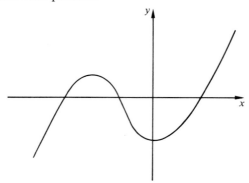

Graphs are not normally used to find the roots of an equation, because even an 'accurate drawing' is not really accurate; but thinking 'graphically' can often provide information about the number of roots of an equation and their approximate location. The idea of continuity now becomes important.

Example 7

Show that the equation $x^3 - 2x^2 + 5x - 7 = 0$ has a positive root, and find its value to the nearest whole number.

Solution

Consider the graph of $f(x) = x^3 - 2x^2 + 5x - 7$. When $x = 0$, $f(x) = -7$, and when x is 'very big' (say 100) $f(x)$ is clearly positive. Since the graph of f is below the x-axis at $x = 0$, and above the x-axis at $x = 100$, (and *assuming that the graph is continuous*) it must cross the x-axis between $x = 0$ and $x = 100$, and so it has a positive root.

Locating the root is largely trial and error; we hunt for a change in the sign of $f(x)$:

x	0	1	2
$f(x)$	-7	-3	3

The table of values shows that there is a root between 1 and 2. To find which integer it is nearer, we evaluate $f(1\frac{1}{2})$. We get $f(1\frac{1}{2}) = -5/8$. Since this is negative, the root lies between $1\frac{1}{2}$ and 2 (see Fig. 2.17). So 2 is the nearest integer to the root. (Note that there may actually be several roots between $1\frac{1}{2}$ and 2; our method merely guarantees that there is at least one.)

Figure 2.17

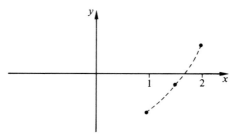

Qu.14 Suppose that $g(x)$ has one root between 7 and 8, and that $g(7)=2$ and $g(8)=-4$. It is tempting to argue that the root is nearer to 7 than to 8 since 2 is nearer to zero than -4. Sketch a graph to show that this argument is not valid.

Qu.15 Sketch two graphs to show that if $f(a)$ and $f(b)$ have the same sign, and f is continuous, then f may or may not have a root between a and b.

Qu.16 (i) Sketch a graph to show that if f is discontinuous at some point between a and b, then f may have no root between a and b even though $f(a)$ and $f(b)$ have different signs.

(ii) Give an actual example of part (i) by specifying a function f and values for a and b.

2.11 Quadratic functions

A function of the form $y=ax^2+bx+c$, where a, b, $c\in R$ and $a\neq0$, is called a quadratic function, and the corresponding equation, $ax^2+bx+c=0$, is called a quadratic equation. Such equations can be solved by factorizing or by the formula

$$x=\frac{-b\pm\sqrt{(b^2-4ac)}}{2a}.$$

The derivation of this formula is given in Section 10.16.

The graph of a quadratic is an example of a *parabola*, and is illustrated in Figs. 2.18 and 2.19. The position of the parabola relative to the axes depends on the particular values of a, b and c.

Figure 2.18

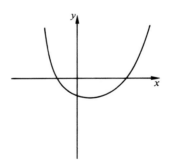

$y=ax^2+bx+c$ in the case $a>0$.

Figure 2.19

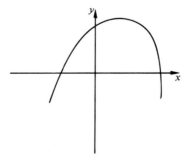

$y=ax^2+bx+c$ in the case $a<0$.

Restricting our attention to the case $a>0$ (since the case $a<0$ is no different), it is clear that a quadratic function may have 0, 1 or 2 roots (Figs. 2.20–22).

Figure 2.20

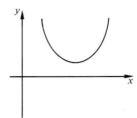

Figure 2.21

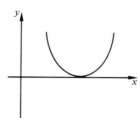

Figure 2.22

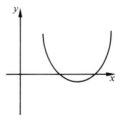

The case of one root (Fig. 2.21) is of interest. It is actually more correct to say that there are two roots which happen to be equal (i.e. Fig. 2.21 may be regarded as a special case of Fig. 2.22). This is illustrated algebraically in the following example.

Example 8
Solve $4x^2 - 12x + 9 = 0$
(i) by factorizing;   (ii) by the formula.
Solution
(i) We get $(2x - 3)(2x - 3) = 0$.
So $x = 1\frac{1}{2}$ (twice!).
(ii) We get $\dfrac{12 \pm \sqrt{0}}{8}$. Taking the + and − separately, we again have $x = 1\frac{1}{2}$ twice.

In Fig. 2.21, we say that the curve has *double contact* with the x-axis, and that the x-axis *touches* the curve (rather than merely cutting it). When a line touches a curve, it is called a *tangent*.
With other types of curve, it is possible to have three or more equal roots, and 'multiple contact' of the curve with the x-axis. The word tangent is used for any multiple contact of a line with a curve, and in

FUNCTIONS AND LIMITS

some cases a line may touch a curve and also cross it at the same point, as in Qu.17. This idea is examined in more detail later.

Qu.17 Factorize $x^3 - 3x^2 + 3x - 1$ by trial and error, showing that it has three equal roots. Plot the graph of this function (on a large scale) for values of x from 0 to 2.

2.12 The discriminant $b^2 - 4ac$

In many problems, we wish to know only how many roots a quadratic equation has, the actual value of the roots being of no interest. It is clear from the formula in Section 2.11 that the equation $ax^2 + bx + c = 0$ (with $a \neq 0$) has:

no roots if $b^2 - 4ac < 0$ (since we cannot take the square root);
equal roots if $b^2 - 4ac = 0$;
two distinct roots if $b^2 - 4ac > 0$.

Because it distinguishes between different types of quadratic in this way, the expression $b^2 - 4ac$ is called the *discriminant* of the equation $ax^2 + bx + c = 0$.

Definition
A function $f: R \to R$ is said to be *positive definite* if and only if

$$\forall x \in R, \ f(x) > 0.$$

Similarly, f is *negative definite* if and only if

$$\forall x \in R, \ f(x) < 0.$$

Graphically, a function is positive definite if and only if its graph lies entirely above the x-axis. The necessary and sufficient conditions for $y = ax^2 + bx + c$ to be positive definite are therefore that the curve is the 'right way up' (as in Fig. 2.18) and that it has no roots; i.e.
    (i) $a > 0$
and   (ii) $b^2 - 4ac < 0$.
These conditions are obtained algebraically in Chapter 10.

Qu.18 State conditions for $ax^2 + bx + c$ to be negative definite.

Example 9
Find the equations of the tangents to the curve $y = x^2 + 2x + 4$ which pass through the origin.
Solution
(Fig. 2.23 shows the tangents that we are seeking, although at this stage we cannot be sure that there are two tangents as shown; the diagram might be misleading.)

Any line through the origin has an equation of the form $y = mx$ (see Chapter 3). Now such a line will, in general, either miss the parabola completely, or cut it at two points. We seek any line that meets the curve at a single point.

Figure 2.23

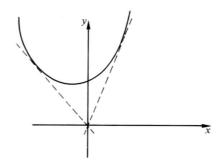

To find where the line $y=mx$ meets the curve, we solve simultaneously the equations

$$\begin{cases} y=mx \\ y=x^2+2x+4 \end{cases}$$
$$\Rightarrow x^2+2x+4=mx$$
$$\Rightarrow x^2+(2-m)x+4=0.$$

Now we require this equation to have equal roots (i.e. $b^2-4ac=0$), so

$$(2-m)^2-16=0$$
$$\Rightarrow (2-m)^2=16$$
$$\Rightarrow 2-m=\pm4$$
$$\Rightarrow m=6 \text{ or } -2.$$

So the required tangents are $y=6x$ and $y=-2x$.

Exercise 2d

1. Show that the equation $x^3+7x^2-221x+790=0$ has a root between -21 and -20. Show that there are two further roots, and locate each between two consecutive integers.

2. Locate between consecutive integers all positive roots of
   (i) $3x^4=5x^3+110$;
   (ii) $2^x-10x^2=0$.

3. (i) What is the relationship between the graphs of the functions $f(x)=x^3-5x^2+7$ and $g(x)=x^3-5x^2+8$?
   (ii) Evaluate $f(4)$ and $f(5)$ and deduce that f has a root between 4 and 5.
   (iii) Explain with the aid of a graph whether the root of g between 4 and 5 will be greater than or less than that of f.

4. For what values of c will the equation $x^3-3x^2+4x+c=0$ have a root between -9 and -10?

5. Find how many roots each of the following equations has. You are not required to find the roots.
   (i) $x^2+x+1=0$;   (ii) $4x^2-2x-1=0$;
   (iii) $9x^2-30x+25=0$;
   (iv) $(k+1)x^2+2kx+(k-1)=0$ where k is a constant.

6. For what values of c does the function $2x^2+5x+c$ have two (distinct) roots?

7. (i) Show that the x-axis is a tangent to the circle $x^2+y^2-10x-4y+25=0$. [*Hint*: consider the intersection of the x-axis with the circle.]

(ii) Find the equation of the other tangent from the origin.

8. The parabola $y = -x^2 + ax + b$ passes through the origin, and has the line $y = 2x + 1$ as a tangent. Find a and b.

9. Prove algebraically that no line through the origin is a tangent to the curve $y = x^2 - 1$. Illustrate this by sketching the curve.

2.13 An introduction to limits

We now give a very informal introduction to the idea of a limit; precise definitions are given in Chapter 39.

Consider first the sequence† of numbers

$$3, \ 2\tfrac{1}{2}, \ 2\tfrac{1}{4}, \ 2\tfrac{1}{8}, \ 2\tfrac{1}{16}, \ \ldots$$

where the dots mean that the sequence continues in the obvious way. Although we never actually reach the number 2, we can get as close to 2 as we like, by going sufficiently far along the sequence. Thus if we are given *any* tiny number, say 10^{-10}, we can find a point in the sequence where we can say that *from this point on, we shall always be within 10^{-10} of 2*. We say that 2 is the *limit* of the sequence, or that the sequence *tends to* 2.

Qu.19 State the limits of the following sequences. If there is no limit, say so.

(i) $16, 8, 4, 2, 1, \tfrac{1}{2}, \tfrac{1}{4}, \ldots$;

(ii) $2, 2\tfrac{1}{2}, 2\tfrac{2}{3}, 2\tfrac{3}{4}, 2\tfrac{4}{5}, \ldots$;

(iii) $1, -\tfrac{1}{3}, \tfrac{1}{9}, -\tfrac{1}{27}, \tfrac{1}{81}, \ldots$;

(iv) $2, -1\tfrac{1}{2}, 1\tfrac{1}{4}, -1\tfrac{1}{8}, 1\tfrac{1}{16}, -1\tfrac{1}{32}, \ldots$;

(v) $1\tfrac{1}{2}, 2\tfrac{1}{3}, 3\tfrac{1}{4}, 4\tfrac{1}{5}, 5\tfrac{1}{6}, \ldots$;

(vi) $1\tfrac{1}{2}, \tfrac{1}{2}, 1\tfrac{1}{4}, \tfrac{1}{4}, 1\tfrac{1}{8}, \tfrac{1}{8}, \ldots$.

Example 10

A regular polygon has n sides, and each vertex is 3 cm from the centre of the polygon. Find the limit of the area of the polygon as $n \to \infty$. (The notation $n \to \infty$ is read as 'n tends to infinity'. It may be taken to mean that n gets bigger than any number we care to name.)

Solution

If we denote the area of the n-sided polygon by A_n, then the numbers $A_3, A_4, A_5, \ldots$ form a sequence of which we wish to find the limit. Complicated trigonometry can be avoided by noting that as $n \to \infty$, the polygon gets nearer and nearer to a circle of radius 3 cm (see Fig. 2.24).

Figure 2.24

$n = 4$ $n = 7$ n 'large'

In the diagram, the shaded area tends to zero as $n \to \infty$, and so the limiting area of the polygon is the area of the circle, $9\pi \text{ cm}^2$.

†Sequences are discussed in more detail in Chapter 8.

The next example introduces the idea of the limit of a function.

Example 11

Find the limit of $\dfrac{3x^2+2x+3}{2x^2-3x+5}$ as $x\to\infty$.

(This may be written as $\displaystyle\lim_{x\to\infty}\dfrac{3x^2+2x+3}{2x^2-3x+5}$.)

Solution

An inelegant method (although a useful mental short cut) is to argue that the dominant terms as x gets large are the terms in x^2. (They become 'much larger' than the terms in x or the constant terms.)

The limit is therefore $\dfrac{3x^2}{2x^2}=\dfrac{3}{2}$.

More correctly, we have

$$\frac{3x^2+2x+3}{2x^2-3x+5}=\frac{3+2/x+3/x^2}{2-3/x+5/x^2}\quad\text{(provided }x\neq0\text{)}.$$

Noting that $2/x$, $3/x^2$, $-3/x$ and $5/x^2$ all tend to zero as $x\to\infty$, we obtain the limit $3/2$.

2.14 One- and two-sided limits

Given a function f, it may be that, as x approaches some number a, the value of $f(x)$ approaches some limit l. (We write $\displaystyle\lim_{x\to a}f(x)=l$.)

Certainly if f is a continuous function, then this limit exists and is equal to $f(a)$. If, however, f has a discontinuity at $x=a$, then $f(a)$ may take a value different from $\displaystyle\lim_{x\to a}f(x)$, or it may not be defined at all.

Consider, for example, the functions

$$g(x)=\begin{cases}x^2+1 & \text{for } x\neq0\\ 2 & \text{for } x=0\end{cases}\qquad\text{(Fig. 2.25)}$$

and

$$h(x)=\frac{1}{x-2}\qquad\text{(Fig. 2.26)}.$$

Figure 2.25

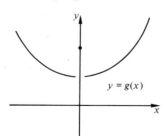

$y = g(x)$

For g: $\displaystyle\lim_{x\to0}g(x)=1$, but $g(0)=2$.

For h: $h(2)$ is not defined, and the limit of $h(x)$ as $x\to2$ depends on whether x approaches 2 from below (we write $x\to2-$) or from above ($x\to2+$). We have

$$\lim_{x\to2-}h(x)=-\infty,\text{ and }\lim_{x\to2+}h(x)=+\infty.$$

Figure 2.26

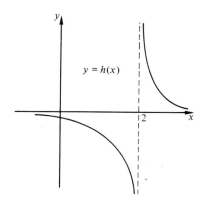

The two limits of h as $x \to 2$ are each *one-sided limits* since each can only be reached by approaching $x = 2$ from one side. The notation $\lim\limits_{x \to a} f(x)$ can be used only if the limit is *two-sided*, i.e. if

$$\lim_{x \to a -} f(x) = \lim_{x \to a +} f(x).$$

Qu.20 Let $f(x) = \dfrac{|x^2 + x|}{x}$. Find $\lim\limits_{x \to 0 -} f(x)$ and $\lim\limits_{x \to 0 +} f(x)$. What is the value of $f(0)$?

Qu.21 Evaluate $\dfrac{\tan x° - 1}{45 - x}$ for $x = 50$, 46 and 45·1, and also for $x = 40$, 44 and 44·9. Find $\lim\limits_{x \to 45} \dfrac{\tan x° - 1}{45 - x}$ to 2 s.f., verifying that the limit is two-sided.

We end this section with a warning. Suppose that f and g are two functions and that as x approaches some number, the values of $f(x)$ and $g(x)$ become 'close' in some sense; then it is tempting to write $f(x) \to g(x)$. This notation is meaningless: it does not say *how* $f(x)$ and $g(x)$ approach one another, and this is usually what matters (see Qu.22). The error can be avoided if it is remembered that the symbol $\to$ must always be followed by some *fixed* number.

Qu.22 Let $f(x) = x^2 + x$ and $g(x) = x^2 + 1$. Find

(i) $\lim\limits_{x \to \infty} \dfrac{f(x)}{g(x)}$;   (ii) $\lim\limits_{x \to \infty} (f(x) - g(x))$.

(Here, f and g are approaching one another in a multiplicative sense, but are diverging from one another in an additive sense.)

Exercise 2e **1.** Find the limits of the following sequences. If no limit exists, say so. (Methods of dealing with some sequences are discussed in later chapters; for the time being, try guessing.)

(i) 1, 2, 4, 8, 16, ...;

(ii) 0·3, 0·33, 0·333, 0·3333, ...;

(iii) 0, 1, $\frac{1}{2}$, $\frac{3}{4}$, $\frac{5}{8}$, $\frac{11}{16}$, ...;

(iv) 1, 10, 2, 9, $2\frac{1}{2}$, $8\frac{1}{2}$, $2\frac{3}{4}$, $8\frac{1}{4}$, ...;

(v) 1, $\frac{2}{3}$, $\frac{4}{9}$, $\frac{8}{27}$, ...

2. (i) Find $\lim\limits_{x \to \infty} \dfrac{1+x^2}{1-x^2}$.

(ii) Comment on the value of $f(x) = \dfrac{\log x}{x-1}$ when $x = 1$.

Evaluate $f(x)$ when x takes the values 2, 1·5, 1·2, 1·1, 1·05, 1·01.

Hence estimate $\lim\limits_{x \to 1+} \dfrac{\log x}{x-1}$. Investigate similarly $\lim\limits_{x \to 1-} \dfrac{\log x}{x-1}$.

3. Find the limits of the following functions as $x \to \infty$:

(i) $\dfrac{2x^2 + x + 3}{3x^2 - 2x - 10}$;   (ii) $\dfrac{x}{1+x}$;

(iii) $\dfrac{x^2 - 3x + 5}{2x^3 + 4x^2 + 6x + 1}$.

4. Estimate

(i) $\lim\limits_{x \to 0+} \dfrac{\sin x^\circ}{x}$;   (ii) $\lim\limits_{x \to 0+} \dfrac{\cos x^\circ - 1}{x^2}$.

5. A block graph is plotted between $x = 0$ and $x = 1$.
(i) Find the total area of the graph when
(a) there are 3 strips, each of width $\frac{1}{3}$, and of heights $\frac{1}{3}$, $\frac{2}{3}$, 1;
(b) there are 4 strips, each of width $\frac{1}{4}$, and of heights $\frac{1}{4}$, $\frac{1}{2}$, $\frac{3}{4}$, 1;
(c) there are 10 strips, each of width 1/10, and of heights 1/10, 2/10, ... 1.
(ii) If the graph has n strips each of width $1/n$, and of heights $1/n$, $2/n$, $3/n$, ... 1, what is the limit of the area as $n \to \infty$?

6. (i) State the limit of the sequence

$$1, \; 1+\tfrac{1}{2}, \; 1+\tfrac{1}{2}+\tfrac{1}{4}, \; 1+\tfrac{1}{2}+\tfrac{1}{4}+\tfrac{1}{8}, \; \ldots .$$

(ii) Show that the sequence

$$1, \; 1+\tfrac{1}{2}, \; 1+\tfrac{1}{2}+\tfrac{1}{3}, \; 1+\tfrac{1}{2}+\tfrac{1}{3}+\tfrac{1}{4}, \; \ldots .$$

tends to infinity, by bracketing thus

$$1+\tfrac{1}{2}+(\tfrac{1}{3}+\tfrac{1}{4})+(\tfrac{1}{5}+\tfrac{1}{6}+\tfrac{1}{7}+\tfrac{1}{8})+(\tfrac{1}{9}+\ldots+\tfrac{1}{16})+\ldots,$$

and demonstrating that each bracket is greater than $\frac{1}{2}$. Complete the argument carefully.

2.15 Countable and uncountable sets

Definition

Let S be an infinite set. We say that S is *countable* if there is a mapping from Z^+ to S which is 1-1 and onto. If no such mapping exists, S is said to be *uncountable*.

If a set S is countable, it is possible to denote the images of 1, 2, 3, ... by, for example, s_1, s_2, s_3, ... Conversely, if the members of S can be listed so that there is a first member of the list, a second member, a third, fourth, and so on (with every member of S included in the list), then S must be countable, because there is the natural mapping from Z^+ to S defined by

$$\forall n \in Z^+, \ f(n) = \text{the } n\text{th member of the list.}$$

We now state two important results which have standard proofs.

Theorem 2.1

(i) Q^+ is countable.

(ii) R^+ is uncountable.

Proof

(i) Imagine that all the positive rationals are cancelled down to their simplest terms, and then arranged in an infinite rectangular array thus:

in the top row, put all the integers;

in the second row, put all the members with denominator 2;

in the third row, put all the members with denominator 3;

and so on.

We then get

1	2	3	4	5	6	. . .
1/2	3/2	5/2	7/2	9/2	11/2	. . .
1/3	2/3	4/3	5/3	7/3	8/3	. . .
1/4	3/4	5/4	7/4	9/4	11/4	. . .

(We must now avoid the temptation to argue that we can 'count' the top row, and then go on to the second row etc.. 'Counting' the top row uses up all the members of Z!) We count the array diagonally as indicated in Fig. 2.27, starting with the top left corner. This 'lists' the

Figure 2.27

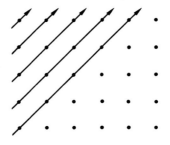

members of Q^+ thus

$$1; \ \tfrac{1}{2}, \ 2, \ \tfrac{1}{3}, \ \tfrac{3}{2}, \ 3, \ \tfrac{1}{4}, \ \tfrac{2}{3}, \ \dots$$

and every member of Q^+ will appear in this list. We can make this list the images of 1, 2, 3, ... in order. Hence result.

(ii) (In proving that R^+ is uncountable, it is not sufficient to show that some particular method of counting fails: this might merely indicate incompetence! Instead, we must show that *any* method of counting will fail. We use R.A.A..)

Suppose that R^+ is countable, so that its members may be listed (as decimals). Construct a decimal less than 1 as follows: choose as the nth digit after the decimal point

    1 if the corresponding digit in the nth member of the list is not a 1;

    2 if the corresponding digit *is* a 1.

We now claim that this decimal does not appear in the original list, since for each $n \in Z^+$, it is different from the nth member of the list. (It differs in the nth decimal place.) The original list is therefore incomplete. Hence result (R.A.A.).

Qu.23 (i) Modify the proof in part (i) above to show that Q is countable.

(ii) What is meant by the phrase *a fortiori*? Explain how, from the result of part (ii) above, R is uncountable *a fortiori*.

Qu.24 (i) Show how the proof of part (i) of the theorem above may be modified to prove that $Z^+ \times Z^+$ is countable.

(ii) Is $Z^+ \times Z^+ \times Z^+$ countable?

Exercise 2f

1. Show that Z is countable.

2. Let S be a countable set. Show that $S \times S$ is countable.

3. Show that $Z^+ \times Z^+ \times Z^+ \times Z^+ \times \dots$ is uncountable. [*Hint*: compare with the proof that R^+ is uncountable.]

4. Let $S_1, S_2, S_3, S_4, \dots$ be countably many countable sets. Investigate whether $S_1 \cup S_2 \cup S_3 \cup S_4 \cup \dots$ is countable.

5. Let $I = [0, 1]$. Find a 1-1 and onto mapping from $I \times I$ to I.

Miscellaneous exercise 2

1. A function f is defined by $f : x \to \dfrac{1}{x+1}$. Write down in similar form expressions for f^{-1} and ff. It is required to find the values of x for which (i) $f = f^{-1}$, (ii) $f = ff$. Show that, in each case, the values of x are given by the equation

$$x^2 + x - 1 = 0.$$

[C (O)]

2. Find the ranges of values of k for which the equation $5x^2 - 2kx + c = 0$ has real roots,

    (i) if $c = 20$,

    (ii) if $c = -20$.

[AEB (O) '80]

3. A function f has domain the set of real numbers. For $0 \leqslant x \leqslant 1$ it is given by the equation

$$f(x) = 1 - x.$$

Given also that f is an even function with period 2, draw its graph over the interval $-3 \leqslant x \leqslant 3$. Write down equations for the function for (i) $-1 \leqslant x \leqslant 0$, (ii) $2 \leqslant x \leqslant 3$.

[SMP]

4. Binary relations R and S are defined on the set $\mathbb{R}$ of all real numbers by

aRb if and only if $a - b \in \mathbb{Q}$, the set of rational numbers,
aSb if and only if $a - b = 0$ or $a - b$ is an irrational number.
Show that *R* is an equivalence relation on $\mathbb{R}$ and that *S* is not.
If *T* denotes the set of equivalence classes with respect to *R* and if
$[x]$ denotes the equivalence class containing the real number *x*,
show that the mapping $f: \mathbb{R} \to T$ defined by $f(a) = [a^2]$ is a mapping
from $\mathbb{R}$ *onto* *T* (i.e. that corresponding to any $[x]$ there is a
number $a \in \mathbb{R}$ such that $f(a) = [x]$.)

[L]

5. (a) A function *f* is defined on the interval [0, 3] as follows:
$$f(x) = 3 - x \text{ if } x \in [0, 2),$$
$$= x - 2 \text{ if } x \in [2, 3].$$
Sketch the graph of *f*, indicating clearly the point of the graph at
any discontinuity. State the range of *f*. Sketch the graphs of the
composite function $f \circ f$ and the inverse function f^{-1}.
(b) Show that the function ϕ defined by

$$\phi(x) = |x - 3| - |x + 3|$$

is an odd function, and sketch its graph.

[W]

6. The domain of the function $f: x \to \dfrac{1}{1-x}$ is the set of all real

numbers not equal to 0 or 1. Determine $f^2(x)$ and show that
$f^3(x) = x$.

$$[f^2(x) = ff(x) \text{ and } f^3(x) = fff(x).]$$

The function *g* has the same domain as *f* and is defined in terms
of *f* by the relation

$$g(x) = 4 + xgf(x).$$

By considering $gf(x)$ and $gf^2(x)$ find $g(x)$ as a rational function of
x.

[JMB (S)]

7. A function *f* is defined on *R* by

$$f: x \to |x + [x]|$$

where $[x]$ indicates the greatest integer less than or equal to *x*,
e.g. $[3] = 3$, $[2 \cdot 4] = 2$, $[-3 \cdot 6] = -4$. Sketch the graph of the function
for $-3 \leqslant x \leqslant 3$. What is the range of *f*? Is the mapping one-one?
The function *g* is defined by $g: x \to |x + [x]|$, $x \in R_+$, $x \notin Z_+$. Find
the rule and domain of the inverse function g^{-1}.

[C (S)]

3 Coordinate geometry I

3.1 Introduction

The reader will already be familiar with the idea of plotting points on a graph, and of drawing a curve to represent an algebraic relationship between two variables x and y. When, for example, we draw the graph of $y = x^2 + x + 1$, then the coordinates of any point on the curve satisfy the equation; conversely, the coordinates of any point not on the curve do not satisfy the equation. In this chapter, we shall examine this relationship between algebra and geometry in more detail. Note that it is almost always helpful to draw a rough sketch when solving a coordinate geometry problem.

3.2 The distance between two points

Example 1

Let P and Q be the points $(1, 2)$ and $(3, -4)$ respectively. Find the distance PQ.

Solution

Completing the triangle PQR (Fig. 3.1) by drawing lines parallel to the axes, we have:

$$PR = 3 - 1 = 2$$
$$RQ = 2 - (-4) = 6.$$

Figure 3.1

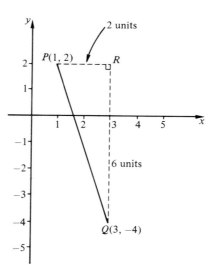

So by Pythagoras's Theorem,

$$PQ = \sqrt{(6^2 + 2^2)} = \sqrt{40}.$$

Qu.1 Find the distance between the following pairs of points:
(i) (2, 3) and (4, 4); (ii) $(-1, 3)$ and $(4, -3)$;
(iii) $(-7, -1)$ and (8, 9).

We can embody this method in a general formula thus: Let P_1 and P_2 be the points (x_1, y_1) and (x_2, y_2) respectively. Then, by the method of Example 1, we have

$$P_1P_2 = \sqrt{\{(x_1-x_2)^2+(y_1-y_2)^2\}}.$$

Note. (1) Sometimes $x_1 - x_2$ will be negative. This does not matter since it is to be squared anyway. The same applies to $y_1 - y_2$.
(2) The formula applies even if some of the four coordinates are negative. (Try it!)

Qu.2 Let P be $(4, -2)$, Q be $(7, -1)$ and R be (5, 5). Find the lengths of PQ, QR, PR, and deduce that angle Q is a right angle.

3.3 The mid-point of a line

Example 2
Find the mid-point of the line-segment PQ where P is the point (3, 4) and Q is (7, 1).

Solution

Figure 3.2

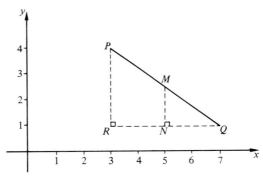

Complete the triangle PQR as shown. Let M be the required mid-point and let MN be parallel to PR. By similar triangles, N is the mid-point of RQ, and so its x-coordinate is the average of the x-coordinates of R and Q, i.e.

$$x\text{-coordinate of } N = \tfrac{1}{2}(3+7) = 5$$
$$\Rightarrow x\text{-coordinate of } M = 5.$$

Similarly, by drawing a horizontal line to meet PR,
y-coordinate of M = average of y-coordinates of P and R

$$= \tfrac{1}{2}(4+1) = 2\tfrac{1}{2}.$$

So M is the point $(5, 2\tfrac{1}{2})$.

Qu.3 Find the mid-points of the lines joining
(i) (4, 9) and (6, 6); (ii) $(2, -1)$ and $(-3, 0)$;
(iii) $(-4, -9)$ and $(1, -6)$.

Generalizing,

> the mid-point of (x_1, y_1) and (x_2, y_2) is
> $$\left(\frac{x_1 + x_2}{2}, \frac{y_1 + y_2}{2}\right).$$

Exercise 3a

1. For each of the following pairs of points, find
   (a) the distance between them, and
   (b) the mid-point of the line joining them.
   (i) $(5, -2)$ and $(-3, -1)$;   (ii) $(-10, 3)$ and $(5, -5)$;
   (iii) $(p+3, q+4)$ and $(p-3, q-4)$;
   (iv) $(a, 0)$ and $(0, b)$.

2. Let A, B, C be the points $(1, -6)$, $(2, 1)$ and $(-3, 6)$ respectively. Mark these points on a rough diagram and prove that triangle ABC is isosceles.

3. Let P, Q, R be the points $(6, 1)$, $(-2, -3)$ and $(3, -4)$ respectively, and let L and M be the mid-points of PQ and PR. Calculate the lengths of QR and LM and verify that $QR = 2LM$.

4. Prove that the circle with centre $C(-1, 5)$ which passes through $G(5, -2)$ also passes through $H(8, 3)$ and $K(1, -4)$.

5. Mark on a diagram the points $A(-2, -3)$, $B(4, -1)$ and $C(3, 1)$. Find by inspection (i.e. by observation and common sense) the coordinates of the point D such that $ABCD$ is a parallelogram. (Make sure that you have the letters in the correct order.) Check your answer by finding
   (i) the lengths AB and CD;
   (ii) the lengths AD and BC;
   (iii) the mid-point of AC and the mid-point of BD.

6. The points P and Q lie on the curve $y = 2x^2 - x - 1$ and have x-coordinates -2 and 3. Find the distance between the points. Find also the mid-point of PQ, and show that it does not lie on the curve.

7. A, B and C lie on the curve $x = y^2 - 6y + 11$ and have y-coordinates $-\frac{1}{2}$, 2 and $4\frac{1}{2}$ respectively. Find the lengths AB, BC and CA, and deduce that AB subtends an angle of $90°$ at C.

8. With P and Q as in question 3, find the coordinates of S and T such that P is the mid-point of SQ and Q is the mid-point of PT.

9. Mark on a diagram the points $A(-2, 1)$ and $B(8, -9)$ and the line $x = 2$. It is required to find a point P on this line such that $\angle APB = 90°$.
   (i) From your diagram, state·how many such points you think there are.
   (ii) Taking P to be the point $(2, \lambda)$, write down Pythagoras's Theorem for triangle APB to obtain an equation in λ.
   (iii) Solve this equation for λ to find all possible positions of P.
   (iv) If M is the mid-point of AB, explain why

   $$\angle APB = 90° \Rightarrow MA = MP.$$

   [*Hint*: think of AB as the diameter of a circle.]
   (v) Check that $MA = MP$ for each of the points P found in (iii).

3.4 The gradient of a straight line

Suppose we take two points $S(2, 1)$ and $T(4, 7)$ and join them by a straight line (Fig. 3.3). Then in moving from S to T, the x-coordinate increases by two, and the y-coordinate by six. These two numbers tell us something about the 'slope' of the line ST.

Figure 3.3

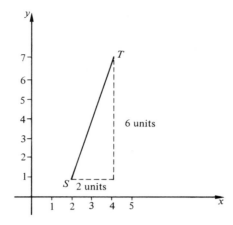

Definition

The *gradient* of a line joining two points is defined by:

$$\text{gradient} = \frac{\text{change in } y\text{-coordinate}}{\text{change in } x\text{-coordinate}}.$$

Thus the gradient of the line joining (x_1, y_1) and (x_2, y_2) is given by

$$\text{gradient} = \frac{y_2 - y_1}{x_2 - x_1}.$$

Qu.4 Let A, B, C, D be any four points on a straight line. By completing the triangles as shown, prove that the gradient of AB and

Figure 3.4

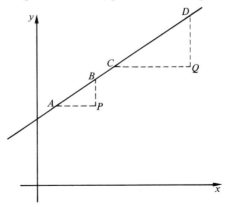

the gradient of CD are equal. (This proves that the gradient of a line depends only on the line, and not on the particular pair of points chosen on it.)

In our example above (Fig. 3.3), the gradient of ST is $6/2 = 3$. This means that y increases by 3 units for every unit that x increases. Note that, in evaluating this gradient, we moved from S to T; if we move from *T* to *S*, we get the same gradient (fortunately) since then

$$\frac{\text{change in } y}{\text{change in } x} = \frac{-6}{-2} = 3.$$

Since the gradient of a straight line is the same between any two points, it follows that if we extend *ST* in both directions, we can write down as many points on the line as we wish, by adding any number to the *x*-coordinate of *S*, and 3 times it to the *y*-coordinate. Here are some such points:

x	2	3	4	5	6	1	0	$2\frac{1}{2}$	2·01
y	1	4	7	10	13	−2	−5	$2\frac{1}{2}$	1·03

Indeed, since the *x*-coordinate of any point on the line can be written as $2+k$ (for suitable *k*), it follows that $(2+k, 1+3k)$ is a typical point on this line. We shall examine this idea in more detail in later chapters.

3.5 The significance of the gradient

Let us now consider how the value of the gradient of a line is related to its 'slope'. (We assume that the graph is drawn with the same scale on each axis. When finding a gradient from a graph with unequal scales, care must be taken to read the 'change in *x*' from the *x*-axis, and the 'change in *y*' from the *y*-axis.)

(i) A positive gradient means that *y* increases as *x* increases, i.e. the line goes up to the right. A negative gradient means that the line goes down to the right (Figs. 3.5–8).

Figure 3.5

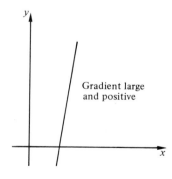

Gradient large and positive

Figure 3.6

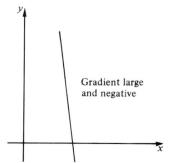

Gradient large and negative

Figure 3.7

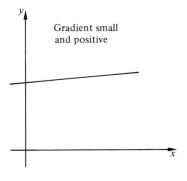

Figure 3.8

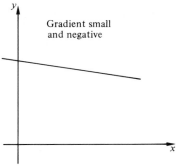

(ii) The magnitude of the gradient tells us how steep the line is. A gradient of 10 or -10 would indicate a steep line, whereas a gradient of $1/10$ or $-1/10$ would correspond to a fairly flat line. More precisely, the gradient is equal to the tangent of the angle that the line makes with the positive x-axis (measured anticlockwise from the axis).

(iii) A gradient of zero means that

$$\frac{\text{change in } y}{\text{change in } x} = 0, \text{ i.e. change in } y = 0.$$

So the line is parallel to the x-axis.

(iv) If a line is parallel to the y-axis (Fig. 3.9), then, in moving from P to Q, we have

$$\frac{\text{change in } y}{\text{change in } x} = \frac{2}{0} \text{ which is meaningless.}$$

In this case, we say that the gradient is infinite.

Figure 3.9

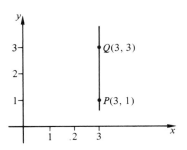

Qu.5 Find the gradients of the lines joining the following pairs of

COORDINATE GEOMETRY I

points, and hence find two more points on each line. Check your answers by rough sketches.

  (i) $(2, 4)$ and $(3, 8)$; (ii) $(-1, -3)$ and $(2, -6)$;

(iii) $(-4, 2)$ and $(-4, 4)$; (iv) $(0, -1)$ and $(4, 4)$;

(v) (a, a^2) and (b, b^2).

Qu.6 (i) Find the acute angle between the x-axis and a line with gradient 2.

(ii) Find the acute angle between two lines whose gradients are $\frac{1}{2}$ and $1\frac{1}{2}$.

(iii) What are the possible gradients of a line making an angle of $45°$ with the x-axis?

[*Hint for Qus. 7, 8, 9*: draw a rough sketch and use the definition of the gradient.]

Qu.7 The line through the point $(-3, -6)$ with gradient 4 passes through the point $(2, k)$. Find k.

Qu.8 The line through $(4, 0)$ with gradient $-2/3$ passes through $(h, 5)$. Find h.

Qu.9 The line through (x_0, y_0) and (x_1, y_1) has gradient m. Express y_1 in terms of x_0, y_0, x_1 and m.

3.6 The equation of a straight line

Consider, as an example, the line with gradient 2 passing through the point $P(4, 1)$ (Fig. 3.10). If (x, y) is any other point on the line, then, since the gradient is 2, we have:

$$\frac{y-1}{x-4}=2$$

$$\Rightarrow\ y-1=2(x-4)$$
$$\Rightarrow\ y=2x-7.$$

Figure 3.10

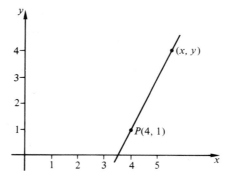

So any point on the line satisfies $y=2x-7$, and this is the equation of the line. (The point P itself does not satisfy $\frac{y-1}{x-4}=2$, but it does satisfy $y-1=2(x-4)$; so all points on the line do in fact satisfy this latter equation.)

    In general, the line with gradient m passing through the point

(x_0, y_0) is

$$\frac{y-y_0}{x-x_0}=m \quad \Rightarrow \quad y-y_0=m(x-x_0)$$

$$\Rightarrow \quad y=mx+(y_0-mx_0).$$

(Note that y_0-mx_0 is a constant.)

It is clear, therefore, that the equation of a line with gradient m can be written in the form $y=mx+c$, where c is a constant. In practice, we can find the equation of a line by the method above, or by the methods used in the next two examples.

Example 3
Find the equation of the line through the point $(-3, 4)$ with gradient $-\frac{1}{2}$.

Solution

The equation is of the form $y=-\frac{1}{2}x+c$. We find c by substituting the given point.

$$4=-\frac{1}{2}(-3)+c \Rightarrow c=2\frac{1}{2}.$$

So the required equation is

$$y=-\frac{1}{2}x+2\frac{1}{2}$$

or

$$2y+x=5.$$

Example 4
Find the equation of the line passing through $(4, 1)$ and $(7, -1)$.

Solution

The gradient of the line is $\dfrac{1-(-1)}{4-7}=-\frac{2}{3}$. So the equation is of the form $y=-\frac{2}{3}x+c$.

Substituting $(4, 1)$ gives

$$1=-\frac{2}{3}\times 4+c$$
$$\Rightarrow \quad c=\frac{11}{3}.$$

The equation is therefore

$$y=-\frac{2}{3}x+\frac{11}{3}$$

or

$$3y+2x=11.$$

(We can check this by substituting the other point, $(7, -1)$.)

We have seen that when the equation of a straight line is written in the form $y=mx+c$, the number m is the gradient of the line. The geometrical significance of the number c can be seen from the fact that c is the value of y when $x=0$; i.e. the point $(0, c)$ lies on the line. So c is the 'height' at which the line cuts the y-axis. This can be of use when drawing a rough sketch of the line.

Qu.10 Use the ideas in the paragraph above to sketch roughly the lines
(i) $y=2x+3$, (ii) $2y=x-4$,

indicating where they cut the y-axis. (Check by accurate drawings.)

Qu.11 (i) Sketch the lines (a) $y=3$; (b) $x=-2$.
(ii) Write down the equations of the lines through the point $(1, -4)$ parallel to
(a) the x-axis; (b) the y-axis.

Finally in this section, the reader should recall that to find the point of intersection of two lines, their equations should be solved simultaneously. This is because the geometrical problem of finding a single point that lies on two lines is equivalent to the algebraic problem of finding a single pair of numbers that satisfy two linear equations. (The reverse process, that of drawing a graph to solve a pair of simultaneous equations, will be familiar from elementary work.)

Qu.12 Find the equations of
(i) the line through $(2, -2)$ with gradient -2;
(ii) the line through $(4, 5)$ and $(7, 1)$.
Find the intersection of these two lines, and check that your answer is reasonable by roughly sketching the lines.

Problems involving the point of intersection of two lines can sometimes be solved without calculating the coordinates of the point of intersection, by using the ideas and method suggested in Qu.13. These ideas are discussed in more detail (in the context of the intersection of planes) in Chapter 25.

Qu.13 Let l_1 be the line $2x+5y=9$, and l_2 the line $3x+2y=6$.
(i) Explain why $(2x+5y-9)+\lambda(3x+2y-6)=0$ (where λ is a constant) is certainly the equation of a straight line. (Call this line l_3.)
(ii) Explain why any pair of values (x, y) satisfying the equations of l_1 and l_2 must also satisfy the equation of l_3.
(iii) Deduce from (i) and (ii) that l_3 is a line through the intersection of l_1 and l_2.
(iv) Find the values of λ that will make l_3
   (a) pass through the origin;
   (b) a vertical line;
   (c) perpendicular to l_2.
(v) Use the method of parts (i) to (iv) to find the equation of the line through the point of intersection of $x+y=5$ and $2x-y=3$ and the point $(4, 2)$.

Exercise 3b **1.** Let P, Q and R be the points $(-2, 1)$, $(3, 3)$ and $(13, 7)$ respectively. Find the gradients of PQ and QR. What does this indicate about the three points? *Write down* the gradient of PR.
2. Find the gradients of the lines joining
   (i) $(4, -5)$ and $(-5, 4)$; (ii) $(0, 0)$ and $(-3, 4)$;
   (iii) (p, p^2) and (q, q^2);
   (iv) $(3, 9)$ and $(3+h, (3+h)^2)$.
3. (i) The line joining $(2, -3)$ and $(-1, p)$ has gradient -1. Find p.
   (ii) The line joining $(-4, 7)$ and $(q, 3)$ has gradient $-\frac{3}{4}$. Find q.

COORDINATE GEOMETRY I

(iii) The line joining (a, b) and $(a+h, c)$ has gradient m. Express c in terms of the other letters.

4. Let A, B, C be the points $(3, 1)$, $(-2, 3)$ and $(-3, -4)$ respectively. Find the equation of the line
   (i) through A with gradient 2;
   (ii) through B with gradient -2;
   (iii) through C with gradient $-\frac{2}{5}$;
   (iv) through B with gradient zero;
   (v) through A with infinite gradient;
   (vi) AB;
   (vii) AD where D is the mid-point of BC;
   (viii) AO where O is the origin.

5. (i) Find the point of intersection of the lines $2x+3y=5$ and $3x+4y=6$, and show that the line $x+2y=4$ also passes through this point.
   (ii) Prove that the lines $3x-y=7$, $2x+3y=5$ and $7x-6y=16$ are concurrent (i.e. all pass through a point).

6. Find the point of intersection of the line through $(-3, -1)$ and $(-1, -7)$ with the line through $(1, 2)$ with gradient λ, in the cases
   (i) $\lambda=1$;   (ii) $\lambda=0$;   (iii) $\lambda=-3$.
   Give a geometrical explanation of (iii).

7. By writing each of the following equations in the form $y=mx+c$, sketch each line roughly, indicating where it cuts the y-axis.
   (i) $y-3x=2$;   (ii) $2y+3x=7$;
   (iii) $x=2y+3$;   (iv) $2(x-y)=3(x+y+1)$.

8. (i) Find the acute angle between the line $y=3x-7$ and the x-axis.
   (ii) Find the acute angle between the line in (i) and the line $2y=x+2$.

9. (i) Show that the equation of the straight line through (x_0, y_0) with gradient m may be written $(y-y_0)=m(x-x_0)$.
   (ii) Show that the equation of the straight line through $A(x_0, y_0)$ and $B(x_1, y_1)$ may be written $\dfrac{y-y_0}{x-x_0}=\dfrac{y_1-y_0}{x_1-x_0}$. (This version of the equation expresses the fact that if $P(x, y)$ is some point on the line, then the gradients of PA and AB are equal.)

10. The lines $3y=2x+8$ and $3y+4x=38$ intersect at A and form two sides of a triangle ABC. B is the point $(-1, 2)$. The median through C has gradient $-\frac{1}{3}$. Find the coordinates of C. (A *median* of a triangle is a line joining a vertex to the mid-point of the opposite side.)

3.7 Parallel and per-
pendicular lines

Since the gradient of a line gives its direction, two lines will be parallel if and only if they have the same gradient. Thus, for example, the lines

$$2y=1-4x \text{ and } y+2x=1$$

are parallel since they both have gradient -2.

Now consider a line which cuts the x-axis and has a positive gradient m_1 (Fig. 3.11). Then, from the definition of the gradient,

$$m_1=\tan\theta_1. \tag{1}$$

Figure 3.11

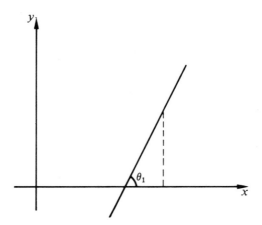

Similarly (Fig. 3.12), for a line with negative gradient m_2,

$$m_2 = -\tan\theta_2. \tag{2}$$

Now suppose that these lines are perpendicular (Fig. 3.13).

Let $PQ = a$ and $PR = b$. Then by elementary trigonometry,

$$\tan\theta_1 = b/a \text{ and } \tan\theta_2 = a/b. \tag{3}$$

Figure 3.12

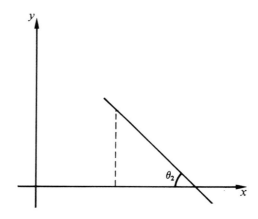

Figure 3.13

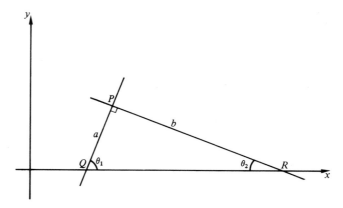

COORDINATE GEOMETRY I

From (1), (2) and (3) we have

$$m_1 m_2 = -\tan\theta_1 \tan\theta_2$$
$$= -(b/a)(a/b)$$
$$= -1.$$

Thus if two lines are perpendicular, the product of their gradients is -1 unless the lines are parallel to the axes; the gradients are then infinite and zero and they cannot be multiplied.

Conversely, suppose that two lines l_1 and l_2 have gradients m_1 and m_2 where $m_1 m_2 = -1$ (i.e. $m_2 = -1/m_1$). By the result above, if a third line l_3 is perpendicular to l_1, it has gradient $m_3 = -1/m_1$. So l_3 and l_2 are parallel, i.e. l_2 is perpendicular to l_1. We have thus proved the following theorem:

Theorem 3.1
Two lines, not parallel to the axes, are perpendicular if and only if their gradients, m_1 and m_2, satisfy $m_1 m_2 = -1$.

Example 5
Find the equations of the lines through the point (2, 4) which are
(i) parallel,   (ii) perpendicular,
to the line $3y + 5x = 6$.
Solution
(i) Any line parallel to $3y + 5x = 6$ can be written in the form $3y + 5x = k$. (Check this: it is a useful short cut.) Substituting the given point gives $k = 22$, so the required equation is $3y + 5x = 22$.
(ii) The given line has gradient $-\frac{5}{3}$, and so a perpendicular line has gradient $\frac{3}{5}$. Substituting (2, 4) into $5y = 3x + k$ gives $k = 14$. So the equation is $5y = 3x + 14$.

Qu.14   Let l be the line $3x + 4y = 8$, and P the point $(-2, 1)$. Find
(i) the equation of the line through P perpendicular to l;
(ii) the point of intersection of l with this perpendicular.
Hence find
(iii) the perpendicular distance of P from l. (See also Example 6 below.)

Example 6
Find the distance of the point $P(h, k)$ from the line l with the equation $ax + by + c = 0$.
Solution
(A slightly simpler method is given in Exercise 25c, question 19.)
Suppose first that $a \neq 0$ and $b \neq 0$, so that the line is not parallel to either axis. From P, we draw lines parallel to the axes to meet l at Q and R (Fig. 3.14).

The y-coordinate of R may be found by substituting $x = h$ into the equation of l to give

$$ah + by + c = 0$$

$$\Rightarrow y = -\frac{ah + c}{b}.$$

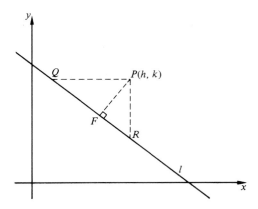

Figure 3.14

So the distance PR is

$$k - \left(-\frac{ah+c}{b} \right) = \frac{ah+bk+c}{b}. \tag{1}$$

Similarly,
$$PQ = \frac{ah+bk+c}{a}. \tag{2}$$

Knowing PQ and PR, we can find the length PF, since, by considering the area of triangle PQR in two ways,

$$\tfrac{1}{2}PQ \cdot PR = \tfrac{1}{2}PF \cdot QR$$

$$\Rightarrow PF = \frac{PQ \cdot PR}{QR}$$

$$\Rightarrow PF = \frac{PQ \cdot PR}{\sqrt{(PQ^2 + PR^2)}}.$$

(This result may also be obtained from similar triangles.) Substituting from (1) and (2),

$$PF = \frac{ah+bk+c}{a} \cdot \frac{ah+bk+c}{b} \div (ah+bk+c) \sqrt{\left(\frac{1}{a^2} + \frac{1}{b^2} \right)}$$

$$\Rightarrow PF = \frac{ah+bk+c}{\sqrt{(a^2+b^2)}}. \tag{3}$$

(To obtain this last line, note that

$$\sqrt{\left(\frac{1}{a^2} + \frac{1}{b^2} \right)} = \sqrt{\left(\frac{a^2+b^2}{a^2 b^2} \right)} = \frac{1}{ab}\sqrt{(a^2+b^2)}.)$$

Now suppose that $a=0$. Then (Fig. 3.15) l is

$$by+c=0$$

or
$$y = -c/b.$$

Then
$$PF = k - (-c/b) = \frac{bk+c}{b}.$$

Since $a=0$, we may write as

$$PF = \frac{ah+bk+c}{\sqrt{(a^2+b^2)}}$$

Figure 3.15

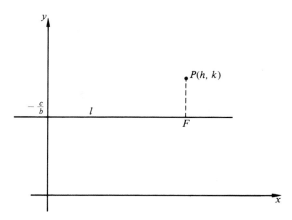

which is the same as (3). Similarly this expression is valid when $b=0$, i.e. it is valid in all cases. Inserting a modulus sign to ensure a non-negative answer,

> the distance of (h, k) from $ax+by+c=0$ is
>
> $$\frac{|ah+bk+c|}{\sqrt{(a^2+b^2)}}.$$

(Note that, strictly, modulus signs should have been used at various points in this solution; they were omitted to avoid confusion. The reader should consider exactly where they are needed, remembering that a 'distance' must be positive.)

Qu.15 Use the formula derived above to find the distance of $(-7, 2)$ from $5x-12y=8$.
Qu.16 Use the result of Example 6 to answer Qu.13 (iii).

3.8 Loci A *locus* (plural *loci*) is a set of points obeying some given condition. For example, if C is a fixed point in a plane, and a is a number, then the locus of P such that $PC=a$ is the circle with centre C and radius a.

Qu.17 If A and B are two fixed points in a plane, describe, with the aid of sketches, the locus of P if
  (i) $AP=BP$;   (ii) $\angle APB=90°$;
(iii) $\angle BAP=90°$;   (iv) $AP=AB$;
 (v) $AP>BP$ (shade the required region);
(vi) the distance of P from the (infinite) line AB is 1 cm.

Example 7
A is the point $(2, 2)$, and a point P moves so that its distance from A is always equal to its distance from the x-axis. Find the equation of the locus of P.
Solution
Let P be the 'general' point (x, y). (The problem is to express the given

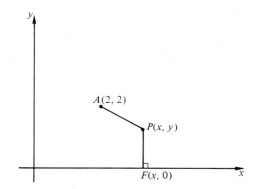

Figure 3.16

geometrical condition on P as an algebraic restriction on x and y.) Let F be the foot of the perpendicular from P to the x-axis.

$$AP = PF$$
$$\Leftrightarrow \sqrt{\{(x-2)^2 + (y-2)^2\}} = |y|$$
$$\Leftrightarrow (x-2)^2 + (y-2)^2 = y^2$$
$$\Leftrightarrow x^2 - 4x + 4 - 4y + 4 = 0$$
$$\Leftrightarrow 4y = x^2 - 4x + 8.$$

This is the required equation.

Exercise 3c

1. Find the equations of the lines through the origin, parallel to
   (i) $y = 2x + 3$; (ii) $2y = 3x - 7$;
   (iii) $5y + 4x = 1$.
2. Find the equations of the lines through $(1, -3)$ parallel to
   (i) $2y - 5x = 7$; (ii) $x + y = -1$;
   (iii) $x = y$; (iv) $y = 3$.
3. and 4. Repeat questions 1 and 2 with 'parallel' replaced by 'perpendicular'.
5. Let P and Q be the points $(1, 4)$ and $(-2, 2)$. Find (i) the midpoint of PQ; (ii) the gradient of PQ. Hence find the equation of the perpendicular bisector of PQ.
6. Let A, B, C, D be the points $(3, 4)$, $(8, -1)$, $(1, -2)$ and $(-4, 3)$ respectively. Show that the lines AC and BD bisect one another at right angles. (Make sure you use the quickest method.) What can you deduce about the quadrilateral $ABCD$?
7. Find the intersection of the line through $A(2, 7)$, parallel to the line joining $B(-2, -1)$ and $C(1, -7)$, with the line through C perpendicular to AB.
8. Let R and S be $(5, 2)$ and $(-3, 4)$, and let P be the point (x, y). Write down expressions, in terms of x and y, for PR^2 and PS^2. Equate these, and deduce that P satisfies $4x - y = 1$. Describe, with reasons, what the locus of P is.
9. Find the perpendicular bisector of AB where A and B are as in question 7, using the method of
   (i) question 5; (ii) question 8.
10. Find the perpendicular bisector of the line joining (a, b) and (c, d).
11. Let P, Q, R, S be the points $(0, 2)$, $(-2, -4)$, $(1, -5)$ and $(3, 1)$

respectively. Show that *PQRS* is a rectangle. (Try to find the best method.)

12. Let *A*, *B*, *C* be the points (0, 3), (6, 5) and (2, −1).
    (i) Find the circumcentre and circumradius of triangle *ABC*. (The circumcircle of three points is the circle passing through them.)
    (ii) Show that the medians are concurrent, and find the centroid (the point where they meet).
    (iii) Show that the altitudes (perpendiculars dropped from vertices) are concurrent, and find the orthocentre (where they meet).

13. Find the distance of the given point from the given line.
    (i) (2, 0), $y = 3x + 2$;
    (ii) (−1, 4), $3y + 2x = 5$;
    (iii) (4, −1), $3y + 2x = 5$.

14. Let *A*, *B*, *C* be (3, 2), (0, −2) and (−1, 0) respectively. Show that *C* lies on the circle with diameter *AB*. (Use a method involving gradients rather than distances.)

15. Let *L* be the line $y = 2x + 3$, and let *G* be the point (4, 2). *P* and *Q* are two points on *L* such that $\angle PGQ = 90°$. Find the *x*-coordinate of *Q* if the *x*-coordinate of *P* is
    (i) 2; (ii) *k*.

16. *A*(3, −1) is one vertex of a rhombus *ABCD*, and the equation of *BD* is $y = 2x + 1$. Find the coordinates of *C*.

17. On the same diagram, sketch the graphs of $2x + 3y = 7$, $2x + 3y = 9$ and $2x + 3y = 12$, and shade the region $9 \leqslant 2x + 3y \leqslant 12$.

18. Sketch the line $2y = x + 7$ and shade the region whose coordinates satisfy $2y \leqslant x + 7$. (Choose two or three 'test' points to determine which side of the line is required.)

19. On separate diagrams, shade
    (i) $\{(x, y): y \leqslant x + 2\}$;
    (ii) $\{(x, y): x + 3y \leqslant 9\}$;
    (iii) $\{(x, y): y - 2 \leqslant x \leqslant 9 - 3y\}$.
    [*Hint*: this is the intersection of two regions.]

20. Find the distance between the parallel lines $8x - 15y - 10 = 0$ and $8x - 15y + 20 = 0$.

21. The points *A* and *B* are (2, 0) and (0, 4) respectively. The point *P* moves so that its distance from *A* is three times its distance from *B*. Find the equation of the locus of *P*.

22. *S* is the point (2, −4), and a point *P* moves so that its distance from *S* is twice its distance from the *y*-axis. Find the equation of the locus of *P*.

23. Find the equation of the locus of a point *P* which moves so that the product of its distance from the lines $x + y = 0$ and $x - y = 0$ is (the constant) c^2.

3.9 The equation of a circle

Definition

A *circle* is defined as the locus of a point whose distance from a fixed point (x_0, y_0) is some fixed number *a*. The point (x_0, y_0) is called the *centre*, and the number *a* the *radius*.

To find the equation of this circle, we express its definition algebraically. Taking a general point (x, y) on the circle, we have

$$\sqrt{\{(x-x_0)^2+(y-y_0)^2\}}=a$$

or
$$(x-x_0)^2+(y-y_0)^2=a^2.$$

This is the equation of the circle.

Example 8
Find the equation of the circle with centre $(2, -3)$ and radius 7.
Solution

$$(x-2)^2+(y+3)^2=7^2$$
$$\Leftrightarrow x^2-4x+4+y^2+6y+9=49$$
$$\Leftrightarrow x^2+y^2-4x+6y-36=0.$$

Qu.18 Find the equations of the circles whose centres and radii are
(i) $(1, 2)$, 4; (ii) $(0, 0)$, a;
(iii) $(-1, -1)$, $\sqrt{2}$.

Example 9
Find all points on the line $y=2x+2$ whose distance from the point $(-6, 1)$ is 5.
Solution
We seek the intersection of the given line with the circle centre $(-6, 1)$ and radius 5. A rough sketch (Fig. 3.17) suggests that there may be

Figure 3.17

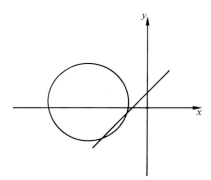

two points, or none (if the line 'misses' the circle), or possibly just one (if the line happens to be a tangent to the circle). We find any such points by solving simultaneously the equations of the circle and the line.

$$(x+6)^2+(y-1)^2=25 \tag{1}$$

$$y=2x+2. \tag{2}$$

Substituting for y from (2) into (1) gives

$$(x+6)^2+(2x+1)^2=25$$
$$\Leftrightarrow 5x^2+16x+12=0$$
$$\Leftrightarrow (5x+6)(x+2)=0$$
$$\Leftrightarrow x=-\tfrac{6}{5} \text{ or } -2.$$

COORDINATE GEOMETRY I

Then from (2)

$$\text{when } x = -\tfrac{6}{5}, \ y = -\tfrac{2}{5}.$$

and
$$\text{when } x = -2, \ y = -2.$$

So the required points are $(-\tfrac{6}{5}, -\tfrac{2}{5})$ and $(-2, -2)$.

(See Chapters 10 and 30 for further work on circles.)

Exercise 3d

1. Find the equation of the circle with centre $(15, -8)$ and radius 17. What is the geometrical significance of the vanishing of the constant term?

2. Find the equation of the set of all points which are a distance 5 from the point $(1, 3)$. Which members of the set lie on
   (i) the x-axis; (ii) the line $y = 5$;
   (iii) the line $y = 2x - 1$?

3. A circle has centre $C(-1, -5)$ and passes through $A(1, -2)$. Find the equation of
   (i) the circle; (ii) the diameter through A;
   (iii) the tangent at A.

4. Let W and Z be the points $(2, -3)$ and $(-3, 9)$. Find
   (i) the mid-point of WZ; (ii) the length of WZ.
   Hence find the equation of the circle with WZ as diameter. (See also questions 5 and 6.)

5. Let A, B and P be $(2, 6)$, $(-1, 2)$ and (x, y) respectively.
   (i) Write down expressions for the gradients of AP and BP.
   (ii) Express algebraically the fact that AP and BP are perpendicular.
   (iii) Simplify your answer to (ii) and explain what the locus of P is.

6. Use the method of question 5 to find the equation of the circle with diameter WZ where W and Z are
   (i) $(-1, 3)$ and $(-4, -6)$; (ii) as in question 4;
   (iii) (x_0, y_0) and (x_1, y_1).

7. On the same axes, sketch the circles $x^2 + y^2 = 9$, $x^2 + y^2 = 16$ and $x^2 + y^2 = 36$. Shade the region $16 \leqslant x^2 + y^2 \leqslant 36$.

8. Shade the region $\{(x, y): (x - 3)^2 + (y + 2)^2 \leqslant 16\}$.

9. (i) If p and q are real numbers satisfying $p^2 + q^2 = 0$, find p and q.
   (ii) What does the graph of $(x - 3)^2 + (y + 1)^2 = 0$ look like?

10. Find the equations of the two circles of radius r which touch the y-axis at the origin.

11. A circle passes through $(0, 2)$ and $(2, 6)$ and its centre lies on the line $y = x - 6$. Find its equation.

12. A and B are the points $(1, -3)$ and $(4, 3)$ respectively. Find the equation of the locus of P if $AP = 2BP$, showing that it is the circle with centre $(5, 5)$ and radius $\sqrt{20}$. [*Hint*: take P to be (x, y).]

13. Find the equation of the circumcircle of $A(-2, -2)$, $B(4, -4)$ and $C(4, 6)$.

14. Find the points of intersection of the line $y + 2x = 5$ and the parabola $y = 3x^2 - 7x - 7$.

15. Find the equations of the circles whose centres and radii are
(i) (5, 2), $\sqrt{20}$;   (ii) $(-10, -3)$, $\sqrt{170}$.
Find also the points of intersection of these circles. [*Hint*: subtract
their equations to obtain a linear equation in x and y, make y the
subject and substitute back into one of the original equations.
Check your final answers by substituting into both equations.]

3.10 Parametric form The usual method of expressing a relationship between x and y is to
give a single equation connecting them. An alternative method is to
express each of x and y in terms of a third variable known as a
parameter. A simple example would be

$$x = 1 + 2t$$
$$y = 2 - t^2.$$

To plot the graph of y against x, we could complete a table thus

t	-3	-2	-1	0	1	2	3
x	-5	-3	-1	1	3	5	7
y	-7	-2	-1	2	1	-2	-7

The values of t can then be ignored when plotting y against x.
In this example, we could easily eliminate the parameter t thus:

$$t = (x - 1)/2,$$

and substituting into the expression for y,

$$y = 2 - \left(\frac{x-1}{2}\right)^2 \Rightarrow 4y = 7 + 2x - x^2.$$

A technique that is sometimes useful when eliminating a parameter
is to divide the two given equations. For example, if $x = \dfrac{t}{1+t^3}$ and
$y = \dfrac{t^2}{1+t^3}$, then $\dfrac{y}{x} = t$, and substituting into the first equation above
gives

$$x = \frac{y/x}{1 + (y/x)^3}$$
$$\Rightarrow x^3 + y^3 = xy.$$

Note that it is often difficult or impossible to eliminate a parameter,
and we shall see later that even when it is not difficult it is often
undesirable. (Parameters are considered further in Chapters 14, 16 and
30.)

Example 10
A curve is given parametrically by $x = at^2$, $y = 2at$, where a is a
constant. P and Q are the points on the curve where $t = t_0$ and $t = t_1$
respectively. Find the equation of the chord PQ. (A *chord* is a straight

line joining two points on a curve.) Find also the x-coordinate of the point where this chord cuts the x-axis.

Solution

The gradient of the line joining $P(at_0^2, 2at_0)$ and $Q(at_1^2, 2at_1)$ is

$$\frac{2at_1 - 2at_0}{at_1^2 - at_0^2} = \frac{2}{t_1 + t_0}.$$

So the equation of PQ is

$$(y - 2at_0) = \frac{2}{(t_1 + t_0)}(x - at_0^2)$$

$$\Rightarrow (t_1 + t_0)(y - 2at_0) = 2(x - at_0^2)$$

$$\Rightarrow (t_1 + t_0)y = 2x + 2at_0 t_1.$$

(Note the symmetry with respect to t_0 and t_1; this is to be expected, since the chord PQ is the same as the chord QP.)

This chord meets the x-axis where $y = 0$, i.e. where

$$2x + 2at_0 t_1 = 0$$

$$\Rightarrow x = -at_0 t_1.$$

Exercise 3e

1. Plot the following graphs for the given ranges of t. (Use any convenient scales.)

   (i) $x = t^2$, $y = t^3$; $-3 \leqslant t \leqslant 3$. (Plot several points with $t \approx 0$ in order to see how the curve behaves near the origin.)

   (ii) $x = \cos t°$, $y = \sin t°$; $0 \leqslant t \leqslant 90$.

   (iii) $x = \dfrac{t}{1+t^2}$, $y = \dfrac{t^2}{1+t^2}$; $-3 \leqslant t \leqslant 3$.

2. Eliminate t between the following:

   (i) $x = 2t + 3$, $y = 1 - t$;

   (ii) $x = t^2 - 1$, $y = -1$;

   (iii) $x = \dfrac{t}{1+t^2}$, $y = \dfrac{t^2}{1+t^2}$;

   (iv) $x = t(1 + t^3)$, $y = t^2(1 + t^3)$.

3. (i) Find the equation of the chord PQ where P and Q are the points on the curve $x = ct$, $y = c/t$ at which $t = p$ and $t = q$. (c is a constant.)

   (ii) If, on the same curve, R is the point where $t = r$, and PQ is perpendicular to QR, show that the gradient of PR is q^2.

4. (i) Find the values of t at which the line $3y + 5x = 1$ cuts the curve

   $$x = \frac{1}{1+t^2}, \quad y = \frac{t}{1+t^2}.$$

   [*Hint*: substitute the expressions for x and y into the equation of the line.]

   (ii) Show that $3x - 4y = 4$ is a tangent to this curve, and find the point of contact. [*Hint*: show that the line has 'double contact' with the curve.]

5. Which point on the curve in question 2 (iii) corresponds to 'very large' t?

1. In a triangle ABC, the coordinates of A and B are $(1, -3)$ and $(9, 3)$ respectively. The equation of AC is $y = 2x - 5$ and the equation of BC is $2y = 15 - x$. Prove that $\angle C = 90°$ and calculate the area of the triangle ABC. Calculate the coordinates of P and Q, the feet of the perpendiculars from the origin O to BC produced and AC respectively. Given that $R(1·8, -2·4)$ is the foot of the perpendicular from O to AB, prove that P, Q and R lie on a straight line.

[JMB (O)]

2. Find the equation of the circle which passes through the point $P(-3, 4)$ and has its centre at the point $Q(-1, 8)$.
Find the equation of the tangent to the circle at P. Show that the line whose equation is $y = 2x$ is a tangent to the circle and find the coordinates of its point of contact, R.
Find the coordinates of the point of intersection, S, of these two tangents and show that $PQRS$ is a square.

[JMB (O)]

3. Three points have coordinates $A(1, 3)$, $B(3, 5)$, $C(1\frac{1}{2}, 4\frac{1}{2})$. The points D and C lie on the same side of AB, the angle DBA is $90°$ and the areas of $\triangle ACB$ and $\triangle ADB$ are equal. Find
(i) the equation of the line through C parallel to AB,
(ii) the coordinates of D.

[C (O)]

4. The point A has coordinates (a, b). B and C are the images of A when A is reflected in the x-axis and y-axis respectively. A moves in such a way that BC always passes through the fixed point $(-2, 3)$. Find the relation between a and b and hence the equation of the path traced by A.

[Ox (O)]

5. A circle touches the line $y = 8$ at the point $(4, 8)$ and also passes through the origin of coordinates. Find the equation of the circle in its simplest form.
Find the equation of the tangent to the circle at its other point of intersection with the x-axis.

[O&C (O)]

6. The points P and Q are $(6, 3)$ and $(14, 9)$ respectively. Find the equations of equal circles, with centres P and Q, which touch each other externally. Find, also, the equations of their common tangents.

[O&C (O)]

7. Find the equation of the circle which passes through the points $A(2, 0)$, $B(10, 4)$ and $C(5, 9)$ and show that it touches the y axis. If the tangents at A and B intersect at D find the co-ordinates of D, the length BD and the angle ADB correct to the nearest degree.

[SUJB]

8. The two circles given by the equations
$x^2+y^2-2x-2y=2$ and $x^2+y^2-8x-10y=a$ (where a is a constant) touch each other externally. Calculate
  (i) the co-ordinates of the centres and the numerical values of the radii of both circles,
  (ii) the value of the constant a,
  (iii) the co-ordinates of the point at which the circles touch.

[W (O)]

9. The point $A(4, 1)$ lies on the line whose equation is $3x-4y-8=0$. A circle touches this line at A and passes through the point $B(5, 3)$. Find the equation of the circle, and show that it touches the y-axis.

  Find also the equation of the line parallel to AB on which the circle cuts off a chord equal in length to AB.

[C]

10. The coordinates of the vertices of the triangle ABC are

$$A(1, 0), \ B(9, -1), \ C(5, 7).$$

Show that the triangle is isosceles and find the length of the altitude AD. Show also that the equation of the circle ABD is

$$x^2+y^2-10x+y+9=0.$$

Find the condition that the line $y=mx$ touches the circle ABD. If the lines $y=m_1 x$ and $y=m_2 x$ are tangents to the circle ABD, show that $|m_1-m_2|=(\sqrt{9360})/35$. Find the angle between the tangents from the origin to the circle ABD.

  (Scale drawing solutions will not be accepted.)

[AEB '78]

4 Surds, indices and logarithms

4.1 Surds   A numerical expression involving a square root, cube root, or higher order root is called a *surd*. Examples are: $\sqrt{2}$, $\sqrt[3]{5}$ and $\sqrt{(1+3\sqrt{2})}$. (The notation $\sqrt{}$ means the *positive* square root.) In this section, we shall be concerned with expressing surds in their simplest form. It does not normally help to replace a surd by a decimal approximation (e.g. $\sqrt{2} \approx 1 \cdot 414$).

Qu.1   (i) Explain how an accurate graph of $y = x^2$ may be used to estimate square roots, showing that a positive number has only one positive square root.
(ii) Show that the squares of $\sqrt{(ab)}$ and $\sqrt{a}\sqrt{b}$ are equal, where a and b are both positive.
(iii) Deduce from (i) and (ii) that $\sqrt{(ab)} = \sqrt{a}\sqrt{b}$.
(iv) If a and b can take any values, show that $\sqrt{(ab)}$ may exist when $\sqrt{a}\sqrt{b}$ does not.

Qu.2   By a similar method, show that $\sqrt{\dfrac{a}{b}} = \dfrac{\sqrt{a}}{\sqrt{b}}$.

Qu.3   Show that $\sqrt[n]{(ab)} = \sqrt[n]{a}\sqrt[n]{b}$ where $n \in Z^+$.

Example 1
Simplify
(i) $\sqrt{4320}$;   (ii) $\sqrt{15} \times \sqrt{12}$;   (iii) $\sqrt{75} - \sqrt{27}$.
Solution
(i) The largest factor of 4320 which is a perfect square is 144, so we write

$$\sqrt{4320} = \sqrt{(144 \times 30)} = \sqrt{144} \times \sqrt{30} = 12\sqrt{30}.$$

(If necessary, the largest square factor may be found by writing the given number in prime factors. Try it.)
(ii) $\sqrt{15} \times \sqrt{12} = (\sqrt{3}\sqrt{5})(\sqrt{3}\sqrt{4}) = 3 \times 2 \times \sqrt{5} = 6\sqrt{5}$.
(iii) $\sqrt{75} - \sqrt{27} = 5\sqrt{3} - 3\sqrt{3} = 2\sqrt{3}$.

Qu.4   Simplify
(i) $\sqrt{45}$;   (ii) $\sqrt{216}$;   (iii) $\sqrt{20} \times \sqrt{80}$;
(iv) $\sqrt{128} - \sqrt{50} + \sqrt{2}$;   (v) $(2\sqrt{3})^2$;
(vi) $(\sqrt{2} + \sqrt{3})^2$;   (vii) $\sqrt[3]{24}$.

In general, when surds are involved in a fraction, it is better to remove them from a denominator, thereby 'rationalizing the denominator'.

Example 2

Rationalize the denominator in the following:

(i) $\dfrac{2}{\sqrt{3}}$;   (ii) $\dfrac{\sqrt{3}-1}{\sqrt{5}-\sqrt{3}}$.

Solution

(i) Multiplying top and bottom by $\sqrt{3}$ gives $\dfrac{2\sqrt{3}}{3}$.

(ii) The method here is important: we use the fact that

$$(\sqrt{5}-\sqrt{3})(\sqrt{5}+\sqrt{3})=(\sqrt{5})^2-(\sqrt{3})^2=5-3.$$

(Note that the *L.H.S.* is the factorization of a 'difference of two squares'.) So, multiplying the numerator and denominator of the given expression by $(\sqrt{5}+\sqrt{3})$, we have

$$\frac{\sqrt{3}-1}{\sqrt{5}-\sqrt{3}}=\frac{(\sqrt{3}-1)(\sqrt{5}+\sqrt{3})}{(\sqrt{5}-\sqrt{3})(\sqrt{5}+\sqrt{3})}$$

$$=\frac{\sqrt{15}-\sqrt{5}+3-\sqrt{3}}{5-3}$$

$$=\frac{\sqrt{15}-\sqrt{5}+3-\sqrt{3}}{2}.$$

Qu.5   Rationalize the denominators in

(i) $\dfrac{4}{\sqrt{5}}$;   (ii) $\dfrac{6}{\sqrt{2}}$;   (iii) $\sqrt{\dfrac{3}{8}}$;

(iv) $\dfrac{1}{\sqrt{3}+\sqrt{2}}$;   (v) $\dfrac{\sqrt{3}}{\sqrt{3}-1}$.

Qu.6   Rationalize the denominator in $\dfrac{1}{\sqrt[3]{2}}$.

Qu.7   (i) Show that

$$x^3-y^3=(x-y)(x^2+xy+y^2)$$

and
$$x^3+y^3=(x+y)(x^2-xy+y^2).$$

(These factorizations are worth learning.)

(ii) Use (i) to rationalize the denominator in

(a) $\dfrac{1}{2-\sqrt[3]{5}}$;   (b) $\dfrac{1}{2+\sqrt[3]{5}}$.

[*Hint*: let $x=2$ and $y=\sqrt[3]{5}$ in each case.]

In the following example, it should be remembered that squaring both sides of an equation can introduce false solutions. (Consider, for example, the equation $2x=6$; squaring gives $4x^2=36$ which has two solutions, only one of which satisfies the original equation.) Each solution must therefore be checked in the original equation.

Example 3
Solve the equations
 (i) $x - \sqrt{(3x+1)} = 1$;
 (ii) $\sqrt{(2x+5)} + \sqrt{(x+6)} = 3$.
Solution
(i) (If we square both sides of the equation as it stands, we still have a square root; so we start by putting the square root on one side on its own.)

$$x - 1 = \sqrt{(3x+1)}.$$

Squaring,

$$(x-1)^2 = 3x+1$$
$$\Rightarrow x^2 - 2x + 1 = 3x + 1$$
$$\Rightarrow x^2 - 5x = 0$$
$$\Rightarrow x(x-5) = 0$$
$$\Rightarrow x = 0 \text{ or } x = 5.$$

Substituting each, we find that $x = 5$ is the only solution.
(iii) (We cannot isolate both square roots, so we shall need to square at two stages.)
Squaring the equation gives

$$(2x+5) + 2\sqrt{(2x+5)}\sqrt{(x+6)} + (x+6) = 9.$$

Multiplying the square roots and collecting terms,

$$2\sqrt{(2x^2 + 17x + 30)} = -(3x+2).$$

Squaring again,

$$4(2x^2 + 17x + 30) = 9x^2 + 12x + 4$$
$$\Rightarrow x^2 - 56x - 116 = 0$$
$$\Rightarrow (x-58)(x+2) = 0$$
$$\Rightarrow x = 58 \text{ or } x = -2.$$

Checking each, we find that $x = -2$ is the only solution.

Qu.8 (i) Explain precisely where the false solution $x = 0$ arises in Example 3(i), by comparing the solution with that of $x + \sqrt{(3x+1)} = 1$. (ii) In Example 3(ii), where does the false solution arise? Of what equation is $x = 58$ a solution?

Exercise 4a
 1. Simplify
   (i) $\sqrt{200}$;   (ii) $\sqrt{90}$;   (iii) $\sqrt{96}$;
   (iv) $\sqrt{240}$;   (v) $\sqrt{5000}$;   (vi) $\sqrt{50\,000}$.
 2. Write in the form $\sqrt{n}$:
   (i) $2\sqrt{3}$;   (ii) $3\sqrt{2}$;   (iii) $5\sqrt{7}$;   (iv) $2\sqrt{2}\sqrt{3}$.
 3. Simplify
   (i) $\sqrt{10}\sqrt{20}$;   (ii) $\sqrt{15}\sqrt{3}$;   (iii) $\sqrt{6}\sqrt{15}\sqrt{10}$;
   (iv) $\sqrt{8} - \sqrt{2}$;   (v) $\sqrt{800} + \sqrt{32} - \sqrt{1152}$.
 4. Let $x = 3\sqrt{2}$, $y = 2\sqrt{2}$ and $z = 2\sqrt{6}$. Simplify

(i) xy; (ii) yx; (iii) x^2; (iv) x^3;

(v) y^5; (vi) $\dfrac{z}{y}$; (vii) $\dfrac{z}{x}$.

5. Expand
   (i) $(1+\sqrt{2})^2$; (ii) $(2-3\sqrt{2})^2$; (iii) $(\sqrt{2}+\sqrt{3})^2$;
   (iv) $(1+\sqrt{2})(1+2\sqrt{2})$; (v) $(\sqrt{6}-\sqrt{2})(\sqrt{3}+2\sqrt{2})$;
   (vi) $(2-\sqrt{3})^3$.

6. Rationalize the denominator in

   (i) $\dfrac{1}{\sqrt{2}}$; (ii) $\dfrac{5}{\sqrt{3}}$; (iii) $\dfrac{4}{\sqrt{2}}$; (iv) $\dfrac{9}{\sqrt{6}}$;

   (v) $\dfrac{3\sqrt{2}}{\sqrt{3}}$; (vi) $\dfrac{1+\sqrt{2}}{\sqrt{10}}$;

   (vii) $\sqrt{\dfrac{1}{2}}$; (viii) $\sqrt{\dfrac{3}{5}}$.

7. Rationalize the denominator in

   (i) $\dfrac{1}{\sqrt{7}-\sqrt{5}}$; (ii) $\dfrac{6}{\sqrt{5}-\sqrt{2}}$;

   (iii) $\dfrac{1}{2\sqrt{3}+\sqrt{7}}$; (iv) $\dfrac{3}{4-\sqrt{3}}$;

   (v) $\dfrac{1}{1+\sqrt{2}}$; (vi) $\dfrac{\sqrt{10}+1}{\sqrt{2}+3}$;

   (vii) $\dfrac{\sqrt{7}-\sqrt{5}}{\sqrt{7}+\sqrt{5}}$; (viii) $\dfrac{1}{\sqrt[3]{9}-\sqrt[3]{3}}$.

8. Solve the equations
   (i) $\sqrt{(x+1)}=3$; (ii) $\sqrt{(x+1)}=-3$.

9. Solve
   (i) $\sqrt{(2x-1)}=x-2$; (ii) $\sqrt{(7-3x)}=x+11$.

10. Solve
    (i) $\sqrt{(x+3)}+1=\sqrt{(2x+7)}$;
    (ii) $\sqrt{(3x+10)}-\sqrt{(2x+5)}=1$;
    (iii) $\sqrt{(2x+6)}-\sqrt{(3x+4)}=\sqrt{(x+2)}$.

11. In this question, we consider various ways of simplifying
    $\sqrt{(22+12\sqrt{2})}$.
    (i) Suppose that $\sqrt{(22+12\sqrt{2})}=a+b\sqrt{2}$. By squaring each side, show that a possible pair of values for a and b satisfies

$$a^2+2b^2=22$$
$$2ab=12.$$

Make a the subject of the second equation and substitute into the first equation to obtain an equation in b. Solve this by noting that it is a quadratic in b^2. Hence show that the required square root is $2+3\sqrt{2}$.

(ii) Suppose that $\sqrt{(22+12\sqrt{2})}=\sqrt{p}+\sqrt{q}$. By squaring each side, show that we may take

$$p+q=22$$
$$pq=72.$$

Hence find p and q and solve the problem.

(iii) Suppose that $\sqrt{(22+12\sqrt{2})}=\sqrt{m}+6\sqrt{n}$. Continue, using the method of (ii) above. (Why was the number 6 chosen?)

12. Find by any of the methods of question 11 the positive square root of

(i) $17-12\sqrt{2}$; (ii) $49+20\sqrt{6}$; (iii) $53+10\sqrt{6}$.

13. In this question, we investigate one method of finding fractions which are near to $\sqrt{3}$.

(i) Verify that the error in the approximation $\sqrt{3}\approx2$ is less than 1.

(ii) Explain how, using (i), we know that

$$(2-\sqrt{3})^2<(2-\sqrt{3}).$$

(iii) Expand $(2-\sqrt{3})^2$ and deduce from (ii) that $\sqrt{3}\approx7/4$ is a better approximation than that in (i).

[*Hint*: by part (ii), $(2-\sqrt{3})^2$ is closer to zero than $(2-\sqrt{3})$ is.]

(iv) By considering $(7-4\sqrt{3})^2$, find a better approximation, and show how the method may be continued to give a sequence of improving approximations.

14. Starting with $\sqrt{2}\approx1$, use the method of question 13 to obtain three further approximations.

15. Find $\lim\limits_{h\to0}\dfrac{\sqrt{(x+h)}-\sqrt{x}}{h}$ by first rationalizing the numerator.

16. Show that if n is a real number greater than 1, then

$$\sqrt{(n-1)}+\sqrt{(n+1)} < 2\sqrt{n}.$$

(Make sure that your proof is written out the right way round. Don't start by assuming what you are trying to prove.)

4.2 Indices In this section, we begin by investigating the meaning of a^n (where $a>0$) for all $n\in Q$. Suppose first that $n\in Z^+$, then we define a^n by

$$a^n=a\times a\times a\times\ldots\times a \text{ (with } n \text{ factors in all).} \tag{1}$$

The number n is called the *index* or *exponent*.

The following three properties of positive integral indices follow immediately from this definition:

(i) $a^n\times a^m=a^{n+m}$;

(ii) $a^n\div a^m=a^{n-m}$ (see note below);

(iii) $(a^n)^m=a^{nm}$.

(Note that we must temporarily impose the condition $n>m$ on rule (ii) above, otherwise a^{n-m} has, as yet, no meaning.)

We now extend our definition of a^n in such a way that these rules continue to hold. This is done in stages 1 to 4 below.

(1) Consider $a^{1/q}$ where $q\in Z^+$. From rule (iii)

$$(a^{1/q})^q=a^1=a$$

and so

$$a^{1/q}=\sqrt[q]{a}. \tag{2}$$

SURDS, INDICES AND LOGARITHMS

(2) Let $p, q \in Z^+$, then

$$a^{p/q} = (a^p)^{1/q} \quad \text{by rule (iii)}$$
$$= \sqrt[q]{(a^p)} \quad \text{from equation (2) above.}$$
$$\text{So } a^{p/q} = \sqrt[q]{(a^p)}. \tag{3}$$

Note that $a^{p/q} = (\sqrt[q]{a})^p$ is sometimes a simpler version to use in calculating.

(3) Now consider a^0. By rule (i), for any s,

$$a^0 \times a^s = a^s.$$

Dividing by a^s gives

$$a^0 = 1. \tag{4}$$

(4) Consider next a^{-s}. By rule (i)

$$a^{-s} \times a^s = a^0 = 1.$$

So
$$a^{-s} = \frac{1}{a^s} \text{ or } \left(\frac{1}{a}\right)^s. \tag{5}$$

Thus a negative sign in an index means 'take the reciprocal'. (The *reciprocal* of a number is its inverse under ordinary multiplication.)

Definition

For any rational s, a^s is defined by equations (1), (3), (4) and (5) above. (Equation (2) is not needed: it is a special case of equation (3).)

Note

We have adopted the only possible definition of a^s that will allow rules (i), (ii) and (iii) to remain true for rational indices; but we have not shown that these rules do actually remain true under the enlarged definition. A rigorous proof that they do is beyond the scope of this chapter.

Example 4

Evaluate

(i) $25^{-1\frac{1}{2}}$; (ii) $\left(\dfrac{27}{8}\right)^{-4/3}$

Solution

(i) $25^{-3/2} = \dfrac{1}{25^{3/2}} = \dfrac{1}{(\sqrt{25})^3} = \dfrac{1}{125}.$

(ii) $\left(\dfrac{27}{8}\right)^{-4/3} = \left(\dfrac{8}{27}\right)^{4/3} = \left(\dfrac{2}{3}\right)^4 = \dfrac{16}{81}.$

Example 5

Simplify

(i) $2^{5/2} - 2^{3/2} + 2^{1/2}$;

(ii) $(x+2)^{3/2} - 3x(x+2)^{1/2} + 2x^2(x+2)^{-1/2}$.

Solution

(i) $2^{5/2} - 2^{3/2} + 2^{1/2} = 2^2\sqrt{2} - 2\sqrt{2} + \sqrt{2} = 3\sqrt{2}.$

(ii) Removing a common factor of $(x+2)^{-1/2}$, we get

$$(x+2)^{-1/2}\{(x+2)^2 - 3x(x+2) + 2x^2\}$$
$$= (x+2)^{-1/2}(4-2x).$$

Qu.9 Write as powers of 2:
(i) 4^n; (ii) 8^{n+1}; (iii) $8^{2n+1} \div 4^{3n-1}$.

Exercise 4b

1. Evaluate
(i) 4^{-1}; (ii) $4^{1/2}$; (iii) $4^{-1/2}$;
(iv) 4^0; (v) $27^{2/3}$; (vi) $\left(\dfrac{8}{125}\right)^{4/3}$;

(vii) $(2\frac{1}{4})^{-3/2}$; (viii) $\left(\dfrac{27}{64}\right)^{-5/3}$; (ix) $\left(\dfrac{27}{64}\right)^0$.

2. Simplify
(i) $3^{7/2} - 3^{5/2} - 3^{3/2} + 3^{1/2}$;
(ii) $8^{1/2} + 2^{1/2}$;
(iii) $50^{1/2} + 18^{3/2} - 2^{5/2}$.

3. Write in the form x^n:
(i) $\sqrt{x}$; (ii) $\sqrt{x^3}$; (iii) $x\sqrt{x}$;
(iv) $\dfrac{1}{x^2\sqrt{x}}$; (v) $\dfrac{x}{\sqrt[3]{x}}$; (vi) $\sqrt[3]{(x\sqrt{x})}$;

(vii) $\sqrt{\dfrac{1}{x^5}}$.

4. Factorize and simplify
(i) $(x+1)^{3/2} + (x+1)^{1/2}$;
(ii) $a^{1/2} + a^{-1/2}$;
(iii) $(2p+3)^{3/2} - (2p+3)^{1/2} + (2p+3)^{-1/2}$.

5. Simplify
(i) $2^{2n} \div 2^n$; (ii) $4^{2n} \div 2^{3n}$; (iii) $81^n \div 27^{n-1}$;
(iv) $50^{n/3} \div 54^{2n/3} \times 6^n \times 9^{n/3} \times 30^{n/3}$.

6. Simplify
(i) $(x^2+1)^{1/2} - x^2(x^2+1)^{-1/2}$;
(ii) $\dfrac{(x-1)^{3/2} + (x-1)^{1/2}}{x}$;
(iii) $(x^2+2)^{1/2} - x^2(x^2+2)^{-1/2} - 2x^2(x^2+2)^{-3/2}$;
(iv) $\dfrac{(x-2)^{5/3} - 3(x-2)^{2/3} + 2(x-2)^{-1/3}}{(x-3)}$.

4.3 The power function a^x

So far, we have defined 2^x only for rational values of x. If we now imagine that we plot the graph of $y = 2^x$ for all rationals, then there are certainly enough points to indicate the general shape of the curve. We now define $y = 2^x$ for *irrational* values of x by choosing values of y that ensure that the graph of $y = 2^x$ is continuous for all real x (Fig. 4.1).

A function of the form $y = a^x$ (where $a > 0$) is known as a *power function* or an *exponential function*.

Figure 4.1

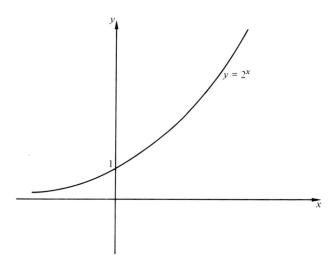

Qu.10 Plot the graphs of $y = a^x$ in the region $|x| < 3$ for
(i) $a = 2$; (ii) $a = \frac{1}{2}$; (iii) $a = 1$.
What is the relationship between the first two graphs?

Parts (i) and (ii) of Qu.10 indicate the general shape of $y = a^x$ for $a > 1$ and $0 < a < 1$ respectively. Note that for any positive a, the graph of $y = a^x$ passes through the point $(0, 1)$. In Qu.11, we see that the function $y = 2^x$ is a very different sort of function from, say, $y = x^2$.

Qu.11 (i) On the same axes, sketch roughly the graphs of $y = 2^x$ and $y = x^2$.

(ii) Find the values of $\dfrac{2^x}{x^2}$ when $x = 4$, $x = 8$ and $x = 16$.

(iii) What happens to $\dfrac{2^x}{x^2}$ as (a) $x \to \infty$; (b) $x \to -\infty$?

(iv) What happens to $\dfrac{2^x}{x^{100}}$ as $x \to \infty$?

4.4 Logarithmic functions

The inverse function of the power function $x \to a^x$ $(a \neq 1)$ is called the *logarithm to the base a* function; it is denoted by $\log_a$. Thus, for example,

$$2^5 = 32, \text{ and so } \log_2 32 = 5.$$

In general,

$$a^p = q \Leftrightarrow \log_a q = p$$

and we say that p is the logarithm of q to the base a. It is conventional that where the base of a logarithm is not stated, it is taken to be 10. Thus $\log x$ means $\log_{10} x$.

In converting between power function and logarithmic notation, it helps to remember that the logarithm is the index. Indeed, we could define $\log_a n$ rather clumsily as 'the power to which a has to be raised to give n'.

The reader may already be familiar with logarithms to base ten and the associated power function $x \to 10^x$ (usually called antilogarithms); another aid in converting between index and logarithmic notation is to remember that the statements $10^2 = 100$ and $\log_{10} 100 = 2$ are equivalent.

Qu.12 Rewrite these statements using logarithmic notation:

(i) $2^6 = 64$; (ii) $10^3 = 1000$; (iii) $8^3 = 512$;

(iv) $9^{1/2} = 3$; (v) $3^{-2} = \frac{1}{9}$; (vi) $(\frac{1}{4})^{3/2} = \frac{1}{8}$.

Qu.13 Rewrite these statements using index notation:
(i) $\log_2 128 = 7$; (ii) $\log_{10} 10\,000 = 4$;
(iii) $\log_8 64 = 2$; (iv) $\log_8 2 = \frac{1}{3}$;
(v) $\log_2 \frac{1}{16} = -4$; (vi) $\log_{4/9} \frac{8}{27} = \frac{3}{2}$.

Qu.14 Find
(i) $\log_3 81$; (ii) $\log_{16} 2$; (iii) $\log_{16} 8$;
(iv) $\log_5 1$; (v) $\log_3 \sqrt{3}$; (vi) $\log_4 \frac{1}{8}$;
(vii) $\log_{1/2} 2$.

Using the fact that a^x and $\log_a x$ are inverse functions (so that their composite is the identity function), we have

(i) $\log_a (a^x) = x$;
(ii) $a^{\log_a x} = x$ (provided $x > 0$).

The graph of a typical logarithmic function (with base greater than 1) is shown in Fig. 4.2. It will be seen that the domain of the function is the positive real numbers only, i.e. we cannot take the logarithm of a negative number. This explains the condition in (ii) above.

Figure 4.2

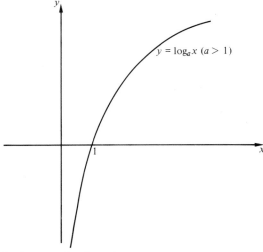

$y = \log_a x \ (a > 1)$

Qu.15 Sketch the graph of $y = \log_{1/2} x$.

4.5 The algebra of logarithms

We saw earlier that indices obey the following three laws:
(i) $a^x \times a^y = a^{x+y}$;

(ii) $a^x \div a^y = a^{x-y}$;

(iii) $(a^x)^y = a^{xy}$.

We can convert these laws into rules for logarithms. Let $a^x = p$ and $a^y = q$, so that $\log_a p = x$ and $\log_a q = y$. Then

(i) $\log_a(pq) = \log_a(a^x a^y)$

$\qquad = \log_a(a^{x+y})$

$\qquad = x + y$

$\qquad = \log_a p + \log_a q.$

So
$$\boxed{\log_a (pq) = \log_a p + \log_a q.}$$

This is the rule that enables us to use 'logs' to multiply two numbers.

(ii) Similarly,

$$\boxed{\log_a \frac{p}{q} = \log_a p - \log_a q.}$$

(iii) $\log_a (p^k) = \log_a (a^x)^k$

$\qquad = \log_a a^{kx}$

$\qquad = kx$

$\qquad = k \log_a p.$

So
$$\boxed{\log_a (p^k) = k \log_a p.}$$

This is the rule used when evaluating powers by logs.

Example 6

Write $3 \log x - \frac{1}{2} \log y$ as a single logarithm.

Solution

$3 \log x - \frac{1}{2} \log y = \log x^3 - \log y^{1/2}$

$$\qquad\qquad = \log \frac{x^3}{\sqrt{y}}.$$

Qu.16 Given that $\log_{10} 2 = 0{\cdot}30103$ and $\log_{10} 3 = 0{\cdot}47712$, evaluate, using the rules derived above,

(i) $\log 6$;　　(ii) $\log 8$;　(iii) $\log 1{\cdot}5$;

(iv) $\log 72$;　　(v) $\log 5$ [*Hint:* you know log 10.];

(vi) $\log \frac{1}{4}$;　　(vii) $\log 375$.

An important rule involving logarithms to different bases is proved in the following example. Note carefully the general shape of the rule: it might be called the 'dominoes rule' or 'chain rule' for logarithms.

Example 7

Show that $\log_a b \times \log_b c = \log_a c$.

Solution

Let $\log_a b = p$ and $\log_b c = q$, so that $a^p = b$ and $b^q = c$.

Then $$c = b^q = (a^p)^q = a^{pq}.$$

So $$\log_a c = pq = \log_a b \times \log_b c$$

as required.

Qu.17 By replacing c by a in this result, show that
$$\log_a b \times \log_b a = 1.$$

Qu.18 (Very easy.) Use the result of Example 7 to show that
$\log_b c = \dfrac{\log_a c}{\log_a b}$. (The result is often used in this form.)

Example 8
Evaluate $\log_3 5$.
First solution
Let $\log_3 5 = x$, then $3^x = 5$.
Taking logarithms to base 10 of each side,

$$\log 3^x = \log 5$$
$$\Rightarrow x \log 3 = \log 5$$
$$\Rightarrow x = \frac{\log 5}{\log 3}$$
$$\approx \frac{0 \cdot 6990}{0 \cdot 4771}$$
$$\approx 1 \cdot 465.$$

Second Solution
Using the result of Qu.18,

$$\log_3 5 = \frac{\log_{10} 5}{\log_{10} 3} = \text{etc.}.$$

Example 9
Solve
 (i) $4^x - 3 \cdot 2^{x+1} = 27$;
 (ii) $\log_2 (x+1) - \log_4(3x-1) = \tfrac{1}{2}$.
Solution
(i) The equation may be written

$(2^2)^x - 3(2 \cdot 2^x) = 27$

or $(2^x)^2 - 6(2^x) = 27$.

Writing $y = 2^x$,

$$y^2 - 6y - 27 = 0$$
$$\Rightarrow y = -3 \text{ or } y = 9.$$

So $2^x = -3$ or $2^x = 9$.
Now, $2^x = -3$ has no solution, and $2^x = 9$ has solution

$$x = \frac{\log 9}{\log 2} \approx 3 \cdot 170.$$

(ii) It is usual to convert all logarithms to the same base. (Here we choose base 4.) The term $\log_2(x+1)$ may be written $\log_4(x+1)/\log_4 2 = 2\ \log_4(x+1)$. So

$$2\ \log_4(x+1) - \log_4(3x-1) = \tfrac{1}{2}$$

$$\Rightarrow\ \log_4 \frac{(x+1)^2}{3x-1} = \frac{1}{2}$$

$$\Rightarrow\ \frac{(x+1)^2}{3x-1} = 4^{1/2} = 2$$

$$\Rightarrow\ (x+1)^2 = 2(3x-1)$$

$$\Rightarrow\ x^2 - 4x + 3 = 0$$

$$\Rightarrow\ x = 1 \text{ or } 3.$$

It is important to check each value in the original equation to ensure that logarithms of negative numbers do not arise. Here, both numbers are genuine solutions.

Exercise 4c

1. Evaluate

(i) $\log_2 16$; (ii) $\log_2 (1/16)$; (iii) $\log_8 (1/4)$;
(iv) $\log_5 (5\sqrt{5})$; (v) $\log_{4/9} (3/2)$; (vi) $\log_4 1$.

2. Simplify

(i) $\log_a a^2$; (ii) $\log_{a^2} (a^3)$;
(iii) $\log_{\sqrt{a}} a^2$; (iv) $\log_{1/a} \sqrt{a}$.

3. If $\log_a x = p$ and $\log_a y = q$, express the following in terms of p and q only.

(i) $\log_a xy$; (ii) $\log_a(x/y)$;

(iii) $\log_a x^2$; (iv) $\log_a \dfrac{x^3}{y^2}$;

(v) $\log_a ax$; (vi) $\log_a(x/ay)$;

(vii) $\log_a \sqrt[3]{\dfrac{x}{y^2}}$.

4. Write the following as single logarithms (i.e. in the form $\log n$).

(i) $\log 14 - \log 7$; (ii) $\log 5 + \log 3$;
(iii) $2 \log 3$; (iv) $5 \log 4 - 3 \log 8$;
(v) $1 + \log 2$; (vi) $\tfrac{1}{2} \log 3 + \log 4$.

5. Evaluate, without using tables or a calculator,

(i) $\log 2 + \log 5$; (ii) $2 \log 20 - \log 4$;
(iii) $\log 3 + \log 7 - \log 21$;
(iv) $\log \sqrt{10}$; (v) $\log \sqrt{1000}$;
(vi) $\dfrac{\log 9}{\log 3}$; (vii) $\dfrac{\log 4}{\log 8}$.

6. Solve the equations

(i) $2^x = 5$; (ii) $(\sqrt{10})^x = 7$; (iii) $(1/3)^x = 10$.

7. Solve the equations

(i) $9^x - 7.3^x = 18$; (ii) $2^{2x+1} = 8(17.2^{x-3} - 1)$;

(iii) $2^x - 3(\frac{1}{2})^{x-1} = 5$; (iv) $2^{x+1} + 2^{-x} = 3$.

8. Solve these equations, checking your answers:

(i) $\log x + \log (2x - 1) = 1$;

(ii) $2 \log_9 x - \frac{1}{2} = \log_9 (5x + 18)$;

(iii) $3 \log_2 x - 6 \log_x 2 = 7$ [*Hint*: let $\log_2 x = y$.];

(iv) $\log_9 x - \log_x 27 = 1$;

(v) $\log (x - 8) = 1 + \log (x - 2)$.

9. On the same axes, sketch the graphs of $y = (2a)^x$ and $y = 2a^x$ for some $a > 1$. Show clearly where the graphs cross, and where each crosses the y-axis.

10. If $\log y = n \log x + c$, show that $y = Ax^n$ for some constant A, explaining clearly the relationship between A and c.

11. Show that $5.2^x = 2^{x+k}$ for suitable k. Find k.

12. There are two simple ways of changing the graph of $y = 2^x$ into the graph of $y = 8.2^x$. One is to change the scale on the y-axis by a factor of 8; what is the other?

13. Simplify

(i) $\log (x^2 - 1) - \log (x - 1)$;

(ii) $\log (2x + 4) - \log (x + 2)$.

14. Write as single logarithms

(i) $2 \log (x + 1) + 3 \log (2x - 3)$;

(ii) $\log (x/2) - \frac{1}{2} \log (x + 1)$.

15. Write as base 3 logarithms:

(i) $\log_9 x$; (ii) $\log_{\sqrt{3}} (2x + 1)$; (iii) $\log_{1/3} x^2$.

16. Solve for x:

(i) $\log_2 x - 3 \log_4 x = 1\frac{1}{2}$;

(ii) $\log_3 (x + 1) - \log_9 (x - 5) = 1\frac{1}{2}$.

(iii) $\log_a (x - 3) = \log_{a^2} (7 - 3x)$.

17. If £100 is invested at 10% compound interest, write down an expression for the value of the investment after n years. How long will it take for the investment to be worth £5000?

18. Say whether each of the following statements about the function $y = \log x$ are true or false. Explain your answers.

(i) The value of y is always increasing as x increases.

(ii) The graph of the function 'flattens off' as x increases.

(iii) There is a maximum value for y.

19. Give a direct proof of the result in Qu.17 (i.e. do not use the result of the preceding example).

4.6 Reduction to linear form

Suppose that in the course of some scientific work, pairs of readings of two variables (x and y say) have been obtained. The data may be plotted on a graph, and it is often possible to do this in such a way that the graph obtained is a straight line, even when the relationship between x and y is not linear. The advantages of obtaining a straight line graph will emerge in the examples below, but one immediate advantage is that we can easily identify any pair of readings that has rather more than its fair share of experimental error.

Example 10

The following readings are believed to obey a law of the form $Q = aP^2 + b$ where a and b are constants.

P	2	4	6	8	9	10
Q	-3.1	0.1	5.5	13.1	17.7	22.8

Use a graphical method to show that this is approximately the case, and estimate
  (i) the values of a and b;
  (ii) the value of Q when $P = 7$;
  (iii) the value of P when $Q = 15$.

Solution

Plotting Q against P will not give a straight line, but by comparing the equation $Q = aP^2 + b$ with the standard form $y = mx + c$, we see that we shall get a straight line by plotting Q against P^2. The gradient will give the value of a, and the intersection with the y-axis gives the value of b. We therefore extend the table to show values of P^2.

P^2	4	16	36	64	81	100
Q	-3.1	0.1	5.5	13.1	17.7	22.8

Plotting these values on a graph, we obtain a straight line, confirming that the law is of the suggested form (Fig. 4.3).

 (i) To find a, we choose two points on the line (not too close together); then

$$a = \text{gradient}$$

$$= \frac{\text{change in } Q}{\text{change in } P^2}$$

$$\approx \frac{12 - 1.2}{60 - 20}$$

$$= 0.27.$$

(Remember to read the change in each variable from the appropriate axis.) The value of b may be read off: $b \approx -4.2$.

 (ii) When $P = 7$, $P^2 = 49$, and from the graph, $Q = 9.0$.
 (iii) When $Q = 15$, from the graph, $P^2 = 71$ and so $P = 8.4$.

Qu.19 Suppose it is known that the readings $(2, -3.1)$ and $(10, 22.8)$ of P and Q are exact (i.e. there is no experimental error). Find the values of a and b by substituting these values of P and Q into $Q = aP^2 + b$ to obtain a pair of simultaneous equations. (In Example 9, the plotted points turn out to be almost exactly collinear, and so the method of Qu.19 could have been used with any pair of points. The problem is, however, that until the graph has been drawn, we cannot judge the accuracy of individual readings.)

Figure 4.3

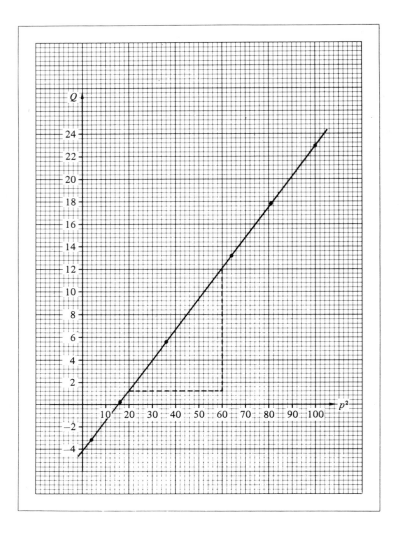

Qu.20 It is believed that the following readings approximately obey a law of the form $S\sqrt{T}=a+bS$. Divide this equation through by S and plot $\sqrt{T}$ against $1/S$ to estimate a and b.

S	3·2	4·5	5·1	7·0	7·9
T	2·3	4·7	5·6	7·6	8·3

Suppose now that x and y are approximately related by a law of the form $y=ax^n$ where a and n are constants. The equation can be reduced to linear form by taking logarithms (to base 10) of each side:

$$\log\ y=\log\ (ax^n)$$
$$\Rightarrow \log\ y=\log\ a+\log\ (x^n)$$
$$\Rightarrow \log\ y=\log\ a+n\log\ x.$$

Comparing with the standard form $y=mx+c$, we see that plotting $\log y$ against $\log x$ should give a straight line. The gradient will give the value of n, and the intercept on the log y axis gives the value of log a from which a can be found.

Note
(1) Care must be taken when plotting negative logarithms obtained from tables. For example, $\bar{2}\cdot3010$ means $-2+0\cdot3010$ and must be plotted as $-1\cdot6990$.
(2) It is possible to obtain graph paper on which one or both axes has a logarithmic scale (as on a slide rule). This avoids the need to look up logarithms.

Qu.21 Let $y=ka^x$ (where k and a are constants). By taking logarithms of each side, write this equation in a form which corresponds to the equation of a straight line. State
 (i) what should be plotted against what;
(ii) the value of the gradient and intercept.

4.7 A note on proportionality

If x and y are connected by a law of the form $y=kx$ (where k is a constant), then we say that y is (*directly*) *proportional to* x or that y *varies as* x; we may also write $y \propto x$.

Similarly, if $y=kx^2$, we say that y varies as x^2, and we may write $y \propto x^2$.

If $y=\dfrac{k}{x}$, we say that y *varies inversely as* x or that y *is inversely proportional to* x. Similarly, y is inversely proportional to the cube root of x would mean $y \propto \dfrac{1}{\sqrt[3]{x}}$ or $y=\dfrac{k}{\sqrt[3]{x}}$ (for some constant k).

Qu.22 (For discussion.) (i) Explain why the amount of an electricity bill is not directly proportional to the number of units used.
(ii) What is meant by saying that gravity obeys 'an inverse square law'?

Exercise 4d
1. The following readings are believed to obey a law of the form $y^2=px+q$.

x	$-9\cdot5$	5	12·5	23	29
y	3·03	3·87	4·24	4·59	4·96

(i) Show graphically that four of the five pairs of readings do obey such a law, and find the values of p and q.
(ii) State which pair of readings is wrong, and use your graph to find
(a) the correct value of y given that the x-coordinate is correct;
(b) the correct value of x if the y-coordinate is correct.
2. The following readings are believed to obey a law of the form $\dfrac{1}{L-1}=a-\dfrac{b}{M}$.

M	1·0	1·2	1·4	1·7	2·0
L	4·33	2·40	1·99	1·75	1·65

(i) What variables may be plotted to obtain a straight line?

(ii) Estimate from such a graph the values of a and b.

(iii) Use your graph to estimate L when $M=1·05$ and M when $L=1·9$.

(iv) Answer (iii) by calculation, using the values of a and b found in (ii).

(v) Show that the equation may be rearranged in such a way that a and b may be found from the graph of $\dfrac{M}{L-1}$ against M.

3. It is believed that z and w obey a law of the form $z = aw^2 + bw + c$, and these readings are known to be exact:

w	0	0·5	1·5	3	5
z	0	1·2	1·8	$-1·8$	-15

Show that this relationship between z and w may be written in the form $z/w = aw + b$ and plot a suitable graph to verify the law and to find a and b.

4. It is believed that two variables, x and y, are related by a law of the form $y = Ax^n$ where A and n are constants. Use a graphical method to estimate A and n from the data:

x	2	3	4	6	10
y	1·2	1·47	1·70	2·08,	2·69

5. Variables P and t are believed to obey a law of the form $P = ka^t$ where k and a are constant. The following readings were taken:

t	1	2	2·5	3	3·5
P	2280	3050	3530	4090	5130

Explain how a graph of $\log P$ against t may be used to estimate k and a, and make such an estimate, ignoring the figure which seems to be incorrect.

6. Two variables, h and d, are related by an equation of the form $d = A(h-10)^n$. Plot a suitable straight line graph to estimate the constants A and n. Use your graph to find h when $d=0·001$.

h	12	15	30	90
d	1·22	0·148	0·00611	0·000252

160	280
0·0000593	0·0000153

1. (i) **Without** the use of tables find, in its simplest form, the value of $80^{2/3} \times 10^{-5/3}$.

(ii) Find, correct to 2 significant figures, the value of x for which $2^{3x} = 15$.

(iii) If $a^2 b^3 = 1$, find the value of $\log_a b$.

[O & C (O)]

2. (i) If x varies directly as y^2, and y varies inversely as the cube root of z, express the relation between z and x in the form $z \propto x^n$. If $z = 3$ when $x = \frac{1}{4}$, find the value of z when $x = 9$.

(ii) If $a^2 = b^3/10$ and $\log_{10} a = \log_{10} b + 1$, find the values of a and b, expressing each answer as a power of 10.

[O & C (O)]

3. (a) Given that $x^y = z$, find

(i) z when $x = 9$ and $y = \frac{1}{2}$;

(ii) z when $x = 64$ and $y = -\frac{1}{3}$;

(iii) x when $y = -\frac{1}{2}$ and $z = 4$; and

(iv) y when $z = 2$ and $x = \frac{1}{2}$.

(b) Express $3^{2y} - 3^{y+1} - 3^y + 3$ in terms of z, where $z = 3^y$. Hence solve the equation

$$3^{2y} - 3^{y+1} - 3^y + 3 = 0.$$

(c) Without using tables evaluate

$$\log 6 + \log 4 + \log 20 - \log 3 - \log 16,$$

where all logarithms are to the base 10.

[W (O)]

4. Solve the equation $4^x - 2^x - 6 = 0$, giving your answer correct to three significant figures.

[AEB (O) '80]

5. Give the value of each of the following, correct to three significant figures where appropriate,

(i) $\log_6 4 + \log_6 9$,

(ii) $\log_2 8 - 2 \log_2 2$,

(iii) $\log_5 3$.

[AEB (O) '79]

6. When a cup of tea is poured out its temperature θ in degrees centigrade t minutes after pouring is given by the formula

$$\theta = 60 \times 2^{-t/15} + 20.$$

(i) Calculate the temperature of the tea immediately after it is poured and also after 10 minutes.

(ii) Calculate after how long the temperature of the tea is
(a) 50°C, (b) 40°C.

[Ox (O)]

7. (a) Given that $\log 2 = 0.301\,030\,0$ and $\log 3 = 0.477\,121\,3$, find, without using tables or calculators,

(i) $\log 16$, (ii) $\log(\frac{1}{9})$, (iii) $\log \sqrt[3]{12}$.

(b) Without using tables or calculators, find the value of

(i) $\dfrac{\log 125}{\log 5}$, (ii) $(\log_7 7) \times (\log_7 49)$.

(c) Given that $\log(x^3 y^2)=9$ and $\log(\frac{x}{y})=2$, find $\log x$ and $\log y$ without using tables or calculators.

(d) Solve the equation $3^x.3^{2x+3}=10$, giving the answer correct to two decimal places.

[JMB (O)]

8. Experiments are made upon two related quantities, x and y, and corresponding values are observed. It is known that the relation between x and y is of the form $y=ax^n$. Explain how the values of a and n may be obtained by plotting $\log y$ against $\log x$.

Water is discharged over a weir and it is found that for different heights, h, of the free surface of the water above the bottom of the weir, the discharge, Q, is given by the following table

h	4	6	8	10	12
Q	650	1740	3640	6360	9790

Show that these observations are consistent with a relation between Q and h of the form $Q=ah^n$ and give estimates, correct to one decimal place, for the values of a and n.

[L]

9. Given that $\log_2(x-5y+4)=0$ and $\log_2(x+1)-1=2\log_2 y$, find the values of x and y.

[AEB '80]

10. Establish the formula

$$\log_y x = \frac{1}{\log_x y}.$$

Solve the simultaneous equations
$$\log_x y + 2\log_y x = 3,$$
$$\log_9 y + \log_9 x = 3.$$

[JMB]

5 Differentiation

5.1 Introduction

We saw in Chapter 2 that if one variable (y) depends on another (x) then we say that y is a function of x, and we may be able to illustrate the relationship between x and y on a graph. If the function is linear (so that the graph is a straight line), then we saw in Chapter 3 that the gradient of the line is a measure of the rate at which y is changing as x changes, i.e. it measures *the rate of change of* y *with respect to* x.

Differentiation is concerned with finding the rate of change of one variable with respect to (w.r.t.) another, in cases when the function connecting them is not necessarily linear. We shall see that such a rate of change can still be represented by a gradient on the associated graph, and we begin by investigating what is meant by the gradient at a particular point on a curve.

5.2 The gradient at a point on a curve: intuitive approach

Consider the curve shown in Fig. 5.1. Intuitively, we would say that the graph is steeper at Q than at P, or that the gradient at Q is greater than the gradient at P. If the curve represented part of the vertical cross-section of a bath, then a spider might be able to stand at P, but would slip if placed at Q.

Figure 5.1

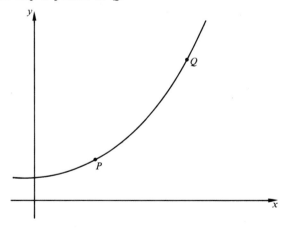

We shall need to convert this intuitive idea of the gradient at a point on a curve into a formal definition, since the definition of the gradient of a straight line is clearly not appropriate for a curve. Whereas the gradient of a straight line is the same at all points of the line, the gradient of a curve will, in general, be different at different points of the curve.

Now suppose that we have a curve $y=f(x)$ and that we wish to estimate its gradient at some particular point P on the curve. One method would be as follows.

Choose some other point Q on the curve (Fig. 5.2), and find the gradient of the chord PQ. This should provide a very rough estimate of the slope at P, provided that Q is not too far from P.

Figure 5.2

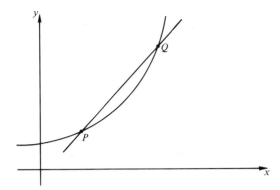

We can obtain better estimates by bringing Q gradually closer to P (Fig. 5.3), and 'in the limit' (i.e. when Q has got 'as close as possible' to P) the chord PQ becomes the tangent at P (Fig. 5.4). The gradient of this tangent is the required value, i.e. it is the same as the gradient of the curve at P.

Figure 5.3

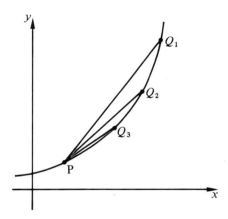

Figure 5.4

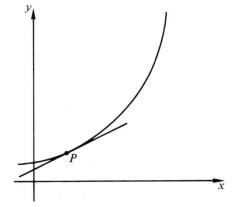

Qu.1 In this question, we attempt to find the gradient at the point $P(3, 9)$ on the curve $y = x^2$. Illustrate each part on a rough sketch.

(i) Let Q be the point on the curve with x-coordinate 5. Find the gradient of PQ.

(ii) Repeat (i) if the x-coordinate of Q is

(a) 4; (b) $3\frac{1}{2}$; (c) 3·1; (d) 3·01; (e) 3·001.

(iii) Suggest a value for the gradient of the curve at P.

(iv) Now try approaching from the left, i.e. take the x-coordinate of Q as

(a) 2; (b) $2\frac{1}{2}$; (c) 2·9; (d) 2·99; (e) 2·999.

Does this suggest the same answer as (iii)?

Generalizing the method of Qu.1, suppose that we wish to find the gradient of $y = f(x)$ at the point P whose x-coordinate is a. Let Q be a nearby point on the curve with x-coordinate $a + h$ where we think of h as being small. Then the coordinates of P and Q are

$$(a, f(a)) \text{ and } (a + h, f(a + h)),$$

and so (see Fig. 5.5)

Figure 5.5

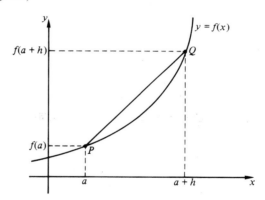

$$\text{gradient of chord } PQ = \frac{f(a + h) - f(a)}{h}.$$

We now wish to move Q 'very close' to P, i.e. we want h to become 'very small'. So we take

$$\text{gradient at } P = \lim_{h \to 0} \frac{f(a + h) - f(a)}{h}. \tag{1}$$

Note

(1) In this limit, we cannot simply put $h = 0$, since this gives $\frac{0}{0}$ (see Section 1.12). Geometrically, putting $h = 0$ would mean that Q and P become the same point, and it is meaningless to speak of the gradient of the line PP.

(2) In Qu.1, we checked that the same value for the gradient at P was obtained whether Q approached from the right or from the left. This checking is implicit in (1) when we write $h \to 0$, which requires that the limit be the same when h approaches zero from above or below. If

different limits are obtained, as in question 5 of Exercise 5a, then the curve does not have a gradient at P.

5.3 The gradient at a point on a curve: formal approach

Definitions

(1) Let P be the point with x-coordinate a on the curve $y=f(x)$. *The gradient of the curve at* P is defined by

$$\text{gradient at } P = \lim_{h \to 0} \frac{f(a+h)-f(a)}{h}.$$

(2) The *tangent* at P is the straight line through P with the same gradient as the curve has at P. (Compare this definition with the use of the word 'tangent' in Section 2.11. They are reconciled in Section 10.6.)

There is no question of these statements' needing proof: they are definitions. But the reasons why we choose to make such definitions are given by the work in the previous section.

Qu.2 Plot the graph of $y=x^2-x+1$ for $-2 \leqslant x \leqslant 3$. Draw by eye the tangent at $(2, 3)$ and hence estimate the gradient of the curve at this point.

Exercise 5a

1. Let P be the point on the parabola $y=3x^2+2x+1$ where $x=2$. Let Q be another point on the curve. Calculate the gradient of the chord PQ when the x-coordinate of Q is
   (i) 4;   (ii) 3;   (iii) 2·1;
   (iv) 2·01;   (v) 2·0001.
   Use your answers to guess the gradient of the curve at P.
2. Check that the gradient of the curve at P in question 1 exists, by repeating the question with positions of Q to the left of P. Take the x-coordinate of Q to be (i) 1·99; (ii) 1·9999. [*Hint*: to find $1·9999^2$, write it as $(2-0·0001)^2$.]
3. Estimate the gradient of the curve $y=\sin x°$ at the point $P(30, 0·5)$ by taking 'Q' to have x-coordinate
   (i) 31;   (ii) 30·1.
   State which of the two estimates you would expect to be more accurate.
4. (i) Plot the curve $y=2^x$ for $-2 \leqslant x \leqslant 3$. Draw by eye the tangent at the point $(0, 1)$ and hence estimate the gradient of the curve at this point.
   (ii) Calculate the gradient of the chord joining $(-0·1, 2^{-0·1})$ and $(0·1, 2^{0·1})$ and show on a diagram why your answer should be approximately the same as that to part (i).
5. Sketch the graph of $f(x)=|x|$. Show that the gradient does not exist at $(0, 0)$ by showing that

$$\lim_{h \to 0-} \frac{f(a+h)-f(a)}{h} \quad \text{and} \quad \lim_{h \to 0+} \frac{f(a+h)-f(a)}{h}$$

are different when $a=0$.

DIFFERENTIATION

6. Estimate the gradient of the tangent to the curve $y = \sqrt{x}$ at the point $P(3, \sqrt{3})$ by calculating the gradient of a 'short' chord with one end at P.

5.4 The derived function

Suppose that $f(x)$ is a function whose gradient exists for some (possibly all) values of x. Then, for those values of x, we may define a new function f' which gives the gradient of f at each point. This new function is called the gradient function or the *derived function* or simply the *derivative*.

Since $f'(a)$ is the gradient of the curve $y = f(x)$ at the point where $x = a$, we have

$$f'(a) = \lim_{h \to 0} \frac{f(a+h) - f(a)}{h}$$

and so, in general, the function f' may be defined thus:

Definition

$$f'(x) = \lim_{h \to 0} \frac{f(x+h) - f(x)}{h}. \tag{2}$$

(The function f is said to be *differentiable* at some point if its gradient exists at that point.)

The relationship between a function f and its derived function f' is illustrated in Fig. 5.6. The values (i.e. heights) plotted on the lower graph are the values of the gradient of the upper graph. To the left of $x = a$, the gradient of f is positive (but decreasing), so the value of f' is positive (and decreasing). At $x = a$, the gradient of f is zero, so f' has a root there, and similarly at $x = b$ and $x = d$. At $x = c$, the graph of f reaches its steepest value between b and d, and so f' 'peaks' at this point.

Figure 5.6

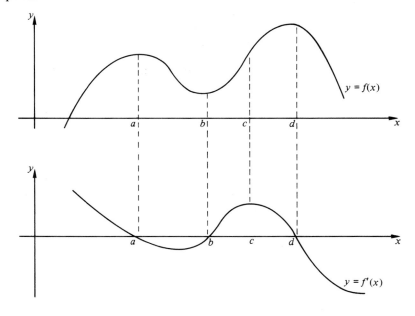

Qu.3 Verify that in Fig. 5.6, the sign of the gradient of f and the sign of f' are always the same.

5.5 Differentiation

Now that we have defined the derivative of a given function, we must turn our attention to the problem of finding it in specific cases. The process of finding a derivative is called *differentiation*, and we say that we *differentiate* $f(x)$ with respect to x to get $f'(x)$. Rather than use the definition (2) each time we wish to differentiate, we shall obtain certain quick rules for differentiating various types of function. These rules often enable us to write down a derivative without any working. Each time we meet a new type of function, we shall use definition (2) to obtain such a 'rule of thumb'. Working from the definition is known as differentiation *from first principles*.

Example 1
Differentiate $f(x) = x^2$ from first principles.
Solution
If $f(x) = x^2$, then $f(x+h) = (x+h)^2$. From the definition,

$$f'(x) = \lim_{h \to 0} \frac{(x+h)^2 - x^2}{h}$$

$$= \lim_{h \to 0} \frac{2xh + h^2}{h}$$

$$= \lim_{h \to 0} (2x + h)$$

$$= 2x.$$

So the derivative of x^2 is $2x$.

Qu.4 Differentiate x^3 from first principles. (It may help to expand $(x+h)^3$ before starting.)

Example 2

Differentiate $\dfrac{1}{x}$ from first principles.

Solution

If $f(x) = \dfrac{1}{x}$, then

$$f'(x) = \lim_{h \to 0} \frac{\frac{1}{x+h} - \frac{1}{x}}{h}$$

$$= \lim_{h \to 0} \frac{1}{h} \left(\frac{x - (x+h)}{x(x+h)} \right)$$

$$= \lim_{h \to 0} \frac{1}{h} \left(\frac{-h}{x^2 + xh} \right)$$

$$= \lim_{h \to 0} \frac{-1}{x^2 + xh}$$

$$= -\frac{1}{x^2}.$$

Qu.5 Differentiate $\sqrt{x}$ from first principles. [*Hint*: it helps at one stage to rationalize the *numerator*.]

Exercise 5b

1. Plot the graph of $y = x^2$ for $|x| \leq 3$. Using the fact that the derivative of x^2 is $2x$, calculate the gradient of the curve at the point where $x = -1\frac{1}{2}$, and hence draw an accurate tangent to the curve at this point.

2. Differentiate from first principles:
   (i) $f(x) = 3x^2$; (ii) $f(x) = 4x^3$; (iii) $f(x) = 6$;
   (iv) $f(x) = x^2 + x$; (v) $f(x) = 3x + 7$;
   (vi) $f(x) = ax^4$ (where a is a constant).

3. Use your answers to question 2 and text Qu.4 to find the gradients of the following curves at the given points:
   (i) $y = 3x^2$ at $(3, 27)$;
   (ii) $y = x^3$ at the point where $x = -4$;
   (iii) $y = 4x^3$ at the point where $y = -32$.

4. Differentiate $\dfrac{1}{\sqrt{x}}$ from first principles. (See the hint for Qu.5.)

5. Differentiate $\sqrt[3]{x}$ from first principles. (Use the hint for Qu.5 and the technique of Qu.7 in Chapter 4.)

6. Plot the graph of $y = x^3$ for $-2 \leq x \leq 2$ and also its derived function (Qu.4). Take equal scales on the two axes for each graph. Check to your own satisfaction that the derived function does give the gradient of $y = x^3$ at each point.

5.6 The differentiation of x^n etc.

In Examples 1 and 2 and Qus. 4 and 5, we saw that

$$f(x) = x^2 \quad \Rightarrow f'(x) = 2x;$$
$$f(x) = x^3 \quad \Rightarrow f'(x) = 3x^2;$$
$$f(x) = x^{-1} \quad \Rightarrow f'(x) = -x^{-2};$$
$$f(x) = x^{1/2} \quad \Rightarrow f'(x) = \tfrac{1}{2}x^{-1/2}.$$

These results suggest that, for any real n, the derivative of x^n is nx^{n-1}. This is indeed the case; but to prove it in general from first principles, we must be able to expand $(x+h)^n$ for any real n, and this requires the Binomial Series (Chapter 28). For the present, we merely assume the result:

Theorem 5.1
For any real number n, the function $f(x) = x^n$ is differentiable, and its derivative is $f'(x) = nx^{n-1}$.

Qu.6 Write down the derivatives of

(i) x^7;   (ii) x^{-3}; (iii) $x^{5/2}$;

(iv) $\sqrt[3]{x}$;   (v) $\dfrac{1}{x^4}$;   (vi) $\dfrac{1}{\sqrt[3]{x^7}}$;

(vii) 1 (write as x^0);   (viii) x.

Theorem 5.2

If $f(x)$ is differentiable and $g(x)=kf(x)$, where k is constant, then $g'(x)=kf'(x)$. (This says that if some function has a constant factor k, then this factor simply appears as a factor in the derivative.)

Proof

Let $g(x)=kf(x)$, then

$$g'(x)=\lim_{h\to 0}\frac{g(x+h)-g(x)}{h}$$

$$=\lim_{h\to 0}\left(\frac{kf(x+h)-kf(x)}{h}\right)$$

$$=k\lim_{h\to 0}\left(\frac{f(x+h)-f(x)}{h}\right)$$

$$=kf'(x).$$

It follows from Theorems 5.1 and 5.2 that the derivative of ax^n is nax^{n-1}.

Qu.7 Write down the derivatives of
   (i) $4x^2$;   (ii) $\frac{1}{6}x^{-3}$;   (iii) $-2x^{-1}$;

(iv) $2\sqrt{x}$;   (v) $\dfrac{7}{3x^2}$;   (vi) $5x$;

(vii) 3 (write as $3x^0$);   (viii) $\dfrac{x^2}{4}$.

Qu.8 Let $f(x)=c$ (where c is a constant). What is the derivative of f? Verify your answer by sketching the graph of f and considering the gradient at each point.

Qu.9 Repeat Qu.8 with the function $f(x)=mx$ where m is constant.

Theorem 5.3

The derivative of $f(x)+g(x)$ is $f'(x)+g'(x)$.

(This theorem says that if some function is the sum of two terms, we can differentiate each term separately. Clearly the theorem can be extended to the sum of three or more terms.)

Proof

Let $s(x)=f(x)+g(x)$, then

$$s'(x)=\lim_{h\to 0}\left(\frac{s(x+h)-s(x)}{h}\right)$$

$$=\lim_{h\to 0}\left(\frac{\{f(x+h)+g(x+h)\}-\{f(x)+g(x)\}}{h}\right)$$

$$= \lim_{h \to 0} \left(\frac{\{f(x+h)-f(x)\} + \{g(x+h)-g(x)\}}{h} \right)$$

$$= f'(x) + g'(x).$$

(Here we are assuming that the limit of a sum is the sum of the individual limits.)

Combining the results of our three theorems:

if
$$f(x) = a_1 x^{n_1} + a_2 x^{n_2} + \dots + a_t x^{n_t}$$

then
$$f'(x) = n_1 a_1 x^{n_1-1} + n_2 a_2 x^{n_2-1} + \dots + n_t a_t x^{n_t-1}$$

which is not as complicated as it looks!

Qu.10 Differentiate
(i) $6x^3 + 3x^2$; (ii) $7x^4 - 7x^{-4}$;
(iii) $5x + 7$; (iv) $mx + c$;
(v) $ax^2 + bx + c$;
(vi) $(x^2 + 2)(x - 3)$ (multiply out first).

Qu.11 (i) Differentiate $\dfrac{x^2 - 3x - 8}{2x}$ by first writing it as

$\frac{1}{2}x - \frac{3}{2} - 4x^{-1}$.

(ii) Similarly differentiate $\dfrac{2x^4 + 3x - 2}{4x^2}$.

Qu.12 Show, by finding a counter-example, that the derivative of $f(x)g(x)$ is not in general $f'(x)g'(x)$ (i.e. the derivative of a product is not the product of the derivatives).

Example 3
Find the gradient of the curve $y = 2x^2 - 3\sqrt{x} + 1$ at the point where $x = 4$.
Solution

$$f(x) = 2x^2 - 3x^{1/2} + 1$$

$$\Rightarrow f'(x) = 4x - \frac{3}{2}x^{-1/2}$$

$$\Rightarrow f'(4) = 16 - \frac{3}{2}\left(\frac{1}{2}\right)$$

$$= 15\tfrac{1}{4}.$$

So the gradient at the given point is $15\frac{1}{4}$.

Qu.13 By differentiating and substituting the appropriate value of x, find the gradients of the following curves at the given points:
(i) $f(x) = x^2 + 2x + 3$, where $x = 5$;
(ii) $f(x) = (x + 3)(x - 1)$, where $x = 1$;
(iii) $f(x) = 4\sqrt{x}$, at the point $(9, 12)$;

(iv) $f(x) = \dfrac{x^2 + 1}{2x}$, where $x = 1$.

5.7 The delta notation

We now introduce an important alternative notation for some of the ideas already encountered in this chapter. In leading up to the definition of the gradient of a curve at a point P, we considered a second 'moveable' point Q near to P. The x-coordinates of P and Q were a and $a+h$, and their y-coordinates were $f(a)$ and $f(a+h)$ respectively. Thus h represented a small change in x, and the corresponding change in y was $f(a+h)-f(a)$. We shall now denote these changes by δx and δy (pronounced delta x and delta y, δ being the Greek letter d. Some books use Δ, a capital delta, in this context.) Thus

$$\delta x = h \text{ and } \delta y = f(a+h)-f(a).$$

Note

(1) It must not be thought that δx means δ times x. When we write δx, it should be regarded as a single symbol meaning 'a small change in x'.

(2) It is important to appreciate that when we are dealing with a particular point on a particular curve, δx can take any value (positive or negative) but the corresponding value of δy is then fixed by $\delta y = f(x+\delta x)-f(x)$.

We can now define the gradient at a point using this new notation. We have (Fig. 5.7)

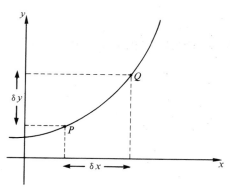

Figure 5.7

gradient of chord $PQ = \dfrac{\delta y}{\delta x}$

and so   gradient of tangent at $P = \lim\limits_{\delta x \to 0} \dfrac{\delta y}{\delta x}$.

This limit is usually denoted by $\dfrac{dy}{dx}$, so that

$$\frac{dy}{dx} = \lim_{\delta x \to 0} \frac{\delta y}{\delta x}.$$

The notation $\dfrac{dy}{dx}$ is thus an alternative to $f'(x)$, and we say that

$\dfrac{dy}{dx}$ is the derivative of y. It must be emphasized that $\dfrac{dy}{dx}$ is a

DIFFERENTIATION

single symbol; it is not a fraction and there is no question of cancelling the d's.

As an illustration, if $y = 3x^2$, then $\dfrac{dy}{dx} = 6x$. We might also write

$$\frac{d}{dx}(3x^2) = 6x.$$

Qu.14 If P is the point $(2, 8)$ on the curve $y = 3x^2 - 2x$, find the value of δy if δx is
(i) 1; (ii) -0.01.
Find $\dfrac{\delta y}{\delta x}$ in case (ii) and compare it with the value of $\dfrac{dy}{dx}$ (at the point P) found by differentiating.

5.8 The significance of the derivative

We have defined the derived function of f by

$$f'(x) = \lim_{h \to 0} \frac{f(x+h) - f(x)}{h}.$$

Although we were led to this definition by considering the gradient at a point on a curve, the definition itself is not geometrical at all and could have been made without reference to graphs or gradients. The real significance of $f'(x)$ is that it provides a measure of *the instantaneous rate of change* of one variable w.r.t. another. Let us consider a specific example.

Suppose that a large tank of water is leaking, and readings taken at 10-second intervals indicate the volume V of water in the tank:

t (seconds)	0	10	20	30
V (litres)	100	80	64	51.

If we wish to estimate the rate at which the water was leaking at the instant $t = 10$, we might take as a first estimate the average rate of leaking between the times $t = 10$ and $t = 20$, i.e.

$$\frac{\text{change in } V}{\text{change in } t} = \frac{64 - 80}{20 - 10} = -1.6.$$

(We might write $\dfrac{\delta V}{\delta t} = -1.6$.)

So between these times, V was decreasing at an average rate of 1·6 litre per second. The number -1.6 corresponds to the gradient of the chord PQ in Fig. 5.8.

A better estimate of the rate of water loss at $t = 10$ might have been obtained if we had known the value of V at say $t = 10.5$. The average rate of flow during the $\frac{1}{2}$-second interval from $t = 10$ to $t = 10.5$ would probably be nearer to the instantaneous value at $t = 10$ than the average during the 10-second interval above. This corresponds to saying that the gradient of the chord PQ is a better estimate of the

Figure 5.8

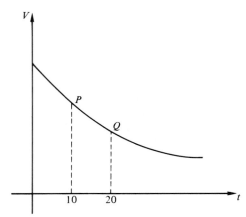

gradient at P if the point Q is 'near' to P (or that $\dfrac{\delta V}{\delta t}$ is a better

approximation to $\dfrac{dV}{dt}$ when the value of δt is small).

Qu.15 Plot accurately the graph of V against t from the values in the table above. Draw by eye the tangent at $t=10$ and hence estimate the rate of water loss at this instant.

Exercise 5c

1. Differentiate the following with respect to x:

(i) $x^4 - x^2 + 7x$;   (ii) $\dfrac{1}{x^2}$;

(iii) $\sqrt[3]{x^2}$;   (iv) $7x^{-2} - 8x^{-3}$;

(v) $4\sqrt{x^3}$;   (vi) $\dfrac{3}{x} - \dfrac{4}{x^3}$;

(vii) $\dfrac{3}{2\sqrt[5]{x^3}}$;   (viii) $\sqrt{(2x^5)} + 9$;

(ix) $(x^2+3)(2x-4)$;   (x) $\dfrac{1}{x^2}(2x+3)$;

(xi) $\sqrt{x}(2+4x-3x^2)$;   (xii) $\dfrac{1+x+x^2}{2x}$;

(xiii) $\dfrac{x^3-2x^2}{3\sqrt{x}}$.

2. Find the gradient of the curve $y=2x^2+3x-3$ at the point $(-2, -1)$.

3. Find the coordinates of the points where the line $y=7$ meets the curve $y=x^2-2x+4$. Find the gradient of the curve at each of these points.

4. Differentiate $f(x)=3x^2-4x+7$ and hence find the coordinates of the point where the gradient is zero.

5. Prove that the gradient at any point on either of the curves $y=\dfrac{1}{x}$ and $y=\dfrac{1}{x^3}$ is negative, and sketch each of these curves on the same axes.

6. Show that for any particular value of x, the curves $y=(x+3)(x+1)$ and $y=x(x+4)$ have the same gradient. Give an explanation, in terms of their graphs, of why this is so.

7. If $\dfrac{dy}{dx}=3x^2+4x+2$, write down a possible function for y.

8. Show that if $y=\dfrac{1}{x}$, then $\dfrac{dy}{dx}+y^2=0$.

9. Find $\dfrac{dy}{dx}$ if
   (i) $2y=1+x^2$;   (ii) $4xy^2=1$.

10. Find the coordinates of the point on the curve $y=x^2+3x-9$ where the gradient is 7.

11. Find the coordinates of the points on the curve $y=x^3-4x^2-3x+1$ where the gradient is zero.

12. Find the coordinates of the points on the curve $y=x^3-3x^2+4x+7$ at which the gradient is 13.

13. Let $y=x^3$. Is it true that the gradient of the chord joining the points whose x-coordinates are 1 and 3 is equal to the gradient of the curve at the point where $x=2$?

14. Prove that the gradient of the chord joining the points on the curve $y=x^3$ whose x-coordinates are $p-1$ and $p+1$ always exceeds the gradient of the curve at (p, p^3) by 1.

15. Find and prove a similar result to that in question 14 for the parabola $y=ax^2+bx+c$ (where a, b and c are constants).

16. The temperature ($\theta°C$) in a certain draught-free corridor is given approximately by $\theta=\dfrac{400-x-x^2}{20}$ where x is the distance in metres from a radiator at one end. Find
   (i) the average rate at which the temperature changes with distance between $x=0$ and $x=5$;
   (ii) the rate of change of temperature w.r.t. distance at the point $x=5$.

6 Applications of differentiation

6.1 A note on polynomial equations

Before considering some of the applications of differentiation, we discuss briefly the solution of simple polynomial equations; the technique is needed later in this chapter.

Consider the equations:

$$2x + 3 = 0; \tag{1}$$
$$x^2 - 5x + 2 = 0; \tag{2}$$
$$3x^3 - 4x^2 + 5x + 1 = 0; \tag{3}$$
$$2x^4 + 7x^3 - 3x^2 - x + 5 = 0. \tag{4}$$

In each case, the L.H.S. of the equation is a *polynomial* (see Chapter 10), and so these are called polynomial equations. The *degree* of a polynomial is the highest power that occurs; so (1) is a 1st degree (or linear) equation, and (2) is a 2nd degree (or quadratic) equation. Equations of degree 3, 4 and 5 are called *cubic*, *quartic* and *quintic* equations respectively.

One method of solving a quadratic equation is to factorize the quadratic into two linear factors. The same method may be used for equations of higher degree, although in practice, it is usually difficult or impossible to find the factors unless further clues are available, as in the examples below.

For quadratic equations which cannot easily be factorized, we may use the method of completing the square (Chapter 10) which leads to 'the formula' for quadratic equations. The question therefore arises as to whether every polynomial equation can be solved algebraically by some standard method. The answer is that the general cubic and quartic equations can be solved, although the methods are not easy. (See Exercise 31e, question 21 for a method for cubics.) Abel and Galois proved that the general equation of degree five or more cannot be solved. This does not mean, of course, that *no* polynomial equation of degree five or more can be solved. For example, $x^5 - 4x^3 = 0$ is trivial.

When algebraic solutions are difficult or impossible, numerical solutions (to any desired degree of accuracy) can always be found (Chapter 29).

Example 1

Show that $x = 4$ is one root of $2x^3 - 5x^2 - 21x + 36 = 0$, and find any other roots.

Solution

We may show that $x = 4$ is a root by direct substitution:

$$2(4^3) - 5(4^2) - 21(4) + 36 = 0.$$

Since $x=4$ is a root, $(x-4)$ must be a factor of the polynomial. (A proof of this relationship between roots and factors is given in Chapter 10.) By inspection, we have

$$2x^3 - 5x^2 - 21x + 36 \equiv (x-4)(2x^2 + 3x - 9). \qquad (5)$$

Here, the term $2x^2$ was found by observing that there is a term $2x^3$ in the product; and the -9 term was found by considering the product of the constant terms. The term $+3x$ is found by using either the term $-5x^2$ or the term $-21x$ on the L.H.S.. Each of these arises in two parts, one of which we know from the terms we have already found. The factorization should be checked when complete.

Factorizing the quadratic factor in (5) gives

$$2x^3 - 5x^2 - 21x + 36 \equiv (x-4)(2x-3)(x+3)$$

and so the original equation is

$$(x-4)(2x-3)(x+3) = 0$$

and the other two roots are $x = \frac{3}{2}$ and $x = -3$.

Qu.1 Prove that $x=3$ is the only root of

$$x^3 - x^2 - x - 15 = 0.$$

Qu.2 Show that $x^4 + x^3 + 3x^2 - 13x + 8$ factorizes as $(x^2 - 2x + 1)(x^2 + 3x + 8)$ and hence solve $x^4 + x^3 + 3x^2 - 13x + 8 = 0$.

It follows from the ideas above that a polynomial equation of degree n may have up to n real roots, but not more. We shall meet and expand upon this idea at several points in this book.

Exercise 6a

Solve the following equations (numbers 1 to 7).
1. $x(x-1)(x+2) = 0$.
2. $(x+1)(2x^2 + 3x - 2) = 0$.
3. $(x^2 - 5x - 6)(x^2 - 5x + 7) = 0$.
4. $x^3 + 5x^2 + 9x + 6 = 0$ given that $x = -2$ is a root.
5. $6x^3 + 19x^2 + 8x - 5 = 0$ given that $x = -1$ is a root.
6. $2x^3 - 3x^2 + 2x - 3 = 0$ given that $x = 3/2$ is a root.
7. $x^3 - 4x^2 - 11x + 30 = 0$.
8. (i) Factorize $x^3 - 1$, and show that $x^3 = 1$ has only one real root.
   (ii) Similarly show that $x^4 = 1$ has only two solutions.
9. (i) Solve $x^4 - 3x^2 - 4 = 0$ by first regarding it as a quadratic equation in x^2.
   (ii) Solve $x^8 + 5x^4 + 6 = 0$.

6.2 Local maxima and minima

Consider the graph shown in Fig. 6.1. The minimum value of y occurs at A, but there is, in a sense, a minimum at C. Similarly, there is a maximum at B, although this is not *the* maximum value of y. Each of A and C is said to be a *local minimum*, and B is a *local maximum*. The idea of a local maximum or minimum is simple enough; the rigorous definition below is rather hard, and may be omitted.

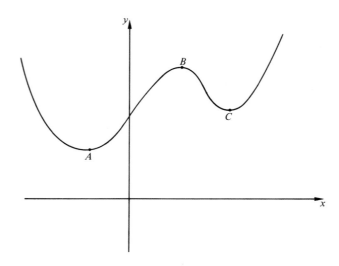

Figure 6.1

Definition

A function f is said to have a *local maximum* at $x=a$ if and only if $\exists \varepsilon > 0$ s.t.

if $a - \varepsilon < x < a + \varepsilon$ and $x \neq a$

then $f(x) < f(a)$.

This says that there is some *neighbourhood* of a such that at every point in the neighbourhood (other than at $x = a$) the value of $f(x)$ is less than the value of $f(a)$. A local minimum is defined similarly.

From now on, we shall omit the word 'local' and refer simply to a maximum or minimum. The term *turning point* will be used for any point which is either a maximum or a minimum.

Qu.3 (i) If f is a continuous function, can it have two maxima with no minimum between?
(ii) Sketch an example of a discontinuous function with two maxima and no minimum between.

6.3 Stationary points

Let f be a differentiable function. It is clear from graphical considerations that

f has a turning point at $x = a$

$\Rightarrow$ the gradient at $x = a$ is zero

and we shall use this fact to find turning points.

Unfortunately, the converse is not true; i.e. if $\dfrac{dy}{dx} = 0$ at some point, we cannot deduce that there is a turning point there. We can show this with a counter-example. Consider the function $y = x^3$ (Fig. 6.2). At $x = 0$, it has neither a maximum nor a minimum, and yet $\dfrac{dy}{dx} = 3x^2$ and so at $x = 0$, $\dfrac{dy}{dx} = 0$.

112   APPLICATIONS OF DIFFERENTIATION

Figure 6.2

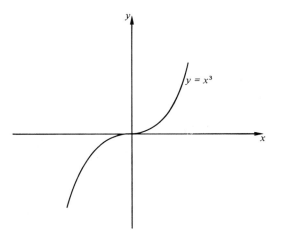

A point where the gradient is zero but which is neither a maximum nor a minimum is a *point of inflexion*. Points where the gradient is zero (maxima, minima and inflexions) are collectively referred to as *stationary points*. (Note that there are other types of inflexion, types where $\dfrac{dy}{dx} \neq 0$ and which are therefore not stationary points. We are not concerned with these here; they are discussed in Section 10.8.)

Qu.4   Find the value of x where the gradient of $y = x^2 - 3x + 4$ is zero and hence find the coordinates of the only stationary point of this curve.

We now consider how, having located the stationary points of a function by putting $\dfrac{dy}{dx} = 0$, we may determine whether each is a maximum, minimum or point of inflexion. Let us examine what happens to the gradient of a curve as we pass through a stationary point. In Fig. 6.3, the numbers refer to the gradient of the curve.

Figure 6.3

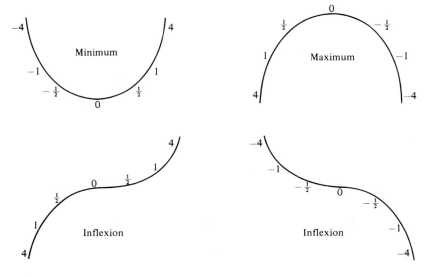

APPLICATIONS OF DIFFERENTIATION

113

These diagrams show that as x increases (i.e. as we move from left to right):

(1) at a minimum, the gradient increases from a negative value to a positive value;

(2) at a maximum, the gradient decreases from a positive value to a negative value;

(3) at an inflexion, the gradient either

(i) decreases to zero and then increases (but is positive on both sides),

or

(ii) increases to zero and then decreases (but is negative on both sides).

Let us now apply these ideas.

Example 2

Investigate the stationary points of the function
$y = x^4 - 5x^3 + 9x^2 - 7x + 3$.

Solution

$$\frac{dy}{dx} = 4x^3 - 15x^2 + 18x - 7.$$

So at a stationary point,

$$4x^3 - 15x^2 + 18x - 7 = 0.$$

By inspection, $x = 1$ is a root, and so $(x - 1)$ is a factor.

Then

$$(x - 1)(4x^2 - 11x + 7) = 0$$

$$\Rightarrow (x - 1)(x - 1)(4x - 7) = 0$$

$$\Rightarrow x = 1 \text{ or } x = \frac{7}{4}.$$

From the original equation

$$x = 1 \Rightarrow y = 1$$

and

$$x = \frac{7}{4} \Rightarrow y = \frac{229}{256}.$$

Now, to determine the nature of the stationary points, we evaluate the gradient of the curve 'near' each point.

(i) When $x = 0 \cdot 9$, $\frac{dy}{dx} = -0 \cdot 034$.

When $x = 1 \cdot 1$, $\frac{dy}{dx} = -0 \cdot 026$.

The gradient is negative on each side of $x = 1$, so there is an inflexion there. (See the last diagram of Fig. 6.3.)

(ii) When $x = 1 \cdot 7$, $\frac{dy}{dx} = -0 \cdot 098$.

When $x = 1 \cdot 8$, $\frac{dy}{dx} = 0 \cdot 128$.

APPLICATIONS OF DIFFERENTIATION

So the gradient changes from a negative value to a positive value at $x = \frac{7}{4}$, and so there is a minimum there.
Summarizing, we have

(i) an inflexion at $(1, 1)$;

(ii) a minimum at $\left(\dfrac{7}{4}, \dfrac{229}{256} \right)$.

Qu.5 Investigate the stationary points of $y = x^3 - 6x^2 - 36x + 4$.

6.4 The second derivative

The method used above, that of finding the gradient on each side of a stationary point, is rather inelegant, and in some cases it can be unclear as to how near to the stationary point we must look in order to be sure that we have not 'trapped' a second stationary point in the interval that we are examining. Fortunately, a better method is generally available.

If y is a function of x, we denote the derivative of y with respect to x by $\dfrac{dy}{dx}$. If we now differentiate again to get the derivative of

$\dfrac{dy}{dx}$, we denote this new function by $\dfrac{d}{dx}\left(\dfrac{dy}{dx}\right)$ or, more usually,

$\dfrac{d^2 y}{dx^2}$. It measures the rate of change (with respect to x) of the

gradient, or the rate of change of the rate of change of y. It is known as the *second derivative*. Higher order derivatives are similarly defined;

thus, for example, $\dfrac{d}{dx}\left(\dfrac{d^2 y}{dx^2}\right)$ is written $\dfrac{d^3 y}{dx^3}$ and is obtained

by differentiating y three times in succession. In our alternative notation, the second derivative of $f(x)$ is written $f''(x)$, the third derivative is $f'''(x)$, and the nth is $f^{(n)}(x)$. The notation y', y'', y''' etc. is also common.

Qu.6 If $y = 3x^4 + 7x^3 + 2$, find $\dfrac{d^2 y}{dx^2}$.

Now if at a stationary point $\dfrac{d^2 y}{dx^2} > 0$, then the rate of change of the

gradient is positive, i.e. the gradient is increasing, and so (see Fig. 6.3)

the stationary point is a minimum. Similarly, if $\dfrac{d^2 y}{dx^2} < 0$, the stationary

point is a maximum. Unfortunately it is not true that if $\dfrac{d^2 y}{dx^2} = 0$ then

the stationary point is an inflexion, since some turning points are so 'flat' that they too have a zero second derivative (see Qu.8 below). So

if, in attempting to classify a stationary point, we get $\dfrac{d^2 y}{dx^2} = 0$, the

method of Example 2 should be used. (In practice, it is often possible to determine the sign of $\dfrac{dy}{dx}$ on each side of the stationary point by inspection, without the direct substitution illustrated in Example 2.)

Summarizing:

$\dfrac{d^2y}{dx^2} > 0 \Rightarrow$ stationary point is a minimum;

$\dfrac{d^2y}{dx^2} < 0 \Rightarrow$ stationary point is a maximum;

$\dfrac{d^2y}{dx^2} = 0 \Rightarrow$ stationary point could be any

of the three types.

Example 3
Investigate the stationary points of

$$y = x^3 - 3x^2 - 24x + 10.$$

Solution

$$\frac{dy}{dx} = 3x^2 - 6x - 24.$$

So at a stationary point,

$$3x^2 - 6x - 24 = 0$$
$$\Rightarrow 3(x + 2)(x - 4) = 0$$
$$\Rightarrow x = -2 \text{ or } 4.$$

Substituting to find y gives stationary points at $(-2, 38)$ and $(4, -70)$.

To determine their nature, we evaluate $\dfrac{d^2y}{dx^2}$ at each point.

$$\frac{d^2y}{dx^2} = 6x - 6,$$

and so

(i) at $x = -2$, $\dfrac{d^2y}{dx^2} = -18$; this is negative, so the stationary point is a maximum;

(ii) at $x = 4$, $\dfrac{d^2y}{dx^2} = 18$; this is positive, so the stationary point is a minimum.

Thus there is a maximum at $(-2, 38)$ and a minimum at $(4, -70)$.

Qu.7 Investigate the stationary points of $y = x^4 - 4x^3 - 20x + 6$.

Qu.8 Sketch the graph of $y = x^4$ near the origin (say for $|x| \leqslant \frac{1}{2}$) and state the nature of the stationary point. Find $\dfrac{d^2y}{dx^2}$ at $x = 0$.

Qu.9 Show that if $y = x^3 - 3x^2 + 4x - 2$, then $\dfrac{d^2y}{dx^2} = 0$ when $x = 1$, but that $x = 1$ is not a stationary point.

Exercise 6b

1. Find the coordinates of the turning point of the curve $y = 2x^2 + 5x + 1$. Use the second derivative to show that the turning point is a minimum.
2. Show that the curve $y = x^3 - 2x^2 + 2x + 1$ has no stationary points.
3. Investigate fully the stationary points of
   (i) $y = 3x^3 - 5x^2 + 3x - 1$;
   (ii) $y = x^3 - 3x^2 + 3x + 4$;
   (iii) $y = x^4 - 2x^3 - 14x^2 + 7$;
   (iv) $y = x^5 + x^4$;
   (v) $y = x^4 + x^3$.
4. Find the minimum value of the gradient of the curve in question 2.
5. Find the minimum value of y if $y = ax^2 + bx + c$, stating any condition that you find it necessary to impose. Deduce that y cannot be zero for any value of x if $b^2 - 4ac < 0$.
6. By finding the turning points of $y = 4x^3 - 54x^2 + 240x - 351$, show that the equation $4x^3 - 54x^2 + 240x - 351 = 0$ has three real roots.
7. Find the greatest and least values of $y = x^3 + 2x^2 - 4x - 1$ in the region $-4 \leqslant x \leqslant 1$.
   [*Hint*: think about the problem graphically.]
8. Find the least value of y if $y = x^4 + 2x^3 - 5x^2$. Deduce that the equation $x^4 + 2x^3 - 5x^2 + 24$ has no real roots.
9. (i) Show that the function $y = 3x^{2/3}$ satisfies the equation $y^2 \dfrac{d^2y}{dx^2} = -6$.

   (ii) Find all functions of the form $y = Ax^n$ satisfying $\dfrac{d^2y}{dx^2} = y^2$ (i.e. find A and n).
10. By considering whether the gradient is increasing or decreasing, state the sign of $\dfrac{d^2y}{dx^2}$ at each of the points labelled in Fig. 6.4.

Figure 6.4

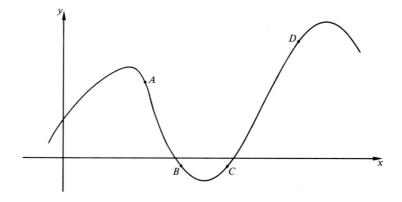

6.5 Sketching polynomial graphs

The process of *sketching* a graph is quite different from that of *plotting* a graph. The distinction is made clear at the beginning of Chapter 16 where the aims and techniques of curve-sketching are discussed. In this section, where we consider only polynomial graphs, it is sufficient to remember that the main purpose of graph-sketching is to illustrate the general shape of the graph and any important features. Accuracy is not important, and it is not normally necessary to find the coordinates of any turning points, although to do so can occasionally be the best way of sketching the graph.

Suppose we wish to sketch the graph of the polynomial

$$y = a_n x^n + a_{n-1} x^{n-1} + \ldots + a_1 x + a_0 \qquad \text{(where } a_n > 0\text{)}.$$

We could begin by listing what we know about the graph. (The reader should check each statement.)
(i) We assume that the graph is continuous, as there does not seem to be any value of x where there is a sudden jump in the value of y.
(ii) As $x \to \infty$, $y \to \infty$ (since $a_n x^n$ becomes the dominant term); and

as $x \to -\infty$, $y \to \infty$ if n is even,

and $y \to -\infty$ if n is odd.

(iii) When $x = 0$, $y = a_0$, so the curve cuts the y-axis at $(0, a_0)$.
(iv) Putting $y = 0$ gives a polynomial equation of degree n which has at most n roots; so the curve cuts the x-axis at no more than n points. Note that, by (ii), if n is odd, the curve, being continuous, must cut the x-axis at least once.
(v) $\dfrac{dy}{dx}$ is a polynomial of degree $n-1$, so there are at most $n-1$ stationary points.

These considerations lead us to the general shapes illustrated in Fig. 6.5.

Figure 6.5

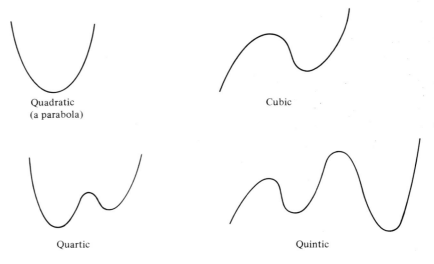

Quadratic
(a parabola)

Cubic

Quartic

Quintic

Note
(1) If the leading coefficient, a_n, is negative, then the graph is 'upside-down', and as $x \to \infty$, $y \to -\infty$.

APPLICATIONS OF DIFFERENTIATION

(2) Two (or more) turning points may 'coalesce' so that, for example, a cubic may look like Fig. 6.6.

Figure 6.6

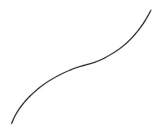

Qu.10   Show, by differentiating, that $y = x^3 + x^2 + 3x - 7$ has no turning points. Find the minimum value of the gradient.

Example 4
Sketch the graph of $y = x^2 + 3x + 7$.
Solution
This is a quadratic with positive leading coefficient, so the graph is a 'right-way-up' parabola. The discriminant, $b^2 - 4ac$, is negative, so the curve does not cross the x-axis. Also, $\dfrac{dy}{dx} = 2x + 3$, so the minimum is at $x = -\frac{3}{2}$. Finally, it cuts the y-axis at $y = 7$. (The value of y at the minimum point could be found if necessary.) The curve is shown in Fig. 6.7.

Figure 6.7

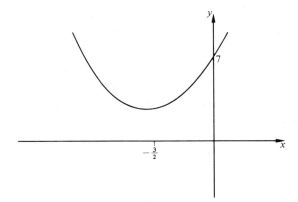

Example 5
Sketch the graph of

$$y = (x - 1)(6x^2 + 2x + 1).$$

Solution
We see that:
(i) the curve is a cubic with a positive leading coefficient;
(ii) it has a root at $x = 1$ and no other root (since $6x^2 + 2x + 1$ has no roots);
(iii) it crosses the y-axis where $y = -1$.
This still leaves two questions:
(iv) does the curve have turning points;

(v) if so, where approximately are the turning points in relation to the axes?

We can answer these as follows:

$$y = (x-1)(6x^2 + 2x + 1)$$
$$= 6x^3 - 4x^2 - x - 1$$
$$\Rightarrow \frac{dy}{dx} = 18x^2 - 8x - 1.$$

This has real roots, so the curve does have turning points. We could now find these turning points, but the following happens to be simpler in this case:

$$x = 0 \Rightarrow \frac{dy}{dx} = -1$$

and

$$x = 1 \Rightarrow \frac{dy}{dx} = 9.$$

Therefore (see Qu.11) the complete curve is as in Fig. 6.8.

Figure 6.8

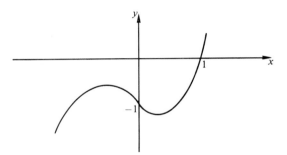

Qu.11 (See Example 5.) By trying to draw other possible positions of the turning points subject to the conditions that $x = 1$ is the only root, and the gradients at $(0, -1)$ and $(1, 0)$ are -1 and 9 respectively, verify that their approximate positions relative to the axes must be as shown.

The sketching of polynomial graphs is discussed further in Chapter 10.

Exercise 6c   Sketch the following curves (nos. 1 to 9), indicating where the curves cut the axes. Find the x-coordinates of any stationary points. You need not necessarily find the corresponding y-coordinates, but the stationary points must be in approximately the correct position relative to the axes.

1. $y = x^2 - 3x + 2$.
2. $y = 3x^2 + 4x - 4$.
3. $y = 2 + x - x^2$.
4. $y = (x-1)(2x-5)$.
5. $y = (x-3)(x^2 - x + 1)$.
6. $y = (x-1)(x^2 + 6x - 2)$.
7. $y = (x-2)^3$.

APPLICATIONS OF DIFFERENTIATION

8. $y=(x-2)^4$.

9. $y=(x-1)^2(x-2)(5x+2)$.

10. By considering the turning points on the graph of $y=x^3+2x^2+x+k$, find the range of values of k for which the equation $x^3+2x^2+x+k=0$ has
   (i) one real root only;
   (ii) two distinct real roots only;
   (iii) three distinct real roots.

11. (See question 10.) Investigate similarly the number of distinct real roots of the equation

$$x^4-10x^3-8x^2+k=0.$$

6.6 Tangents and normals

In the last chapter, we defined the tangent to a curve at a point P, and saw that it has the same gradient as the curve at P. We now define a normal to a curve.

Definition

The *normal* to a curve at a point P on the curve is the straight line through P perpendicular to the tangent at P. (See Fig. 6.9.)

Figure 6.9

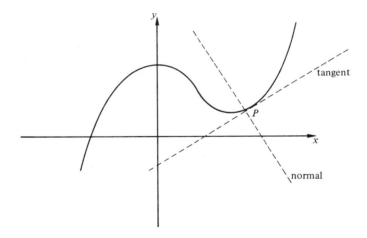

Example 6

Find the equation of the tangent to the curve $y=(2x+3)^2$ at the point $(-\frac{1}{2},4)$.

Solution

$$y=(2x+3)^2$$
$$=4x^2+12x+9$$
$$\Rightarrow \frac{dy}{dx}=8x+12.$$

When $x=-\frac{1}{2}$, $\dfrac{dy}{dx}=8(-\frac{1}{2})+12=8$.

So the gradient of the required tangent is 8, and its equation is of the form $y = 8x + c$. But the tangent must pass through $(-\frac{1}{2}, 4)$, and so

$$4 = 8(-\frac{1}{2}) + c,$$

so

$$c = 8.$$

The tangent at $(-\frac{1}{2}, 4)$ is therefore $y = 8x + 8$.

Example 7
Find the equation of the normal to the curve $y = x^3 - 6x^2 + 10x + 1$ at the point where $x = 3$.
Solution
Call the point in question P. Substituting $x = 3$ into the equation of the curve gives $y = 4$; so P is the point $(3, 4)$. Now

$$y = x^3 - 6x^2 + 10x + 1$$

$$\Rightarrow \frac{dy}{dx} = 3x^2 - 12x + 10.$$

When $x = 3$, $\frac{dy}{dx} = 1$, so the gradient of the curve at P is 1. The gradient of the normal at P is therefore -1 (using $m_1 m_2 = -1$; see Section 3.7). The normal therefore has an equation of the form $y = -x + c$. Substituting $(3, 4)$ gives $c = 7$. The required equation is therefore $y + x = 7$.

Qu.12 Find the equations of the tangent and normal to $y = x^3 + 3x^2 - 1$ at the point $(-1, 1)$.

Now consider the problem of finding any other points where the tangent or normal to a curve at P cuts the curve. If the curve is a cubic, i.e. is of the form $y = ax^3 + bx^2 + cx + d$, then a general line $y = mx + k$ meets it at the points whose coordinates satisfy the simultaneous equations

$$y = ax^3 + bx^2 + cx + d$$
$$y = mx + k.$$

Subtracting these equations leaves us with a cubic equation to solve. If, however, the line is known to be the normal to the curve at the point $P(x_0, y_0)$, then P is itself one of the points where the line meets the curve, and so $x = x_0$ is one root of the cubic equation.

Example 8
Find where the normal in Example 7 meets the curve again.
Solution
We found that the normal at the point where $x = 3$ is $y + x = 7$. This meets the curve where

$$\begin{cases} y = -x + 7 \\ y = x^3 - 6x^2 + 10x + 1. \end{cases}$$

Subtracting,

$$x^3 - 6x^2 + 11x - 6 = 0.$$

But the line *must* meet the curve where $x=3$ (since it is the normal there); so we can factorize to give

$$(x-3)(x^2-3x+2)=0$$

$$\Rightarrow (x-3)(x-1)(x-2)=0$$

$$\Rightarrow x=1 \text{ or } 2 \text{ or } 3.$$

The normal therefore meets the curve again where $x=1$ and where $x=2$. Substituting into the equation of the normal (or of the curve) gives

$$x=1 \Rightarrow y=6$$

and
$$x=2 \Rightarrow y=5.$$

So the points where the normal meets the curve again are (1, 6) and (2, 5).

In the example above, we used the fact that the normal to a curve at a point P meets the curve at P. In the case of a tangent at P, we know that it has *double contact* at P (see Section 2.11 and Section 10.6). This fact is used in the next example.

Example 9
Find where the tangent to $y=x^4-x^3-3x^2+2x-7$ at the point $P(2, -7)$ meets the curve again.
Solution

$$y=x^4-x^3-3x^2+2x-7$$

$$\Rightarrow \frac{dy}{dx}=4x^3-3x^2-6x+2.$$

When $x=2$, $\frac{dy}{dx}=10$, so the tangent is of the form $y=10x+c$.
Substituting (2, -7) gives $c=-27$. So the tangent is $y=10x-27$. This meets the curve where

$$\begin{cases} y=x^4-x^3-3x^2+2x-7 \\ y=10x-27 \end{cases}$$

$$\Rightarrow x^4-x^3-3x^2+2x-7=10x-27$$

$$\Rightarrow x^4-x^3-3x^2-8x+20=0.$$

Now since $x=2$ is a double root, $(x-2)^2$ must be a factor; i.e. (x^2-4x+4) is a factor. The other quadratic factor can be found by inspection. (Again the first and last terms are easy: see Example 1.)

$$(x^2-4x+4)(x^2+3x+5)=0.$$

But x^2+3x+5 has no roots; so the tangent does not meet the curve again.

Qu.13 Find where the tangent to $y=x^3+2x^2-7x-5$ at the point (-2, 9) meets the curve again.

1. Find the equation of the tangent to the curve $y = 3x^2 - 2x + 2$ at the point where $x = -4$. Find where this tangent cuts the x-axis.

2. Find the equation of the normal to $y = 2x^2 + 5x - 10$ at the point $(-2, -12)$. Find the coordinates of the point where this normal cuts the curve again.

3. At what point on the curve $y = x^2 - 5x - 1$ is the gradient -2? Find the equation of the tangent at this point.

4. Find the equation of the normal to the curve $y = 2x^2 - 3x + 3$ which has gradient 1.

5. Find the equations of the tangents to $y = x^2 - 3x - 4$ at the points where the curve cuts the x-axis.

6. Find the equation of the tangent to $y = 1 - 2x^2$ at the point $(-1, -1)$. Find also the equation of the tangent to the curve which is perpendicular to this first tangent.

7. The tangents to the curve $y = (x-1)(2x-3)$ at the points where $x = 2$ and $x = p$ are perpendicular. Find the value of p, the equations of the two tangents, and their point of intersection.

8. Find the equations of the tangent and normal to $y = 1/x$ at the point $(2, \frac{1}{2})$.

9. Find the equations of the tangent and normal to $y = x^{2/3}$ at the point (p^3, p^2).

10. Find the equation of the tangent to the curve $y = x^3 - x^2 - 35x - 50$ at the point where $x = -3$. Find also the coordinates of the point where the tangent meets the curve again.

11. Find the x-coordinates of the points where the tangent to the curve $y = x^4 - 4x^3 - 5x^2 + 41x - 29$ at $x = 2$ meets the curve again.

12. Repeat question 10 for the tangent to the curve $y = 2x^3 - 21x^2 + 59x - 20$ at the point where $x = 5$.

13. Find the equation of the normal to the curve $y = x^3 - 10x^2 + 32x - 29$ at the point $(3, 4)$. Find where this normal meets the curve again.

14. Repeat question 13 for the normals to
   (i) $y = 4x^3 + 28x^2 + 64x + 43$ at the point $(-2\frac{1}{2}, -4\frac{1}{2})$;
   (ii) $y = x^3 + x^2 + 3x + 5$ at the point $(-2, -5)$.

15. Show that the equation of the tangent to the curve $y = ax^2 + bx + c$ at the point where $x = 2$ is $y = (4a+b)x - 4a + c$. Find also the equation of the tangent at the point where $x = 1$. Show that these two tangents intersect at a point where $x = \frac{3}{2}$.

16. Prove that the tangents to $y = x^2$ at the points where $x = p$ and $x = q$ intersect on the line $y = pq$. Hence show that any pair of perpendicular tangents to this curve intersect on the line $y = -\frac{1}{4}$.

17. Find the equation of the tangent to $y = x^2$ at (p, p^2). Show algebraically that if this tangent must pass through the point (h, k), then there are two possible values of p provided that $k < h^2$. Explain the geometrical significance of this condition. Find the equations of the two tangents passing through $(3, -7)$.

18. Where does the normal to the curve $y^2 = 4ax$ (where $a > 0$ is a constant) at the point $(at^2, 2at)$ cut the x-axis? Deduce that, with the exception of the case $t = 0$, no normal cuts the x-axis in the region $x \leqslant 2a$.

19. Find, in terms of t, the y-coordinate of the point where the normal

to $y = x^3$ at (t, t^3) cuts the y-axis. Show that no normal to the curve cuts the positive y-axis below $y = \dfrac{4\sqrt{3}}{9}$.

20. Find the points of intersection of the curves $y = 1 + \dfrac{1}{x}$ and $y = 3 - \dfrac{1}{x^2}$. For the intersection lying in the region $x > 0$, find the angle at which the curves intersect.

6.7 Problems involving maxima and minima

Example 10

A rectangular area is to be enclosed on three sides by a fence, an existing hedge forming the fourth side. Find the greatest area of the rectangle if 100 m of fencing is available.

Solution

Let the width of the rectangle be x metre (Fig. 6.10). Then the length is $(100 - 2x)$ metre, so that

$$A = x(100 - 2x)$$
$$= 100x - 2x^2.$$

Figure 6.10

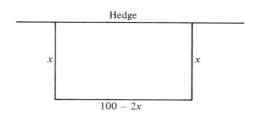

To find the value of x that gives a maximum area, we put $dA/dx = 0$.

So
$$\frac{dA}{dx} = 100 - 4x = 0$$
$$\Rightarrow \quad x = 25.$$

Thus the greatest area is $25 \times 50 = 1250 \text{ m}^2$.
(The reasons why we can tell that this area is a maximum rather than a minimum are discussed in Qu.15.)

An essential feature of the example above and of its solution is that we were able to express the quantity that we were maximizing (the area) in terms of a single variable (the width, x). The same feature occurs in Example 11.

Qu.14 Solve Example 10 by taking y to be the *length* of the rectangle, and first expressing A in terms of y.

Qu.15 Show that the area in Example 10 is a maximum rather than a minimum, by each of the following methods:
(i) sketching the graph of A against x;
(ii) finding $\dfrac{d^2 A}{dx^2}$.

(iii) drawing diagrams of the field to show that when x is either 'very small' or 'almost 50' the area is very small.

Example 11

A cylinder, closed at both ends, is to have a volume of 2000π cm². What should its dimensions be if its surface area (S) is to be as small as possible?

Solution

For a closed cylinder of radius r and height h,

$$S = 2\pi r^2 + 2\pi rh. \tag{1}$$

(Without further information, we should now be faced with the difficulty that S is a function of two variables; but the extra constraint of a fixed volume enables us to express one variable in terms of the other.)

Now
$$V = 2000\pi$$

$$\Rightarrow \pi r^2 h = 2000\pi$$

$$\Rightarrow h = \frac{2000\pi}{\pi r^2} = \frac{2000}{r^2}. \tag{2}$$

Substituting into (1), we get

$$S = 2\pi r^2 + \frac{4000\pi}{r}$$

$$\Rightarrow \frac{dS}{dr} = 4\pi r - \frac{4000\pi}{r^2}.$$

So for a minimum,

$$4\pi r - \frac{4000\pi}{r^2} = 0$$

$$\Rightarrow r^3 = 1000$$

$$\Rightarrow r = 10 \text{ cm}$$

and, from (2), $h = 20$ cm.

Qu.16 (See Qu.15.) Use any method to show that the solution of Example 11 is a minimum rather than a maximum.

Exercise 6e

1. If the perimeter of a rectangle is to be 40 cm, prove that its area will be a maximum when it is a square.
2. If the area of a rectangle is A, show that the minimum value of the perimeter is $4\sqrt{A}$.
3. A rectangular box, without a top, and of volume 120 cm³ is to have a square base of side x cm.
   (i) Express the total surface area of the box in terms of x only, and hence find the value of x which gives a minimum surface area.
   (ii) Find the value of x which makes the total length of the twelve edges a minimum.

4. An equilateral triangle of side a is to have a rectangle inscribed in it as shown (Fig. 6.11). Obtain, in terms of x, an expression for the area of the rectangle, and hence find the maximum value of this area.

Figure 6.11

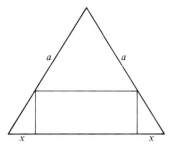

5. A cylinder, closed at one end only, is to have a surface area of 100π cm^2. Find its maximum volume.

6. An open box is to be made from a rectangular piece of cardboard measuring 20 cm by 10 cm, by cutting from each corner a square of side x cm and then turning up the sides. Find the value of x which gives a maximum volume.

7. A hollow right circular cone rests over a cylinder of height a and radius a, so that the base of the cone lies in the same plane as the base of the cylinder. Find the minimum volume of the cone. (Take the radius of the cone to be $a+x$. This avoids the need to differentiate a quotient, a technique which is dealt with in Chapter 7.)

6.8 Simple kinematics

Kinematics is the study of motion (without any particular reference to the *cause* of the motion). In this section, we consider the motion of a particle (i.e. an object whose size is negligible) moving in one dimension.

Suppose that a particle P moves along a straight line so that its position measured (in metres) from a fixed point O, at time t seconds after it starts, is given by $x=f(t)$.

Qu.17 Suppose that $x=t^2-8t+12$.
  (i). Where is the particle at time $t=1$? Where is it at $t=3$?
  (ii) Where was the particle initially (i.e. where did it start)?
  (iii) At what time(s) is the particle at O?
  (iv) How far does the particle travel in the two seconds from $t=6$ to $t=8$?
  (v) What is its average velocity between $t=6$ and $t=8$? (The units of velocity are *metres per second*, written ms^{-1}.)
  (vi) When is the particle at the point $x=21$? Discuss the significance of the negative root.
  (vii) What is the *net* distance travelled (also called the *displacement*) between times $t=1$ and $t=7$?
  (viii) Sketch the graph of x against t. (Note that x is the *dependent* variable and so should be plotted on the vertical axis.)

Now suppose that we wish to find the average velocity of P between two times t_0 and t_1. At time t_0 the position of P is given by $f(t_0)$, and at time t_1 the position of P is given by $f(t_1)$. So the net distance travelled by P is $f(t_1) - f(t_0)$ and the time that passes is $t_1 - t_0$.

So $$\text{average velocity} = \frac{f(t_1) - f(t_0)}{t_1 - t_0}. \qquad (1)$$

6.9 Velocity at an instant

We can certainly talk of the average velocity over an interval of time. Can we talk of the velocity *at an instant* (e.g. at $t = 3$)? Does this have a meaning? To answer these questions, it is helpful to consider how we might attempt to measure the velocity of a particle at the instant $t = 3$.

We could argue that a good approximation would be the average velocity between $t = 3$ and $t = 3 \cdot 1$. Better would be the average velocity between $t = 3$ and $t = 3 \cdot 0001$. This leads us to the following definition:

Definition

If the position of a particle is given by $x = f(t)$, then its velocity at the instant $t = t_0$ is

$$\lim_{h \to 0} (\text{average velocity between times } t_0 \text{ and } t_0 + h)$$

or $$\lim_{h \to 0} \frac{f(t_0 + h) - f(t_0)}{h} \qquad (2)$$

(using (1) in the previous section).

The expression (2) should be familiar. It is identical to the definition of the derivative of $f(t)$ at $t = t_0$. Thus we have the following important result: the velocity of a particle whose position is $f(t)$ is $f'(t)$. Alternatively we may say that if x is the position of a particle, then $\frac{dx}{dt}$ is its velocity.

Qu.18 Find the velocity of the particle in Qu.17 at the instant $t = 3$. What is the significance of the sign of your answer?

6.10 Distance-time graphs

By considering the graph of x against t for the particle P we can give a geometrical interpretation of the ideas in the previous section.

Between t_0 and t_1 the average velocity is

$$\frac{\text{distance gone}}{\text{time taken}} = \frac{f(t_1) - f(t_0)}{t_1 - t_0}$$

which is also the gradient of the chord joining the points P and Q on the curve. The velocity when $t = t_0$ is defined as the limit of the average velocity as Q approaches P, and this is the gradient of the curve at P.

APPLICATIONS OF DIFFERENTIATION

Figure 6.12

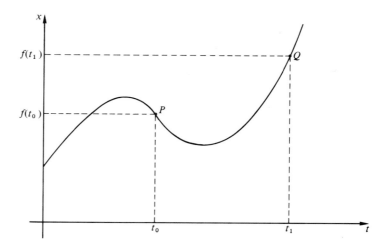

6.11 Acceleration

If the position of our particle P is given by $x = f(t)$, then the velocity is given by $v = \dfrac{dx}{dt}$. The *acceleration* of the particle is a measure of how fast the velocity is changing. Thus the average acceleration between t_0 and t_1 is

$$\frac{v(t_1) - v(t_0)}{t_1 - t_0}$$

and, using the limiting argument which was used to define velocity at an instant, we have

acceleration at time t_0 is $v'(t_0)$.

Thus, in general, the acceleration is given by

$$a = \frac{dv}{dt} = \frac{d}{dt}\left(\frac{dx}{dt}\right) = \frac{d^2 x}{dt^2}.$$

The acceleration of a particle is perhaps one of the simplest representations of the idea of a second derivative. It is measured in *metres per second per second*, or *metres per second squared*, written ms^{-2}.

Example 12

The position of a particle P at time t is given by
$x = t^3 - 15t^2 + 63t + 21$. Find
(i) the velocity and acceleration when $t = 1$;
(ii) the time(s) when the particle is instantaneously at rest;
(iii) the average velocity between $t = 1$ and $t = 4$.

Solution

(i) $v = \dfrac{dx}{dt} = 3t^2 - 30t + 63$, so

$$t = 1 \Rightarrow v = 36 \,\mathrm{ms}^{-1};$$

and $a = \dfrac{d^2 x}{dt^2} = 6t - 30$, so

$$t = 1 \Rightarrow a = -24 \,\mathrm{ms}^{-2}.$$

(ii) The particle is at rest when $v=0$.

$$3t^2-30t+63=0$$
$$\Rightarrow 3(t-3)(t-7)=0$$
$$\Rightarrow t=3 \text{ or } 7.$$

(iii) The average velocity between times $t=1$ and $t=4$ is

$$\frac{\text{distance gone}}{\text{time taken}}=\frac{97-70}{4-1}=9 \text{ ms}^{-1}.$$

Example 13
A stone is thrown vertically into the air and its height above the ground t seconds later is given by

$$h=2+7t-4\!\cdot\!9t^2.$$

Find the maximum height reached by the stone.
Solution
We give two arguments which are essentially the same.
(i) From the work on maxima and minima, h is a maximum when $\dfrac{dh}{dt}=0$. So

$$7-9\!\cdot\!8t=0$$

$$\Rightarrow t=\frac{7}{9\!\cdot\!8}.$$

Substituting, $h=4\!\cdot\!5$.
So the maximum height reached is $4\!\cdot\!5$ m.
(ii) At the instant when the stone reaches the top of its flight, its velocity is instantaneously zero. Then

$$v=0 \ \Rightarrow \ \frac{dh}{dt}=0$$

$$\Rightarrow 7-9\!\cdot\!8t=0$$

etc. as above.

Exercise 6f   (Distance is measured in metres, and time in seconds.)
   1. The position of a particle at time t is given by
$$x=(t-1)(t-3)(t-4).$$
   Find
      (i) the times at which the particle passes through the origin;
      (ii) the velocity and acceleration at each of these times;
      (iii) the times at which the particle is at rest;
      (iv) the minimum velocity of the particle, and when this occurs.
   2. The velocity of a particle is given by
$$v=2t^2-3t+1.$$
   Find
      (i) its initial velocity;
      (ii) its velocity when $t=1$;

(iii) its acceleration when $t=1$;

(iv) the minimum velocity.

3. The position of a particle at time t is given by

$$x = t^3(1-t)^2.$$

(i) Show that it starts from rest.

(ii) Show that when it returns to the origin, it again comes to rest there.

(iii) Find its greatest distance from the origin between the times $t=0$ and $t=1$.

(iv) Sketch a distance-time graph.

4. Show that a particle whose position is given by

$$x = 2t^3 - 3t^2 + 4t - 1$$

never comes to rest.

5. The height above the ground of a balloon is given by

$$h = \frac{1}{1000} t^2 (40-t)(41-t) \text{ for } 0 \leqslant t \leqslant 40.$$

Show that it reaches its greatest height after about $20\frac{1}{4}$ seconds. Find the speed at which it crashes.

6. A stone is thrown downwards from the top of a cliff, its height above the sea t seconds later being given by

$$h = 60 - 5t - 5t^2.$$

Find

(i) the speed with which it is thrown;

(ii) when it hits the sea;

(iii) the height of the cliff.

7. A particle moves according to the law

$$x = k\sqrt{t}$$

(where k is a positive constant). Show that

(i) the velocity varies inversely as the distance gone;

(ii) there is no limit to the distance that the particle will eventually go;

(iii) the particle's velocity is always decreasing.

8. For what positive values of t is the particle in question 1 slowing down?

9. Between the times $t=1$ and $t=3$, a particle's position is given by $x = t + \frac{1}{t}$. Find its average velocity during this interval. At what instant is it actually moving at this average velocity?

10. A particle's position at time t is given by

$$x = 2t^3 - 3t^2 - 12t + 4.$$

Find

(i) its greatest velocity between $t=0$ and $t=2\frac{1}{2}$;

(ii) its greatest speed during the same interval.

(Note: velocity is directed, i.e. it has a sign. Speed is the modulus of the velocity, and so is always positive.)

1. Find the equation of the tangent at the point $P\left(3p, \dfrac{3}{p}\right)$ on the curve $y = \dfrac{9}{x}$. The tangent meets the x-axis at A and the y-axis at B, and O is the origin. Find the area of the triangle OAB, showing that it is independent of p. The normal to the curve at P meets the x-axis at N. In the case when $p = 2$, find
   (i) the coordinates of N,
   (ii) the area of triangle APN.

   [JMB (O)]

2. Find the equation of the tangent to the curve

   $$y = x^3 - 7x^2 + 14x - 8$$

   at the point where $x = 1$.
   Find the x co-ordinate of the point at which the tangent is parallel to the tangent at $x = 1$.

   [C (O)]

3. Two cyclists A and B are moving, one on each of two straight roads, with uniform speeds towards the cross-roads where the two roads intersect at right angles. When they are first observed A is 300 m and B is 400 m from the crossroads; A has a speed of 16 m/s and B has a speed of 12 m/s. Prove that their distance apart t seconds later is given by the expression

   $$s^2 = 2 \cdot 5 \times 10^5 - 1 \cdot 92 \times 10^4 t + 4 \times 10^2 t^2.$$

   By considering the minimum value of s^2, or otherwise, find when the cyclists are the shortest distance apart and show that this distance is 140 m.

   [Ox (O)]

4. Draw a rough sketch of the two graphs $y = 8x^2$ and $y = 1/x$ and calculate the co-ordinates of the point A where they intersect. Find the equation of the tangent to $y = 1/x$ at A and calculate the co-ordinates of the point B where this tangent meets $y = 8x^2$ again. Find also the equation of the tangent to $y = 8x^2$ at A and calculate the co-ordinates of the point C where this tangent meets $y = 1/x$ again.

   [Ox (O)]

5. Prove that the curve

   $$y = x^3 + 2x^2 + 9x + 8$$

   has neither maximum nor minimum points on it.
   Calculate

   (i) the value of x for which $\dfrac{d^2 y}{dx^2} = 0$;

   (ii) the value of $\dfrac{dy}{dx}$ for this value of x.

   [AEB (O) '80]

6. An open water tank, with vertical sides, stands on a horizontal square base of side x metres. The total exterior surface area (sides and base) is 75 m^2.

Find expressions for its height and for its volume in terms of x. Calculate the dimensions of the tank if it is to be capable of holding as much water as possible.

[SUJB (O)]

7. A curve passes through the point $(2, 3\frac{1}{2})$ and its gradient at the point $P(x, y)$ is $1 - \dfrac{1}{x^2}$. Find the equation of the curve.

   Find also the equation of the normal at the point $(2, 3\frac{1}{2})$ and the co-ordinates of another point on the curve at which the tangent will be parallel to the tangent at $(2, 3\frac{1}{2})$.
   Obtain the equation of this second tangent.

[W (O)]

8. The tangents to the curve $y = 12/x$ at the points $P(4, 3)$ and $Q(-3, -4)$ meet at R. Obtain the equations of the tangents and prove that the line joining R to the origin passes through the middle point of PQ.

[O & C (O)]

9. A particle P moves along a straight line which passes through a fixed point O. At a time t second after the particle has passed O the distance OP is $(18t + 15t^2 - 4t^3)$cm. Find

   (i) the velocity and acceleration of P when it passes O;
   (ii) the time that has elapsed and the distance travelled when P comes to instantaneous rest;
   (iii) the time that has elapsed when P returns to O, giving the answer to the nearest tenth of a second.

[O & C (O)]

10. The point P on the curve $x^2 = 4y$ has coordinates $(2t, t^2)$. Find the equation of the normal at P to the curve.

    The normal at P passes through the point $A(-\frac{3}{16}, 2\frac{13}{16})$. Write down an equation for t. Show that there are three points on the curve at which the normals pass through A. Show that one of these points has coordinates $(-2, 1)$, and find the coordinates of the other two points.

[Ox (O)]

11. The area of the sheet metal used in the manufacture of a closed cylindrical can is $400\,\text{cm}^2$. Find, to the nearest cm^3, the largest possible volume of the can.

[AEB '80]

7 Products, quotients and composite functions

7.1 Introduction

So far, we have seen how to differentiate only those functions which are made up of sums of terms of the form ax^n (where a and n are constants). In this chapter, we return to the definition of the derived function in order to obtain further rules for differentiating. We use both the delta notation and the f' notation.

7.2 Products

Let $f(x) = u(x)v(x)$. (This means that the value of $f(x)$ is obtained simply by multiplying the values of $u(x)$ and $v(x)$. Examples would be $f(x) = (x+2)(x^2-3x)$ and $f(x) = 2x \sin x$.) Then

$$f'(x) = \lim_{h \to 0} \frac{f(x+h) - f(x)}{h}$$

$$= \lim_{h \to 0} \frac{1}{h} \{u(x+h)v(x+h) - u(x)v(x)\}.$$

The next step may seem unusual: we add an extra term and also subtract it.

$$f'(x) = \lim_{h \to 0} \frac{1}{h} \{u(x+h)v(x+h) - u(x)v(x+h) + u(x)v(x+h) - u(x)v(x)\}$$

$$= \lim_{h \to 0} v(x+h) \left(\frac{u(x+h) - u(x)}{h} \right) + \lim_{h \to 0} u(x) \left(\frac{v(x+h) - v(x)}{h} \right). \qquad (1)$$

Now as $h \to 0$,

$$v(x+h) \to v(x),$$

$$\frac{u(x+h) - u(x)}{h} \quad \to \quad u'(x)$$

and

$$\frac{v(x+h) - v(x)}{h} \quad \to \quad v'(x).$$

So, taking limits in (1) gives the result:

> if $f(x) = u(x)v(x)$, then
> $f'(x) = u'(x)v(x) + v'(x)u(x)$.

This important result is generally remembered in words rather than as a formula: 'to differentiate a product, differentiate the first and multiply by the second, plus differentiate the second and multiply by the first'.

(In taking the limit of (1) in the proof above, we assumed that the limit of a product is the product of the individual limits. This result is proved in Chapter 39.)

We now derive the product rule using the delta notation. Let

$$y = uv, \qquad (2)$$

where u and v (and hence y) are functions of x. Then an increase, δx, in x gives rise to increases δu and δv in u and v, and, in turn, a corresponding increase, δy, in y. Now y is the product of u and v for all values of x, so the new values of y, u and v satisfy

$$y + \delta y = (u + \delta u)(v + \delta v)$$
$$\Rightarrow \quad y + \delta y = uv + v\delta u + u\delta v + \delta u \delta v. \qquad (3)$$

Subtracting (2) from (3)

$$\delta y = v\delta u + u\delta v + \delta u \delta v$$

$$\Rightarrow \quad \frac{\delta y}{\delta x} = v\frac{\delta u}{\delta x} + u\frac{\delta v}{\delta x} + \delta u\frac{\delta v}{\delta x}.$$

Taking limits as $\delta x \to 0$,

$$\frac{dy}{dx} = v\frac{du}{dx} + u\frac{dv}{dx} + 0\frac{dv}{dx}$$

or

$$\boxed{\frac{d}{dx}(uv) = v\frac{du}{dx} + u\frac{dv}{dx}.}$$

Qu.1 Use the product rule to differentiate the following. Check each answer by multiplying out and then differentiating.

  (i) $(x^2 + 1)(3x + 2)$;   (ii) $(2x^3 - x^2 + 1)(x^2 + 3)$;

(iii) $(x - 1)(x + 1)$;   (iv) $x^2 . x^3$.

The reader may be wondering what use the product rule is, when it is just as easy to 'multiply out' first. The answer is that later, when we have seen how to differentiate a variety of functions, the product rule is essential for differentiating products such as $x \sin x$ which cannot be 'multiplied out'.

7.3 Quotients   We now derive a rule for differentiating the quotient of two functions.

Let $f(x) = \dfrac{u(x)}{v(x)}$, then

$$f'(x) = \lim_{h \to 0} \frac{1}{h}\{f(x + h) - f(x)\}$$

$$= \lim_{h \to 0} \frac{1}{h}\left\{\frac{u(x + h)}{v(x + h)} - \frac{u(x)}{v(x)}\right\}$$

$$= \lim_{h \to 0} \frac{1}{h}\left\{\frac{u(x + h)v(x) - u(x)v(x + h)}{v(x + h)v(x)}\right\}$$

$$= \lim_{h \to 0} \frac{1}{h} \left[\frac{\{u(x+h) - u(x)\}v(x) - u(x)\{v(x+h) - v(x)\}}{v(x+h)v(x)} \right]$$

$$= \lim_{h \to 0} \frac{1}{v(x+h)v(x)} \left\{ v(x) \frac{u(x+h) - u(x)}{h} - u(x) \frac{v(x+h) - v(x)}{h} \right\}.$$

Taking limits,

$$f'(x) = \frac{1}{\{v(x)\}^2} \{v(x)u'(x) - u(x)v'(x)\}.$$

Thus

> if $f(x) = \dfrac{u(x)}{v(x)}$, then
>
> $f'(x) = \dfrac{u'(x)v(x) - v'(x)u(x)}{\{v(x)\}^2}.$

In words, this says: 'to differentiate a quotient, differentiate the top and multiply by the bottom, minus differentiate the bottom and multiply by the top, all over the denominator squared'.

In the delta notation, we have

$$y = \frac{u}{v} \quad \text{and} \quad y + \delta y = \frac{u + \delta u}{v + \delta v}.$$

Subtracting,

$$\delta y = \frac{u + \delta u}{v + \delta v} - \frac{u}{v}$$

$$= \frac{(u + \delta u)v - u(v + \delta v)}{(v + \delta v)v}$$

$$= \frac{v\delta u - u\delta v}{(v + \delta v)v}$$

$$\Rightarrow \quad \frac{\delta y}{\delta x} = \frac{1}{(v + \delta v)v} \left\{ v \frac{\delta u}{\delta x} - u \frac{\delta v}{\delta x} \right\}$$

$$\Rightarrow \quad \frac{dy}{dx} = \frac{1}{v^2} \left(v \frac{du}{dx} - u \frac{dv}{dx} \right).$$

So

> $$\frac{d}{dx}\left(\frac{u}{v}\right) = \frac{v \dfrac{du}{dx} - u \dfrac{dv}{dx}}{v^2},$$

which is the same result as above.

Qu.2 Use the quotient rule to differentiate the following. Simplify the numerator of the answer in each case.

(i) $\dfrac{x^2}{x^2 + 1}$; (ii) $\dfrac{x+1}{x+2}$; (iii) $\dfrac{1}{2x^2 + 3}$; (iv) $\dfrac{x^4}{x^3}$; (v) $\dfrac{x^m}{x^n}$.

Check your answers to (iv) and (v) by dividing first.

Exercise 7a

1. Differentiate the following using the product rule, and also by multiplying out the question first. Show that your two answers are the same by simplifying each.

   (i) $(1 - \sqrt{x})(1 + \sqrt{x^3})$;

   (ii) $(1 + x + x^2)\left(x - \dfrac{1}{x}\right)$;

   (iii) $\left(1 + \dfrac{1}{x} + \dfrac{2}{x^2}\right)(1 + x + 2x^2)$.

2. Differentiate

   (i) $\dfrac{x}{1 + x^2}$;   (ii) $\dfrac{x^2 + 2x - 3}{x^2 - 2x + 4}$;

   (iii) $\dfrac{2x^3}{3(x^3 + 2)}$;   (iv) $\dfrac{3x^3}{2x^2 + 5}$;

   (v) $\dfrac{1 + 2\sqrt{x}}{1 + \sqrt{(2x)}}$.

3. Find the maximum value of $y = \dfrac{x - 2}{x^2 - x + 2}$. What happens to y as $x \to \infty$?

4. Use a calculus method to show that for $x \geqslant 0$, the value of $y = \dfrac{x + 1}{x + 2}$ is always increasing. What is $\lim\limits_{x \to \infty} \dfrac{x + 1}{x + 2}$? Sketch $y = \dfrac{x + 1}{x + 2}$ for $x \geqslant 0$.

5. (i) Let u, v and w be functions of x. By writing uvw as $u(vw)$, i.e. by considering it initially as the product of u and vw, use two applications of the product rule to show that

$$\frac{d}{dx}(uvw) = vw\frac{du}{dx} + uw\frac{dv}{dx} + uv\frac{dw}{dx}.$$

   (ii) Find the corresponding result for $\dfrac{d}{dx}(uvwz)$. (This result is generalized in question 8 of Exercise 8f.)

7.4 Composite functions (function of a function)

We now derive a rule for differentiating those functions which are most simply regarded as being the composite of two or more functions.

Suppose that $f(x) = v(u(x))$, i.e. f is u followed by v, and suppose that our increase h in x gives rise to an increase k in $u(x)$ so that

$$u(x + h) - u(x) = k.$$

Then,

$$f'(x) = \lim_{h \to 0} \frac{f(x + h) - f(x)}{h}$$

$$= \lim_{h \to 0} \frac{v(u(x+h)) - v(u(x))}{h}$$

$$= \lim_{h \to 0} \left(\frac{v(u(x) + k) - v(u(x))}{k} \cdot \frac{k}{h} \right) \tag{1}$$

$$= \lim_{h \to 0} \left(\frac{v(u(x) + k) - v(u(x))}{k} \cdot \frac{u(x+h) - u(x)}{h} \right).$$

Taking limits of each factor separately, and noting that, for the first factor, we may replace $\lim\limits_{h \to 0}$ by $\lim\limits_{k \to 0}$ since $h \to 0 \Rightarrow k \to 0$, we get

$$f'(x) = v'(u(x)) \cdot u'(x). \tag{2}$$

Note

(This note is quite hard and may be omitted. It is included because the argument above is not rigorous without it.) At (1) above, we inserted a factor $\frac{k}{k}$. We can only do this if we can avoid the possibility that $k = 0$ (for $h \neq 0$). We shall be unable to avoid this if k repeatedly takes the value zero as $h \to 0$. But in this case:

(i) $\frac{k}{h}$ is repeatedly zero, so $\frac{k}{h} \to 0$, i.e. $u'(x) = 0$;

and (ii) $\frac{v(u(x) + k) - v(u(x))}{h}$ is repeatedly zero, i.e. $f'(x) = 0$.

So the final result (2) remains true.

Using the delta notation, the proof is deceptively simple. If $y = f(x) = v(u(x))$, then a change δx in x gives rise to changes δu in u and δy in y. Since $\frac{\delta y}{\delta u}$ etc. are fractions,

$$\frac{\delta y}{\delta x} = \frac{\delta y}{\delta u} \times \frac{\delta u}{\delta x}.$$

Taking limits of each side, and noting that $\delta x \to 0 \Rightarrow \delta u \to 0$, we get

$$\boxed{\frac{dy}{dx} = \frac{dy}{du} \times \frac{du}{dx}.} \tag{3}$$

Qu.3 Verify that this result (3) says the same as (2).

Note

(1) The rule for differentiating a function of a function is often called the *Chain Rule*.

(2) Written in the form (3) above, the chain rule makes it even more tempting to regard $\frac{dy}{dx}$ etc. as fractions; this temptation must be

resisted. The chain rule can, however, be remembered by noting the similarity to fractions.

(3) For triple composite functions, $y = g(v(u(x)))$, we have similarly

$$\frac{dy}{dx} = \frac{dy}{dv} \times \frac{dv}{du} \times \frac{du}{dx}.$$

In the f' notation,

$$f'(x) = g'(v(u(x))) . v'(u(x)) . u'(x).$$

The following example shows the use of the chain rule, although, in practice, it is not necessary to show any working as this can usually be done mentally as in Example 2.

Example 1

Differentiate $(x^2 - 3x + 5)^4$.

Solution

Let $y = (x^2 - 3x + 5)^4$. We take u to be the 'inner' function, i.e. $u = x^2 - 3x + 5$; then $y = u^4$. (Thus u is a function of x, and y is a function of u; i.e. u is 'half-way' between x and y.)

By the chain rule

$$\frac{dy}{dx} = \frac{dy}{du} \times \frac{du}{dx}$$

$$= 4u^3 . (2x - 3)$$
$$= 4(x^2 - 3x + 5)^3 . (2x - 3).$$

The next example shows how, in practice, no working is needed.

Example 2

Differentiate $\dfrac{1}{(3x^2 + 2)^5}$.

Solution

Although the given function is written as a quotient, there is no need to use the quotient rule; we write

$$y = (3x^2 + 2)^{-5}.$$

(This is of the form 'something' to the power -5; differentiating gives '-5(something)$^{-6}$, times differentiate the something'.)

Then $$\frac{dy}{dx} = -5(3x^2 + 2)^{-6} . 6x$$
$$= -30x(3x^2 + 2)^{-6}.$$

Qu.4 Differentiate the following, showing working as in Example 1:
(i) $(2x + 3)^{10}$; (ii) $(3x^4 + 5)^{-2}$; (iii) $\sqrt{(x^2 + 1)}$.

Qu.5 Write down the derivatives of the following, doing all working mentally as in Example 2. (In some parts it may help to rewrite the question first.)

(i) $\dfrac{1}{x + 1}$;   (ii) $\dfrac{3}{2x + 1}$;

(iii) $(3x+4)^7$;   (iv) $\dfrac{1}{(2x^2-3)^5}$;

(v) $\sqrt{(4x^3-3)}$.

Suppose that y is a function of x, and z is a function of y, then by the chain rule,

$$\frac{dz}{dx}=\frac{dz}{dy}\times\frac{dy}{dx}.$$

Putting $z=x$,

$$\frac{dx}{dx}=\frac{dx}{dy}\times\frac{dy}{dx}.$$

But $\dfrac{dx}{dx}=1$, so

$$\frac{dx}{dy}\times\frac{dy}{dx}=1$$

or

$$\boxed{\frac{dx}{dy}=1\bigg/\frac{dy}{dx}.}$$

This is really a statement about the derivatives of mutually inverse functions, since when we write $\dfrac{dx}{dy}$, we are thinking of x as a function of y. We shall meet this idea again in Chapter 14.

7.5 A harder example

So far in this chapter, we have met the rules for differentiating products, quotients and composite functions. Some functions require a combination of these rules.

Example 3

Differentiate $y=\dfrac{\sqrt{(x^2+2x)}}{(3x-4)^3}$.

Solution

The function is a quotient, and so the quotient rule is used; but when differentiating the numerator and denominator, the chain rule will be needed. (The reader may wish to differentiate the numerator and denominator first.) We get

$$\frac{dy}{dx}=\frac{\frac{1}{2}(x^2+2x)^{-1/2}(2x+2)(3x-4)^3-9(3x-4)^2(x^2+2x)^{1/2}}{(3x-4)^6}$$

$$=(x^2+2x)^{-1/2}\left\{\frac{(x+1)(3x-4)-9(x^2+2x)}{(3x-4)^4}\right\}$$

$$=-\frac{6x^2+19x+4}{(3x-4)^4\sqrt{(x^2+2x)}}.$$

PRODUCTS, QUOTIENTS AND COMPOSITE FUNCTIONS

An alternative method of differentiating functions like the one above is given in Section 15.6.

Differentiate numbers 1 to 12.

1. $\sqrt{(1+x^2)}$. **2.** $(2x^2-3x+1)^5$.
3. $\sqrt[5]{(3x-7)^4}$. **4.** $(x^2+3x-2)^{10}$.
5. $\sqrt{(x^2-2x+1)}$.

6. $\dfrac{3}{2\sqrt[3]{(x^2-1)}}$. (There is no need to regard this as a quotient.)

7. $4\sqrt[3]{(x^3+2)^2}$. **8.** $\dfrac{3-x+x^2}{\sqrt{(1+x)}}$.

9. $2x\sqrt{(x+1)}$. **10.** $\dfrac{\sqrt{(x+1)}}{x^2+1}$.

11. $\dfrac{(x^2-3)^{10}}{2x+1}$. **12.** $\left(\dfrac{x}{x+1}\right)^{4/5}$.

13. Let $x=2y^2-3y-1$. Find $\dfrac{dy}{dx}$ at the point $(13, -2)$. [*Hint*: find $\dfrac{dx}{dy}$ first.]

14. Make y the subject of the formula $x^2+y^2=4$, and hence find $\dfrac{dy}{dx}$. Find the equation of the normal to this curve at the point where $x=p$ and y is positive. Show that the origin lies on this normal. (Why should you have expected it to?)

15. The position of a particle at time t is given by

$$x=12\sqrt{(t^2+16)}.$$

Find the average acceleration between $t=0$ and $t=3$. Find also the acceleration when $t=3$.

16. Two straight roads intersect at right angles at X. At time $t=0$, particle A is at X moving due East, and particle B is on the other road 100 m from X moving towards X. The speeds of the particles are $3\,\text{ms}^{-1}$ and $2\,\text{ms}^{-1}$ respectively. Find
(i) the distance from X of each particle at time t $(t>0)$;
(ii) the distance between the particles then;
(iii) the time at which the particles are closest together.

17. Find the equation of the tangent to the curve $y=\dfrac{1}{1+x^2}$ at the point where $x=p$. Find where this tangent cuts the y-axis. Find the values of p for which this intersection with the y-axis is furthest from the origin.

18. If x, v and a denote the position, velocity and acceleration of a particle moving in one dimension, show that $a=v\dfrac{dv}{dx}$.

19. Differentiate $y=\dfrac{2x+3}{x-4}$

(i) as a quotient;
(ii) by first showing that it may be written as $y=2+11(x-4)^{-1}$.

PRODUCTS, QUOTIENTS AND COMPOSITE FUNCTIONS

8 Series I

The use of the terms *sequence, progression* and *series* varies from one book to another, but the most usual definitions are as follows:

Definitions

(1) A *sequence* is a set of numbers in some fixed order, so that there is a first number, a second number, a third and so on. If the sequence terminates (so that for some $n \in Z^+$, the nth number is the last), it is said to be a *finite sequence*; otherwise it is an *infinite sequence*. The numbers in a sequence are called the *terms* of the sequence.

(2) A *progression* is an alternative word for a sequence. It is used particularly when the terms of a sequence are related in some way, i.e. when there is a pattern in the sequence. In this book, the use of the word is restricted to arithmetic and geometric progressions (Section 8.3 and Section 8.5).

(3) A *series* is the sum of the terms of a sequence.

We saw in Chapter 2 that an infinite sequence may tend to some (finite) limit a; in this case, we say that the sequence *converges* to a.

Consider now the infinite sequence whose nth term is denoted by u_n, i.e. the sequence

$$u_1, u_2, u_3, \ldots..$$

Each of the series

$$s_1 = u_1$$
$$s_2 = u_1 + u_2$$
$$s_3 = u_1 + u_2 + u_3$$
$$s_4 = u_1 + u_2 + u_3 + u_4$$

(and so on)

is called a *partial sum* of the sequence $\{u_n\}$. The numbers $s_1, s_2, s_3, \ldots$ form a new sequence which may or may not converge. If this sequence of partial sums *does* converge to some number s, then the more terms we take in the series

$$u_1 + u_2 + u_3 + \ldots$$

the closer we get to s. For short, we may write

$$s = u_1 + u_2 + u_3 + \ldots$$

and we say that the *infinite series* $u_1 + u_2 + u_3 + \ldots$ *converges to s*, or that s is the sum of the series.

As an example, consider the sequence

$$2, 1\tfrac{1}{2}, 1\tfrac{1}{4}, 1\tfrac{1}{8}, \ldots.$$

Here, the sequence converges to 1, but the corresponding series does not converge, since the partial sums

$$2, 3\tfrac{1}{2}, 4\tfrac{3}{4}, 5\tfrac{7}{8}, \ldots$$

eventually get larger than any number we care to name. (We say that the series tends to infinity.)

Qu.1 (i) For each of the following sequences, state whether or not the sequence converges, giving its limit where appropriate.

(a) $2, 1, \tfrac{1}{2}, \tfrac{1}{4}, \tfrac{1}{8}, \ldots$;
(b) $3, 3\tfrac{1}{2}, 3\tfrac{3}{4}, 3\tfrac{7}{8}, 3\tfrac{15}{16}, \ldots$;
(c) $1, -1, 1, -1, 1, \ldots.$

(ii) Find the first four partial sums of each of the sequences above. State whether each of the corresponding series converges, giving its limit where appropriate.

Qu.2 (For discussion.) (i) Can anything be said about the convergence of a sequence $u_1, u_2, u_3, \ldots$ if the corresponding series $u_1 + u_2 + u_3 + \ldots$ converges?

(ii) If a sequence converges, can you say anything about the convergence of the series? (Be careful. Make sure that Exercise 2e, question 6(ii) does not provide a counter-example to what you decide.)

Qu.3 Let $u_1, u_2, u_3, \ldots$ be a sequence whose partial sums are $s_1, s_2, s_3, \ldots.$ Simplify

(i) $s_5 - s_4$;   (ii) $s_n - s_{n-1}$;
(iii) $s_{10} - 2s_9 + s_8$;   (iv) $s_n + u_{n+1}$;
(v) $(s_{n+1})^2 - (s_n)^2 - (u_{n+1})^2.$

8.2 Expressing a sequence

In this section, we consider infinite sequences that have some definite pattern (as distinct from, for example, a sequence of random numbers). There are three common ways in which the terms of a sequence may be given:

(1) Where the pattern is clear, it is sufficient to give the first few terms of the sequence. Thus to write

$$1, 2, 4, 8, 16, \ldots$$
or
$$2, 4, 6, 8, 10, \ldots$$

does, in effect, give the whole sequence.

(2) The nth term of the sequence may be given explicitly in terms of n. Thus, for example, if we write $u_n = n^2 + 2$, this tells us every member of the sequence. It begins $3, 6, 11, 18, 27, \ldots.$

We may combine our first two methods and write

$$3, 6, 11, 18, \ldots, (n^2 + 2), \ldots.$$

The term $(n^2 + 2)$ would then be referred to as the 'general term'.

(3) We may give a *recurrence relation* connecting the terms of the sequence. One or more initial values are also needed. Thus

$$\begin{cases} u_1 = 5 \\ u_{n+1} = 2u_n - 3 \end{cases}$$

together define the whole sequence. We have

$$u_1 = 5$$
$$u_2 = 2u_1 - 3 = 2 \times 5 - 3 = 7$$
$$u_3 = 2u_2 - 3 = 2 \times 7 - 3 = 11$$
$$u_4 = 2u_3 - 3 = 2 \times 11 - 3 = 19$$

and so on.

Some recurrence relations connect each term with the preceding two or more terms. Two or more initial values are then needed.

Of the three methods above, the second has the distinct advantage that it allows us immediately to write down any member (e.g. the 100th) of the sequence.

Qu.4 Write out the first five terms of each of the following sequences:

(i) $u_n = 3n - 2$;

(ii) $u_n = \dfrac{n+1}{n^2}$;

(iii) $\begin{cases} u_{n+1} = u_n + 3, \\ u_1 = 4; \end{cases}$

(iv) $\begin{cases} u_{n+1} = (n+1)u_n, \\ u_1 = 1. \end{cases}$

Qu.5 The *Fibonacci Sequence* is defined by $u_{n+2} = u_{n+1} + u_n$ together with the initial values $u_1 = u_2 = 1$. Write out the first 10 terms of the sequence. (The reader may like to find out as much as possible about this remarkable sequence. Space does not allow a discussion here.)

Qu.6 Verify that these all define the same sequence:

(i) 1, 3, 6, 10, 15, ...;
(ii) $u_n = \frac{1}{2}n(n+1)$;
(ii) $u_{n+1} = u_n + n + 1$ and $u_1 = 1$;
(iv) $u_{n+1} = (n+1)^2 - u_n$ and $u_1 = 1$.

Qu.7 Find the 100th term of the sequence in Qu.6.

Example 1

Express in terms of n
(i) the nth member, and
(ii) the $(n+1)$th member, of the sequence

$$(1 \times 2 \times 4), \ -(2 \times 3 \times 5), \ (3 \times 4 \times 6), \ -(4 \times 5 \times 7), \ldots.$$

Solution

(i) First we must make sure that the nth term has the correct sign. A factor of $(-1)^n$ will give alternating signs, the first being negative, the second positive and so on. We want exactly the opposite signs, and so we use a factor of $(-1)^{n+1}$. Now we note that each term is the product of three factors, the lowest factor being equal to the number of the term (e.g. the lowest of the three factors in the 4th term is 4). In each case, the other two factors exceed the lowest by 1 and 3. Thus the nth term is

$$(-1)^{n+1}n(n+1)(n+3).$$

(ii) the $(n+1)$th term can now be found by replacing n by $n+1$; this gives

$$(-1)^{n+2}(n+1)(n+2)(n+4)$$

or, more simply,

$$(-1)^n(n+1)(n+2)(n+4).$$

Exercise 8a   **1.** Write out the first six terms of each of these sequences.

(i) $u_n = 2n + 1$;   (ii) $u_n = 2n - 1$;

(iii) $u_n = 1 - \dfrac{1}{n}$;

(iv) $u_n =$ the nearest whole number to $\dfrac{n}{3}$;

(v) $u_n = (-1)^n$;   (vi) $u_n = (-1)^n n^2$;

(vii) $u_n = (-1)^{n-1} 2n$;   (viii) $u_n = 5 + \left(-\dfrac{1}{2}\right)^n$.

2. Write out the first five terms of each of these sequences.

(i) $\begin{cases} u_{n+1} = 2u_n, \\ u_1 = 1; \end{cases}$ (ii) $\begin{cases} u_{n+1} = \dfrac{1}{2}\left(u_n + \dfrac{2}{u_n}\right), \\ u_1 = 1; \end{cases}$

(iii) $\begin{cases} u_{n+1} = \dfrac{1}{2}\left(u_n + \dfrac{2}{u_n}\right), \\ u_1 = \sqrt{2}; \end{cases}$

(iv) $\begin{cases} u_{n+1} = -\dfrac{1}{2}u_n, \\ u_1 = 1; \end{cases}$ (v) $\begin{cases} u_{n+1} = (-1)^n \dfrac{1}{2}u_n, \\ u_1 = 1. \end{cases}$

3. Consider the sequence $2, 1\frac{1}{2}, 1\frac{1}{4}, 1\frac{1}{8}, \ldots$.
(i) State the limit of this sequence.
(ii) Write down the values of the partial sums $s_1, s_2, \ldots, s_5$.
(iii) Does the series $2 + 1\frac{1}{2} + 1\frac{1}{4} + \ldots$ converge?

4. For each of the sequences in questions 1 and 2, state
(i) the limit of the sequence if it tends to a limit;
(ii) whether the corresponding series converges.

5. Express in terms of n:
(i) the nth positive even number;
(ii) the nth odd number;
(iii) the nth member of the sequence $1, 4, 9, 16, 25 \ldots$;
(iv) the $(n+1)$th member of the sequence in (iii);
(v) the cube of the nth odd number.

6. Express in terms of n the $(n+1)$th member of the sequence whose nth member is

(i) n^2; (ii) $2n-5$;

(iii) $3n^2-2$; (iv) $\dfrac{n}{n+1}$;

(v) $(2n+1)^2$; (vi) $\dfrac{1}{6}n(n+1)(2n+1)$.

7. Express in terms of n the nth member of each of these sequences:
  (i) $2, 4, 6, 8, \ldots$;
  (ii) $4, 6, 8, 10, 12, \ldots$;
  (iii) $1, 2, 4, 8, 16, \ldots$;
  (iv) $1, \frac{1}{3}, \frac{1}{9}, \frac{1}{27}, \ldots$;
  (v) $-3, 5, -7, 9, -11, \ldots$;
  (vi) $3, -6, 12, -24, 48, \ldots$;
  (vii) $1, 9, 25, 49, 81, 121, \ldots$;
  (viii) $1\times2, 2\times3, 3\times4, 4\times5, \ldots$;
  (ix) $2, 12, 30, 56, 90, \ldots$.

8. Express in terms of n the $(n+1)$th member of each of the sequences in question 7.

9. (i) With the notation for partial sums used in the text, suppose that for a certain sequence $s_1=7$, $s_2=10$ and $s_3=15$. Find u_1, u_2 and u_3.

(ii) Suppose that in general $s_n=n^2+6$. Express s_{n-1} in terms of n, and hence find u_n in terms of n. Explain why your result is not valid for $n=1$.

10. For each of these sequences, defined in terms of their partial sums, find the values of u_1, u_2, u_3 and u_n. (Use the method of question 9.)

(i) $s_n=3n$; (ii) $s_n=n^3$;

(iii) $s_n=\dfrac{n}{n+1}$; (iv) $s_n=2^n-1$.

8.3 Arithmetic progressions

An Arithmetic Progression (A.P.) is a sequence in which each term exceeds the previous one by a fixed (possibly negative) number. Examples are:

$$10, 13, 16, 19, 22, \ldots;$$
$$-12, -10\tfrac{1}{2}, -9, -7\tfrac{1}{2}, \ldots;$$
$$10, 6, 2, -2, \ldots.$$

The number by which the terms increase is called the *common difference* and is often denoted by d. In the examples above, d takes the values 3, $1\frac{1}{2}$ and -4 respectively.

The first term of an A.P. is usually denoted by a. In general, therefore,

$$u_1 = a$$
$$u_2 = a+d$$
$$u_3 = a+2d$$
$$u_4 = a+3d \quad \text{etc.},$$

and the nth term is

$$u_n = a + (n-1)d.$$

Qu.8 Write out the first four terms and the 100th term of each of the arithmetic progressions determined by
(i) $a=1$, $d=2$;   (ii) $a=13$, $d=4$;
(iii) $a=7\frac{1}{2}$, $d=1$;   (iv) $a=12$, $d=-3\frac{1}{4}$.

Qu.9 For each of the following arithmetic progressions, state the values of a and d, and the value of the 20th term.
(i) 1, 4, 7, 10, ...;
(ii) 6, $8\frac{1}{2}$, 11, $13\frac{1}{2}$, ...;
(iii) 10, 4, -2, -8, ...;
(iv) -8, $-8\frac{1}{2}$, ...

Qu.10 Use the formula $u_n = a + (n-1)d$ to find the number of terms in the following arithmetic progressions:
(i) 10, 12, 14, ..., 100;
(ii) 6, 10, 14, ..., 362;
(iii) 12, 7, 2, -3, ..., -5008.

An A.P. with first term a and common difference d may be defined by the recurrence relation $\begin{cases} u_1 = a \\ u_{n+1} = u_n + d. \end{cases}$

8.4 Summing an A.P. If $u_1, u_2, u_3, \ldots$ is an A.P., then the infinite series $u_1 + u_2 + u_3 + \ldots$ does not converge (except in the trivial case $a = d = 0$, when $\forall\, r \in Z^+$, $u_r = 0$). In fact, this series tends to $+\infty$ or $-\infty$ depending on the sign of d (or a, if $d=0$).

Now suppose that we wish to sum the first n terms of an A.P.. Let the first term be a and the common difference be d. Then

$$s_n = a + (a+d) + (a+2d) + \ldots + (a+(n-1)d).$$

Denoting the last term by l, we have

$$s_n = a + (a+d) + (a+2d) + \ldots + (l-2d) + (l-d) + l, \tag{1}$$

and, reversing the order of the terms on the R.H.S.,

$$s_n = l + (l-d) + (l-2d) + \ldots + (a+2d) + (a+d) + a. \tag{2}$$

Adding each pair of terms in (1) and (2),

$$2s_n = (a+l) + (a+l) + \ldots + (a+l)$$
$$\Rightarrow 2s_n = n(a+l)$$

since there are n terms of $(a+l)$.

So
$$s_n = \tfrac{1}{2}n(a+l).$$

But $l = u_n = a + (n-1)d$, so substituting for l,

$$s_n = \tfrac{1}{2}n\{2a + (n-1)d\}.$$

Example 2

Sum the following arithmetic series:

 (i) $2+5+8+\ldots+122$;

(ii) $25+21+17+\ldots$ (to 20 terms).

Solution

(i) We have $a=2$ and $d=3$. The number of terms can be found from

$$2+(n-1)\times 3=122$$
$$\Rightarrow n=41.$$

So
$$s_{41}=\tfrac{1}{2}n(a+l)$$
$$=\tfrac{1}{2}\times 41\times(2+122)$$
$$=2542.$$

(ii) Here, $a=25$, $d=-4$ and $n=20$. Since l is not given, we use the formula $s_n=\tfrac{1}{2}n\{2a+(n-1)d\}$:

$$s_{20}=\tfrac{1}{2}\times 20\times\{50+19\times(-4)\}$$
$$=10\times(-26)=-260.$$

Example 3

The fourth term of an A.P. is 13, and the 7th term is -5. Find the sum of the first 16 terms.

Solution

Let the first term be a and the common difference be d. We are given that

$$a+3d=13$$
and
$$a+6d=-5.$$

Solving these simultaneously gives $d=-6$ and $a=31$.

So
$$s_{16}=8(62-90)=-224.$$

Example 4

The sum of the first 12 terms of an A.P. is 33, and the 3rd term exceeds twice the sixth term by 1. Find the first term and common difference.

Solution

Expressing the two pieces of information algebraically,

$$s_{12}=33 \Rightarrow \tfrac{1}{2}\times 12(2a+11d)=33$$
$$\Rightarrow 12a+66d=33$$
$$\Rightarrow 4a+22d=11, \tag{1}$$

and
$$u_3=2u_6+1 \Rightarrow a+2d=2(a+5d)+1$$
$$\Rightarrow a+8d=-1. \tag{2}$$

Solving (1) and (2) gives $a=11$ and $d=-1\tfrac{1}{2}$.

Example 5

How many terms of the series

$$26+29+32+\ldots$$

must be taken for the sum to exceed 5000?

Solution

Here, $a = 26$ and $d = 3$; we attempt to find n such that $s_n = 5000$.

$$\tfrac{1}{2}n\{52 + (n-1) \times 3\} = 5000$$
$$\Rightarrow 3n^2 + 49n - 10\,000 = 0.$$

Solving by the formula gives a positive root of approximately 50·1. Thus for the sum to exceed 5000 we need 51 terms.

Finally in this section, we give a warning about the various formulae which we have derived (and which should be learnt). Remember that they apply to an A.P. whose first term is a and whose common difference is d; so care must be taken on a question like the following.

Qu.11 Find expressions for
(i) the $(n+1)$th term, and
(ii) the sum of the first $n+1$ terms, of the A.P.
$$2a + d,\ 2a + 1,\ 2a - d + 2,\ 2a - 2d + 3,\ \ldots.$$
[*Hint*: here 'a' is $2a + d$, 'd' is $1 - d$ and 'n' is $n + 1$.]

Exercise 8b

1. Find the number of terms in each of these arithmetic progressions, and the sum of the terms.
   (i) 10, 17, 24, ..., 696;
   (ii) 23, $27\frac{1}{2}$, 32, ..., 806;
   (iii) 42, 33, 24, ..., -435;
   (iv) $p + 7q,\ p + 5q,\ p + 3q,\ \ldots,\ p - 129q$.
2. Find the sum of the first 40 terms of each of the sequences in question 1.
3. (i) Find the sum of the odd numbers from 1 to 99 inclusive.
   (ii) Find the sum of the even numbers from 2 to 100 inclusive.
   (iii) *Write down* the sum of the series

   $$-1 + 2 - 3 + 4 - \ldots - 99 + 100.$$

   [*Hint*: think of the terms in pairs.]
   (iv) Verify that your answers to (i) and (ii) differ by your answer to (iii). Explain why this should be.
4. The 10th term of an A.P. is 17, and the 16th term is 44. Find
   (i) the first two terms; (ii) the nth term;
   (iii) the $2n$th term.
5. The sum of the first two terms of an A.P. is 22, and the third term is 8. Find the 4th term.
6. The 5th term of an A.P. is 12 times the 9th term, and 5 times the 8th term exceeds the 3rd term by 5. Find the sum of the first ten terms.
7. (i) Which is the first term of the A.P.

   $$23, 27, 31, \ldots$$

   to exceed 5000?
   (ii) How many terms of the corresponding series must be taken for the sum to exceed 5000?

8. Find the sum of all the positive terms of the series

$$105 + 101\tfrac{1}{4} + 97\tfrac{1}{2} + \ldots.$$

9. What can be said about an A.P. if the sum of the 5th and 8th terms is equal to the sum of the 6th and 7th terms, and the sum of the 10th and 14th terms is twice the 12th term?

10. The first term of an A.P. is -136, and the common difference is 7. Find the sum of
  (i) the first 50 terms;
  (ii) the first 100 terms;
  (iii) the 51st to the 100th term inclusive.

11. The 7th term of an A.P. exceeds the 4th term by 9, and the 6th term is 5. How many terms must be added to give a total of 1168?

12. The sum of the first 16 terms of an A.P. is -632, and the sum of the next 10 terms is -525. Find the 100th term.

13. The nth term of an A.P. is p and the $(n+2)$th term is q. *Write down* the $(n+1)$th term.

14. Obtain an expression for s_n, the sum of the first n terms of an A.P., in terms of n, d and l only (where d and l are the common difference and last term respectively).

8.5 Geometric progressions

A Geometric Progression (G.P.) is a sequence in which each term is a fixed multiple of the previous term. Examples are:

$$3, 6, 12, 24, 48, \ldots;$$
$$5, -15, 45, -135, 405, \ldots;$$
$$3, 2, \tfrac{4}{3}, \tfrac{8}{9}, \tfrac{16}{27}, \ldots.$$

The factor by which each term changes is called the *common ratio* and is usually denoted by r. In the examples above, r takes the values 2, -3 and $\tfrac{2}{3}$ respectively.

Again denoting the first term by a, we have

$$u_1 = a$$
$$u_2 = ar$$
$$u_3 = ar^2$$
$$u_4 = ar^3 \quad \text{etc.}$$

Clearly the nth term is

$$\boxed{u_n = ar^{n-1}.}$$

We may also define this G.P. by the recurrence relation

$$\begin{cases} u_1 = a \\ u_{n+1} = ru_n. \end{cases}$$

Qu.12 Write down the first four terms of the geometric progressions whose first terms and common ratios are
  (i) $a=2$, $r=3$;   (ii) $a=7$, $r=-2$;
  (iii) $a=11$, $r=\tfrac{1}{3}$;   (iv) $a=36$, $r=-\tfrac{3}{4}$;
  (v) $a=7$, $r=1$.

Qu.13 For each of these geometric progressions, state the common ratio and evaluate the 10th term.
- (i) $10, 5, 2\frac{1}{2}, 1\frac{1}{4}, \ldots$;
- (ii) $1, -2, 4, -8, \ldots$;
- (iii) $6, -6, 6, -6, \ldots$;
- (iv) $625, 250, 100, 40, \ldots$.

Qu.14 Use the formula $u_n = ar^{n-1}$ to find the number of terms in these geometric progressions:
- (i) $384, 192, 96, \ldots, \frac{3}{256}$;
- (ii) $0 \cdot 06, -0 \cdot 18, 0 \cdot 54, \ldots, 43 \cdot 74$;
- (iii) $1\frac{1}{2}, 1, \frac{2}{3}, \frac{4}{9}, \ldots, (\frac{2}{3})^k$;
- (iv) $128, 512, 2048, \ldots, 2^{3p+1}$ where p is even.

8.6 Summing a finite G.P.

Let s_n be the sum of the first n terms of the G.P. with first term a and common ratio r. Then

$$s_n = a + ar + ar^2 + \ldots + ar^{n-2} + ar^{n-1} \qquad (1)$$

$$\Rightarrow rs_n = ar + ar^2 + ar^3 + \ldots + ar^{n-1} + ar^n. \qquad (2)$$

Subtracting (2) from (1),

$$s_n - rs_n = a - ar^n,$$

the majority of the terms cancelling. Thus

$$\boxed{s_n = \frac{a(1 - r^n)}{1 - r} \qquad \text{provided } r \neq 1.}$$

Changing the signs of both numerator and denominator gives

$$\boxed{s_n = \frac{a(r^n - 1)}{r - 1} \qquad \text{provided } r \neq 1}$$

which is slightly more convenient if $r > 1$. Clearly for $r = 1$, $s_n = na$.

Qu.15 Sum these geometric series, using the formula above. Leave answers in index form.

- (i) $2 + 6 + 18 + \ldots$ (20 terms);
- (ii) $81 - 54 + 36 - 24 + \ldots$ (10 terms);
- (iii) $1 + \frac{1}{3} + \frac{1}{9} + \ldots + \frac{1}{729}$;
- (iv) $x^3 - 2x^5 + 4x^7 - 8x^9 + \ldots + 2^{48}x^{99}$.

Example 6
Find all geometric progressions such that the sum of the first two terms is 12 and the 6th term is 16 times the 10th term.

Solution
With the usual notation, we have

$$a + ar = 12 \qquad (1)$$

$$ar^5 = 16ar^9. \qquad (2)$$

From (2),
$$r^4 = \frac{1}{16}$$

$$\Rightarrow r = \pm\tfrac{1}{2}.$$

Substituting each into (1),

$$r = \tfrac{1}{2} \quad \Rightarrow \quad a = 8$$
$$r = -\tfrac{1}{2} \quad \Rightarrow \quad a = 24.$$

So there are two geometric progressions satisfying the data; they are

$$8, \ 4, \ 2, \ 1, \ \ldots$$
and
$$24, \ -12, \ 6, \ -3 \ \ldots .$$

The next example illustrates a useful technique, that of *dividing* a pair of simultaneous equations.

Example 7

The sum of the second and third terms of a G.P. is 280, and the sum of the 5th and 6th terms is 4375. Find the 4th term.
Solution
We have

$$ar + ar^2 = 280$$
and
$$ar^4 + ar^5 = 4375.$$

Factorizing the L.H.S. of each, and dividing the equations,

$$\frac{ar^4(1+r)}{ar(1+r)} = \frac{4375}{280}$$

$$\Rightarrow r^3 = \frac{4375}{280} = \frac{125}{8}$$

$$\Rightarrow r = \frac{5}{2}.$$

Substituting back gives $a = 32$. So the 4th term is

$$32\left(\frac{5}{2}\right)^3 = 500.$$

Example 8

How many terms of the geometric series

$$15 - 30 + 60 - 120 + \ldots$$

must be taken for the sum to exceed 3000?
Solution
Using the formula for the sum of a G.P., we have

$$\frac{15\{1 - (-2)^n\}}{1 + 2} > 3000$$

$$\Rightarrow 1 - (-2)^n > 600$$

$$\Rightarrow -(-2)^n > 599.$$

Clearly n must be odd, and so by inspection, the least value of n satisfying this condition is $n=11$. So 11 terms are needed.

Qu.16 What is the least *even* integer n such that $s_n > 3000$ in Example 8?

Qu.17 The first term of a G.P. is $a+1$, and the common ratio is r^2. Write down expressions for
(i) the 4th term; (ii) the nth term;
(iii) the $2n$th term; (iv) the sum to $n+1$ terms.

Exercise 8c

(Where necessary in this exercise, answers may be left in index form.)

1. Find the 10th term and the sum to 20 terms of these geometric progressions:
   (i) $1, \frac{2}{3}, \frac{4}{9}, \frac{8}{27}, \ldots$;
   (ii) $16, -8, 4, -2, \ldots$;
   (iii) $-48, 36, -27, \frac{81}{4}, \ldots$;
   (iv) $p^3q, p^2q^2, pq^3, q^4, \ldots$;
   (v) $x^3, x, \dfrac{1}{x}, \dfrac{1}{x^3}, \ldots$.

2. The kth term of the G.P.

$$1, 1\cdot2, 1\cdot44, \ldots$$

   is the first to exceed 10. Find k. [*Hint*: you will need logarithms.]

3. The 4th term of a G.P. is 20 and the 7th term is -540. Find the first term.

4. The sum of the first two terms of a G.P. is -8, and the sum of the second and third terms is $1\cdot6$. Find the first three terms.

5. The third term of a G.P. exceeds the first by 1, and the 4th term exceeds the third by $\frac{1}{2}$. Find the first 4 terms.

6. The sum of the first three terms of a G.P. is 6, and the first term is 2. Find the first three terms of every G.P. satisfying these conditions.

7. The fourth term of a G.P. is 4 times the 8th term, and the sum of the first and third terms is 6. Find the first three terms of each of the two possible geometric progressions, and the difference between their sums to 10 terms.

8. Find the sum of the first 11 terms of the geometric progressions
   (i) $\frac{1}{32}, \frac{1}{16}, \frac{1}{8}, \ldots$; (ii) $32, 16, 8, \ldots$.
   Explain the relationship between your answers.

9. A man pays an insurance company a premium of £100 at the beginning of each year. His money is guaranteed to grow at 5% compound interest. Find the value of the policy immediately after his 8th payment.

10. (i) Show that the square of the $(n+1)$th term of a G.P. is equal to the product of the nth and $(n+2)$th terms.
    (ii) If the 10th term of a G.P. is p and the 12th term is q, what is the 11th term?

11. By multiplying out the brackets, simplify
    (i) $(1-x)(1+x+x^2+ \ldots +x^n)$;

(ii) $(1+x)(1-x+x^2-x^3+ \ldots +(-x)^n)$.

Explain the connection with the work in this chapter.

12. A snail is 1 m from the top of a wall. Each day he climbs $\frac{2}{3}$ of the remaining distance.

  (i) How far in metres does he travel on each of the first three days?

  (ii) Write out a series for the total distance travelled in n days.

  (iii) How far will be left after each of the first three days?

  (iv) How far will be left after n days?

  (v) Verify that the sum of the series in (ii) plus the answer to (iv) is 1 metre.

13. At the beginning of 1980, a man invested £10 000 at a fixed rate of 5% p.a. compound interest. At the end of 1980, and at the end of each subsequent year, he withdraws £1000. Show that, immediately after the nth withdrawal, his remaining investment is worth

$$£\{10\,000 \times 1 \cdot 05^n - 1000(1 \cdot 05^{n-1} + 1 \cdot 05^{n-2} + \ldots + 1 \cdot 05 + 1)\}.$$

Show that if the value of the investment is zero, then $1 \cdot 05 = 2^n$. Hence find the number of (complete) withdrawals that can be made.

14. In question 13, how much money can be withdrawn each year if the balance is exactly reduced to zero at the 10th withdrawal?

15. A man takes out a £20 000 25-year mortgage when the mortgage rate is 15%. What is his annual repayment? (Assume that the repayments are annual, starting at the end of the first year, and that the interest is 'compounded' annually.)

16. Sum the series

$$1 + r + r^3 + r^4 + r^6 + r^7 + \ldots + r^{28} + r^{30} + r^{31}$$

(i) by inserting brackets to 'pair' the terms so that the series is a G.P.;

(ii) by inserting the missing terms, summing, and then subtracting the series of missing terms.

Show that your two answers are the same.

17. (i) Find all values of a, b, c such that

$$10, a, b, c, 810$$

is a G.P..

(ii) If $0 < a < b$ and a, x_1, x_2, b is a G.P., express x_1 and x_2 in terms of a and b.

(iii) If u_1, u_2, u_3, $\ldots$ is a G.P. with all terms positive, what can be said about the sequence

$$\log_{10} u_1, \log_{10} u_2, \log_{10} u_3, \ldots ?$$

8.7 Summing infinite geometric progressions

Let $$s_\infty = a + ar + ar^2 + ar^3 + \ldots.$$
Denoting the partial sum $a + ar + \ldots + ar^{n-1}$ by s_n as usual, we have

$$s_\infty = \lim_{n \to \infty} s_n$$

$$= \lim_{n \to \infty} \frac{a(1-r^n)}{1-r} \quad \text{(provided } r \neq 1\text{).}$$

We must therefore investigate what happens to r^n as $n \to \infty$. Clearly, as $n \to \infty$,

$$|r| < 1 \;\Rightarrow\; r^n \to 0,$$
$$r > 1 \;\Rightarrow\; r^n \to \infty,$$
$$r < -1 \;\Rightarrow\; r^n \text{ oscillates with } |r^n| \text{ ever increasing,}$$

and $\qquad r = -1 \;\Rightarrow\; r^n$ alternates between 1 and -1.

It follows that

$$\text{for } |r| < 1, \quad \lim_{n \to \infty} \frac{a(1-r^n)}{1-r} = \frac{a}{1-r}.$$

So $\qquad\qquad a + ar + ar^2 + \ldots = \dfrac{a}{1-r} \quad \text{(for } |r| < 1\text{),}$

which we may write as

$$\boxed{\; s_\infty = \frac{a}{1-r} \qquad \text{(for } |r| < 1\text{).} \;}$$

For all other values of r, the series does not converge (unless, of course, $a = 0$).

Qu.18 Find

(i) $1 + \frac{1}{3} + \frac{1}{9} + \frac{1}{27} + \ldots$;

(ii) $1 - \frac{1}{3} + \frac{1}{9} - \frac{1}{27} + \ldots$.

Example 9

Find the sum of the infinite G.P.

$$3x^2 + 6x + 12 + \frac{24}{x} + \ldots$$

stating the values of x for which this sum exists.

Solution

The first term is $3x^2$ and the common ratio is $\dfrac{2}{x}$. Substituting into the formula,

$$s_\infty = \frac{3x^2}{1 - 2/x} = \frac{3x^3}{x-2}.$$

Now the series will converge (to this sum) only if $\left|\dfrac{2}{x}\right| < 1$ i.e. only if $|x| > 2$.

Qu.19 For what values of x do the following converge?

(i) $1 + 2x + 4x^2 + 8x^3 + \ldots$;

(ii) $1 + \dfrac{x}{2} + \dfrac{x^2}{4} + \dfrac{x^3}{8} + \ldots$;

(iii) $1 - x^2 + x^4 - x^6 + \ldots$.

Qu.20 By first writing the recurring decimal $0\cdot2\dot{7}$ in the form

$$\frac{27}{100}+\frac{27}{10^4}+\frac{27}{10^6}+\dots,$$

write it as a vulgar fraction. (Discuss how this method is related to that in Section 1.4.)

8.8 Problems involving both arithmetic and geometric progressions

The next example illustrates a number of important techniques.

Example 10

The second, fourth and ninth terms of an arithmetic progression are in geometric progression, and the third term is 21. Find the first two terms of the A.P. and the common ratio of the G.P..

Solution

Let the A.P. have first term a and common difference d. Then we are given that $a+d$, $a+3d$, $a+8d$ are in geometric progression. So

$$\frac{a+3d}{a+d}=\frac{a+8d}{a+3d}(=\text{common ratio})$$

$$\Rightarrow \quad (a+3d)^2=(a+d)(a+8d)$$
$$\Rightarrow \quad a^2+6ad+9d^2=a^2+9ad+8d^2$$
$$\Rightarrow \quad d^2-3ad=0$$
$$\Rightarrow \quad d=0 \text{ or } d=3a.$$

We are also given that $a+2d=21$. If $d=0$, then $a=21$ and the A.P. is

$$21, 21, \dots.$$

If $d=3a$, then $7a=21$; so $a=3$, $d=9$ and the A.P. is

$$3, 12, \dots.$$

The common ratio of the G.P. is $\dfrac{a+3d}{a+d}$, so r takes the values 1 or $2\frac{1}{2}$ respectively.

Exercise 8d

1. State whether each of the following infinite geometric series is convergent. Find the sum to infinity where appropriate.
   (i) $1+\frac{1}{2}+\frac{1}{4}+\frac{1}{8}+\dots$;
   (ii) $1-\frac{1}{2}+\frac{1}{4}-\frac{1}{8}+\dots$;
   (iii) $6+4+\frac{8}{3}+\dots$;
   (iv) $10-2+0\cdot4-0\cdot08+\dots$;
   (v) $\frac{1}{3}+\frac{1}{2}+\frac{3}{4}+\dots$;
   (vi) $\frac{1}{2}-\frac{1}{2}+\frac{1}{2}-\frac{1}{2}+\dots$;
   (vii) $1-2+4-8+\dots$.

2. Find the sum of these infinite geometric series, stating the range of values of x for which each converges.
   (i) $1-3x+9x^2-27x^3+\dots$;
   (ii) $1+4x^2+16x^4+64x^6+\dots$;
   (iii) $x+x(1-x)+x(1-x)^2+x(1-x)^3+\dots$;
   (iv) $x^3+x^2+x+1+\dots$.

3. Write down the first few terms of the infinite geometrical series whose first term is 1 and whose sum is

(i) $\dfrac{1}{1-x}$;　　(ii) $\dfrac{1}{1+x}$.

4. The sum to infinity of a certain G.P. is 10 and the first term is 2. Find the 2nd term.

5. (i) The sum to infinity of a G.P. is 9, and the second term is 2. Show that there are two such geometric progressions, and find the first term of each.

(ii) Show that no G.P. has a sum to infinity of 1 and a second term of -2.

6. Show that (with the exception of the trivial case in which every term is zero) no G.P. can have the sum of the first two terms equal to the sum to infinity.

7. Derive the sum to infinity of

$$S = a + ar + ar^2 + \ldots$$

by showing directly that $S - rS = a$. Explain why this proof breaks down for $|r| \geqslant 1$.

8. A sequence is defined by $u_1 = 3$, $u_{n+1} = \frac{1}{2}u_n$.

(i) Write down the first few terms, and hence express u_n in terms of n only.

(ii) Find the sum to infinity.

A new sequence $\{v_n\}$ is defined by $v_n = u_{2n-1} + u_{2n}$.

(iii) Find v_1, v_2, v_3 and v_4.

(iv) Find the sum to infinity of this new sequence. Compare your answer with that of (ii).

9. Find the difference between the sums to 10 terms of the series beginning $2 + 4 + \ldots$ when it is an A.P. and when it is a G.P..

10. The third, sixth and tenth terms of an A.P. are in geometrical progression, and the first term is 14. Find all possible values of the third term and the corresponding common ratios of the G.P..

11. The 2nd, 5th and 8th terms of an A.P. are in geometric progression. Discuss.

12. The first, third and fourth terms of a G.P. are in arithmetic progression. Find all possible values of the common ratio.

13. Find an expression for the sum to n terms of the series

$$a + (a + rd) + (a + r^2d) + (a + r^3d) + \ldots.$$

14. (See also questions 15 and 16.) Find the sum of the infinite series

$$1 + 2x + 3x^2 + 4x^3 + \ldots$$

by first writing it as the sum of infinitely many infinite series thus

$$
\begin{aligned}
1 + x + x^2 + x^3 + x^4 + \ldots \\
+ x + x^2 + x^3 + x^4 + \ldots \\
+ x^2 + x^3 + x^4 + \ldots \quad \text{etc.}
\end{aligned}
$$

State the range of values of x for which the series converges.

15. Find the sum of the series in question 14 by differentiating each side of

$$1 + x + x^2 + x^3 + \ldots = \frac{1}{1-x}.$$

(A discussion of the conditions under which an infinite series may be differentiated term by term is beyond the scope of this book.)

16. Sum the series in question 14 by calling it S and considering $S - xS$, itself an infinite series.

8.9 The sigma notation

Sigma is the Greek letter corresponding to S. A capital sigma, written Σ, is used to denote a summation. Its precise use is best illustrated by an example.

$\displaystyle\sum_{k=10}^{k=15} 3k^2$ means 'the sum of the values of $3k^2$ as k takes each integer value from 10 to 15 inclusive'.

Thus

$$\sum_{k=10}^{k=15} 3k^2 = 3(10)^2 + 3(11)^2 + \ldots + 3(15)^2$$
$$= 2865.$$

Since k was the only variable in the summation, we could have written simply $\displaystyle\sum_{10}^{15} 3k^2$. The numbers 15 and 10 are called the upper and lower limits of the summation. Further examples are:

$$\sum_{0}^{10} 2^n = 1 + 2 + 4 + \ldots + 1024 = 2047$$

and

$$\sum_{r=1}^{\infty} \frac{x^r}{r} = x + \frac{x^2}{2} + \frac{x^3}{3} + \ldots.$$

Qu.21 Write the following series without using the Σ notation (i.e. give the first few terms and the last term). Do not evaluate the series.

(i) $\displaystyle\sum_{0}^{10} r^3;$ (ii) $\displaystyle\sum_{1}^{100} \frac{1}{k};$

(iii) $\displaystyle\sum_{20}^{30} \sqrt{r};$ (iv) $\displaystyle\sum_{k=0}^{n} (a + kd);$ (v) $\displaystyle\sum_{\lambda=0}^{\infty} ar^{\lambda}.$

Qu.22 By first writing out the first few terms, evaluate

(i) $\displaystyle\sum_{1}^{100} r;$ (ii) $\displaystyle\sum_{3}^{20} (2k + 3);$ (iii) $\displaystyle\sum_{1}^{\infty} \frac{3}{2^{\lambda}};$ (iv) $\displaystyle\sum_{r=1}^{r=10} 1.$

Qu.23 Let $u_1, u_2, u_3, \ldots$ be a sequence.

(i) How may $\displaystyle\sum_{r=1}^{100} u_r - \sum_{r=1}^{50} u_r$ be written more simply?

(ii) Verify that $a \displaystyle\sum_{r=1}^{n} u_r = \sum_{r=1}^{n} au_r.$

(iii) If v_1, v_2, v_3, ... is another sequence, is it true that

$$\sum_{i=1}^{n} (u_i + v_i) = \sum_{i=1}^{n} u_i + \sum_{i=1}^{n} v_i?$$

Qu.24 (For discussion.) Verify that

(i) $\displaystyle\sum_{0}^{10}(3r+5) = \sum_{2}^{12}(3r-1)$;   (ii) $\displaystyle\sum_{0}^{n} r^3 = \sum_{1}^{n} r^3$.

The following example shows how to express a series using the sigma notation. The reader should first look again at Example 1 of this chapter.

Example 11

Express

$$\frac{1}{1.2} + \frac{1}{3.4} + \frac{1}{5.6} + \ldots + \frac{1}{99.100}$$

using the sigma notation.

Solution

The 'typical term' is of the form $\dfrac{1}{(2r-1)(2r)}$. Thus the series may be written as $\displaystyle\sum \frac{1}{2r(2r-1)}$ for suitable limits which may be found by inspection. The series is therefore $\displaystyle\sum_{1}^{50} \frac{1}{2r(2r-1)}$.

Qu.25 Write the series

$$\frac{1}{1.2} + \frac{1}{2.3} + \frac{1}{3.4} + \ldots + \frac{1}{99.100}$$

using Σ notation.

Exercise 8e   **1.** Evaluate

(i) $\displaystyle\sum_{1}^{3} r^2$;   (ii) $\displaystyle\sum_{1}^{3}(r^2+1)$;   (iii) $\displaystyle\sum_{1}^{10}(-2)^r$;   (iv) $\displaystyle\sum_{1}^{8}(-1)^r r^2$;

(v) $\displaystyle\sum_{1}^{100} r^2 - \sum_{1}^{99} r^2$;   (vi) $\displaystyle\sum_{0}^{6}(r+1)(r+2)$;   (vii) $\displaystyle\sum_{0}^{7} r(r+1)$;   (viii) $\displaystyle\sum_{0}^{\infty}\left(-\frac{2}{3}\right)^k$.

2. Write down the following series without using the sigma notation, i.e. give the first few terms and, where appropriate, the last term.

(i) $\displaystyle\sum_{r=0}^{r=100} x^r$;   (ii) $\displaystyle\sum_{k=1}^{k=\infty} \frac{x^k}{k^2}$;   (iii) $\displaystyle\sum_{r=2}^{20} \frac{x^{r-1}}{r(r+1)}$.

3. Write in sigma notation:
    (i) $1+3+5+ \ldots +101$;
    (ii) $1.2+4.5+7.8+ \ldots +100.101$ [*Hint*: find first the rth member of the sequence 1, 4, 7, ...];

(iii) $\frac{1}{2}+\frac{2}{3}+\frac{3}{4}+\ldots+\frac{49}{50}$;

(iv) $\frac{1}{2}+\frac{3}{4}+\frac{5}{6}+\ldots+\frac{49}{50}$;

(v) $1+\frac{1}{2}+\frac{1}{4}+\frac{1}{8}+\ldots$;

(vi) $1-\frac{1}{2}+\frac{1}{4}-\frac{1}{8}+\ldots$;

(vii) $-x+\frac{x^2}{2}-\frac{x^3}{3}+\ldots$;

(viii) $1+2x+3x^2+4x^3+\ldots$;

(ix) $1-4x+9x^2-16x^3+\ldots$;

(x) $\frac{2x}{3.4}+\frac{3x^2}{4.5}+\frac{4x^3}{5.6}+\ldots$.

(xi) $\frac{1}{2.4}-\frac{3}{4.6}+\frac{5}{6.8}-\frac{7}{8.10}+\ldots$.

8.10 Mathematical induction

In this section, we shall use the notation $P(n)$ to denote a *statement* about the number n. Suppose, for example, that $P(n)$ is the statement 'n is a perfect square'. Then $P(1)$, $P(4)$ and $P(9)$ would be true statements, while $P(2)$, $P(3)$ and $P(5)$ would be false.

Leaving this example, let $P(n)$ be some other statement, and suppose that we know that

$$\left.\begin{array}{ll} & \text{(i) } P(1) \text{ is true,} \\ \text{and} & \text{(ii) for all } k\in Z^+, \ P(k)\Rightarrow P(k+1). \end{array}\right\} \tag{1}$$

What can we deduce? Well, $P(1)$ is true, and from (ii), $P(1)\Rightarrow P(2)$, so $P(2)$ is true. But from (ii) again, $P(2)\Rightarrow P(3)$, so $P(3)$ is true. Continuing this process, it is clear that $P(n)$ will be true for every positive integer n.

It follows that if we wish to prove that some statement $P(n)$ is true for all positive integers, we have to prove only the two statements (1) above. This fact is sometimes called the *Principle of Induction*.

Note

When proving the second of the two statements in (1) above, it is important to remember that we are not trying to prove that $P(k)$ is true; we are merely trying to show that *if* $P(k)$ is true, then $P(k+1)$ is also true. It helps to think of k as some particular number for which $P(n)$ happens to be true.

Example 12

Prove that for all $n\in Z^+$, 4^n-1 is divisible by 3.

Solution

Let $P(n)$ be the statement '(4^n-1) is divisible by 3'.

(i) $P(1)$ is true, since 4^1-1 is divisible by 3.

(ii) Now *suppose* that for some value of k, $P(k)$ is true, i.e. that 4^k-1 is divisible by 3. We wish to show that $P(k+1)$ is true, i.e. that $4^{k+1}-1$ is divisible by 3. Well,

$$\begin{aligned} 4^{k+1}-1 &= (4^k-1)+(4^{k+1}-4^k) \\ &= (4^k-1)+4^k(4-1) \\ &= (4^k-1)+(3\times 4^k). \end{aligned}$$

Thus if $4^k - 1$ is divisible by 3, then so is $4^{k+1} - 1$ (since both terms on the R.H.S. are divisible by 3); i.e. $P(k) \Rightarrow P(k+1)$.
The required result now follows by induction.

Example 13
Prove that

$$\sum_{1}^{n}(3r-2)^2 = \tfrac{1}{2}n(6n^2 - 3n - 1).$$

Solution
Let $P(n)$ be the statement

$$\sum_{1}^{n}(3r-2)^2 = \tfrac{1}{2}n(6n^2 - 3n - 1).$$

(i) Putting $n=1$, the L.H.S. is the single term 1, and the R.H.S. $= 1$. So $P(1)$ is true.
(ii) Now suppose that $P(k)$ is true, i.e. that

$$\sum_{1}^{k}(3r-2)^2 = \tfrac{1}{2}k(6k^2 - 3k - 1). \qquad (1)$$

[We must try to show that $P(k+1)$ is true, i.e. that

$$\sum_{1}^{k+1}(3r-2)^2 = \tfrac{1}{2}(k+1)\{6(k+1)^2 - 3(k+1) - 1\}$$

or

$$\sum_{1}^{k+1}(3r-2)^2 = \tfrac{1}{2}(k+1)(6k^2 + 9^k + 2).]$$

Adding the next term to each side of (1) gives

$$\sum_{1}^{k}(3r-2)^2 + (3k+1)^2 = \tfrac{1}{2}k(6k^2 - 3k - 1) + (3k+1)^2$$

$$\Rightarrow \sum_{1}^{k+1}(3r-2)^2 = \tfrac{1}{2}(6k^3 - 3k^2 - k) + (9k^2 + 6k + 1)$$

$$= \tfrac{1}{2}(6k^3 + 15k^2 + 11k + 2)$$

$$\Rightarrow \sum_{1}^{k+1}(3r-2)^2 = \tfrac{1}{2}(k+1)(6k^2 + 9k + 2),$$

which is the statement $P(k+1)$.
Thus we have shown that $P(1)$ is true, and that $P(k) \Rightarrow P(k+1)$. Hence $P(n)$ is true for all $n \in Z^+$ by induction.

Qu.26 Prove that $1 + 3 + 5 + \ldots + (2n-1) = n^2$. [*Hint*: when trying to reach $P(k+1)$, write down this statement in rough so that you know what you are trying to obtain. The technique is again to add the next term—in this case $(2k+1)$—to each side of $P(k)$.]
Qu.27 (i) Show that if $P(n)$ is the statement

$$1 + 3 + 5 + \ldots + (2n-1) = n^2 + 1,$$

then $P(k) \Rightarrow P(k+1)$.
(ii) Explain the contradiction with Qu.26.

Each of Examples 12 and 13 illustrates the main limitation of the method of proof by induction, namely that it is only a method of proof. In Example 13, we could not have used induction if the question had said merely 'Find an expression for $\sum\limits_{1}^{n}(3r-2)^2$'. Some other method—perhaps an inspired guess—is needed first to obtain the required expression.

Induction is often the appropriate method of proving results which seem obvious; but the obviousness must not be allowed to obscure the need for an inductive argument. The following example also illustrates how induction may begin at $n=2$ rather than at $n=1$. In general, if the induction begins at $n=r$, then the result will have been proved for $n \geqslant r$ only.

Example 14
Prove that the product of any number of odd numbers is odd.
Solution
The proof is by induction on n, the number of odd numbers, and is in two stages.
(i) We prove first that the product of *two* odd numbers is odd. Let a and b be odd numbers, then there exist integers p and q such that $a=2p+1$ and $b=2q+1$.

Then
$$ab=(2p+1)(2q+1)$$
$$=4pq+2p+2q+1$$
$$=2(2pq+p+q)+1.$$

So ab is odd.
(ii) Now suppose that the result has been proved for the product of any k odd numbers, and let $a_1, a_2 \ldots a_{k+1}$ be a set of $k+1$ odd numbers. Then

$$a_1 \times a_2 \times \ldots \times a_{k+1}=(a_1 \times a_2 \times \ldots \times a_k) \times a_{k+1}.$$

Now by the inductive hypothesis, the bracket is an odd number b say; and the product $b \times a_{k+1}$ is odd by part (i). So $a_1 \times a_2 \times \ldots \times a_{k+1}$ is odd, i.e. the result is true for $n=k+1$.

We have therefore shown that the result is true for $n=2$, and that if it is true for $n=k$ then it is also true for $n=k+1$. Therefore it is true for $n \geqslant 2$ by induction.

Exercise 8f   **1.** Prove the following by induction:

(i) $1+2+3+\ldots+n=\dfrac{1}{2}n(n+1)$;

(ii) $1+4+7+\ldots+(3n-2)=\dfrac{3}{2}n^2-\dfrac{1}{2}n$;

(iii) $\dfrac{1}{2}+\dfrac{1}{4}+\dfrac{1}{8}+\ldots+\dfrac{1}{2^n}=1-\dfrac{1}{2^n}$;

(iv) $1+4+9+\ldots+n^2=\dfrac{1}{6}n(n+1)(2n+1)$;

(v) $\sum\limits_{1}^{n}r^3=\dfrac{1}{4}n^2(n+1)^2$;   (vi) $\sum\limits_{1}^{n}\dfrac{1}{r(r+1)}=\dfrac{n}{n+1}$;

(vii) $\sum_{r=0}^{n-1} t^r = \dfrac{1-t^n}{1-t}$ (for $t \neq 1$);   (viii) $\sum_{1}^{n} r.2^{r-1} = (n-1).2^n + 1$.

2. Prove by induction that for all positive integers n
   (i) $5^n - 1$ is divisible by 4;
   (ii) $n^5 - n$ is divisible by 5;
   (iii) $7^n + 3n - 1$ is divisible by 9. [*Hint*: at one stage, it may help to write $7^{k+1} + 3(k+1) - 1$ as $7(7^k + 3k - 1) - 18k + 9$.];
   (iv) $5^n + 12n - 1$ is divisible by 16;
   (v) $9.4^{2n} - 5^{n-1}$ is divisible by 11;
   (vi) $n^3 - n$ is divisible by 6. [*Hint*: show, in a separate induction, that $n^2 + n$ is divisible by 2.].

3. Prove by induction that for $n \in Z$ with $n \geq 5$, $2^n > n^2$.

4. Prove that if $y = \dfrac{1}{1-x}$ then
$$\frac{d^n y}{dx^n} = \frac{1 \times 2 \times 3 \times \ldots \times n}{(1-x)^{n+1}}.$$

5. Prove that for $n \in Z^+$
   (i) $\begin{pmatrix} 0 & -1 \\ -1 & 0 \end{pmatrix}^{2n-1} = \begin{pmatrix} 0 & -1 \\ -1 & 0 \end{pmatrix}$;

   (ii) $\begin{pmatrix} 0 & -1 \\ 1 & 0 \end{pmatrix}^{4n-1} = \begin{pmatrix} 0 & 1 \\ -1 & 0 \end{pmatrix}$.

6. Prove that the maximum number of points of intersection of n straight lines lying in a plane is $\frac{1}{2}n(n-1)$.

7. Suppose that $f : R \rightarrow R$ is such that $\forall \ a_1, a_2 \in R$
$$f(a_1 + a_2) = f(a_1) + f(a_2).$$
   (i) Show that $f(a_1 + a_2 + a_3) = f(a_1) + f(a_2) + f(a_3)$.
   (Every step in your argument must be clear.)
   (ii) Prove (rigorously) that
$$f(a_1 + a_2 + \ldots + a_n) = f(a_1) + f(a_2) + \ldots + f(a_n)$$
   for any $n \in Z^+$.

8. Let $y_n = u_1 u_2 u_3 \ldots u_n$ where each of the u_r are functions of x. Prove by induction that
$$\frac{dy_n}{dx} = \sum_{r=1}^{n} \left(\frac{y_n}{u_r} \frac{du_r}{dx} \right).$$

9. (i) Evaluate $\begin{pmatrix} 1 & 1 \\ 0 & 1 \end{pmatrix}^n$ for $n = 1$, 2 and 3, and hence suggest a general formula. Prove your formula by induction.

   (ii) Repeat (i) for $\begin{pmatrix} 1 & \frac{1}{2} \\ 0 & \frac{1}{2} \end{pmatrix}^n$.

1. (a) The sum of seven consecutive terms of an arithmetic progression of positive terms is 147 and the product of the first and last of these terms is 297. Calculate the common difference of the arithmetic progression.

(b) A geometric progression has first term a and common ratio r. If S denotes the sum of the first n terms of this progression and if R denotes the sum of the reciprocals of these terms, show that

$$S = a^2 r^{n-1} R.$$

[JMB (O)]

2. In a geometric progression of positive terms the second and fourth terms are 4 and $\frac{1}{4}$ respectively.

Find

(i) the sum to infinity of the terms of the progression,

(ii) the least number of terms needed to obtain a sum to within 0·1 of the sum to infinity.

[AEB (O) '79]

3. (a) Write down an expression for the nth term of a Geometric Progression whose first term is a and whose common ratio is r.

A G.P., all terms of which are positive, is such that the fourth term is 16 times greater than the eighth term. Find the common ratio. Given that the third term is 5, find the first term and the sum to infinity of the progression.

(b) The first three terms of an Arithmetic Progression are 2, x and $4\frac{1}{2}$, in that order. Find x and the sum of the first twenty-five terms of the progression.

[SUJB (O)]

4. (i) The sum of the 4th and 11th terms of an arithmetical progression is half of the 33rd term. The sum of the first 16 terms is 228. Find the first term and the common difference.

(ii) The first two terms of a geometric progression are 2, 3. Find which is the first term of the progression to exceed 4000.

[O & C (O)]

5. A sequence $\{u_n\}$ is defined by

$$u_1 = u_2 = 1,$$
$$u_{n+1} = au_n + bu_{n-1}, \quad n \geqslant 2.$$

Given that a and b are positive, prove by induction, or otherwise, that $u_{n+3} > a^{n+1}$ for all $n \geqslant 0$. Deduce that $\{u_n\}$ is not convergent if $a > 1$.

[JMB]

6. (a) Prove that the sum of all the integers between m and n inclusive $(m, n \in Z_+, n > m)$ is $\frac{1}{2}(m+n)(n-m+1)$.

Find the sum of all the integers between 1000 and 2000 which are *not* divisible by 5.

(b) A geometric series has first term 2 and common ratio 0·95. The sum of the first n terms of the series is denoted by S_n and the sum to infinity is denoted by S. Calculate the least value of n for which $S - S_n < 1$.

[C]

7. The first term of a geometric series is 2 and the second term is x.

State the set of values of x for which the series is convergent. Show that when convergent the series converges to a sum greater than 1.

If $x = \frac{1}{2}$, find the smallest positive integer n such that the sum of the first n terms differs from the sum to infinity by less than 2^{-10}.

[L]

8. (i) Express the recurring decimal $0 \cdot 1\dot{4}\dot{8}$ as an infinite geometric series. Sum the geometric series and express your result as a fraction in its lowest terms.

$$[0 \cdot 1\dot{4}\dot{8} \equiv 0 \cdot 148148 \dots]$$

(ii) Given that $\sum_{r=1}^{n} a_r = pn^2 + qn$ for each integer, n, where p and q are constants, show that the terms of the sequence $a_1, a_2, \dots$ form an arithmetic progression.

[L (S)]

9. An arithmetical progression and a geometrical progression have the same positive first term. The common difference r of the arithmetical progression is equal to the common ratio of the geometrical progression. The sum of the first 8 terms of the former is 5 times the sum of the first and third terms of the latter. The fifth term of the geometrical progression is $-r$. Find r and the first term of either progression.

Show that the geometrical progression converges and find its sum to infinity.

[L]

10. A cheetah starts with a leap of 4 m to chase a deer. Each leap of the cheetah is shorter by 10 cm than the preceding leap. The deer has a start of $21\frac{2}{5}$ m and runs away with uniform leaps of 2 m each, taking 5 leaps to every 6 of the cheetah and starting at the same moment as the cheetah. Suppose that the cheetah takes n leaps to overtake the deer. Show that the distance gone by the cheetah is

$$\tfrac{1}{2}n[8 - \tfrac{1}{10}(n-1)] \text{ metres}$$

and simplify this expression. Show also that when the deer is caught it will be a distance $(21\frac{2}{5} + \frac{5}{3}n)$ metres from the cheetah's starting point and hence calculate how far the cheetah has to go in order to catch the deer.

[Ox (O)]

11. If $T_n = a^{n-1}$, $a \neq 1$, and $S_n = T_1 + T_2 + \dots + T_n$, find, in terms of a and n, in their simplest form,

   (i) $T_1 + T_2 + T_3 + \dots + T_n$,
   (ii) $T_1 T_2 T_3 \dots T_n$,
   (iii) $S_1 + S_2 + S_3 + \dots S_n$.

[AEB '80]

12. (i) An inexperienced operator sets a programmable calculator to perform repeatedly the following loop of operations:

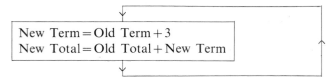

New Term = Old Term + 3
New Total = Old Total + New Term

starting with both Total and Term equal to 2, but forgetting to tell it when to stop. Write down the first six values of Term.

After two minutes the operator stops the machine and Term reads 3599. Assuming that the last loop has been completed, exactly what will Total read and how many additions per second has the machine been performing?

(ii) A ball is thrown vertically up from the ground to a height of 25 m, falls back and bounces repeatedly. After each successive impact the ball rises to four-fifths of the height of the previous bounce. Write down the distances travelled between the start and the first impact and between successive impacts up to the fourth. Deduce and simplify a formula for the total distance travelled by the ball from the first throw to the nth impact.

[MEI (O)]

9 Permutations and combinations; the Binomial Theorem

9.1 Counting the members of a set

In the first part of this chapter, we shall consider various problems concerning the number of ways of arranging or selecting objects. It is a feature of such problems that there may be several correct methods of solving them and several (plausible) incorrect ones. The reader should therefore make a habit of considering alternative solutions and of searching for the error if two different answers are obtained.

Since the problems that we shall be concerned with are essentially problems of counting, it is helpful to remember that there are only two mistakes that can be made when counting the members of a set: counting some members more than once, and failing to count some members at all.

Qu.1 A lecture is to be attended by all sixth-formers studying Mathematics or Physics. If r pupils study Maths and s study Physics, why is $r+s$ not necessarily the number of pupils who should be at the lecture?

Let $n(S)$ denote the number of elements in the set S; then if A and B are sets,

$$n(A\cup B)=n(A)+n(B)-n(A\cap B). \tag{1}$$

This formula would give the correct answer to Qu.1, since the extra term of $-n(A\cap B)$ removes those elements which have been counted twice, so that the net effect is that they are counted once only.

Qu.2 By repeated application of (1) above, show that

$$n(A\cup B\cup C)=n(A)+n(B)+n(C)-n(A\cap B)$$
$$-n(B\cap C)-n(C\cap A)+n(A\cap B\cap C).$$

[*Hint*: start by writing the L.H.S. as $n((A\cup B)\cup C)$.]

Qu.3 Verify the formula in Qu. 2 by drawing a Venn Diagram and checking that each of the seven regions of $A\cup B\cup C$ is counted once only overall.

Qu.4 Find the corresponding expression for $n(A\cup B\cup C\cup D)$.

9.2 The multiplicative rule

The multiplicative rule is the fundamental way of dealing with counting problems that can be analysed or broken down into two or more stages. It is best illustrated by an example.

Suppose we throw a die† and toss a coin, and wish to find the number of possible outcomes of this composite 'experiment'. We could list them all: $(1, H), (1, T), (2, H), (2, T), \ldots, (6, T)$. More simply, we could say that for each of the six possible scores for the die, there are two possible results for the coin. The answer is therefore $6 \times 2 = 12$ outcomes. We could illustrate this method on a *tree diagram* (Fig. 9.1).

Figure 9.1

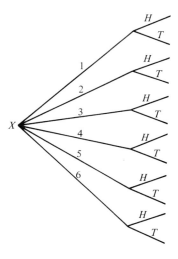

The diagram begins at X and shows six branches corresponding to the six possible results from the coin. Whichever of these results occurs, there are two possible results for the coin; so each of the six branches splits into two, giving twelve possible routes in all.

The multiplicative rule says that if an experiment may be divided into two (or more) stages, and if the number of outcomes at the second stage is fixed and does not depend on the outcome of the first stage, then the number of outcomes for the whole experiment is the product of the numbers of outcomes of the two stages separately; i.e. if the first stage has s outcomes and the second stage has t outcomes then the whole experiment has st outcomes.

Qu.5 The tree diagram above corresponds to looking at the die and then the coin. Draw the diagram that corresponds to looking at the coin and then the die.

Qu.6 Let R be the set of outcomes when a die is thrown, and let S be the set of outcomes when a coin is tossed. Verify that the set of outcomes of the joint experiment corresponds to the Cartesian product $R \times S$.

Qu.7 Find the number of possible outcomes when two dice are thrown. (Assume that the dice are distinguishable—perhaps different colours—so that $(3, 5)$ is a different outcome from $(5, 3)$.)

Qu.8 (For discussion.) What is the answer to Qu.7 if the dice are indistinguishable (so that $(3, 5)$ and $(5, 3)$ are together counted as one outcome)? Why exactly does the multiplicative rule not apply here?

---

† 'Die' is the singular of the plural word 'dice'.

Example 1

Four people take part in a simple raffle for a first and second prize. Their initials, P, Q, R and S, are written on four balls which are put in a bag, and two balls are drawn consecutively without replacement (i.e. the first is not replaced before the second is drawn). Find the number of possible results.

Solution

The first ball can be chosen in 4 ways, and, whichever ball is chosen first, there are 3 possible second choices. By the multiplicative rule, the total number of results is $4 \times 3 = 12$.

Qu.9 Draw a clearly labelled tree diagram to illustrate this solution.

Qu.10 Solve Example 1 if selection is made *with replacement* (so that a ball may be chosen twice).

Qu.11 Suppose that the balls in Example 1 had been labelled P, P, Q, R, and two were drawn without replacement. Find the number of outcomes by simply listing them. Show why the multiplicative rule cannot be used here, by drawing a tree diagram to illustrate the solution. (Be careful to show only one letter P on the first draw; if you show two, you will find yourself counting some outcomes twice.)

9.3 Permutations of distinguishable objects

A *permutation* is an arrangement or ordering of a number of objects. (A more precise definition was given in Exercise 2a, question 12.) In this section, we consider only sets of distinguishable objects, i.e. sets of objects such that if any two are interchanged, we get a different arrangement which would be counted separately.

Example 2

In how many ways can the letters A, B, C, D be arranged?

Solution

The letter that is to go first may be chosen in 4 ways. The second letter may then be chosen in 3 ways (whichever one was chosen first). The third letter may be chosen in 2 ways, and there is then only one letter that can go last. The number of permutations is therefore $4 \times 3 \times 2 \times 1 = 24$.

Qu.12 List the 24 permutations of A, B, C, D.

Qu.13 How many arrangements are there of
(i) 5 letters;   (ii) 6 letters?

Qu.14 Verify that Example 2 may be argued thus: The letter A may be placed in 4 possible positions (1st, 2nd, 3rd or 4th). Then B can go in 3 places, etc.

Extending the method of Example 2 and Qu.13, we see that the number of arrangements of n distinguishable objects is

$$n \times (n-1) \times (n-2) \times \ldots \times 3 \times 2 \times 1.$$

Notation

The number $n \times (n-1) \times \ldots \times 3 \times 2 \times 1$ is denoted by $n!$ which is pronounced 'n factorial' or 'factorial n' or 'n shriek'. Thus $4! = 4 \times 3 \times 2 \times 1 = 24$, and $5! = 120$.

Qu.15 (i) If n is 'very large', which of 10^n, n^{10}, $n!$ is greatest? Which is smallest?
(ii) Write down the next number in the sequence

$$120, 24, 6, 2, 1.$$

Write each of the six numbers in factorial form. What should we take as the value of $0!$?
Qu.16 (i) Write out in full $\frac{12!}{8!}$, and evaluate it.
(ii) Express $10 \times 9 \times 8$ using factorials only.

Now suppose that we have n distinguishable objects of which we wish to arrange only r ($r < n$). The method is a truncated version of the method above.

Example 3
Ten runners compete in an Olympic final. In how many ways can the three medals be awarded?
Solution
The winner (gold) can be chosen in 10 ways, leaving 9 choices for the silver and then 8 for the bronze. So the answer is $10 \times 9 \times 8 = 720$ ways.

In general, if we wish to arrange r objects out of n, we can choose the first in n ways, the second in $(n-1)$ ways and so on, the rth being chosen in $(n-r+1)$ ways. So the total number of ways of arranging r objects from n is

$$n(n-1)(n-2) \ldots (n-r+1).$$

(Note that there are r factors here.) This number may be written

$$\frac{n!}{(n-r)!}$$

Notation
The number of ways of arranging r objects from n is denoted by ^nP_r.
We have seen above that $^nP_r = \dfrac{n!}{(n-r)!}$. Note that $^nP_n = n!$ since $0! = 1$
(see Qu.16).

9.4 Problems involving restrictions

Example 4
Three men and three women are to sit in six seats arranged in a line. In how many ways can this be done if
 (i) there are no restrictions;
 (ii) no two men and no two women may sit together;
(iii) the only restriction is that Mr X and Mr Y must not be adjacent?
Solution
(i) With no restrictions, they may sit in $6!$ ways $= 720$ ways.
(ii) Numbering the seats 1 to 6, the men must sit either in seats 1, 3, 5 or in seats 2, 4, 6. So
(a) choose whether the men sit in the odd- or even-numbered seats.

This can be done in two ways.

(b) Arrange the men in these seats (3!=6 ways).

(c) Arrange the women in the remaining seats (3!=6 ways).

So the total number of arrangements is $2 \times 6 \times 6 = 72$ ways. (Another method is suggested in Qu.18.)

(iii) (Two features of the following solution should be noted:

(1) It is in fact simpler to find the number of ways of arranging the six people so that X and Y *are* adjacent, and then to subtract this number from the total of 720.

(2) In calculating the number of ways in which X and Y are adjacent, it is better to deal with them first rather than to try to seat the other four in such a way as to leave two adjacent gaps.)

We proceed by seating X and Y.

(a) First choose two seats for them. This can be done in 5 ways (1 and 2, or 2 and 3 etc.).

(b) Then decide which way round they sit (i.e. whether X or Y is in the lower-numbered seat). This can be done in 2 ways.

(c) Finally put the other four people in the empty seats. This can be done in 4!=24 ways.

Thus the number of arrangements with X and Y adjacent is $5 \times 2 \times 24 = 240$. So the number with X and Y *not* together is $720 - 240 = 480$.

Qu.17 *Write down* how many of these 480 ways have X sitting in a lower-numbered seat than Y.

Qu.18 Solve part (ii) of Example 4 by arguing thus: seat 1 may be filled in 6 ways, then seat 2 may be filled in ....

Exercise 9a

(Answers should be left in factorial form where this avoids heavy calculation.)

1. Evaluate

(i) 6!; (ii) $\dfrac{9!}{7!}$; (iii) $\dfrac{10!}{7! \ 3!}$; (iv) $\dfrac{5! \times 4!}{6! \times 3!}$; (v) $\dfrac{100! \times 49!}{98! \times 50!}$.

2. (i) Write down the L.C.M. of 8! and 6!.

(ii) Write down the H.C.F. of 5!, 8! and 10!.

3. Factorize

(i) $10! - 9!$; (ii) $16! - 14!$;

(iii) $(n+1)! - 2(n-1)!$; (iv) $(n+1)! - n! + (n-1)!$.

4. (i) Write down the L.C.M. of $n!$ and $(n+1)!$.

(ii) Express $\dfrac{1}{n!} + \dfrac{1}{(n+1)!}$ as a single fraction.

(iii) Express $\dfrac{n!}{(n-r)!} + \dfrac{n!}{(n-r+1)!}$ as a single fraction.

5. Simplify $\dfrac{n!}{3!(n-3)!} + \dfrac{n!}{4!(n-4)!}$.

6. Evaluate

(i) $^5P_3$; (ii) $^6P_6$; (ii) $^{100}P_2$.

7. On a menu, there are three soups, four main courses and three desserts. How many different three-course meals may be ordered?

8. How many three-lettered 'words' can be formed from the letters A, B, C, D, E, F, G, H, if repeats
   (i) are not allowed; (ii) are allowed?

9. (i) Each day of a 5-day week, a boy takes an apple or an orange or a banana to school. In how many different ways can he organize a week's fruit-eating?
   (ii) What would be the answer to (i) if he were not allowed to take the same fruit every day?
   (iii) What is the answer to (i) if he may not take the same fruit on any two consecutive days?

10. For a photograph, four ladies sit in a line, and four gentlemen stand behind them. In how many ways can they arrange themselves?

11. A room has five separate lights, each with its own switch. In how many ways can the room be lit if at least one light must be on?

12. A child's drawing of a man has a body, two arms, two legs and a head. In how many ways can it be coloured if three colours are available, but the body must be a different colour from the other five members?

13. (i) How many different 6-digit numbers can be formed using each of the digits 1, 2, 3, 5, 7, 9 once only?
    (ii) How many of these numbers will be even?
    (iii) How many will be odd?

14. Four-digit numbers are to be formed using the digits 1, 3, 5, 6, 7, 8 (no digit to be repeated). How many can be formed if they are to be
    (i) even;   (ii) over 8000;   (iii) over 8000 and even;
    (iv) over 8000 or even (or both)? [*Hint*: use the expansion of $n(A \cup B)$.]

15. The five volumes of an encyclopaedia are replaced at random. How many possible arrangements are there if it is known that volume 2 is in the middle, and volumes 1 and 5 are adjacent to one another?

16. In how many ways can the letters of the word WINDY be arranged? In how many of these is the W
    (i) at the front;   (ii) at the end;   (iii) in the middle?

17. Write down the expansion of $n(A \cup B)$. How many of the arrangements of the letters REAPING either begin with the R or end with the G, or both?

18. How many 6-digit numbers can be formed from the digits 1, 2, 3, 4, 5, 6 if the 1 must not come first and the 6 must not come last? [*Hint*: one possible way is to use the method of question 17 to find the number of ways in which the 1 does come first or the six last.]

19. The 12 cards comprising the Ace, King and Queen of each suit are to be put in a line. In how many ways can this be done if
    (i) cards of the same suit are to be kept together;
    (ii) cards of the same denomination are to be kept together?

20. In how many ways can 4 boys and 5 girls sit in a line if
    (i) no two girls are adjacent;
    (ii) all the girls sit together and all the boys sit together?

21. In a small railway compartment, four seats face four others.

(i) In how many ways can five men sit down?

(ii) Repeat (i) if Mr X and Mr Y must sit next to one another.

(iii) Repeat (i) if Mr A and Mr B must sit on opposite sides.

22. The 13 spades from a pack of cards are to be arranged in a line so that the K, Q and J are separated from one another. Find, by each of the following methods, the number of ways in which this can be done.

(i) Let A be the event 'K is next to Q', B be 'Q is next to J' and C be 'J is next to K'. Draw a Venn Diagram and decide which region corresponds to the answer. Use the result of Qu.2 in the text, noting that $n(A \cap B \cap C) = 0$.

(ii) Place the other ten cards in order first; then put the K in one of the gaps etc.

23. Use the method of question 22(ii) to find the number of ways in which all 52 cards in a pack may be put in a line with the 4 Aces separated.

24. Repeat question 20(i) with 4 boys and 4 girls. (Be careful.)

9.5 Permutations involving indistinguishable objects

Two objects are indistinguishable if, for the purposes of the problem under consideration, merely interchanging them does not give a different arrangement. An example will make this clear.

Example 5

In how many ways can the letters A, A, B, C be arranged in a line?

Solution

Suppose that we temporarily mark the two A's so that they are distinguishable, calling them A_1 and A_2; then the answer is $4! = 24$. But when we remove the suffices, we find that each 'word' appears twice; for example, BA_1A_2C and BA_2A_1C both become $BAAC$. Our answer of 24 has counted each genuine arrangement twice, so the number of distinct arrangements is only $\frac{24}{2} = 12$.

Example 6

How many 'words' can be made from the letters $AAABBCDE$ if all the letters must be used in each word?

Solution

With suffices (as in Example 5) the answer would be $8!$. We must now decide how many times each word is repeated when suffices are removed. Consider any arrangement, for example

$$A_2 B_1 D B_2 A_3 C E A_1.$$

We could change the A's about in 6 ways and the B's in 2 ways without altering the word obtained when the suffices are removed. Thus each word appears $6 \times 2 = 12$ times in the $8!$ original arrangements. The answer is therefore $\frac{8!}{12} (= 3360)$ distinct arrangements.

Qu.19 How many words can be made in Example 6 if the three A's must be kept together? [*Hint*: treat the 3 A's together as a single letter.]

We can generalize the method of Examples 5 and 6. Suppose we have n objects of m different types: r_1 of one type, r_2 of another, and so on (so that $\sum_{k=1}^{m} r_k = n$) where objects of the same type are indistinguishable; then the number of arrangements is $\dfrac{n!}{r_1!r_2!\dots r_m!}$.

9.6 Roundabouts and necklaces

Example 7

A roundabout has eight identical seats. In how many ways can eight children ride on it?

Solution

Imagine first that the roundabout is stationary. (We may then regard the seats as being labelled by the eight compass directions N, NE, E etc.) The children may now be seated in 8! ways. But when the roundabout is free to move, we find that we have counted each genuine arrangement 8 times, since, for example, the following arrangements are really the same:

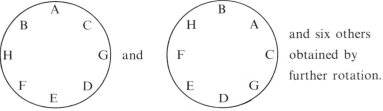

and six others obtained by further rotation.

The correct answer is therefore $\dfrac{8!}{8} = 7!$

Qu.20 Let S be the set of 8! arrangements of 8 children on a stationary roundabout. Let R be a relation (see Section 1.7) defined on S thus: let a, $b \in S$, then aRb if and only if the arrangement a can be turned into b merely by rotating the roundabout.
(i) Show that R is an equivalence relation.
(ii) State the number of elements in each equivalence class.
(iii) Verify that in Example 7 we are in effect required to count the equivalence classes.
(iv) Deduce the answer to Example 7.

Qu.21 Verify that Example 7 may be solved by counting only those arrangements in which one particular child (A say) is in the 'North' seat.

In problems involving circular necklaces of beads, it must be remembered that we can turn the necklace over, thus reducing the number of genuinely distinct arrangements even further. (We assume that the necklace has no knot or clasp which could be taken as a reference point.)

Qu.22 Four beads coloured red, blue, yellow and green are used to make a circular necklace. How many necklaces can be made using all four beads? (It may help to find (i) the number of apparently different necklaces if no movement is allowed, and (ii) the number of necklaces that become equivalent when they may be rotated or turned over.)

Qu.23 Draw the three necklaces in Qu.22, and verify that they may be characterized by specifying which of the three other colours is not adjacent to the red.

Qu.24 Four beads of different colours are threaded on a string which is knotted at each end with identical knots. Find the number of different strings that can be made.

Problems involving necklaces where some of the beads are indistinguishable are generally quite difficult, and no standard method is practicable.

Qu.25 Find, by drawing them all, the number of different circular necklaces that can be made using 3 black and 2 white beads.

Exercise 9b

1. How many different arrangements are there of all the letters in
   (i) HIDDEN;   (ii) RARER;
   (iii) COMMITTEE;   (iv) SUCCESSES?

2. In how many of the arrangements in question 1(i) are the two D's together?

3. How many different numbers may be formed using all the digits in 11222345? How many of these will be odd?

4. How many 'towers' can a child build with 4 red bricks, 3 green and 2 blue if all the bricks must be used?

5. Every year, five schools each send three athletes to compete against one another in a race. The results sheet shows only which positions each school obtained (i.e. names of individuals do not appear). How many different results sheets are possible?

6. How many arrangements are there of the letters $AABBCDE$ if
   (i) there are no constraints;
   (ii) the A's must be together;
   (iii) the A's must be together and so must the B's;
   (iv) either the A's are together or the B's are together or both;
   (v) the A's are separated and so are the B's. [*Hint*: use two of the previous answers.]

7. A string of beads consists of 8 different beads threaded on a string which is knotted at each end. How many distinguishable strings can be made?

8. Repeat question 7 in the case where three of the beads are identical.

9. A roundabout at a fair has 8 identical horses. In how many ways can seven children sit on it?

10. Repeat question 9 for
    (i) 6 children; (ii) 3 children; (iii) 1 child.

11. In how many different ways can 3 girls and 5 boys sit on a roundabout with 8 seats if
    (i) the girls all sit together;
    (ii) no two girls sit together?

9.7 Combinations   The word *combination*, as opposed to permutation, indicates that a subset is to be *chosen* from a set of objects, rather than arranged in some order.

Example 8
Three men are to be chosen from seven to form a committee. In how many ways can this be done? (Contrast this with Example 3 in which three runners were to be put in order; no order is required in this question.)

Solution

Call the men A, B, ... G, and suppose that we *are* required to put them in order. This can be done in $^7P_3 = 7 \times 6 \times 5 = 210$ ways.

If we now ignore the order of the three men, we find that we have counted each selection six times since, for example, we have counted BDG and DGB (and four other arrangements) separately. The correct number of choices is thus only $\frac{210}{6} = 35$.

Qu.26   Find the number of choices of 3 men from 5, and check by listing them all.

Let us now generalize. Suppose we wish to choose r objects from n. We can *arrange* r from n in ^nP_r ways; but each *combination* now appears $r!$ times. So the number of choices is $\dfrac{^nP_r}{r!} = \dfrac{n!}{r!(n-r)!}$

Notation
The number of ways of choosing r objects from n distinguishable objects is denoted by ^nC_r or $\dbinom{n}{r}$. We have shown above that

$$^nC_r = \frac{n!}{r!(n-r)!}$$

Note

(1) ^nC_r may be written $\dfrac{n(n-1)(n-2)\ldots(n-r+1)}{1.2.3\ldots r}$ (after cancelling the $(n-r)!$). The quickest way of evaluating ^nC_r in particular cases is therefore:

    for the numerator, write down r numbers starting at n and decreasing;

    for the denominator, write down r numbers starting at 1 and increasing.

For example, $^{25}C_4 = \dfrac{25.24.23.22}{1.2.3.4} = \text{etc.}$

(2) The number of ways of choosing no objects from n ought to be 1. This is indeed the case if we remember the convention that $0! = 1$. Similarly $^nC_n = 1$.

Theorem 9.1

$$^nC_r = {}^nC_{n-r}.$$

(So, for example, $^{12}C_8 = {}^{12}C_4$ and $^{10}C_7 = {}^{10}C_3$.)

First proof

We know that $^nC_r = \dfrac{n!}{r!(n-r)!}$. The value of $^nC_{n-r}$ may be obtained by replacing r by $n-r$ in this expression. So

$$^nC_{n-r} = \frac{n!}{(n-r)!\{n-(n-r)\}!} = \frac{n!}{(n-r)!r!} = {}^nC_r.$$

Second proof

If we wish to choose $(n-r)$ objects from n, we may do this by selecting the r that are *not* to be chosen. Thus to each subset of size $(n-r)$, there corresponds a subset of size r, namely its complement. So the number of ways of choosing $(n-r)$ from n is the same as the number of ways of choosing r from n.

Qu.27 List all the subsets of $\{A, B, C, D, E\}$ having two elements, and also all subsets with three elements. Indicate clearly the correspondence between the two lists suggested by the second proof above.

Qu.28 Each of n light bulbs in a row may be on or off.

(i) In how many ways may the row be lit?

(ii) In how many of these will exactly r bulbs be on?

(iii) By considering the fact that the number of bulbs that are on must be 0 or 1 or 2 ... or n, show that

$$\sum_{r=0}^{n} {}^nC_r = 2^n.$$

Example 9

Three boys and two girls are to be chosen from a class of ten boys and twelve girls. In how many ways can this be done?

Solution

The boys may be chosen in $^{10}C_3$ ways, and the girls in $^{12}C_2$ ways. So, by the multiplicative rule, the answer is

$$^{10}C_3 \times {}^{12}C_2 = \frac{10.9.8}{1.2.3} \times \frac{12.11}{1.2} = 7920 \text{ ways.}$$

Exercise 9c

1. Evaluate
   (i) $^5C_2$; (ii) $^5C_3$; (iii) $^{10}C_9$; (iv) $^8C_4$; (v) $^{100}C_{98}$.

2. Rewrite without using the ^nC_r notation, and simplify
   (i) $^nC_2$; (ii) $^nC_{n-1}$; (iii) $^{n+1}C_{n-1}$; (iv) $^{n+3}C_n$; (v) $^nC_0$.

3. In how many ways can 8 tenors be split into 5 'firsts' and 3 'seconds'.

4. In how many ways can a committee of 3 men and 4 women be chosen from 6 men and 5 women?

5. A cricket team is to comprise 1 wicket-keeper, 4 bowlers and 6

batsmen. How many different teams could be selected from 3 wicket-keepers, 6 bowlers and 10 batsmen?

6. Repeat question 3 if one of the 8 can only sing 'second'.

7. Three children are to be chosen from 24. In how many ways can this be done? How many of these choices will not include John Smith.

8. In how many ways can 7 children be split into two teams, one of 3 and one of 4, if two particular children insist on being in the same team?

9. (i) Ten points lie in a plane with no three in line. How many lines can be drawn passing through two of the points?
   (ii) If no four of the points are concyclic, how many circles can be drawn through three of the points?

10. In an experiment to test a new drug, 12 of the 24 volunteers are given the new drug, 8 are given an old drug, and 4 are given dummy tablets. In how many ways can this be done?

11. How many different bridge hands (13 cards) containing exactly 6 spades can be dealt?

9.8 Further examples

The next three examples need careful thought if wrong methods are to be avoided.

Example 10

A committee of 4 people is to be chosen from 5 men and 4 women. In how many ways can this be done if at least 2 women must be chosen?

Solution

We consider separately the cases in which there are exactly two women chosen, exactly three, and exactly four.

If we choose exactly 2 women, then we must choose two men. This gives $^4C_2 \times {}^5C_2 = 6 \times 10 = 60$ ways.

Choosing 3 women and 1 man gives $^4C_3 \times {}^5C_1 = 20$ ways; and choosing 4 women can be done in just one way. So in all there are $60 + 20 + 1 = 81$ ways.

Qu.29 Criticize the following method for Example 10. Choose two women first ($^4C_2$ ways); then choose any two people from the remaining seven ($^7C_2$ ways). The answer is therefore $^4C_2 \times {}^7C_2 = 6 \times 21 = 126$ ways. [*Hint*: consider why a selection such as $M_2W_1W_2W_4$ would be counted more than once by this method.]

Example 11

Six boys are to be split into 3 pairs for judo practice. In how many ways can this be done?

Solution

Name the boys A to F. We can choose A's partner in five ways. The partner of any one of the remaining four boys can be chosen in three ways, and this completes the pairing. Thus there are $5 \times 3 = 15$ ways.

Qu.30 (i) What is wrong with the following solution to Example 11? Choose a pair of boys ($^6C_2$ ways), then choose a pair from the remaing four ($^4C_2$ ways). Answer $^6C_2 \times {}^4C_2 = 90$ ways.

(ii) Explain how this argument can be adjusted to give a valid method.

Example 12

A bag contains 8 balls. Three are white and the others are different colours. How many choices of 4 balls can be made?

Solution

There is no quick method: we must consider separately the cases in which three, two, one or no whites are chosen.

Three whites and one colour may be chosen in $1 \times 5 = 5$ ways.

Two whites and two colours may .be chosen in $1 \times {}^5C_2 = 10$ ways.

One white and three colours may be chosen in $1 \times {}^5C_3 = 10$ ways.

Four colours may be chosen in ${}^5C_4 = 5$ ways.

In all there are 30 ways.

We end this section with an important problem (Example 13) to which there are two standard and elegant solutions. It involves putting *indistinguishable* balls into *distinguishable* boxes. Note that the simplest way to make boxes distinguishable is to put them (or imagine them) in a line, so that they become 'numbered' by their position in the line.

Qu.31   In how many ways can r *distinguishable* balls be put into n distinguishable boxes? (You are allowed to put more than one ball in a box.)

Example 13

In how many ways can r indistinguishable balls be put into n distinguishable boxes? (Try this before reading on.)

First solution

A particular way of distributing the balls may be illustrated by a diagram of r balls and $(n-1)$ lines dividing the balls into n sections. Thus in Fig. 9.2, there are 3 balls in box 1, none in box 2, 1 in box 3, and so on. The problem is therefore identical to that of arranging r balls and $(n-1)$ lines. This may be done in ${}^{r+n-1}C_r$ ways.

Figure 9.2

Second solution (Hard.)

Consider the coefficient of t^r in the expansion of

$$(1+t+t^2+t^3+\ldots)(1+t+t^2+t^3+\ldots)\ldots(1+t+t^2+t^3+\ldots) \quad (1)$$

where there are n brackets. A term t^r can be obtained by multiplying t^{p_1} from the first bracket, t^{p_2} from the second bracket, and so on, provided that $\sum_{i=1}^{n} p_i = r$. We may let this term correspond to putting p_1 balls in box 1, p_2 in box 2 etc.. The total number of such distributions of balls is measured by the number of ways of obtaining a term t^r, i.e. by the coefficient of t^r.

Now (1) above may be written $\left(\dfrac{1}{1-t}\right)^n$ where we regard the 'dummy' t as satisfying $|t| < 1$. We therefore seek the coefficient of t^r in the expansion of $(1-t)^{-n}$. This may be found most simply using the Binomial Series (Chapter 28), but the following method gives the required answer.

$$(1-t)^{-1} = \sum_{k=0}^{\infty} t^k.$$

Differentiating $n-1$ times with respect to t,

$$(n-1)! \, (1-t)^{-n} = \sum_{k=n-1}^{\infty} \frac{k!}{(k-n+1)!} t^{k-n+1}$$

$$\Rightarrow (1-t)^{-n} = \frac{1}{(n-1)!} \sum_{k=n-1}^{\infty} \frac{k!}{(k-n+1)!} t^{k-n+1}.$$

Now we wish to pick out the term in t^r from the R.H.S.. This can be done by putting $k = r + n - 1$. We get the term

$$\frac{1}{(n-1)!} \frac{(r+n-1)!}{r!} t^r = {}^{r+n-1}C_r t^r.$$

So the number of ways is ${}^{r+n-1}C_r$.

Qu.32 In how many ways can 5 identical balls be put into 3 numbered boxes?

Exercise 9d (Harder questions)

1. How many bridge hands contain a six-card suit (i.e. exactly six, not six or more)? Explain why the answer is not simply four times the answer to Exercise 9c, question 11.
2. How many four-digit numbers can be made from a selection of 11223456?
3. Three married couples enter a railway compartment in which four seats face four others. In how many ways can they sit down if
   (i) each man sits opposite his wife;
   (ii) no man sits opposite his wife?
4. How many distinguishable cubes are there in which each face is
   (i) either black or white;
   (ii) black, white or red?
5. How many distinguishable necklaces are there with
   (i) 2 red and 4 other beads of different colours;
   (ii) 2 red, 2 green and 2 other different colours;
   (iii) 2 red and 4 green?
6. Explain why the number of ways in which 10 boys can be split into two teams of 5 is not ${}^{10}C_5$, whereas they can be split into a team of 4 and a team of 6 in ${}^{10}C_4$ ways.
7. A committee of 7 is to be chosen from 5 men and 6 women. In how many ways can it be done if there must be at least 3 men and at least 2 women?
8. How many choices of 3 letters can be made from
   (i) $AABBCDE$;   (ii) $AAABCDE$?
9. In how many ways can the ten volumes of an encyclopaedia be rearranged so that they are all in the correct position except for

(i) 1 volume;   (ii) 2 volumes;

(iii) 3 volumes;   (iv) 4 volumes?

10. In how many ways can 5 people be chosen from 6 men and 5 women if at least 1 man must be chosen? (Use the quickest method.)

11. (i) A class of n children is to be split into three groups: p are to do music, q art, and r games (where $p+q+r=n$). By first choosing those who are to do music, show that the number of ways of dividing the class is $\dfrac{n!}{p!\,q!\,r!}$.

(ii) Explain why this answer is the same as the number of ways of arranging p red, q white and r black balls.

12. In how many ways can 5 indistinguishable balls be put into 3 indistinguishable boxes? (Listing is the best way.)

13. There are n people at a party. If everybody knows precisely three other people, show that n is even. (Assume that acquaintanceship is mutual, i.e. A knows $B \Leftrightarrow B$ knows A.)

14. (i) 2^n teams take part in a knock-out competition. The complete draw (for all rounds) is made before any matches are played. Find the number of distinct draws. (The order in which two teams appear is to be disregarded, i.e. there is no 'home' team.)

(ii) Show that the answer to (i) is always odd.

9.9 Vandermonde's Theorem and Pascal's Triangle

Theorem 9.2   (Vandermonde)

$$^{n}C_r = {}^{n-1}C_r + {}^{n-1}C_{r-1}.$$

First proof

$$\text{R.H.S.} = \frac{(n-1)!}{r!\,(n-r-1)!} + \frac{(n-1)!}{(r-1)!\,(n-r)!}$$

$$= \frac{(n-1)!\,(n-r) + (n-1)!\,r}{r!\,(n-r)!}$$

$$= \frac{(n-1)!\,n}{r!\,(n-r)!}$$

$$= \frac{n!}{r!\,(n-r)!} = {}^{n}C_r = \text{L.H.S.}$$

Second proof

Suppose we have n objects and wish to choose r. Call one of the objects X. If we do not choose X, then the r must be chosen from the remaining $n-1$. This can be done in $^{n-1}C_r$ ways. If we do choose X, then we must choose $r-1$ from the $n-1$. This can be done in $^{n-1}C_{r-1}$ ways. Together these give all choices of r from n. So

$$^{n}C_r = {}^{n-1}C_r + {}^{n-1}C_{r-1}.$$

Qu.33 (i) Extend the second proof above by considering whether the choice of r objects includes neither, one, or both of two particular objects X and Y, to show that

$$^nC_r = {}^{n-2}C_r + 2({}^{n-2}C_{r-1}) + {}^{n-2}C_{r-2}$$

provided $r \geqslant 2$.

(ii) Extend it further by partitioning the n objects into subsets of size k and $n-k$. Deduce that

$$^nC_r = {}^kC_r + {}^{n-k}C_1{}^kC_{r-1} + {}^{n-k}C_2{}^kC_{r-2} + \ldots + {}^{n-k}C_r$$

(where we adopt the convention that $^aC_b = 0$ if $b > a$).

Vandermonde's Theorem has one very important application which enables us to obtain very rapidly a large number of values of ^nC_r. Consider a triangle of number thus:

$$^0C_0$$
$$^1C_0 \qquad ^1C_1$$
$$^2C_0 \qquad ^2C_1 \qquad ^2C_2$$
$$^3C_0 \qquad ^3C_1 \qquad ^3C_2 \qquad ^3C_3$$
$$^4C_0 \qquad ^4C_1 \qquad ^4C_2 \qquad ^4C_3 \qquad ^4C_4$$
$$^5C_0 \qquad ^5C_1 \qquad ^5C_2 \qquad ^5C_3 \qquad ^5C_4 \qquad ^5C_5$$

By Vandermonde's Theorem, each number in the triangle is the sum of the two numbers just above it. So each row can be constructed from the row above (together with the knowledge that there is a 1 at each end). We therefore have the triangle

$$
\begin{array}{ccccccccccc}
 & & & & & 1 & & & & & \\
 & & & & 1 & & 1 & & & & \\
 & & & 1 & & 2 & & 1 & & & \\
 & & 1 & & 3 & & 3 & & 1 & & \\
 & 1 & & 4 & & 6 & & 4 & & 1 & \\
1 & & 5 & & 10 & & 10 & & 5 & & 1 \\
\end{array}
$$

This is known as Pascal's Triangle. It is particularly useful when a whole row of numbers is required, as in the Binomial Theorem below. To find a single value of ^nC_r, it is quicker to use direct calculation.

Qu.34 Write down the next three rows of Pascal's Triangle.

9.10 The Binomial Theorem

Consider the problem of expanding $(a+b)^n$ where $n \in Z^+$. We know that

$$(a+b)^2 = a^2 + 2ab + b^2,$$

and by continued multiplication,

$$(a+b)^3 = a^3 + 3a^2b + 3ab^2 + b^3$$

and $(a+b)^4 = a^4 + 4a^3b + 6a^2b^2 + 4ab^3 + b^4$

and so on. Clearly the process becomes tedious for large indices; the Binomial Theorem enables such expansions to be found fairly easily.

Theorem 9.3. (The Binomial Theorem.)

For $n \in Z^+$,

$$(a+b)^n = a^n + {}^nC_1 a^{n-1}b + {}^nC_2 a^{n-2}b^2 + \ldots + {}^nC_{n-1} ab^{n-1} + b^n.$$

(For the sake of neatness, the coefficients of the terms a^n and b^n could be written ${}^nC_0$ and ${}^nC_n$ respectively.)

Proof

Note first that when a number of brackets are to be multiplied together, the terms obtained are precisely those that can be obtained by multiplying together a choice of one term from each bracket.

Now

$$(a+b)^n = (a+b)(a+b) \ldots (a+b) \text{ with } n \text{ brackets.}$$

Selecting the a from each bracket given a^n. Selecting b from one bracket and a from each of the others gives a term $a^{n-1}b$. But there are ${}^nC_1$ ways of choosing which bracket is to provide the b. So in all we have ${}^nC_1$ such terms, i.e. ${}^nC_1 a^{n-1}b$.

Continuing the process, the general term is obtained by choosing b from r brackets and a from the other $n-r$. This can be done in ${}^nC_r$ ways, giving the term ${}^nC_r a^{n-r}b^r$. Finally we note that, proceeding in this way, we do include every possible term. Hence the result.

Qu.35 Use the Binomial Theorem to write down the expansion of $(x+y)^5$, obtaining the coefficients from the appropriate row of Pascal's Triangle. Check your answer by multiplying this out.

Qu.36 Prove the Binomial Theorem by induction. [*Hint*: consider $(a+b)^{k+1} = (a+b)(a+b)^k$. Use the inductive hypothesis to expand $(a+b)^k$, multiply, collect terms and use Vandermonde's Theorem.]

Example 14

Expand $\left(2 - \dfrac{x}{4}\right)^{10}$ as far as the term in x^3.

Solution

Thinking of this as $\left(2 + \left(-\dfrac{x}{4}\right)\right)^{10}$ we have

$$2^{10} + 10 \cdot 2^9 \left(-\frac{x}{4}\right) + 45 \cdot 2^8 \left(-\frac{x}{4}\right)^2 + 120 \cdot 2^7 \left(-\frac{x}{4}\right)^3 + \ldots$$

$$= 1024 - 1280x + 720x^2 - 240x^3 + \ldots.$$

(Note that the signs alternate.)

Example 15

Evaluate $(1 \cdot 998)^6$ to 6 decimal places.

Solution

$$(1.988)^6 = (2 - 0.002)^6$$
$$= 2^6 - 6.2^5(0.002) + 15.2^4(0.002)^2 - 20.2^3(0.002)^3 + \ldots$$
$$= 64 - 0.384 + 0.00096 - 0.00000128 + \ldots$$

(where the next term clearly does not affect the sixth decimal place)

$$= 63.616959 \text{ (6 d.p.)}.$$

Example 16

Find the coefficient of y^3 in the expansion of $(2 - y)(1 + 3y)^7$.

Solution

The term in y^3 arises in two parts: by taking the 2 in the first bracket with the y^3 term in $(1 + 3y)^7$, and by taking the $-y$ in the first bracket with the y^2 term in $(1 + 3y)^7$. Now

$$(2 - y)(1 + 3y)^7 = (2 - y)(\ldots + {}^7C_2(3y)^2 + {}^7C_3(3y)^3 + \ldots)$$
$$= (2 - y)(\ldots + 189y^2 + 945y^3 + \ldots).$$

The term in y^3 is therefore

$$(2 \times 945 - 189)y^3 = 1701y^3.$$

Qu.37 Find the first four terms (i.e. up to the term in x^3) in the expansion of $(1 + x - x^2)^5$ in ascending powers of x. [*Hint*: start by writing it as $\{1 + (x - x^2)\}^5$.]

Qu.38 (i) By considering $(1 + 1)^n$ show that $\displaystyle\sum_{r=0}^{n} {}^nC_r = 2^n$.

(See also Qu.28.)

(ii) By a similar method, show that

$${}^nC_0 - {}^nC_1 + {}^nC_2 - {}^nC_3 + \ldots + (-1)^n \, {}^nC_n = 0.$$

(iii) Use the results of (i) and (ii) to show that

$${}^nC_0 + {}^nC_2 + {}^nC_4 + \ldots = 2^{n-1}$$

where the last term on the L.H.S. is ${}^nC_{n-1}$ if n is odd and ${}^nC_n$ if n is even.

Exercise 9e

1. Use the Binomial Theorem to expand
   (i) $(x + y)^6$;   (ii) $(x - y)^4$;   (iii) $(x + 2)^3$;   (iv) $(2x + 1)^5$;

   (v) $\left(2x - \dfrac{y}{2}\right)^4$;   (vi) $(2 - 3x^2)^5$;   (vii) $\left(x - \dfrac{1}{x}\right)^8$;   (viii) $\left(x - \dfrac{3}{x}\right)^6$.

2. Expand the following in ascending powers of x as far as the term in x^4:

   (i) $(1 + x)^{10}$;   (ii) $\left(2 - \dfrac{x}{4}\right)^8$;   (iii) $(1 + 3x^2)^{40}$.

3. Use the Binomial Theorem to evaluate
   (i) $(1.02)^4$ exactly;   (ii) $(1.002)^{10}$ to 7 d.p.;
   (iii) $(0.998)^{10}$ to 7 d.p.

4. Evaluate

(i) $(1{\cdot}99)^5$ exactly; (ii) $(3{\cdot}0002)^5$ to 10 d.p.;

(iii) $(2{\cdot}003)^5 - (1{\cdot}997)^5$ to 8 d.p.

5. Find the coefficient of

(i) x^3 in the expansion of $\left(3-\dfrac{x}{9}\right)^{10}$;

(ii) t^4 in the expansion of $\left(t+\dfrac{2}{t}\right)^{12}$;

(iii) t^{-3} in the expansion of $\left(4t-\dfrac{1}{2t}\right)^{9}$;

(iv) p^2 in the expansion of $\left(p-\dfrac{3}{p}\right)^{7}$;

(v) t^3 in the expansion of $(1-t)^{10}(1+2t)$.

6. Find the constant terms in the expansions of

(i) $\left(2y+\dfrac{1}{3y}\right)^{6}$; (ii) $\left(r^2-\dfrac{2}{r^3}\right)^{10}$.

7. Expand the following as far as the term in x^3 (in ascending powers):

(i) $(1-x^2)^{10}$; (ii) $\left(1-\dfrac{x}{2}\right)^{5}\left(1+\dfrac{x}{2}\right)^{5}$ (think first);

(iii) $(1+x)(1-2x)^6$; (iv) $(x-2)(3+x)^4$.

8. Write down the first three terms in the expansion of $(2a-5b)^8$. By substituting values for a and b, estimate the value of $(19{\cdot}95)^8$.

9. Given that the first two terms of the expansion of $(1+ax)^5 + (1-ax)^5$ are $2+5x^2$, find the two possible values of a.

10. Expand the following as far as the term in x^3:

(i) $(1+x+x^2)^7$; (ii) $(1-x+2x^2)^5$; (iii) $(2-3x-x^2)^4$.

11. Expand and simplify, leaving in surd form:

(i) $(1+\sqrt{2})^4$; (ii) $(\sqrt{2}-3)^3$; (iii) $(\sqrt{5}-1)^4$;

(iv) $\left(\sqrt{2}-\dfrac{1}{\sqrt{2}}\right)^{5}$; (v) $(1+w)^6$ where $w=\sqrt[3]{2}$.

12. The first three terms in the expansion of $(1+px+qx^2)^8$ are $1-12x+47x^2$. Find p and q and the next term in the expansion.

13. Let $(5+2x)^{10}=a_0+a_1x+a_2x^2+\ldots+a_{10}x^{10}$. Write down an expression for a_r in terms of r. Show that $\dfrac{a_{r+1}}{a_r}=\dfrac{2(10-r)}{5(r+1)}$ and that the smallest integral value of r for which this expression is less than unity is $r=3$. Which is the largest of the a_i and what is its value?

14. Use the method of question 13 to find the largest coefficient in the expansion of $(3x+2)^{12}$. (Leave powers in your answer.)

15. Use Vandermonde's Theorem twice to show that

$$\binom{n}{r}=\binom{n-2}{r}+2\binom{n-2}{r-1}+\binom{n-2}{r-2}.$$

16. Show that $^{2n}C_n = \sum_{r=0}^{n} (^nC_r)^2$.

17. Show that

(i) $^nC_r(n-r) = n(^{n-1}C_r)$;

(ii) $^{n-r}C_{i-r} \times {^nC_r} = {^nC_i} \times {^iC_r}$.

18. (i) Write down an expression for the value of £100 after it has been invested at 4% compound interest for 10 years.

(ii) Use the Binomial Theorem to evaluate this amount to the nearest pound.

19. Find the coefficient of x^2y^3 in the expansion of $(1+x+2y)^8$.

20. Write out the first few terms in the expansion of $(x+h)^n$ where $n \in Z^+$. Hence differentiate x^n from first principles in the case $n \in Z^+$.

Miscellaneous exercise 9

1. Obtain the binomial expansion of $(1+2x)^7$ in ascending powers of x, as far as the term in x^3.

If $(1+2x)^7(a+bx+cx^2)$, where a, b, c are constants, is expanded in ascending powers of x, the first three terms are $1+11x+44x^2$. Calculate the values of a, b and c.

[AEB (O) '80]

2. (a) How many *odd* numbers, greater than 500 000, can be made from the digits 2, 3, 4, 5, 6, 7, without repetitions? How many *odd* numbers, less than 600 000, can be made from these digits without repetitions?

(b) Ten beads of different colours are arranged on a ring. If no two rings are the same, what is the greatest number of possible arrangements? (A ring can be turned over.)

[SUJB (O)]

3. (a) Write down and simplify the coefficient of x^4 in the expansion of $(3-2x)^7$.

(b) Expand $(1-x)^{10} - (1-2x)^5$ as far as x^3 and show that the result is $5x^2 - 40x^3$.

Hence, find the value of $(0.99)^{10} - (0.98)^5$ correct to **four** decimal places.

[W (O)]

4. One representative from each of four Western countries attends a conference with the Sheikh of each of four oil-producing Middle East countries.

At the opening session the Western delegates were invited to choose on which side of the square table they would sit, and the Sheikhs then sat opposite them. In how many different ways can the eight delegates be seated?

In how many different ways can the delegates to this conference be split into two sub-committees, if each sub-committee is made up of two delegates from each group?

[MEI (O)]

5. A double-decker bus arrives at a bus stop with 10 vacant seats, 6 of which are downstairs and the remaining 4 upstairs. 5 people are waiting at the stop to get on. Those already on the bus remain in their seats.

(i) Find the number of different possible seating arrangements of the 5 new passengers.

(ii) Given that all 5 must sit downstairs find the number of different seating arrangements which are now possible.

(iii) Given that of the 5 new passengers 2 particular people must sit downstairs and one particular person must sit upstairs, find the number of different possible seating arrangements.

Later on, during the journey, the bus has on it 15 passengers. The bus company wishes to ask a sample of 5 of the passengers questions about their travelling habits.

(iv) Find the number of ways in which such a sample can be chosen.

(v) Given that of the 15 passengers 9 are downstairs and 6 are upstairs, and that at least 3 of the sample of 5 must be downstairs, find the number of ways of choosing the sample.

[JMB (O)]

6. (a) Factorise completely

$$(x+y)^6 - (x-y)^6.$$

(b) Find the number of ways in which n boys $(n > 3)$ can be arranged if two given boys are always to be separated,

(i) when the boys stand in a straight line,

(ii) when the boys stand in a circle, clockwise and anti-clockwise arrangements being regarded as different.

[C]

7. (i) Ten school-children form a 'crocodile' for church attendance, i.e. they walk in pairs along the road. Find the number of ways of forming the crocodile,

(a) if order *within* the pairs does not matter, but if the order *of* the pairs is important,

(b) if neither order *of* the pairs nor order *within* the pairs matters.

(ii) Prove that, if p and n are positive integers, then

$$\frac{(pn)!}{(p!)^n(n!)}$$

is an integer.

[C (S)]

10 Polynomials, identities etc.

10.1 Definitions   A *polynomial* is an expression of the form

$$a_n x^n + a_{n-1} x^{n-1} + \ldots + a_2 x^2 + a_1 x + a_0$$

where the a_i are real† numbers called *coefficients*.

The *degree* of a polynomial is the largest value of r such that $a_r \neq 0$; i.e. it is the highest power of x that occurs. The coefficient corresponding to this highest power is called the *leading coefficient*.

Polynomials of degree 0, 1, 2, 3, 4 and 5 are often referred to as constant, linear, quadratic, cubic, quartic and quintic polynomials respectively. The *zero polynomial* is the constant 0; its degree is not defined.

Polynomials may be denoted by capital letters, e.g. $P(x)$; and other letters may be used in place of x as 'dummies'. Thus $2t^2 + t - 5$ is a quadratic polynomial in t. It is assumed that the reader is familiar with the elementary algebra of polynomials, namely their addition and multiplication.

Qu.1   Let $P(x)$ be $2x^2 + 3x - 4$ and $Q(x)$ be $5 - 3x$. Write down
(i) the polynomial $2P(x)$;
(ii) the degree of $P(x)$;
(iii) the leading coefficient of $Q(x)$;
(iv) the degree of the product $P(x)Q(x)$;
(v) the leading coefficient of $P(x)Q(x)$.

Qu.2   (i) Let $P(x)$ and $Q(x)$ have degrees m and n respectively. What is the degree of $P(x)Q(x)$?
(ii) Show that the result of (i) would not continue to be true if a value were given to the degree of the zero polynomial.
(iii) If, in (i), $m > n$, what is the degree of $P(x) + Q(x)$?

10.2 Identical polynomials   **Definition**
Two polynomials $P(x)$ and $Q(x)$ are said to be *identical* or *identically equal* if their corresponding coefficients are equal. We then write $P(x) \equiv Q(x)$. So

$$a_0 + a_1 x + a_2 x^2 + \ldots + a_n x^n$$
$$\equiv b_0 + b_1 x + b_2 x^2 + \ldots + b_n x^n$$

if and only if $a_0 = b_0$, $a_1 = b_1$, etc.. Clearly two identical polynomials have the same degree.

---

† More generally, the coefficients may be complex numbers (Chapter 31) or members of any commutative ring (Chapter 33); but we shall consider only *real* polynomials.

Example 1

If

$$a(x-1)(x-2)+b(x-3)+c \equiv 3x^2+4x-1$$

find a, b and c.

Solution

(An alternative method is given after Theorem 10.1.) Expanding the L.H.S. and collecting terms,

$$ax^2+(b-3a)x+(2a-3b+c) \equiv 3x^2+4x-1.$$

Equating corresponding coefficients, we get the three equations

$$a=3$$
$$b-3a=4$$
$$2a-3b+c=-1.$$

This set of simultaneous equations gives the solution $a=3$, $b=13$, $c=32$.

Qu.3 If $(x+b)(x-3) \equiv ax^2+5x+c$, find a, b, c.

Qu.4 If $p(x-2)(x+3) \equiv qx+r$, find p, q, r.

We may regard a polynomial $P(x)$ as a function of x, so that the notation $P(2)$ means the value of $P(x)$ when $x=2$.

Theorem 10.1

$$P(x) \equiv Q(x) \Rightarrow \forall\, a \in R,\ P(a)=Q(a).$$

(Note that we must not write $P(a) \equiv Q(a)$: $P(a)$ and $Q(a)$ are numbers, not polynomials.)

The theorem says that when any number is substituted into each of two identical polynomials, the same value is obtained. This follows directly from the definition of identical polynomials.

Qu.5 State in words the converse of Theorem 10.1. Do you think it is true? (This is answered later in this chapter by Theorem 10.6.)

We can use Theorem 10.1 to provide an alternative solution to Example 1:

$$a(x-1)(x-2)+b(x-3)+c \equiv 3x^2+4x-1.$$

By the theorem, we may substitute any number for x:

$$x=1 \Rightarrow -2b+c=6 \tag{1}$$

$$x=2 \Rightarrow -b+c=19 \tag{2}$$

$$x=3 \Rightarrow 2a+c=38. \tag{3}$$

Solving (1) and (2) simultaneously gives $b=13$, $c=32$, and substituting into (3) gives $a=3$.

In this solution, *any* three values of x could have been substituted; the values 1, 2 and 3 were chosen because, in each case, one of the three unknowns vanishes. Three equations are needed since there are three unknowns. In general, there is no reason why the methods of 'equating coefficients' and 'substituting values' should not be mixed.

Qu.6 If

$$a(x+1)(x-2)+b(x+1)(x+3)+c(x-2)(x+3)\equiv x^2-x+2,$$

find a, b and c by the method of the alternative solution to Example 1 above. [*Hint*: decide which are the most convenient three values to substitute.]

Qu.7 (For discussion.) Write out the best solution to Example 1 that you can find, combining the methods of the two given solutions.

Qu.8 By substituting $x=-2$ into each side, show that it is impossible to find values of a, b and c such that

$$a(x^2+x-2)+b(x+2)+c(x^2-4)\equiv x^2+x+1.$$

(The reader may like to investigate how the method of 'equating coefficients' fails.)

10.3 Algebraic long division

Algebraic long division is a method of dividing one polynomial by another. As with ordinary division, which it closely resembles, there may be a remainder. We shall assume that the two polynomials have each been written in *descending powers of* x.

Example 2

Divide $4x^6-4x^5+x^4+6x^2-16x+2$ by $2x^2-3x+4$.

Solution

In the dividend, the coefficient of x^3 is zero; a gap is left corresponding to this missing term. At each stage, the next term in the quotient (answer) is found by dividing the leading term of the divisor (in this case $2x^2$) into the leading term of what is left of the dividend.

$$
\begin{array}{r}
2x^4+\ x^3-2x^2-\ 5x\ -\tfrac{1}{2}\ \text{rem}\ (\tfrac{5}{2}x+4) \\
2x^2-3x+4\ \overline{\big)\ 4x^6-4x^5+\ x^4\qquad\ +\ 6x^2-16x+2} \\
4x^6-6x^5+8x^4 \\
\hline
2x^5-7x^4 \\
2x^5-3x^4+\ 4x^3 \\
\hline
-4x^4-\ 4x^3+\ 6x^2 \\
-4x^4+\ 6x^3-\ 8x^2 \\
\hline
-10x^3+14x^2-16x \\
-10x^3+15x^2-20x \\
\hline
-x^2+\ 4x+2 \\
-x^2+\ \tfrac{3}{2}x-2 \\
\hline
\tfrac{5}{2}x+4
\end{array}
$$

The division in Example 2 gives us that

$$
\begin{aligned}
&4x^6-4x^5+x^4+6x^2-16x+2\\
&\qquad\equiv(2x^2-3x+4)(2x^4+x^3-2x^2-5x-\tfrac{1}{2})+(\tfrac{5}{2}x+4).
\end{aligned}
$$

In general, if we divide $P(x)$ by $S(x)$ to give a quotient $Q(x)$ and a remainder $R(x)$, then

$$P(x) \equiv S(x)Q(x) + R(x).$$

It is clear from the method of algebraic long division that if $R(x)$ is not the zero polynomial, then the degree of R will be less than the degree of S. More precisely, we have the following theorem, the proof of which is outlined in questions 10 and 11 of Exercise 10a.

Theorem 10.2

Given polynomials $P(x)$ and $S(x)$, where S is not the zero polynomial, then there exist unique polynomials $Q(x)$ and $R(x)$ such that

$$P(x) \equiv S(x)Q(x) + R(x)$$

where either the degree of R is less than the degree of S or $R(x) \equiv 0$. (This is sometimes called the *Euclidean* property of polynomials.)

Note that the proof of Theorem 10.2 outlined in Exercise 10a is in two parts: question 10 shows the existence of Q and R, and question 11 shows their uniqueness. Such a two-part proof is typical of many 'existence and uniqueness' theorems.

Qu.9 Use algebraic long division to find Q and R when

$$P(x) \equiv 2x^6 - x^5 + 3x^4 + 7x^3 + 2x + 3, \text{ and}$$
$$S(x) \equiv x^2 - 2x + 5.$$

It is also possible to divide polynomials when they are expressed in *ascending* powers of x. At each stage, the next term in the quotient is found by examining the terms of *lowest* degree. As we see in the example below, the division may not terminate naturally; it can then be stopped at any stage, and the remainder at that stage stated. If the range of values of x is restricted in such a way that, for those values of x, the remainder tends to zero as the division is continued, then the quotient may be regarded as an infinite series, and the remainder omitted completely. (See Qus. 11 and 12.)

Example 3

Divide $2 + 4x + x^2$ by $1 + 3x$ in ascending powers of x as far as the term in x^3.

Solution

$$
\begin{array}{r}
2 - 2x + 7x^2 - 21x^3 \quad \text{rem}(63x^4) \\
\hline
1+3x \,\big|\, 2 + 4x + x^2 \\
2 + 6x \\
\hline
-2x + x^2 \\
-2x - 6x^2 \\
\hline
7x^2 \\
7x^2 + 21x^3 \\
\hline
-21x^3 \\
-21x^3 - 63x^4 \\
\hline
63x^4 \\
\end{array}
$$

Qu.10 Continue this division two stages further, obtaining the quotient as far as the term in x^5, and stating the remainder at this stage.

Qu.11 (i) Divide the polynomial 1 by $1-x$ as far as the term in x^4. State the remainder.

(ii) What is the remainder when the division is continued as far as the term in x^n?

(iii) For what values of x does the remainder tend to zero as $n \to \infty$?

(iv) Find the sum of the infinite G.P. formed by the quotient (for those values of x in (iii)) and compare your answer with the original question.

Qu.12 Repeat Qu.11 with $P(x) \equiv 2$, $S(x) \equiv 1 + 2x$.

Exercise 10a

1. This question illustrates an efficient method of substituting a value of x into a polynomial.

(i) Use a calculator to work through the flow diagram below with the given data. State the value of S printed. (Assume that each time the instruction 'Read A' is reached, the next value of A is read in.)

Figure 10.1

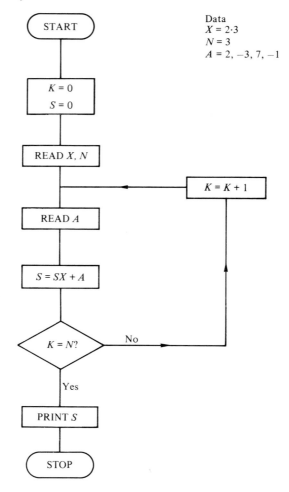

Data
$X = 2 \cdot 3$
$N = 3$
$A = 2, -3, 7, -1$

POLYNOMIALS, IDENTITIES ETC.

(ii) Work through the flow diagram again, leaving X as X (rather than putting $X = 2 \cdot 3$) to obtain a polynomial. State this polynomial.

(iii) It is required to calculate the value of $x^4 - 2x^3 + x^2 - 7x - 9$ when $x = 1 \cdot 2$. State the input values that would be needed in the flow diagram.

(iv) Write down the polynomial in (iii), inserting brackets to show the order in which the calculation is carried out by the method of the flow diagram.

(v) Use this method to calculate the value of the polynomial when $x = 1 \cdot 2$.

2. (i) Find values of a and b such that

$$a(2x - 1) + b(x + 3) \equiv x + 17.$$

(ii) If $a = 3$ and $b = -1$, solve the equation

$$a(2x - 1) + b(x + 3) = x + 17.$$

3. If $\qquad a(x - 1)(x + 2) + b(x - 5) + c(x - 3) \equiv 3x^2 - 2x + 3,$

find a, b and c by
(i) equating coefficients;
(ii) substituting suitable values for x.

4. If $\quad ax^2 + 3x - 4 \equiv bx(x + 1) + c(x - 2)$, find a, b and c.

5. Find out as much as possible about p, q and r if

$$p(x + 3)^2 + q(x + 3) + r \equiv px^2 + qx + r.$$

6. Repeat question 3 for

$$a(x + 2)(x - 1) + b(x - 3) + c \equiv 2x^2 + 3x - 4.$$

7. Divide $P(x)$ by $S(x)$, stating the quotient and remainder, when
(i) $P(x) \equiv 2x^3 + 2x^2 - 7x + 2$, $S(x) \equiv 2x^2 + x - 3$;
(ii) $P(x) \equiv 2x^4 + 5x^3 - 2x^2 - 37x - 28$, $S(x) \equiv x^2 + 4x + 7$;
(iii) $P(x) \equiv x^3 - 4x^2 - 3x + 10$, $S(x) \equiv x - 4$;
(iv) $P(x) \equiv 4x^4 - x - 1$, $S(x) \equiv 2x - 1$.

8. (i) Find, by division, the quotient and remainder when $2x^3 - 3x^2 - 6x - 3$ is divided by $2x + 3$.

(ii) Divide the same polynomials in ascending powers of x as far as the term in x^2. State the quotient and remainder at that stage.

9. Divide $2x^5 - x^4 - 17x^3 + 9x^2 + 27x - 20$ by $2x^2 + 3x - 5$ in
(i) descending powers of x; (ii) ascending powers of x.

10. (This question and question 11 outline a rigorous proof of Theorem 10.2 in the text.) Given polynomials $P(x)$ and $S(x)$, it is certainly possible to find pairs of polynomials $Q(x)$ and $R(x)$ such that

$$P(x) \equiv S(x)Q(x) + R(x). \tag{1}$$

(We could, for example, take $Q \equiv 0$ and $R \equiv P$.)

Now suppose that from all possible pairs of Q and R satisfying (1), we choose a particular pair, $Q_1(x)$ and $R_1(x)$, with $R_1(x)$ having the least possible degree. Suppose nevertheless that deg $R_1(x) \geqslant$ deg $S(x)$. We aim to obtain a contradiction.

Let
$$R_1(x) \equiv \sum_{k=0}^{m} r_k x^k \text{ and } S(x) \equiv \sum_{k=0}^{n} s_k x^k$$

where r_m and s_n are non-zero, and $m \geqslant n$.

Consider
$$Q_2(x) \equiv Q_1(x) + \frac{r_m}{s_n} x^{m-n}$$

and
$$R_2(x) \equiv R_1(x) - \frac{r_m}{s_n} x^{m-n} S(x).$$

Show that (i) Q_2 and R_2 satisfy (1),
(ii) deg $R_2 <$ deg R_1,
thus contradicting the assumption that deg R_1 was the least possible. (It follows that $\exists\, Q$ and R satisfying (1), with deg $R <$ deg S.)

11. (Here we show the *uniqueness* of the Q and R whose existence is established in question 10.) Suppose that there are two distinct pairs Q_1, R_1 and Q_2, R_2 of polynomials satisfying

$$P \equiv SQ + R \text{ with deg } R < \text{deg } S.$$

By subtracting the pair of identities

$$P(x) \equiv S(x)Q_1(x) + R_1(x)$$
and
$$P(x) \equiv S(x)Q_2(x) + R_2(x)$$

and considering the degree of $R_1(x) - R_2(x)$, complete the *reductio ad absurdum* proof.

12. When the polynomial $P(x)$ is divided by $S(x)$, the quotient is $Q(x)$ and the remainder $R(x)$. What will be
(i) the quotient, (ii) the remainder,
when $P(x)$ is divided by $2S(x)$? [*Hint*: it may help to set out the problem in the form of Theorem 10.2 in the chapter.]

13. Show that 125 000 027 and 124 999 973 are not prime. [*Hint*: see Qu.7 of Chapter 4.]

10.4 The Remainder Theorem and its corollaries

Theorem 10.3 (The Remainder Theorem.)
If a polynomial $P(x)$ is divided by the linear polynomial $(x - \alpha)$, then the remainder is the number $P(\alpha)$.

Proof
By Theorem 10.2, we may write

$$P(x) \equiv (x - \alpha)Q(x) + R(x) \qquad (1)$$

where $R(x)$ is some constant. By Theorem 10.1, we may substitute any value of x into (1). Putting $x = \alpha$, we have $P(\alpha) = R$. So R is the number $P(\alpha)$.

Example 4
Find the remainder when $x^3 + 2x^2 - 5x - 4$ is divided by
(i) $(x + 2)$; (ii) $(2x - 1)$.

Solution

(i) By the Remainder Theorem, the remainder may be obtained by substituting $x = -2$. This gives

$$-8 + 8 + 10 - 4 = 6.$$

(ii) Writing $x^3 + 2x^2 - 5x - 4 \equiv (2x - 1)Q(x) + R$ and putting $x = \frac{1}{2}$, we have

$$R = \tfrac{1}{8} + \tfrac{1}{2} - 2\tfrac{1}{2} - 4 = -5\tfrac{7}{8}.$$

Qu.13 Find the remainder when $2x^3 - x^2 - 2x + 4$ is divided by
(i) $x - 3$; (ii) $x + 1$.

Qu.14 Prove that if $P(x)$ is divided by $(\lambda x - \mu)$, the remainder is $P(\mu/\lambda)$. Use this result to solve Example 4(ii).

Qu.15 Find the remainder when the polynomial in Qu.13 is divided by
(i) $2x - 3$; (ii) $3x + 1$.

Example 5

Find the remainder when $x^6 - x^5 + 2$ is divided by
(i) $(x + 1)(x - 2)$; (ii) $(x + 1)^2$.

Solution

(i) The divisor is quadratic, so the remainder will be linear (or constant). A convenient 'trick' is to write this remainder in the form $\lambda(x + 1) + \mu(x - 2)$. Thus

$$x^6 - x^5 + 2 \equiv Q(x)(x + 1)(x - 2) + \lambda(x + 1) + \mu(x - 2).$$

Then
$$x = -1 \Rightarrow \quad 4 = -3\mu$$

and
$$x = 2 \quad \Rightarrow 34 = 3\lambda.$$

So
$$\lambda = \tfrac{34}{3} \text{ and } \mu = -\tfrac{4}{3}.$$

The remainder is therefore

$$\tfrac{34}{3}(x + 1) - \tfrac{4}{3}(x - 2)$$
$$= 10x + 14.$$

(ii) Again the remainder is linear (or constant). We write it in the form $\lambda(x + 1) + \mu$. So

$$x^6 - x^5 + 2 \equiv Q(x)(x + 1)^2 + \lambda(x + 1) + \mu. \tag{1}$$

Putting $x = -1$ gives $\mu = 4$.

Now differentiate (1) and then put $x = -1$:

$$6x^5 - 5x^4 \equiv Q'(x)(x + 1)^2 + 2Q(x)(x + 1) + \lambda$$
$$\Rightarrow \lambda = -11$$

So the remainder is

$$-11(x + 1) + 4$$
$$= -11x - 7.$$

Qu.16 Solve each part of Example 5 by taking the remainder as $ax + b$ (but using essentially the same method.)

Theorem 10.4 (The Factor Theorem.)
Let $P(x)$ be a polynomial and let $\alpha \in R$, then

$$P(\alpha) = 0 \Leftrightarrow (x - \alpha) \text{ is a factor of } P(x).$$

Proof
By the Remainder Theorem,

$P(\alpha) = 0$
$\Leftrightarrow$ the remainder when $P(x)$ is divided by $(x - \alpha)$ is zero
$\Leftrightarrow (x - \alpha)$ is a factor of $P(x)$.

This important relationship between the factors of a polynomial and its roots will already be familiar to the reader from having solved quadratic equations and other polynomial equations (Chapter 6) by factorizing.

Qu.17 Use the 'formula' to solve the quadratic equation $x^2 - 4x - 1 = 0$. Hence express $(x^2 - 4x - 1)$ in the form $(x - \alpha)(x - \beta)$.

Theorem 10.5 (The Extended Factor Theorem.)
If $P(x)$ has distinct† roots $\alpha_1, \alpha_2, \alpha_3, \ldots \alpha_n$, then

$$P(x) \equiv (x - \alpha_1)(x - \alpha_2) \ldots (x - \alpha_n)Q(x)$$

for some polynomial $Q(x)$.
Proof
By the Factor Theorem, since α_1 is a root of $P(x)$,

$$P(x) \equiv (x - \alpha_1)P_1(x)$$

for some polynomial $P_1(x)$. Putting $x = \alpha_2$, we see that

$$P(\alpha_2) = 0 \Rightarrow (\alpha_2 - \alpha_1)P_1(\alpha_2) = 0$$
$$\Rightarrow P_1(\alpha_2) = 0 \text{ (since } \alpha_2 - \alpha_1 \neq 0)$$
$$\Rightarrow P_1(x) \equiv (x - \alpha_2)P_2(x)$$

for some polynomial $P_2(x)$. So

$$P(x) \equiv (x - \alpha_1)(x - \alpha_2)P_2(x).$$

We may extend this process (by induction) until the desired result is obtained, the polynomial $P_n(x)$ being renamed $Q(x)$. (The reader may care to supply the details of the induction.)

Corollary
A polynomial of degree n cannot have more than n distinct roots.
Proof
(We use the method of proof by contradiction.) Suppose that $P(x)$ is of degree n and has distinct roots $\alpha_1, \alpha_2, \ldots \alpha_{n+1}$; then using the first n of these in Theorem 10.5, we have

$$P(x) \equiv (x - \alpha_1)(x - \alpha_2) \ldots (x - \alpha_n)Q(x).$$

But since the degree of $P(x)$ is n, $Q(x)$ must be a constant, i.e.

$$P(x) \equiv a(x - \alpha_1)(x - \alpha_2) \ldots (x - \alpha_n)$$

† 'Distinct' here means that no two of the roots are equal.

for some $a \in R$.

Substituting $x = \alpha_{n+1}$ gives

$$0 = a(\alpha_{n+1} - \alpha_1)(\alpha_{n+1} - \alpha_2) \dots (\alpha_{n+1} - \alpha_n).$$

Since no two of the α_i are equal, it follows that $a = 0$, so that $P(x) \equiv 0$, which contradicts our assumption that the degree of P is n. The required result is thus proved.

We can now prove the converse of Theorem 10.1 of this chapter.

Theorem 10.6 (Converse of Theorem 10.1.)
If $P(x)$ and $Q(x)$ are two polynomials such that, for all $a \in R$, $P(a) = Q(a)$, then $P(x) \equiv Q(x)$. (Remember that we defined the identity sign in terms of equal coefficients.)
Proof
Consider the polynomial $P(x) - Q(x)$; it will take the value zero for all real values of x. In particular, if we suppose that $P(x) - Q(x)$ has degree n, then we can choose $n + 1$ distinct roots and apply the corollary of Theorem 10.5 to obtain a contradiction. So $P(x) - Q(x)$ cannot have degree n for any non-negative integer n; and so $P(x) - Q(x) \equiv 0$, which gives $P(x) \equiv Q(x)$.

Theorem 10.6 and Theorem 10.1 together prove that the two statements

(i) $P(x)$ and $Q(x)$ have the same coefficients,
(ii) $\forall\, a \in R,\ P(a) = Q(a)$

are equivalent. We may now take the sign $\equiv$ to indicate either of the statements.

Qu.18 Prove that if two polynomials $P(x)$ and $Q(x)$, each of degree n, are equal for $n + 1$ distinct values of x, then they are equal for all values of x. [*Hint*: this is almost the same result as Theorem 10.6.]

Example 6
Show that, if no two of a, b, c are equal, then

$$\frac{(x-a)(x-b)}{(c-a)(c-b)} + \frac{(x-a)(x-c)}{(b-a)(b-c)} + \frac{(x-b)(x-c)}{(a-b)(a-c)} \equiv 1.$$

Solution
The L.H.S. of the identity is a polynomial whose degree is at most two; call it $P(x)$. Substituting $x = a$, $x = b$ and $x = c$ gives

$$P(a) = 1, \ P(b) = 1 \text{ and } P(c) = 1.$$

Thus the (at-most-quadratic) polynomial $P(x) - 1$ vanishes for three distinct values of x and so (Qu.18) is identically zero. So

$$P(x) - 1 \equiv 0$$

which gives

$$P(x) \equiv 1.$$

10.5 Further use of the identity sign

The identity sign $\equiv$ is sometimes used for functions which are not polynomials. Clearly the 'equal coefficients' meaning is not then appropriate, and so the 'equal for each value of x' meaning is used instead. It is understood that the sign means that the values of the functions are equal only for those values of x for which they are defined. Thus, for example, we may write

$$1 + \frac{x}{1-x} \equiv \frac{1}{1-x}$$

and it is understood that $x \neq 1$.

Care must be taken not to apply the theorems of this chapter to functions which are not polynomials. It would, for example, be simple enough to find two functions f and g such that $f(x) = g(x)$ for infinitely many values of x, but such that $f(x) \not\equiv g(x)$.

Qu.19 (For discussion.) Find two such functions. [*Hint*: it may help to make one of the functions periodic.]

10.6 Repeated roots

We have seen that a polynomial $P(x)$ has a root α if and only if $(x - \alpha)$ is a factor of $P(x)$. If $(x - \alpha)^2$ is a factor of $P(x)$, then we may say that $(x - \alpha)$ is a *repeated factor* of $P(x)$, and that α is a *repeated root* of $P(x)$. In general, if $P(x)$ may be written in the form

$$P(x) \equiv (x - \alpha)^n Q(x)$$

where $(x - \alpha)$ is not a factor of $Q(x)$, then we say that α is a root of multiplicity n. Thus, for example, the polynomial

$$3x^2(x-1)^3(x-2)$$

has a double root at $x = 0$, a triple root at $x = 1$, and a single root at $x = 2$. The total number of roots of a polynomial (counting multiple roots the appropriate number of times) cannot exceed the degree of the polynomial.

The next theorem gives a necessary and sufficient condition for $P(x)$ to have a repeated root.

Theorem 10.7

α is a repeated root of $P(x) \Leftrightarrow \alpha$ is a root of both $P(x)$ and its derivative $P'(x)$.

(This theorem can, of course, be stated—and is proved below—in terms of factors rather than roots.)

Proof

The proof is in two parts corresponding to the two directions of the implication sign.

(i) $\Rightarrow$

If α is a repeated root of $P(x)$, then

$$P(x) \equiv (x - \alpha)^2 Q(x) \quad \text{for some } Q(x).$$

Differentiating with respect to x by the product rule,

$$P'(x) \equiv 2(x - \alpha)Q(x) + (x - \alpha)^2 Q'(x)$$
$$\equiv (x - \alpha)\{2Q(x) + (x - \alpha)Q'(x)\}.$$

So α is a root of $P'(x)$, and is also certainly a root of $P(x)$.

(ii) $\Leftarrow$

Suppose that α is a root of both $P(x)$ and of $P'(x)$, then

$$P(x) \equiv (x-\alpha)H(x) \qquad \text{for some } H(x).$$

Differentiating,

$$P'(x) \equiv H(x) + (x-\alpha)H'(x)$$
$$\Rightarrow H(x) \equiv P'(x) - (x-\alpha)H'(x).$$

Now both terms on the R.H.S. are divisible by $(x-\alpha)$, so $(x-\alpha)$ is a factor of $H(x)$, i.e. $H(x) \equiv (x-\alpha)G(x)$ for some $G(x)$. Then

$$P(x) \equiv (x-\alpha)H(x) \equiv (x-\alpha)^2 G(x),$$

so that α is a repeated root of $P(x)$.

Qu.20 (For discussion.) Find a counter-example to show that

$$\alpha \text{ is a root of } P'(x) \;\not\Rightarrow\; \alpha \text{ is a repeated root of } P(x).$$

Example 7

Solve $2x^3 - 39x^2 + 72x + 1296 = 0$ given that it has a repeated root.

Solution

By Theorem 10.7, the repeated root is also a root of the derivative of this polynomial, i.e. it is a root of $6x^2 - 78x + 72 \equiv 6(x-1)(x-12)$. So the repeated root is either 1 or 12. By substitution, 1 is not a root at all, so 12 is the double root. The complete factorization is then

$$(x-12)^2(2x+9) = 0,$$

the remaining factor being found from consideration of the x^3 term and the constant term. Thus $x = 12$ or $-\frac{9}{2}$.

Theorem 10.7 may be used to reconcile (at least for polynomial graphs) the two apparently different meanings of a *tangent* to a curve. In Chapter 2 we said that a tangent to the curve $y = f(x)$ at the point P was

(i) a line through P having double (or more) contact at P with the curve.

In Chapter 5 we defined a tangent to be

(ii) the line through P with the same gradient as the curve $y = f(x)$ at P.

We now show that these definitions are equivalent.

Let $f(x)$ be a polynomial, and let $y = g(x)$ be the equation of a straight line meeting the curve $f(x)$ at the point P. (The function $g(x)$ is therefore a polynomial of the form $mx + c$.) Let the x-coordinate of P be α. Since both curves pass through P, $f(\alpha) = g(\alpha)$, and so α is a root of the polynomial $f(x) - g(x)$. Then

the graphs of f and g have double contact at P

$\Leftrightarrow f(x) - g(x)$ has a double root at $x = \alpha$

$\Leftrightarrow \dfrac{d}{dx}\{f(x) - g(x)\}$ has a root at $x = \alpha$

$$\Leftrightarrow f'(\alpha) - g'(\alpha) = 0$$
$$\Leftrightarrow f'(\alpha) = g'(\alpha)$$
$$\Leftrightarrow f \text{ and } g \text{ have the same gradient at } P.$$

This establishes the equivalence of the two possible definitions for polynomial curves.

1. Use the Remainder Theorem to find the remainder when
   (i) $x^3 - 7x^2 - 2x - 4$ is divided by $x - 3$;
   (ii) $x^4 + x^2 + 2$ is divided by $x + 1$;
   (iii) $2x^4 - 3x^3 + x^2 - 7x + 2$ is divided by $x - 2$.

2. Use the Remainder Theorem (as extended in Qu.14 in the chapter) to find the remainder when
   (i) $x^3 + 2x^2 + 3x - 5$ is divided by $2x - 1$;
   (ii) $3x^4 - x^3 - 5x^2 + 3$ is divided by $3x + 2$;
   (iii) $x^4 - 1$ is divided by $10x - 9$.

3. When a polynomial $P(x)$ is divided by $S(x)$, the quotient is $Q(x)$ and the remainder $R(x)$. Find, in terms of $Q(x)$, $R(x)$ and λ, (a) the quotient, and (b) the remainder when
   (i) $P(x)$ is divided by $\lambda S(x)$;
   (ii) $\lambda P(x)$ is divided by $S(x)$.

4. (i) When the polynomial $2x^4 - 7x^3 + 5x^2 - 3x + c$ is divided by $x - 1$, the remainder is 5. Find c.
   (ii) When $x^3 + ax^2 - 3x + 15$ is divided by $x - 2$, the remainder is 1. Find a.

5. When $x^3 + ax^2 + bx - 7$ is divided by $x + 1$, the remainder is -3; when it is divided by $x + 2$, the remainder is 1. Find a and b.

6. By finding roots, factorize the following as far as possible:
   (i) $x^2 - 5x + 3$;   (ii) $x^2 - 3x + 5$;   (iii) $x^3 - 3x^2 - 2x + 4$;
   (iv) $x^3 - 1$;   (v) $x^3 - 2x^2 + x$.

7. Use the method of Example 5 to find the remainder when
   (i) x^{10} is divided by $(x - 1)(x - 2)$;
   (ii) $x^6 - 2x^5$ is divided by $x^2 - 4x + 3$.

8. Use the method of Example 5 to find the remainder when
   (i) $x^7 + 3x^6$ is divided by $(x + 1)^2$;
   (ii) $8x^6 - x^2$ is divided by $4x^2 - 4x + 1$.

9. (Alternative proof of the Remainder Theorem.)
   (i) Verify that if $k \in Z^+$, then
   $$x^k - a^k \equiv (x - a)(x^{k-1} + x^{k-2}a + x^{k-3}a^2 + \dots + a^{k-1}).$$

   (ii) Show that if $P(x) \equiv \sum_{r=0}^{n} c_r x^r$, then
   $$P(x) - P(a) \equiv \sum_{r=0}^{n} c_r(x^r - a^r).$$

   (iii) Show that $P(x) - P(a)$ has a factor of $x - a$, and so may be written in the form $Q(x)(x - a)$. [*Hint*: use parts (i) and (ii).]
   (iv) Deduce the Remainder Theorem.

10. Let $L(x)$ be a linear polynomial, and let a and b be real numbers with $a \neq b$. Suppose that we wish to find λ and μ satisfying
    $$L(x) \equiv \lambda(x - a) + \mu(x - b).$$

(i) By substituting $x = a$ and $x = b$, find expressions for λ and μ.

(ii) Rewrite the identity with λ and μ replaced by these expressions. Explain clearly how you know that the identity is now true for all values of x. [*Hint*: each side is linear.]

(Note that it follows from this question that, given a and b with $a \neq b$, any linear polynomial may be expressed in the form $\lambda(x - a) + \mu(x - b)$.)

11. Find the constant term in the remainder when $P(x)$ is divided by $(x - a)(x - b)$.

12. (i) By writing $P(x) \equiv (x - a)^2 Q(x) + \lambda(x - a) + \mu$, show that the remainder when $P(x)$ is divided by $(x - a)^2$ is $P'(a)(x - a) + P(a)$.

(ii) What is the remainder when $P(x)$ is divided by $(x - a)^3$?

13. For what values of $k \in Z^+$ is $x^k + a^k$ divisible by $x + a$? What is then the quotient?

14. Show that

$$\frac{(x + a + b)(x + a + c)}{(a - b)(a - c)} + \frac{(x + b + c)(x + b + a)}{(b - c)(b - a)} +$$

$$+ \frac{(x + c + a)(x + c + b)}{(c - a)(c - b)} \equiv 1.$$

15. Show that

$$y = \frac{y_1(x - x_2)(x - x_3)}{(x_1 - x_2)(x_1 - x_3)} + \frac{y_2(x - x_3)(x - x_1)}{(x_2 - x_3)(x_2 - x_1)} +$$

$$+ \frac{y_3(x - x_1)(x - x_2)}{(x_3 - x_1)(x_3 - x_2)}$$

is the only quadratic polynomial passing through the points (x_1, y_1), (x_2, y_2), (x_3, y_3). Write down, in similar form, the cubic passing through these three points and the point (x_4, y_4).

16. (i) Use the formula in question 15 to find the equation of the parabola through the points $(1, -3)$, $(2, 1)$, $(4, 27)$.

(ii) Solve this also by supposing that the equation is $y = ax^2 + bx + c$ and obtaining a set of three simultaneous equations in a, b and c.

17. Solve the following equations given that each has a repeated root:
(i) $16x^3 + 20x^2 - 32x + 9 = 0$;
(ii) $x^4 + 2x^3 - 27x^2 + 108 = 0$.

18. (i) Suppose that $P(x)$ is divided by $S(x)$ to give a quotient $Q(x)$ and a remainder $R(x)$. Show that if P and S have a common algebraic factor, then R also has this factor (unless $R \equiv 0$).

(ii) Verify that, in the flow diagram below, any algebraic factor common to the input polynomials P and S will be a factor of S at every stage. (This is *Euclid's algorithm* for the H.C.F. of two polynomials—see question 13 of Exercise 1b. In practice, at each stage P and S may be multiplied or divided by any suitable constants to avoid fractions and unnecessarily large integers.)

(iii) Given that $P(x) \equiv x^4 - 12x^3 - 171x^2 + 1242x - 1944$ has a repeated root, find this root. [*Hint*: consider also $P'(x)$ and use the idea of part (i).]

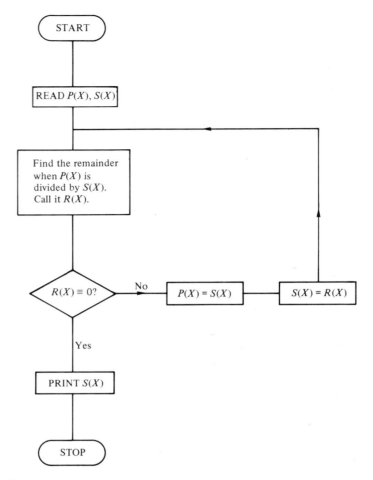

Figure 10.2

START

READ $P(X)$, $S(X)$

Find the remainder when $P(X)$ is divided by $S(X)$. Call it $R(X)$.

$R(X) \equiv 0?$ — No → $P(X) = S(X)$ — $S(X) = R(X)$

Yes

PRINT $S(X)$

STOP

19. Show that

$$P(x)Q(x) \equiv 0 \;\Rightarrow\; P(x) \equiv 0 \text{ or } Q(x) \equiv 0.$$

20. Determine whether polynomials form a group under
(i) addition; (ii) multiplication.

10.7 Graphs and repeated roots

We have already seen (Theorem 10.7) that at a repeated root the gradient of a polynomial graph is zero. Suppose that α is a double root of some polynomial $P(x)$ so that $y = P(x)$ may be written $y = (x - \alpha)^2 Q(x)$ where $Q(x)$ does not have $(x - \alpha)$ as a factor. Then, as x passes through the value α, the sign of $Q(x)$ does not change, since $Q(\alpha) \neq 0$; also $(x - \alpha)^2$ is positive on each side of $x = \alpha$. Thus the sign of y is the same on each side of $x = \alpha$, and the graph of y near to $x = \alpha$ is as shown in Fig. 10.3 or Fig. 10.4. Exactly the same considerations apply to any root of even multiplicity.

Consider, by contrast, a triple root at $x = \alpha$, so that $y = (x - \alpha)^3 Q(x)$. Here, the sign of y *does* change as x passes through the value α. The graph near $x = \alpha$ is then as in Fig. 10.5 or Fig. 10.6. The same considerations apply to any root of odd multiplicity.

POLYNOMIALS, IDENTITIES ETC.

Figure 10.3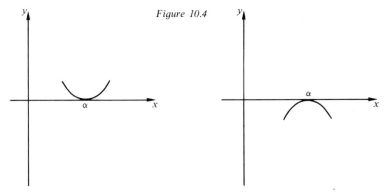

Figure 10.4

α

α

Double roots

Figure 10.5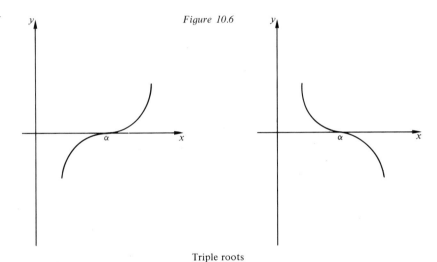

Figure 10.6

α

α

Triple roots

Qu.21 Plot accurately on the same axes the graphs of $y=x^2$, $y=x^3$, $y=x^4$ and $y=x^5$ for $-1 \leqslant x \leqslant 1$. (Choose the largest sensible scales that will fit on the graph paper.)

Example 8
Sketch the graph of

$$y = x^2(x-2)^3(3-x).$$

Solution
We note that
(i) this is a sixth degree polynomial with negative leading coefficient (so both 'tails' are downwards);
(ii) there is a double root at $x=0$, a triple root at $x=2$, and a single root at $x=3$;
(iii) the curve crosses the y-axis at the origin.
So the curve is as in Fig. 10.7.

POLYNOMIALS, IDENTITIES ETC.

203

Figure 10.7

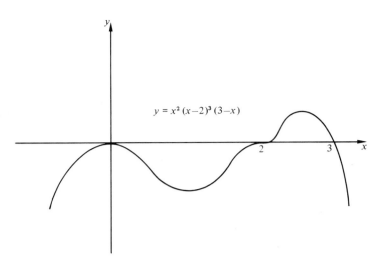

$$y = x^2 (x-2)^3 (3-x)$$

10.8 A note on points of inflexion

We now extend our use of the phrase 'point of inflexion' by the following definition which applies to any suitably smooth curve, and not merely to polynomial graphs.

Definition

A curve $y=f(x)$ is said to have a *point of inflexion* at $x=a$ if $f''(a)=0$ and $f''(x)$ changes sign as x passes through $x=a$.

This means that the gradient of the curve stops increasing and starts decreasing (or vice versa), i.e. the gradient has a local maximum or minimum at a point of inflexion. Another interpretation is that it is a point where a curve changes from a 'right-hand bend' to a 'left-hand bend' (or vice versa). A car being driven along the curve would instantaneously have its wheels straight at a point of inflexion.

Figs. 10.8–11 illustrate some points of inflexion (arrowed).

Figure 10.8

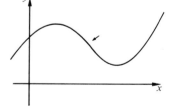

Figure 10.9

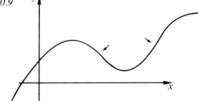

Figure 10.10

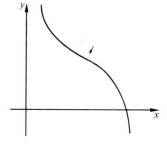

Figure 10.11

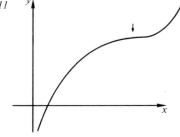

POLYNOMIALS, IDENTITIES ETC.

Finding points of inflexion is another available aid to sketching a polynomial graph.

Qu.22   If $y = x^3 + 3x^2 + 5x - 7$, find the value of x at which $\dfrac{d^2y}{dx^2} = 0$, and show that this is a point of inflexion. Sketch this cubic.

Qu.23   (For discussion.) Decide whether the following statement is equivalent to the definition of an inflexion given above: a point of inflexion is a point where a curve crosses its own tangent at that point.

Exercise 10c

Sketch the following graphs (questions 1 to 10). You are not required to find the coordinates of turning points.

1. $y = (x - 1)^2$.
2. $y = (x - 1)^3$.
3. $y = x(x - 1)^2$.
4. $y = x^2(x - 1)^2$.
5. $y = (x - 1)^3(x - 2)$.
6. $y = (x + 2)^3(x - 3)^2$.
7. $y = x(x - 3)^3(5 - x)$.
8. $y = (2x + 1)^4(x - 3)^2$.
9. $y = (2 - x)^5(3 + x)$.
10. $y = (x - 1)(x - 2)(x + 2)^2(x + 3)^3$.
11. Sketch the graph of $y = (x - 2)^3 + 8$ by first considering the graph of $y = (x - 2)^3$.
12. Find the x-coordinates of any points of inflexion of
    (i) $y = 2x^3 - 3x^2 + x - 2$;
    (ii) $y = 3x^4 + 16x^3 + 30x^2 - 7x + 1$;
    (iii) $y = 3x^4 + 8x^3 + 8x^2 - 19x$.
    Sketch the graph of (iii).
13. Prove that any polynomial of odd degree greater than 2 has at least one point of inflexion.
14. (For discussion.) Is it possible to find a polynomial curve which has
    (i) exactly two points of inflexion and no turning points;
    (ii) exactly three points of inflexion and no turning points?
    (Discuss these both algebraically and graphically.)

10.9   Roots and coefficients

We have seen that a polynomial of degree n cannot have more than n roots. In Chapter 31, we shall see that, when we extend our number system to the set of complex numbers, any nth degree polynomial has exactly n roots (counting repeated roots the appropriate number of times). The work below assumes that polynomials have the full number of roots, and it is valid whether or not the roots are all real numbers. We begin by considering quadratic polynomials.

Suppose that $ax^2 + bx + c$ (where $a \neq 0$) has roots α and β. Then

$$ax^2 + bx + c \equiv a(x - \alpha)(x - \beta)$$

$$\Rightarrow \quad x^2 + \frac{b}{a}x + \frac{c}{a} \equiv x^2 - (\alpha + \beta)x + \alpha\beta.$$

Equating coefficients,

$$\alpha + \beta = -\frac{b}{a} \text{ and } \alpha\beta = \frac{c}{a}.$$

Thus we have the important results:
(i) The sum of the roots of the quadratic polynomial $ax^2 + bx + c$ (or of the quadratic equation $ax^2 + bx + c = 0$) is $-\frac{b}{a}$.

(ii) The product of the roots is $\frac{c}{a}$.

Any quadratic may therefore be written in the form

$$x^2 - (\text{sum of roots})x + (\text{product of roots}).$$

Qu.24 *Write down*
(i) the sum,   (ii) the product,
of the roots of $3x^2 - 2x - 4 = 0$. Check your answers by direct calculation.

Qu.25 (i) Write down the quadratic equation whose roots (α and β) are the solutions of the pair of simultanous equations

$$\alpha + \beta = 1\tfrac{1}{3}$$
$$\alpha\beta = -2.$$

(ii) Solve the pair of simultaneous equations above (by elimination or substitution) taking the solution as far as obtaining a quadratic equation. (Compare with part (i).)

There are two standard methods of solving problems like the following example; one method is given here, the other is the substitution method (Section 10.10).

Example 9
Find the quadratic equation whose roots exceed the roots of $x^2 + px + q = 0$ by 2.
Solution
Let the roots of $x^2 + px + q = 0$ be α and β; then

$$\alpha + \beta = -p \text{ and } \alpha\beta = q.$$

We seek the equation whose roots are $(\alpha + 2)$ and $(\beta + 2)$. For this new equation, we have:

$$\begin{aligned}
\text{sum of roots} &= (\alpha + 2) + (\beta + 2) \\
&= (\alpha + \beta) + 4 \\
&= -p + 4; \\
\text{product of roots} &= (\alpha + 2)(\beta + 2) \\
&= \alpha\beta + 2(\alpha + \beta) + 4 \\
&= q - 2p + 4.
\end{aligned}$$

The required equation may be written

$$x^2 - (\text{sum of roots})x + (\text{product of roots}) = 0,$$

and is therefore

$$x^2 + (p - 4)x + (q - 2p + 4) = 0.$$

Example 10

Find the quadratic equation whose roots are $\alpha^2 - \beta$ and $\beta^2 - \alpha$ where α and β are the roots of $2x^2 - 3x - 4 = 0$.

Solution

We have $\alpha + \beta = \frac{3}{2}$ and $\alpha\beta = -2$.

The sum of the new roots is

$$(\alpha^2 - \beta) + (\beta^2 - \alpha)$$
$$= \alpha^2 + \beta^2 - (\alpha + \beta)$$
$$= (\alpha + \beta)^2 - 2\alpha\beta - (\alpha + \beta)$$
$$= \frac{9}{4} + 4 - \frac{3}{2}$$
$$= 4\frac{3}{4}.$$

The product of the new roots is

$$(\alpha^2 - \beta)(\beta^2 - \alpha)$$
$$= \alpha^2\beta^2 + \alpha\beta - (\alpha^3 + \beta^3)$$
$$= (\alpha\beta)^2 + \alpha\beta - \{(\alpha + \beta)^3 - 3\alpha\beta(\alpha + \beta)\}$$
$$= 4 - 2 - (\frac{27}{8} + 9)$$
$$= -10\frac{3}{8}.$$

The required equation is therefore

$$x^2 - 4\frac{3}{4}x - 10\frac{3}{8} = 0$$

or

$$8x^2 - 38x - 83 = 0.$$

Qu.26 Find the equation whose roots are the cubes of the roots of $ax^2 + bx + c = 0$.

10.10 The substitution method

Many problems like those in the previous section can be solved by the substitution method which is illustrated in the following alternative solution to Example 9.

Solution (to Example 9).

The equation $x^2 + px + q = 0$ has roots $x = \alpha$ and $x = \beta$. The required equation has roots $y = \alpha + 2$ and $y = \beta + 2$. (We use y as the variable for the new equation to avoid confusion.) So $y - 2 = \alpha$, or $y - 2 = \beta$. Therefore the variable $(y - 2)$ satisfies the original equation, i.e.

$$(y - 2)^2 + p(y - 2) + q = 0.$$

Expanding,

$$y^2 - 4y + 4 + py - 2p + q = 0$$

or

$$y^2 + (p - 4)y + (q - 2p + 4) = 0.$$

Qu.27 Solve Qu.26 by the substitution method.

[*Hint*: show first that $ay^{2/3} + by^{1/3} = -c$ and cube each side to get $a^3y^2 + b^3y + 3aby(ay^{2/3} + by^{1/3}) = -c^3$. Substitute from the first equation into the second to obtain the required equation.]

POLYNOMIALS, IDENTITIES ETC.

Definitions

(1) An expression involving α and β is said to be *symmetric* in α and β if and only if it is unchanged when α and β are interchanged. For example, $\alpha^2\beta + \beta^2\alpha + \alpha\beta$ and $\dfrac{\beta}{\alpha} + \dfrac{\alpha}{\beta}$ are both symmetric, whereas $\alpha^2 - \beta^2$ and $\alpha^2\beta + \alpha\beta^3$ are not.

(2) An expression involving α, β and γ is said to be *symmetric* in α, β and γ if it is unchanged by the transposition (interchange) of any two of α, β, γ. The expression will then be unchanged by *any* permutation (rearrangement) of α, β and γ, including the two *cyclic* permutations

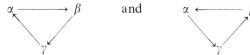 and

since these are each the composite of two simple transpositions. A similar definition of *symmetric* applies to expressions involving four or more symbols. Thus

$$(\alpha-\beta)^2 + (\beta-\gamma)^2 + (\gamma-\alpha)^2$$

and

$$\frac{\alpha}{\beta\gamma\delta} + \frac{\beta}{\alpha\gamma\delta} + \frac{\gamma}{\alpha\beta\delta} + \frac{\delta}{\alpha\beta\gamma} - \frac{4}{\alpha\beta\gamma\delta}$$

are symmetric, whereas

$$\alpha\beta^2\gamma^3 + \beta\gamma^2\alpha^3 + \gamma\alpha^2\beta^3$$

and

$$\alpha\beta + \beta\gamma + \gamma\delta + \delta\alpha$$

are not.

(3) The term *skew-symmetric* is used for an expression which changes sign under a simple transposition. Qu.29(ii) is one example.

Qu.28 What two terms may be added to $\alpha\beta + \beta\gamma + \gamma\delta + \delta\alpha$ to make it symmetric?

Qu.29 Verify that neither of the following expressions is symmetric in p, q and r, although both are unchanged by the two cyclic permutations of p, q and r.

(i) $pq^2 + qr^2 + rp^2$;   (ii) $(p-q)(q-r)(r-p)$.

Notation

A modified version of the sigma notation is sometimes used for symmetric functions. The summation is understood to be of all terms of the same form as the one given. If we have three symbols α, β and γ, then, for example,

$$\begin{aligned}
\Sigma\alpha &= \alpha + \beta + \gamma, \\
\Sigma\alpha^2 &= \alpha^2 + \beta^2 + \gamma^2, \\
\Sigma\alpha\beta &= \alpha\beta + \beta\gamma + \gamma\alpha, \\
\Sigma\alpha\beta^2 &= \alpha\beta^2 + \beta\alpha^2 + \alpha\gamma^2 + \gamma\alpha^2 + \beta\gamma^2 + \gamma\beta^2.
\end{aligned}$$

(Note the different numbers of terms in these last two.) Similarly, with four symbols α, β, γ and δ,

$$\Sigma\alpha \quad = \alpha + \beta + \gamma + \delta,$$
$$\Sigma\alpha\beta \quad = \alpha\beta + \alpha\gamma + \alpha\delta + \beta\gamma + \beta\delta + \gamma\delta,$$
$$\Sigma\alpha\beta\gamma = \alpha\beta\gamma + \alpha\beta\delta + \alpha\gamma\delta + \beta\gamma\delta.$$

Qu.30 Write out in full the following expressions which are symmetric in α, β and γ.

(i) $\sum \dfrac{1}{\alpha}$;　　(ii) $\sum \dfrac{\alpha}{\beta}$;　　(iii) $\sum \alpha^2 \beta^2$;　(iv) $\sum \alpha^2 \beta^3$.

Qu.31 Find the number of terms in the following (which are symmetric in α, β, γ and δ):

(i) $\sum \dfrac{\alpha\beta}{\gamma}$;　(ii) $\sum \alpha^2 \beta\gamma$;　(iii) $\sum \alpha\beta^2\gamma^3$;　(iv) $\sum \dfrac{\alpha}{\beta^2}$.

Qu.32 Simplify the following expressions in α, β and γ. (Do not simply write out every term; try to write down merely enough terms to be able to see what is happening.)

(i) $\left(\sum \alpha \right)^2 - \sum \alpha^2$;　　　　(ii) $\sum \alpha^2 \sum \alpha - \sum \alpha^2 \beta$;

(iii) $\sum \alpha \left(\sum \alpha^2 - \sum \alpha\beta \right)$;　　(iv) $\sum \alpha\beta - \sum \alpha\gamma$.

Definitions
(1) The *elementary symmetric functions* in α and β are the two functions $\alpha + \beta$ and $\alpha\beta$.
(2) The elementary symmetric functions in α, β and γ are $\Sigma\alpha$, $\Sigma\alpha\beta$ and $\alpha\beta\gamma$.
(3) The elementary symmetric functions in α, β, γ and δ are $\Sigma\alpha$, $\Sigma\alpha\beta$, $\Sigma\alpha\beta\gamma$ and $\alpha\beta\gamma\delta$.
(And so on.)

It can be shown that any symmetric function may be expressed in terms of the appropriate elementary symmetric functions.

The following example illustrates a useful method of checking an identity involving symmetric functions, namely that of counting the total number of terms on each side of the identity.

Example 11
If α, β, γ and δ are four numbers, express $\Sigma\alpha^2\beta$ in terms of the symmetric functions of α, β, γ, δ.
Solution
All the required terms are contained in $(\Sigma\alpha\beta)(\Sigma\alpha)$, but this also gives unwanted terms of the form $\alpha\beta\gamma$, each such term appearing three times. This suggests that

$$\sum \alpha^2 \beta = \left(\sum \alpha\beta \right) \left(\sum \alpha \right) - 3\sum \alpha\beta\gamma. \tag{1}$$

As a check, we have:

$\Sigma\alpha^2\beta$ contains $4\times3=12$ terms;

$\Sigma\alpha\beta$ contains 6 terms;

$\Sigma\alpha$ contains 4 terms;

$\Sigma\alpha\beta\gamma$ contains 4 terms.

Counting terms in (1) gives

$$12=6\times4-3\times4$$

which is correct. Note that this is not an *absolute* check; it guarantees only that we have the right number of terms, not that we have the correct terms.

Qu.33 Use this method to check your answers to Qu.32.

10.12 Roots and coefficients of cubics and quartics
Suppose that the cubic equation $ax^3+bx^2+cx+d=0$ has roots α, β and γ. Then

$$ax^3+bx^2+cx+d\equiv a(x-\alpha)(x-\beta)(x-\gamma).$$

Dividing by a, expanding the R.H.S. and equating coefficients,

$$\sum\alpha \ =-\frac{b}{a}$$

$$\sum\alpha\beta =\frac{c}{a}$$

$$\alpha\beta\gamma \ =-\frac{d}{a}.$$

Thus any cubic may be written in the form

$$x^3-(\text{sum of roots})x^2+(\text{sum of products in pairs})x$$
$$-(\text{product of roots})=0.$$

Similarly, if the quartic equation $ax^4+bx^3+cx^2+dx+e=0$ has roots α, β, γ and δ, then

$$\sum\alpha \ = \ -\frac{b}{a},$$

$$\sum\alpha\beta \ = \ \frac{c}{a},$$

$$\sum\alpha\beta\gamma \ = \ -\frac{d}{a},$$

$$\alpha\beta\gamma\delta \ = \ \frac{e}{a}.$$

(And similarly for equations of higher degree.)

Example 12
If the roots of $x^3-7x^2+3x+9=0$ are α, β and γ, find the equation with roots $(\alpha-\beta\gamma)$, $(\beta-\alpha\gamma)$, $(\gamma-\alpha\beta)$.

Solution

We have

$$\sum \alpha = 7,$$
$$\sum \alpha\beta = 3,$$
$$\alpha\beta\gamma = -9.$$

For the new roots, we have

$$\begin{aligned}
\text{sum of roots} &= (\alpha - \beta\gamma) + (\beta - \alpha\gamma) + (\gamma - \alpha\beta) \\
&= \Sigma\alpha - \Sigma\alpha\beta \\
&= 7 - 3 \\
&= 4;
\end{aligned}$$

$$\begin{aligned}
\text{product in pairs} &= (\alpha - \beta\gamma)(\beta - \alpha\gamma) + (\alpha - \beta\gamma)(\gamma - \alpha\beta) + (\beta - \alpha\gamma)(\gamma - \alpha\beta) \\
&= \Sigma\alpha\beta - \Sigma\alpha^2\beta + \Sigma\alpha^2\beta\gamma \\
&= \Sigma\alpha\beta - (\Sigma\alpha\Sigma\alpha\beta - 3\alpha\beta\gamma) + \alpha\beta\gamma\Sigma\alpha \\
&= 3 - (7.3 + 27) - 9.7 \\
&= 3 - 48 - 63 \\
&= -108;
\end{aligned}$$

$$\begin{aligned}
\text{product of roots} &= (\alpha - \beta\gamma)(\beta - \alpha\gamma)(\gamma - \alpha\beta) \\
&= \alpha\beta\gamma - \Sigma\alpha^2\beta^2 + \Sigma\alpha^3\beta\gamma - (\alpha\beta\gamma)^2 \\
&= \alpha\beta\gamma - \{(\Sigma\alpha\beta)^2 - 2\alpha\beta\gamma\Sigma\alpha\} + \alpha\beta\gamma\{(\Sigma\alpha)^2 - 2\Sigma\alpha\beta\} - (\alpha\beta\gamma)^2 \\
&= -9 - (9 + 126) - 9(49 - 6) - 81 \\
&= -612.
\end{aligned}$$

The required equation is therefore

$$x^3 - 4x^2 - 108x + 612 = 0.$$

Qu. 34 With α, β, γ as in Example 12, use the method of the example to find the equation with roots $(\alpha + \beta)$, $(\beta + \gamma)$, $(\gamma + \alpha)$.

The next example illustrates an important technique which makes it possible to use the substitution method.

Example 13

Let $x^4 - 6x^3 - 8x^2 + 4x + 3 = 0$ have roots α, β, γ, δ. Find the equation with roots $\alpha\beta\gamma$, $\alpha\beta\delta$, $\alpha\gamma\delta$ and $\beta\gamma\delta$.

Solution

The new roots may be written as $\dfrac{\alpha\beta\gamma\delta}{\delta}$, $\dfrac{\alpha\beta\gamma\delta}{\gamma}$, $\dfrac{\alpha\beta\gamma\delta}{\beta}$, $\dfrac{\alpha\beta\gamma\delta}{\alpha}$, i.e. as $\dfrac{3}{\delta}$, $\dfrac{3}{\gamma}$, $\dfrac{3}{\beta}$ and $\dfrac{3}{\alpha}$. If the new variable y may take these values, then $\dfrac{3}{y}$ takes the values α, β, γ, δ, i.e. $\dfrac{3}{y}$ satisfies the original equation†. So

$$\left(\frac{3}{y}\right)^4 - 6\left(\frac{3}{y}\right)^3 - 8\left(\frac{3}{y}\right)^2 + 4\left(\frac{3}{y}\right) + 3 = 0$$

† More briefly we may say that if y takes the values $\dfrac{3}{\delta}$, $\dfrac{3}{\gamma}$, $\dfrac{3}{\beta}$, $\dfrac{3}{\alpha}$, then it takes the values $\dfrac{3}{x}$; thus $y = \dfrac{3}{x}$, so $x = \dfrac{3}{y}$, and this is our substitution.

$$\Rightarrow \frac{27}{y^4} - \frac{54}{y^3} - \frac{24}{y^2} + \frac{4}{y} + 1 = 0$$

$$\Rightarrow y^4 + 4y^3 - 24y^2 - 54y + 27 = 0.$$

Qu.35 With α, β, γ, δ as in Example 13, find the equation whose roots are $(\alpha + \beta + \gamma - \delta)$, $(\alpha + \beta - \gamma + \delta)$, $(\alpha - \beta + \gamma + \delta)$, $(-\alpha + \beta + \gamma + \delta)$.
[*Hint:* the new roots are $(\Sigma\alpha) - 2\alpha$, $(\Sigma\alpha) - 2\beta$ etc. Show that $\dfrac{6-y}{2}$ satisfies the original equation.]

Example 14
Prove that if the roots of $x^3 - px^2 + qx - r = 0$ are in arithmetic progression, then $2p^3 = 9pq - 27r$.
First solution
Let the roots be $a - d$, a and $a + d$. Then

$$\text{sum of roots} = 3a = p \tag{1}$$

$$\text{sum of products in pairs} = 3a^2 - d^2 = q \tag{2}$$

$$\text{product of roots} = a^3 - ad^2 = r. \tag{3}$$

We eliminate a and d between these three equations.

$$a \times (2) - (3): \quad 2a^3 = aq - r.$$

Substituting for a from (1),

$$2\left(\frac{p}{3}\right)^3 = \frac{p}{3}q - r$$

$$\Rightarrow 2p^3 = 9pq - 27r.$$

Second solution
Let the roots be α, β, γ. Since they are in A.P., either $2\alpha = \beta + \gamma$ or $2\beta = \alpha + \gamma$ or $2\gamma = \alpha + \beta$.

So
$$(2\alpha - \beta - \gamma)(2\beta - \alpha - \gamma)(2\gamma - \alpha - \beta) = 0$$
$$\Rightarrow (3\alpha - p)(3\beta - p)(3\gamma - p) = 0$$
$$\Rightarrow 27\alpha\beta\gamma - 9p\Sigma\alpha\beta + 3p^2\Sigma\alpha - p^3 = 0$$
$$\Rightarrow 27r - 9pq + 3p^3 - p^3 = 0$$
$$\Rightarrow 2p^3 = 9pq - 27r.$$

(This second proof may be reversed to prove the converse result.)

10.13 Sums of powers of roots Let α, β, γ be the roots of $ax^3 + bx^2 + cx + d = 0$, and let s_n denote the sum of the nth power of the roots, so that

$$s_n = \alpha^n + \beta^n + \gamma^n.$$

Then
$$s_0 = \alpha^0 + \beta^0 + \gamma^0 = 3;$$

$$s_1 = \Sigma\alpha = -\frac{b}{a};$$

$$s_2 = \Sigma\alpha^2$$

$$= (\Sigma\alpha)^2 - 2\Sigma\alpha\beta$$

$$= \frac{b^2}{a^2} - 2\frac{c}{a}$$

$$= \frac{b^2 - 2ac}{a^2}.$$

Further values of s_n may now be obtained as follows.
Since α, β and γ are roots,

$$\begin{aligned}
a\alpha^3 + b\alpha^2 + c\alpha + d &= 0 \\
a\beta^3 + b\beta^2 + c\beta + d &= 0 \\
a\gamma^3 + b\gamma^2 + c\gamma + d &= 0.
\end{aligned}$$

Adding,

$$as_3 + bs_2 + cs_1 + 3d = 0$$

which gives s_3 if s_1 and s_2 are known. Similarly, if the original
equation is multiplied through by x^{n-3} to give

$$ax^n + bx^{n-1} + cx^{n-2} + dx^{n-3} = 0$$

then α, β and γ are roots of this equation. Substituting and adding as
before gives

$$as_n + bs_{n-1} + cs_{n-2} + ds_{n-3} = 0.$$

So any s_n can be found from the recurrence relation

$$s_n = -\frac{1}{a}(bs_{n-1} + cs_{n-2} + ds_{n-3}).$$

Qu.36 Find s_0, s_1, s_2, and hence find s_3 and s_4 for the equation
$x^3 + 2x^2 - 3x - 5 = 0$.

The method described above may be applied to polynomials of any
degree.

Qu.37 Let α and β be the roots of $x^2 - 3x - 1 = 0$. By finding a
suitable recurrence relation for s_n, find $\alpha^6 + \beta^6$.

Example 15
If α, β and γ are the roots of $2x^3 - 5x^2 - 6x + 2 = 0$, find $\dfrac{1}{\alpha^4} + \dfrac{1}{\beta^4} + \dfrac{1}{\gamma^4}$.

Solution
Extending the notation above, we seek s_{-4}. Now $s_0 = 3$, $s_1 = 2\frac{1}{2}$ and

$$s_{-1} = \Sigma\frac{1}{\alpha} = \Sigma\frac{\beta\gamma}{\alpha\beta\gamma} = \frac{1}{\alpha\beta\gamma}\Sigma\beta\gamma = 3.$$ Writing the recurrence relation

$$2s_n - 5s_{n-1} - 6s_{n-2} + 2s_{n-3} = 0$$

in the form

$$s_{n-3} = 3s_{n-2} + \frac{5}{2}s_{n-1} - s_n$$

and using it successively with $n=1$, 0 and -1, gives $s_{-2}=14$, $s_{-3}=46\frac{1}{2}$ and $s_{-4}=171\frac{1}{2}$.

So $$\frac{1}{\alpha^4}+\frac{1}{\beta^4}+\frac{1}{\gamma^4}=171\frac{1}{2}.$$

Example 16
Solve the simultaneous equations
$$x+y+z=5$$
$$x^2+y^2+z^2=69$$
$$x^3+y^3+z^3=287.$$

Solution
Let x, y, z be the roots of some cubic equation $t^3+at^2+bt+c=0$.
Then
$$\Sigma x=5 \quad \Rightarrow \; a=-5$$
$$\Sigma x^2=69 \; \Rightarrow \; \Sigma xy=\tfrac{1}{2}\{(\Sigma x)^2-\Sigma x^2\}$$
$$=\tfrac{1}{2}(25-69)$$
$$\Rightarrow \quad b=-22.$$

and $$s_3+as_2+bs_1+3c=0$$
$$\Rightarrow \; c=-\tfrac{1}{3}(287-5\times 69-22\times 5)=56.$$

The cubic is then
$$t^3-5t^2-22t+56=0.$$

By inspection (see note below), $t=2$ is a root, so
$$(t-2)(t^2-3t-28)=0$$
$$\Rightarrow (t-2)(t+4)(t-7)=0.$$
$$\Rightarrow t=2 \text{ or } -4 \text{ or } 7.$$

So $$\{x,\,y,\,z\}=\{-4,\,2,\,7\}.$$

Note
'Inspection' was used to solve the cubic, but the condition $x^2+y^2+z^2=69$ guarantees that all three roots have a modulus less than $\sqrt{69}$, so the problem is not too difficult. Also, it is certainly easier to search for one root at a time than to try to solve the three original equations by inspection, when all three values must be found at the same time.

Exercise 10d

1. Let $p=\alpha+\beta$ and $q=\alpha\beta$. Express each of the following in terms of p and q only:

   (i) $\alpha^2+\beta^2$;   (ii) $\alpha^2\beta+\beta^2\alpha$;   (iii) $\alpha^3+\beta^3$;

   (iv) $\dfrac{1}{\alpha}+\dfrac{1}{\beta}$;   (v) $\dfrac{1}{\alpha^2}+\dfrac{1}{\beta^2}$;   (vi) $(\alpha^3-2\beta)(\beta^3-2\alpha)$.

2. Let α and β be the roots of $2x^2-7x-5=0$. By expressing the sum and product of the new roots in terms of $\alpha+\beta$ and $\alpha\beta$, find the equation whose roots are
   (i) α^2 and β^2;   (ii) 2α and 2β;   (iii) $\alpha-1$ and $\beta-1$.

3. Solve each part of question 2 by the substitution method.

4. and 5. Repeat questions 2 and 3 when α and β are the roots of $x^2 - px + q = 0$.

6. Let α and β be the roots of $x^2 - 3x + 5 = 0$. Find the equation whose roots are
   (i) $\alpha + 2\beta$ and $\beta + 2\alpha$;
   (ii) $\alpha^2 + \beta^2$ and $\alpha + \beta$;
   (iii) $\alpha^3 - \beta^2$ and $\beta^3 - \alpha^2$.
   Show that α and β are not real numbers.

7. For each of the following quadratics, state whether there are no real roots, two positive roots, two negative roots or one positive and one negative root.
   (i) $x^2 - 7x + 3 = 0$;   (ii) $x^2 - 7x - 3 = 0$;
   (iii) $x^2 + 2x + 4 = 0$;   (iv) $x^2 + 2x - 4 = 0$;
   (v) $3x^2 + 5x + 1 = 0$;   (vi) $1 - x - x^2 = 0$;
   (vii) $5x - 3x^2 - 2 = 0$.

8. Prove that if one of the roots of $x^2 - px + q = 0$ is twice the other then $2p^2 = 9q$. Prove or disprove the converse.

9. Prove that the roots of $x^2 - 2px + q$ differ by 2 if and only if $p^2 = q + 1$.

10. Let the roots of $3x^3 - 6x^2 - 2x + 9 = 0$ be α, β and γ. By expressing the sum, product, and sum of the products in pairs of the new roots in terms of $(\alpha + \beta + \gamma)$, $(\alpha\beta + \beta\gamma + \gamma\alpha)$ and $\alpha\beta\gamma$, find the equation whose roots are
    (i) 3α, 3β, 3γ;   (ii) $\alpha - 4$, $\beta - 4$, $\gamma - 4$;
    (iii) $1/\alpha$, $1/\beta$, $1/\gamma$;   (iv) $\alpha\beta$, $\beta\gamma$, $\gamma\alpha$;
    (v) $\alpha + \beta$, $\beta + \gamma$, $\gamma + \alpha$;   (vi) $1/\alpha\beta$, $1/\beta\gamma$, $1/\gamma\alpha$.

11. Repeat question 10 in the case where α, β and γ are the roots of $ax^3 + bx^2 + cx + d = 0$.

12. and 13. Repeat questions 10 and 11 using the substitution method.

14. Prove that if one of the roots of $x^3 - px^2 + qx - r = 0$ is the sum of the other two, then $p^3 = 4pq - 8r$.

15. Find a relationship between p, q and r given that the roots of $x^3 - px^2 + qx - r = 0$ are in geometric progression.

16. Find the cubic equation whose roots are α, β and $\alpha\beta$, where α and β are the roots of $ax^2 + bx + c = 0$.

17. Let α, β, γ and δ be the roots of $x^4 - 7x^3 - 2x^2 - x + 8 = 0$. Find the equation whose roots are
    (i) $\alpha/2$, $\beta/2$, $\gamma/2$, $\delta/2$;
    (ii) $\alpha\beta\gamma$, $\beta\gamma\delta$, $\gamma\delta\alpha$, $\delta\alpha\beta$;
    (iii) α^2, β^2, γ^2, δ^2.

18. Find the quartic equation whose roots are the two roots of $ax^2 + bx + c = 0$ and the two roots of $px^2 + qx + r = 0$. [*Hint*: there is a simple method.]

19. Find the value of λ and the roots of the equation $x^3 - 45x^2 + \lambda x - 3240 = 0$, given that the roots are in arithmetic progression.

20. Given that the sum of two of the roots of
$$x^4 + ax^3 + bx^2 + cx + d = 0$$

is zero, show that

(i) $\dfrac{c}{a}$ is a root of $x^2 - bx + d = 0$;

(ii) the quartic factorizes as $\left(x^2 + ax + \dfrac{da}{c}\right)\left(x^2 + \dfrac{c}{a}\right)$.

21. (i) If α, β and γ are the roots of $x^3 - 2x^2 + 3x + 1 = 0$, find $\Sigma\alpha^3$ and $\Sigma\dfrac{1}{\alpha^3}$.

(ii) Use your answers to find the equation whose roots are the cubes of the roots of the given equation.

22. If α and β are the roots of $x^2 - 2x - 2 = 0$, find $\alpha^{10} + \beta^{10}$.

23. Solve
$$x + y + z = -7$$
$$x^2 + y^2 + z^2 = 115$$
$$x^3 + y^3 + z^3 = -631.$$

24. Solve
$$x + y + z = 4\tfrac{1}{2}$$
$$xy + yz + zx = -10$$
$$\frac{1}{x} + \frac{1}{y} + \frac{1}{z} = \frac{4}{15}.$$

25. (i) Show that the equation

$$18x^4 - 21x^3 - 94x^2 - 21x + 18 = 0$$

may be written as

$$18(y^2 - 2) - 21y - 94 = 0$$

where $y = x + \dfrac{1}{x}$.

(ii) Find the two possible values of y and hence find the four roots of the original equation.

(iii) Solve, by the same method,

$$50x^4 + 115x^3 - 654x^2 + 115x + 50 = 0.$$

(This is a standard method of solving a *palindromic* quartic equation, i.e. one whose coefficients form a symmetrical pattern about the centre term.)

26. (See question 25.) Find a method for solving

$$12x^4 - 8x^3 - 39x^2 + 8x + 12 = 0.$$

10.14 Partial fractions   Consider the problem of expressing

$$\frac{4}{x-3} - \frac{3}{x+2}$$

as a single fraction. It is simple enough:

$$\frac{4}{x-3} - \frac{3}{x+2} \equiv \frac{4(x+2) - 3(x-3)}{(x-3)(x+2)}$$

$$\equiv \frac{x+17}{(x-3)(x+2)}.$$

Now consider the problem in reverse. We often wish to express an algebraic fraction as the sum (or difference) of several fractions, each having a linear denominator. (When this is not possible, we aim to make the denominators as simple as possible.) This process is known as putting the expression into *partial fractions*.

Example 17

Express $\dfrac{2x+1}{2x^2+5x-12}$ in partial fractions.

Solution

$$\frac{2x+1}{2x^2+5x-12} \equiv \frac{2x+1}{(2x-3)(x+4)}$$

$$\equiv \frac{A}{2x-3} + \frac{B}{x+4} \quad \text{say.}$$

Multiplying by $(2x-3)(x+4)$,

$$(2x+1) \equiv A(x+4) + B(2x-3).$$

(We may now either substitute for x or equate coefficients.)

Put $x = -4$: $\quad -7 = -11B \Rightarrow B = \frac{7}{11}$.

Put $x = \frac{3}{2}$: $\quad -4 = 5\frac{1}{2}A \quad \Rightarrow A = \frac{8}{11}$.

Thus $\quad \dfrac{2x+1}{2x^2+5x-12} \equiv \dfrac{8}{11(2x-3)} + \dfrac{7}{11(x+4)}.$

Qu.38 Check this answer by putting the R.H.S. over a common denominator.

Qu.39 Find A, B and C such that

$$\frac{5x^2-2x-1}{(x-1)(x-2)(x-3)} \equiv \frac{A}{x-1} + \frac{B}{x-2} + \frac{C}{x-3}.$$

Although the method of Example 17 is short enough, it may be condensed into a simple rule, called the *cover-up rule*, which allows the answer to be written down with a minimum of working. The rule, which is proved after the example below, states that:

> to find the numerator corresponding to a particular factor in the denominator, 'cover up' that factor in the original expression and substitute into the rest of the expression the value of x that would make the factor zero.

Example 18

Express $\dfrac{2x-3}{(3x-1)(x+2)}$ in partial fractions.

Solution

If

$$\frac{2x-3}{(3x-1)(x+2)} \equiv \frac{A}{3x-1} + \frac{B}{x+2}$$

then the value of A may be found by substituting $x = \frac{1}{3}$ into

$\dfrac{2x-3}{(3x-1)(x+2)}$. This gives $A = -\dfrac{2\frac{1}{3}}{2\frac{1}{3}} = -1$. Similarly, B is found by

putting $x = -2$ into $\dfrac{2x-3}{(3x-1)(x+2)}$. We get $B = \dfrac{-7}{-7} = 1$.

So
$$\frac{2x-3}{(3x-1)(x+2)} \equiv -\frac{1}{3x-1} + \frac{1}{x+2}.$$

Qu.40 Use the cover-up rule to solve
(i) Example 17;   (ii) Qu.39.

An important limitation of the cover-up rule is that it is only really useful when the denominator of the given expression factorizes into distinct linear factors. Repeated factors and quadratic factors are discussed in the next section. We now outline a proof of the cover-up rule.

Suppose that
$$\frac{P(x)}{(\lambda x - \mu)Q(x)} \equiv \frac{A}{(\lambda x - \mu)} + f(x)$$

where $Q(x)$ does not have a factor $(\lambda x - \mu)$, and $f(x)$ represents all the other terms that would appear on the R.H.S.. Then

$$\frac{P(x)}{Q(x)} \equiv A + (\lambda x - \mu)f(x).$$

Putting $x = \mu/\lambda$, gives

$$A = \frac{P(\mu/\lambda)}{Q(\mu/\lambda)},$$

which is the value given by the cover-up rule.

The next example shows a method of dealing with an expression in which the degree of the numerator is as great as, or greater than the degree of the denominator.

Example 19

Express $\dfrac{x^3 - 3x^2 - x + 5}{x^2 - 5x - 6}$ in partial fractions.

Solution

Dividing, we get

$$
\begin{array}{r}
x + 2 \\
x^2-5x-6 \enclose{longdiv}{x^3 - 3x^2 - x + 5} \\
\underline{x^3 - 5x^2 - 6x } \\
2x^2 + 5x + 5 \\
\underline{2x^2 - 10x - 12} \\
15x + 17.
\end{array}
$$

So
$$\frac{x^3 - 3x^2 - x + 5}{x^2 - 5x - 6} \equiv x + 2 + \frac{15x + 17}{x^2 - 5x - 6}$$
$$\equiv x + 2 + \frac{15x + 17}{(x+1)(x-6)}$$
$$\equiv x + 2 - \frac{2}{7(x+1)} + \frac{107}{7(x-6)}$$

by the cover-up rule.

10.15 Partial fractions with repeated roots or quadratic factors

Suppose that we are required to put into partial fractions an expression of the form $\dfrac{P(x)}{(x-\alpha)^2(x-\beta)}$ (where $P(x)$ is of degree less than three). By considering the reverse process, it is clear that this expression could have arisen from a sum of terms of the form

$$\frac{A}{(x-\alpha)^2} + \frac{B}{(x-\alpha)} + \frac{C}{(x-\beta)}.$$

Qu.41 (i) (This question shows why, in general, all three terms are needed.)
(a) Suppose that

$$\frac{P(x)}{(x-\alpha)^2(x-\beta)} \equiv \frac{D}{x-\alpha} + \frac{E}{x-\beta}.$$

By multiplying to remove denominators, show that $P(x)$ is divisible by $(x-\alpha)$. (This is not true for general $P(x)$. If it were true, we should have cancelled a factor of $(x-\alpha)$ at the beginning.)
(b) Suppose that

$$\frac{P(x)}{(x-\alpha)^2(x-\beta)} \equiv \frac{F}{(x-\alpha)^2} + \frac{G}{(x-\beta)}.$$

Show that when $P(x)$ is divided by $(x-\alpha)^2$, the remainder is a constant multiple of $(x-\beta)$. (This also is not true in general.)
(ii) Verify that $\dfrac{Hx + K}{(x-\alpha)^2}$ may be written in the form $\dfrac{H}{(x-\alpha)} + \dfrac{L}{(x-\alpha)^2}$.
(This explains why the numerator corresponding to $(x-\alpha)^2$ need only be a constant.)

Similarly, a triple factor $(x-\alpha)^3$ will require terms

$$\frac{A}{(x-\alpha)^3} + \frac{B}{(x-\alpha)^2} + \frac{C}{x-\alpha}$$

on the R.H.S..

Example 20

Express $\dfrac{x^3 + 4x - 8}{(x^2 - 4)^2}$ in partial fractions.

Solution
Let
$$\frac{x^3+4x-8}{(x-2)^2(x+2)^2} \equiv \frac{A}{(x-2)^2} + \frac{B}{(x-2)} + \frac{C}{(x+2)^2} + \frac{D}{(x+2)}.$$

Then
$$x^3+4x-8 \equiv A(x+2)^2 + B(x-2)(x+2)^2 + C(x-2)^2 +$$
$$+ D(x-2)^2(x+2).$$

Putting $x=2$: $16A=8 \Rightarrow A=\frac{1}{2}$.

Putting $x=-2$: $16C=-24 \Rightarrow C=-\frac{3}{2}$.

Equating coefficients of x^3: $B+D=1$.   (1)

Putting $x=0$: $4A-8B+4C+8D=-8$
$$\Rightarrow D-B=-\frac{1}{2}.$$ (2)

Solving (1) and (2) gives $D=\frac{1}{4}$ and $B=\frac{3}{4}$. So the required expression is
$$\frac{1}{2(x-2)^2} + \frac{3}{4(x-2)} - \frac{3}{2(x+2)^2} + \frac{1}{4(x+2)}.$$

Qu.42 Express $\dfrac{3x+2}{(x-2)^2(2x+1)}$ in partial fractions.

Qu. 43 (For discussion.) Investigate the effect of trying the cover-up rule in the case of repeated factors.

Now suppose that we wish to put into partial fractions an expression in which one of the factors in the denominator is a quadratic which does not factorize into linear factors. We may proceed as in the example below. Note that we cannot expect the numerator corresponding to the quadratic factor to be a constant.

Example 21
Express $\dfrac{5x^2-7x-8}{(x^2-x+2)(x-3)}$ in partial fractions.

Solution
Let
$$\frac{5x^2-7x-8}{(x^2-x+2)(x-3)} \equiv \frac{Ax+B}{x^2-x+2} + \frac{C}{x-3}.$$

Then
$$5x^2-7x-8 \equiv (Ax+B)(x-3) + C(x^2-x+2).$$

Putting $x=3$: $16=8C \Rightarrow C=2$.

Equating coefficients of x^2: $5=A+C \Rightarrow A=3$.

Putting $x=0$: $-8=-3B+2C \Rightarrow B=4$.

So

$$\frac{5x^2 - 7x - 8}{(x^2 - x + 2)(x - 3)} \equiv \frac{3x + 4}{x^2 - x + 2} + \frac{2}{x - 3}.$$

Exercise 10e   Express in partial fractions:

1. $\dfrac{3x - 5}{(x - 1)(x - 3)}$;

2. $\dfrac{2x + 35}{x^2 - x - 20}$;

3. $\dfrac{x - 2}{(x + 1)(x - 1)}$;

4. $\dfrac{11x - 21}{(2x - 3)(x - 3)}$;

5. $\dfrac{15x - 11}{3x^2 - 5x}$;

6. $\dfrac{10x + 1}{3(4x^2 - 1)}$;

7. $\dfrac{x^2 + 5x}{(x - 1)(x + 2)(x + 3)}$;

8. $\dfrac{6x^2 + 133x + 67}{5(2x + 7)(x - 3)(x + 1)}$;

9. $\dfrac{-x^2 + 3x + 7}{(x + 2)^2(x - 1)}$;

10. $\dfrac{x + 4}{(x + 1)^2}$;

11. $\dfrac{4x^2 + 24x + 37}{(2x + 5)^2(x + 3)}$;

12. $\dfrac{-x^2 + 4x - 2}{(x - 1)^3}$;

13. $\dfrac{x^4 + 14x^3 + 72x^2 + 146x + 103}{(x + 3)^3(x + 1)^2}$;

14. $\dfrac{2x^2 + 4x - 1}{(x^2 + 2)(x + 1)}$;

15. $\dfrac{4x^2 - 7x + 18}{(x^2 + 2x + 3)(x - 4)}$;

16. $\dfrac{4x - 10}{x^3 - 4x^2 + 10x}$;

17. $\dfrac{2x^3 + x^2 - 2x + 1}{(x^2 + 1)(x + 1)^2}$;

18. $\dfrac{2x^2 - 3x + 7}{x^2 - 1}$;

19. $\dfrac{6x^3 - 17x^2 - 31x + 23}{(2x + 3)(x - 4)}$;

20. $\dfrac{2x^3 - 2x^2 + 1}{x - 1}$.

21. By first expressing it in partial fractions, differentiate

$$y = \frac{3x - 1}{(x - 3)(x + 1)}.$$

10.16 Completing the square   **Qu.44**   What number must be added to $x^2 - 6x$ to make it a perfect square?

The process known as completing the square is simple and useful. The problem is to express a general quadratic $ax^2 + bx + c$ in the form $a(x + d)^2 + e$.

Example 22
Complete the square for $x^2 - 4x + 7$.
Solution

If $x^2 - 4x + 7 \equiv (x + d)^2 + e$

then $x^2 - 4x + 7 \equiv x^2 + 2dx + (d^2 + e)$.

POLYNOMIALS IDENTITIES ETC.   221

Equating coefficients gives $d = -2$ and $e = 3$.

So
$$x^2 - 4x + 7 \equiv (x-2)^2 + 3.$$

Qu.45 Repeat the process of Example 22 for
(i) $x^2 + 10x - 5$;   (ii) $x^2 - 7x + 12$.

Note that, provided the coefficient of x^2 is unity, the value of 'd' in $(x+d)^2 + e$ is half the coefficient of x. It is therefore sometimes possible to *write down* the result of completing the square, the necessary arithmetic being done mentally.

Qu.46 *Write down* the result of completing the square for
(i) $x^2 - 8x + 19$;   (ii) $x^2 - 5x - 3$.

Example 23
Complete the square for $3x^2 - 2x + 1$.
Solution
Perhaps the simplest method is

$$3x^2 - 2x + 1 \equiv 3\left(x^2 - \frac{2}{3}x + \frac{1}{3} \right)$$

$$\equiv 3\left\{ \left(x - \frac{1}{3} \right)^2 + \frac{2}{9} \right\}$$

$$\equiv 3\left(x - \frac{1}{3} \right)^2 + \frac{2}{3}.$$

Qu.47 Complete the square for
(i) $2x^2 - x + 1$;   (ii) $4x^2 - 4x + 1$;
(iii) $1 - x - x^2$ (take out a factor of -1 first).

The following examples illustrate some of the applications of completing the square.

Example 24
Find the minimum value of $3x^2 - 6x - 2$, and the corresponding value of x.
Solution
$$3x^2 - 6x - 2 \equiv 3\left(x^2 - 2x - \frac{2}{3} \right)$$

$$\equiv 3\left\{ (x-1)^2 - \frac{5}{3} \right\}$$

$$\equiv 3(x-1)^2 - 5.$$

Now the term $3(x-1)^2$ cannot be negative, and has a minimum value of zero when $x = 1$. So $3(x-1)^2 - 5$ has a minimum value of -5 when $x = 1$.

Qu.48 Solve Example 24 by a calculus method.

Qu.49 By completing the square, show that $5x^2 - 3x + 1$ is positive definite (i.e. positive for all x), and state its minimum value.

Qu.50 Find the maximum value of $-3 + 2x - 4x^2$ and the value of x for which it occurs. Hence sketch the curve.

Example 25

Find the centre and radius of the circle

$$x^2 + y^2 - 3x + 5y = 2.$$

Solution

Completing the square for x and y separately, we have

$$x^2 + y^2 - 3x + 5y = 2$$
$$\Rightarrow (x - 1\tfrac{1}{2})^2 + (y + 2\tfrac{1}{2})^2 = 10\tfrac{1}{2}.$$

So the centre is $(1\tfrac{1}{2}, \ -2\tfrac{1}{2})$ and the radius is $\sqrt{10\tfrac{1}{2}}$.

Example 26

Solve the equation $x^2 - 4x - 3 = 0$.

Solution

$$x^2 - 4x - 3 = 0 \ \Rightarrow \ (x - 2)^2 - 7 = 0$$
$$\Rightarrow \ x - 2 = \pm\sqrt{7}$$
$$\Rightarrow \ x = 2 \pm \sqrt{7}.$$

Qu.51 (The formula for a quadratic equation.) Show, by completing the square, that if $ax^2 + bx + c = 0$ (where $a \neq 0$) then

$$\left(x + \frac{b}{2a}\right)^2 + \frac{c}{a} - \frac{b^2}{4a^2} = 0.$$

Hence show that

$$x + \frac{b}{2a} = \frac{\pm\sqrt{(b^2 - 4ac)}}{2a}$$

and deduce the well known formula for x.

We can now derive the condition for a quadratic function to be positive definite. If $y = ax^2 + bx + c$, then

$$y = a\left(x + \frac{b}{2a}\right)^2 + c - \frac{b^2}{4a}.$$

So y will be positive definite if and only if

$$a > 0 \text{ and } c - \frac{b^2}{4a} > 0$$

i.e. if and only if

$$a > 0 \text{ and } b^2 - 4ac < 0.$$

Qu.52 Derive the conditions for y to be negative definite (see Section 2.12). (Note that, when multiplying each side of an inequality by a negative number $(4a)$, the inequality sign must be reversed.)

The following two questions indicate a method of determining the number of real roots of a cubic equation.

Qu.53 (i) Show, by differentiating, that the maximum and mimimum values of $y = x^3 + px + q$ occur at $x = \pm \sqrt{\dfrac{-p}{3}}$, and show that the corresponding values of y are $q \pm \frac{2}{3} p \sqrt{\dfrac{-p}{3}}$.

(ii) How many roots will the equation $x^3 + px + q = 0$ have if these two values of y have different signs?

(iii) Show that the equation will have 1, 2 or 3 real roots according as $4p^3 + 27q^2$ is positive, zero or negative. (The expression $4p^3 + 27q^2$ is called the *discriminant* of the cubic.)

(iv) What happens to the method of (i) if $p > 0$? Is the result in (iii) still valid?

Qu.54 (i) Show that the substitution $x = y - 4$ reduces the equation $x^3 + 12x^2 + 39x + 38 = 0$ to the form discussed in Qu.53. Hence determine the number of real roots of this equation.

(ii) What substitution would be used on $x^3 + ax^2 + bx + c$ to reduce it to a form with no x^2 term?

Exercise 10f

1. By completing the square, find the minimum values of the following functions, and the corresponding values of x:
   (i) $x^2 + 2x - 3$;   (ii) $x^2 - 2x + 5$;
   (iii) $x^2 - 8x + 10$;   (iv) $x^2 + 5x + 20$;
   (v) $2x^2 - 8x - 9$;   (vi) $3x^2 + 4x - 7$;
   (vii) $6x^2 - x$.

2. By completing the square, find the maximum value of
   (i) $2 + x - x^2$;   (ii) $3x - x^2$;
   (iii) $7 - 4x - 5x^2$;   (iv) $1 + 5x - 5x^2$.

3. (i) Complete the square for $y = x^4 + 4x^2 + 10$. What is the minimum value of y?
   (ii) Repeat for $y = x^4 - 4x^2 + 10$.

4. By completing the square, solve
   (i) $x^2 - 6x + 3 = 0$;   (ii) $x^2 + 13x + 36 = 0$;
   (iii) $2x^2 - x - 5 = 0$.

5. Find the centre and radius of each of the following circles:
   (i) $x^2 + y^2 - 4x + 2y - 4 = 0$;
   (ii) $2x^2 + 2y^2 + 10x - 14y + 31 = 0$.

6. Describe the graph of
   (i) $x^2 + y^2 - 6x - 8y + 25 = 0$;
   (ii) $x^2 + 2y^2 + 4x - 12y + 23 = 0$.

7. S is the circle $x^2 + y^2 + 10x - 2y + 24 = 0$, and P is the point $(-2, -2)$. Find
   (i) the centre and radius of S;
   (ii) the distance of P from S;
   (iii) the length of the tangent from P to S.

8. A and B are the points $(0, 2)$ and $(2, 0)$ respectively. A point P

moves such that $AP=2BP$. Show that the locus of P is a circle, and find its centre and radius.

9. Repeat question 8 where $AP=\lambda BP$, discussing separately what happens when
(i) $\lambda=1$;　(ii) $\lambda\rightarrow1+$;
(iii) $\lambda\rightarrow1-$. (When a circle is specified in this way, it is sometimes known as a *circle of Apollonius*.)

10. Determine the number of real roots of

$$32x^3-176x^2-2762x+17081=0.$$

Miscellaneous exercise 10

1. An equilateral triangle ABC has its vertices at the points $A(1,0)$, $B(0,\sqrt{3})$ and $C(-1,0)$. The point P moves so that

$$PA^2+PB^2+PC^2=k^2.$$

Show that the locus of P is a circle for all values of k and that the position of the centre of the circle is independent of k. Find the value of k^2 if the circle passes through A.

Show that the radius, r, of any one of these circles is given by

$$r^2=\tfrac{1}{3}(k^2-4).$$

[SUJB (O)]

2. (a) If α and β are the roots of the equation $2x^2-5x-2=0$, find the equation whose roots are α/β and β/α.
(b) Find the values of a and b if the expression $x^4+3x^3-15x^2+ax+b$ is exactly divisible by $x+2$ and $x-1$.
Using these values of a and b, solve completely the equation

$$x^4+3x^3-15x^2+ax+b=0.$$

[W (O)]

3. Find the values of a and b and factorize the expression x^3+ax^2+bx+6 in each of the following separate cases:
(i) it is divisible by $(x-1)$ and there is a remainder of 12 when it is divided by $(x+1)$;
(ii) it is divisible by $(x-1)^2$.

[O & C (O)]

4. (a) Given the identity

$$x^4+x^2+x+1\equiv(x^2+A)(x^2-1)+Bx+C,$$

determine the numerical values of A, B and C.
By giving x a suitable value, or otherwise, find the remainder when $100\,010\,101$ is divided by 9999.
(b) Given that $\dfrac{1}{x^2(x-1)}$ may be expressed in partial fractions of the form $\dfrac{P}{x}+\dfrac{Q}{x^2}+\dfrac{R}{x-1}$, find the numbers P, Q and R.

[C]

5. (a) The remainder when $x^2+3x+20$ is divided by $x-a$ is twice the remainder when it is divided by $x+a$. Find the possible values of a.
(b) Use the factor theorem to find the value of k for which $a+2b$

is a factor of

$$a^4 + 32b^4 + a^3 b(k + 3).$$

[C (O)]

6. (a) Given that α and β are the roots of the equation $x^2 - 5x + 8 = 0$, find, without attempting to solve the equation, the values of

(i) $\dfrac{1}{\alpha} + \dfrac{1}{\beta}$, (ii) $\alpha^2 + \beta^2$.

(b) Find the range of values of k for which the roots of the equation

$$x^2 + 2x - k = 0$$

are real.

[AEB (O) '79]

7. The expression $2x^3 + ax^2 + bx + 6$ is exactly divisible by $(x - 2)$ and on division by $(x + 1)$ gives a remainder of -12.
Calculate the values of a and b and factorise the expression completely.

[AEB (O) '80]

8. Given that $f(x) \equiv x^4 + px^3 + qx^2 - 40x + 16$ is a perfect square and that p is positive, find the values of the constants p and q.
Find the remainder when $f(x)$ is divided by $(x + 2)$.

[JMB]

9. (a) Solve the equation $2x^3 - 3x^2 - 59x + 30 = 0$, given that the roots are in arithmetic progression.
(b) The roots of the equation $2x^3 + x^2 - 5x + 3 = 0$ are α, β, γ. Find
(i) $\alpha^2 + \beta^2 + \gamma^2$, (ii) $\alpha^3 + \beta^3 + \gamma^3$,
(iii) a cubic equation, with numerical coefficients, which has roots $\alpha + 1$, $\beta + 1$, $\gamma + 1$.

[C]

10. Given that α and β are the roots of the equation

$$x^2 - px + q = 0,$$

prove that $\alpha + \beta = p$ and $\alpha\beta = q$.
Prove also that
(a) $\alpha^{2n} + \beta^{2n} = (\alpha^n + \beta^n)^2 - 2q^n$,
(b) $\alpha^4 + \beta^4 = p^4 - 4p^2 q + 2q^2$.
Hence, or otherwise, form the quadratic equation whose roots are the fourth powers of those of the equation $x^2 - 3x + 1 = 0$.

[L]

11. (a) Show that, if the equations

$$x^2 + bx + c = 0, \quad x^2 + px + q = 0$$

have a common root, then

$$(c - q)^2 = (b - p)(cp - bq).$$

(b) Using the Remainder Theorem, or otherwise, factorise the expression

POLYNOMIALS IDENTITIES ETC.

$$(x-y)^3+(y-z)^3+(z-x)^3$$

into linear factors.

(c) The polynomial equation $f(x)=0$ has roots a and b. These roots are not repeated and there are no roots between a and b. Show that, if

$$f(x)=(x-a)(x-b)g(x),$$

then $g(a)$ and $g(b)$ have the same sign.

[MEI]

12. (a) If the roots of the equation $54x^4-81x^3+18x^2+20x-8=0$ are α, α, α and β find them.

(b) If $m^2>n>0$ and the equation $ax^2+2bx+c=0$ has real roots, show that the equation $ax^2+2mbx+nc=0$ has real roots. Find the condition that these two equations have a common root.

[AEB (S) '80]

13. If α, β, γ, δ are the roots of the equation

$$x^4-6x^3+8x^2+4x-4=0,$$

prove that

$$\Sigma\alpha^2=20,$$
$$\Sigma\alpha^2\beta=60,$$
$$\Sigma\alpha^3=60,$$
$$\Sigma\alpha^4=192,$$

where in each equation Σ denotes the sum of all the terms of the same type as the one given.

[Ox]

14. The quadratic equation $2x^2+b^2x+c=1$, where b and c are real constants, has two distinct real roots in the range $-1\leqslant x\leqslant1$. By considering the graph of $y=2x^2+b^2x+c-1$, or otherwise, prove that

$$b^2-1\leqslant c<\tfrac{1}{8}b^4+1<3.$$

[JMB (S)]

11 Trigonometric functions

11.1 Radians   Angles are often measured in degrees; for mathematical purposes, the *radian* is a more convenient unit. (The main reason for this preference will appear in Chapter 13.) The radian is defined thus:

Definition
If the length of an arc of a circle is equal to the radius of the circle, then the arc subtends an angle of 1 *radian* at the centre (Fig. 11.1). The symbol ^{c} is sometimes used for radians.

Figure 11.1

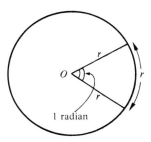

Comparison of Fig. 11.1 with an equilateral triangle (in which the *chord* length would be r) suggests that 1 radian is slightly less than 60°. In fact, we have:

2π radians gives an arc length of $2\pi r$ (the circumference),

so 2π radians $= 360°$,

π radians $= 180°$,

$\dfrac{\pi}{2}$ radians $= 90°$,

$\dfrac{\pi}{3}$ radians $= 60°$,

$\dfrac{\pi}{4}$ radians $= 45°$, etc.;

and   1 radian $= \dfrac{360°}{2\pi} \approx 57 \cdot 3°$.

From now on, all angles will be measured in radians unless otherwise stated.

Qu.1   Write the following angles in degrees:

(i) $\dfrac{\pi}{6}$;   (ii) $\dfrac{4\pi}{3}$;   (iii) 3π;   (iv) 2.

Qu.2 Write the following angles in radians:
(i) 120°;   (ii) 135°;   (iii) 1°;   (iv) 720°.

Qu.3 A circle has radius r and a sector of it subtends an angle θ (radians) at the centre. Show that
(i) arc length of sector $= r\theta$;
(ii) area of sector $= \frac{1}{2}r^2\theta$.

$$\left[\textit{Hint}: \text{ the fraction of the circle taken up by the sector is } \frac{\theta}{2\pi}. \right]$$

Qu.4 A sector of a circle subtends an angle of 3 radians at the centre, and has an area of 6 cm². Find its total perimeter.

11.2 Definitions of the trigonometric functions

The definitions of the trigonometric functions given below will not resemble the 'opposite over hypotenuse' type of definition used for elementary work, although the two approaches are, of course, consistent (see Qu.10).

We examine first the sine and cosine functions, defining, as an example, sin 2 and cos 2.

Consider a unit circle (i.e. a circle of radius 1) with centre the origin O, and let OP be a radius of the circle at an angle of 2 radians (measured, as is conventional in this work, in an anticlockwise sense from the positive x-axis) (Fig. 11.2).

Figure 11.2

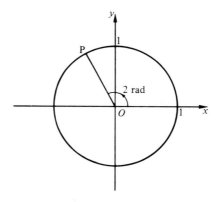

Then cos 2 and sin 2 are defined by
$$\cos\ 2 = \text{the } x\text{-coordinate of } P;$$
$$\sin\ 2 = \text{the } y\text{-coordinate of } P.$$

Qu.5 On graph paper, draw a 'unit circle' (taking, say, 10 cm as 1 unit) and make an accurate version of Fig. 11.2. Read off to 1 d.p.:
(i) cos 2;   (ii) sin 2.

We now define $\sin\theta$ and $\cos\theta$ for all real values of θ.

Definition
If a unit line OP is drawn from the origin O, making an angle θ with the positive x-axis, and if the coordinates of P are (λ, μ) then $\cos\theta = \lambda$ and $\sin\theta = \mu$.

Note

(1) This definition applies to all real values of θ (i.e. the domain of the sine and cosine functions is R). If, for example, $\theta = 9$, then the position of P is as shown in Fig. 11.3; and if $\theta = -1.4$, then P is as in Fig. 11.4 (i.e. negative angles are measured clockwise).

Figure 11.3

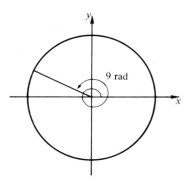

Figure 11.4

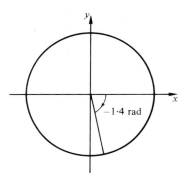

(2) If (λ, μ) is any point on the unit circle, then $-1 \leqslant \lambda \leqslant 1$ and $-1 \leqslant \mu \leqslant 1$. So, for all θ, $-1 \leqslant \cos\theta \leqslant 1$ and $-1 \leqslant \sin\theta \leqslant 1$; i.e. the range of each function is the set $\{y \in R: |y| \leqslant 1\}$.

(3) Since the length $OP = 1$, we have, by Pythagoras,

$$\lambda^2 + \mu^2 = 1.$$

So, for all θ,

$$\cos^2\theta + \sin^2\theta = 1.$$

(This is one of the 'Pythagorean' identities discussed in Section 11.10.)

(4) We may write $\sin\theta^\circ$ or $\cos\theta^\circ$, but the degrees symbol must not be omitted when degrees are meant.

Qu.6 From the definitions of the sine and cosine functions, find

(i) $\sin \pi$; (ii) $\cos \pi$; (iii) $\sin \dfrac{3\pi}{2}$, (iv) $\cos 2\pi$; (v) $\sin \dfrac{5\pi}{2}$;

(vi) $\cos 9\pi$; (vii) $\sin\left(\dfrac{-\pi}{2}\right)$; (viii) $\cos\left(\dfrac{-7\pi}{2}\right)$; (ix) $\sin 270^\circ$;

(x) $\cos(-450^\circ)$.

TRIGONOMETRIC FUNCTIONS

Qu.7 (i) If, in the definition of sin and cos, the point P corresponds to the angle θ, and the point P' corresponds to the angle $-\theta$, what is the geometrical relationship between P and P'?

(ii) If the coordinates of P are (λ, μ), what are the coordinates of P'?

(iii) Deduce that for all θ, $\cos(-\theta) = \cos\theta$, and state the corresponding result for sines.

Qu.8 Use the method of Qu.7 to determine which of the following statements are true for all values of θ.

(i) $\cos\theta = \cos(\theta + 2\pi)$;   (ii) $\sin\theta = -\sin(\theta + \pi)$;

(iii) $\sin(\pi - \theta) = \sin\theta$;   (iv) $\sin\theta = \cos\left(\dfrac{\pi}{2} - \theta\right)$.

Qu.9 (i) If $\sin\theta = 1$, what are the possible values of θ?

(ii) If $\cos\theta = 0$, what are the possible values of θ?

Qu.10 Verify that for $0 \leqslant \theta \leqslant \dfrac{\pi}{2}$, the definitions of sin and cos agree with the 'elementary' definitions (i.e. those involving opposite, adjacent and hypotenuse).

We can now define four more trigonometric functions. They are the tangent (tan), cotangent (cot), secant (sec) and cosecant (cosec) functions.

Definitions

(1) $\tan\theta = \dfrac{\sin\theta}{\cos\theta}\left(=\dfrac{\mu}{\lambda}$ in the notation of our earlier definition$\right)$.

(2) $\cot\theta = \dfrac{\cos\theta}{\sin\theta}\left(=\dfrac{\lambda}{\mu}\right)$.   (3) $\sec\theta = \dfrac{1}{\cos\theta}\left(=\dfrac{1}{\lambda}\right)$.

(4) $\mathrm{cosec}\,\theta = \dfrac{1}{\sin\theta}\left(=\dfrac{1}{\mu}\right)$.

These definitions are not valid for all values of θ, since, for example, $\tan\dfrac{\pi}{2} = \dfrac{\sin(\pi/2)}{\cos(\pi/2)} = \dfrac{1}{0}$ which is meaningless. Where the definitions given are not valid, the functions remain undefined.

Qu.11 Find the values of the following. Write 'undefined' where appropriate.

(i) $\tan\pi$;   (ii) $\sec\dfrac{\pi}{2}$;   (iii) $\mathrm{cosec}\,\dfrac{3\pi}{2}$;

(iv) $\cot\dfrac{-\pi}{2}$;   (v) $\sec 2\pi$;   (vi) $\mathrm{cosec}\,0$;

(vii) $\tan(-540°)$;   (viii) $\cot 270°$.

Qu.12 By first expressing them in terms of $\sin\theta$ and $\cos\theta$ only, simplify

(i) $\dfrac{1}{\cot\theta}$;   (ii) $\dfrac{\sin\theta\cot\theta}{\cos\theta}$;   (iii) $\dfrac{\sec\theta}{\tan\theta}$;   (iv) $\dfrac{\cos\theta + \sin\theta\tan\theta}{\sec\theta}$.

11.3 Quadrants and signs

The reader may be familiar with the use of trigonometric tables† for dealing with angles between 0 and $\frac{\pi}{2}$. As a first step towards extending the use of these tables to angles of any size, we now classify angles according to which of the four *quadrants* they lie in (see Fig. 11.5).

Figure 11.5.

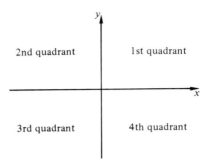

Angles between 0 and $\frac{\pi}{2}$ lie in the first quadrant; angles between $\frac{\pi}{2}$ and π lie in the second quadrant, and so on. An angle of $500°$ would, for example, lie in the second quadrant. The classification applies also to negative angles, so that an angle of $-\frac{2\pi}{3}(=-120°)$ lies in the third quadrant. Angles of 0, $\frac{\pi}{2}$, π, etc. are not classified, and in the work that follows, they must be dealt with separately.

Qu.13 In which quadrants do the following angles lie?
(i) $50°$;   (ii) $100°$;   (iii) $150°$;

(iv) $300°$;   (v) $\frac{2\pi}{3}$;   (vi) $-\frac{\pi}{6}$;

(vii) $-300°$;   (viii) 9.

We now consider the signs of the six trigonometric functions in each quadrant. These can be obtained directly from the definitions, and the reader should check them. The results may be summarized thus:

	quadrant			
	1st	2nd	3rd	4th
sine and cosecant	$+$	$+$	$-$	$-$
cosine and secant	$+$	$-$	$-$	$+$
tangent and cotangent	$+$	$-$	$+$	$-$

(Note that the six ratios fall into mutually reciprocal pairs.)

---

†It is assumed that the reader has a calculator; but the work here, and in the next few sections, is just as important as when only tables were available. It may help temporarily to restrict the use of a calculator to angles between 0 and $90°$.

The following diagram acts as a useful mnemonic for these results:

S	A
T	C

In the 'C' quadrant, Cos is positive and sin and tan are negative.
In the 'A' quadrant, All are positive.
In the 'S' quadrant, Sin is positive and cos and tan are negative.
In the 'T' quadrant, Tan is positive and sin and cos are negative.
This is sometimes known as the CAST rule. (Note that the word begins in the 4th quadrant.)

11.4 Related angles

Consider again the unit line OP used in the definition of the sine and cosine functions. Let Q, R and S be the images of P under reflection in the x-axis, y-axis and origin respectively (Fig. 11.6). Now the coordinates of P, Q, R and S are identical except for changes in sign. (In each case, the x-coordinate is $\pm\lambda$ and the y-coordinate is $\pm\mu$.) It follows that the four angles corresponding to P, Q, R and S have exactly the same sine, cosine, tangent etc. except for the signs.

We shall call these four angles *related angles*.

Figure 11.6

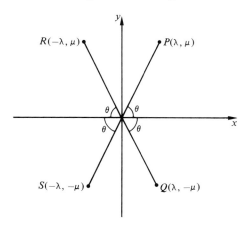

Qu.14 Which three angles between 0 and 360° are 'related' to 25°?

11.5 The use of tables for angles of any size†

Example 1
Find $\tan 290°$.
Solution
First find the *sign* using the CAST rule: 290° is in the 4th quadrant where tangents are *negative*.

Then find the related angle in the first quadrant: this is 70° (Figs. 11.7a and 11.7b).
From tables,

$$\tan\ 70° = 2{\cdot}7475,$$

so

$$\tan 290° = -2{\cdot}7475.$$

†See footnote on page 232.

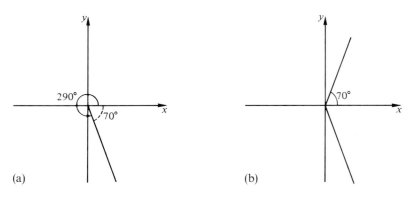

Figure 11.7

(a)   (b)

Example 2

Find $\operatorname{cosec}\left(\dfrac{13\pi}{5}\right)$.

Solution

$\dfrac{13\pi}{5}$ is in the same position as $\dfrac{13\pi}{5}-2\pi=\dfrac{3\pi}{5}=108°$. This is in the 2nd quadrant where sin (and hence cosec) is *positive*. The related angle in the first quadrant is $72°$. From tables,

$$\operatorname{cosec}\ 72°=1\!\cdot\!0515,$$

so

$$\operatorname{cosec}\ 108°=1\!\cdot\!0515,$$

i.e.

$$\operatorname{cosec}\left(\dfrac{13\pi}{5}\right)=1\!\cdot\!0515.$$

Qu.15 Find
   (i) $\sin 200°$;   (ii) $\cos(-40°)$;   (iii) $\cot 100°$;
   (iv) $\tan 260°$;   (v) $\sec(-100°)$;   (vi) $\operatorname{cosec}(-500°)$;
(vii) $\cos 5000°$.
(Note: if using a calculator, then, for practice, restrict its use to angles between 0 and $90°$ in this question.)

**11.6 Special angles
(30°, 45°, 60°)**   The trigonometric ratios of $30°$, $45°$ and $60°$ can easily be found without the use of tables or calculator, and they should be memorized.

In Fig. 11.8, ABC is an equilateral triangle of side 2, and D is the mid-point of BC. By Pythagoras, $AD=\sqrt{3}$, and so, from triangle ADC,

$$\sin 30°=\cos 60°=\frac{1}{2};$$

$$\sin 60°=\cos 30°=\frac{\sqrt{3}}{2};$$

$$\tan 30°=\cot 60°=\frac{1}{\sqrt{3}};$$

$$\tan 60°=\cot 30°=\sqrt{3}.$$

Qu.16 Write down the values of
(i) $\sec 30°$;   (ii) $\operatorname{cosec} 30°$;   (iii) $\sec 60°$;   (iv) $\operatorname{cosec} 60°$.

Figure 11.8

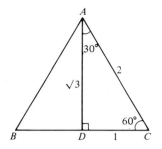

In Fig. 11.9, $\angle B = 90°$, $AB = BC = 1$. By Pythagoras, $AC = \sqrt{2}$, and so

$$\sin 45° = \cos 45° = \frac{1}{\sqrt{2}};$$

$$\tan 45° = \cot 45° = 1;$$

$$\sec 45° = \csc 45° = \sqrt{2}.$$

Figure 11.9

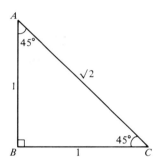

Qu.17 Without using tables or calculator, find the value of
(i) $\tan 240°$; (ii) $\sec 135°$; (iii) $\csc 330°$.

Exercise 11a **1.** Convert the following angles to degrees:

(i) $\dfrac{3\pi}{4}$; (ii) $\dfrac{5\pi}{3}$; (iii) $\dfrac{7\pi}{2}$;

(iv) 10π; (v) $\dfrac{5\pi}{8}$; (vi) $\dfrac{13\pi}{9}$.

2. Convert to radians:

(i) $225°$; (ii) $240°$; (iii) $210°$;

(iv) $270°$; (v) $22\frac{1}{2}°$; (vi) $600°$.

3. In which quadrants do the following angles lie?

(i) $\dfrac{3\pi}{4}$; (ii) $\dfrac{7\pi}{3}$; (iii) 3; (iv) 5;

(v) $-\dfrac{2\pi}{3}$; (vi) -4; (vii) 10.

4. For each of the following, find the related angle in the first quadrant:

    (i) $120°$;   (ii) $\dfrac{7\pi}{4}$;   (iii) $250°$;

    (iv) $\dfrac{8\pi}{5}$;   (v) $\dfrac{11\pi}{3}$;   (vi) $350°$;

    (vii) $260°$.

5. Use tables (or a calculator) to find

    (i) $\sin 140°$;   (ii) $\cos 260°$;

    (iii) $\sin 292°$;   (iv) $\cos 163°$;

    (v) $\sin 117° 12'$;   (vi) $\cos 232° 46'$.

6. Find, without using tables or a calculator, and leaving in surd form where appropriate,

    (i) $\sin 135°$;   (ii) $\sin \dfrac{2\pi}{3}$;   (iii) $\cos 330°$;   (iv) $\cos \dfrac{7\pi}{4}$;

    (v) $\sin 960°$;   (vi) $\cos \dfrac{17\pi}{6}$.

7. Using tables (or a calculator) only where necessary, find

    (i) $\tan 135°$;   (ii) $\cot 240°$;   (iii) $\sec 350°$;

    (iv) $\operatorname{cosec}(-130°)$;   (v) $\tan(-25°)$;   (vi) $\cot \dfrac{5\pi}{8}$;

    (vii) $\sec\left(-\dfrac{3\pi}{4}\right)$;   (viii) $\operatorname{cosec}(-1500°)$.

8. If the point $P(\lambda, \mu)$ on the unit circle centre O is rotated by $\dfrac{\pi}{2}$ anticlockwise about O to the point P', what are the coordinates of P'? Deduce expressions for

    (i) $\sin\left(\theta+\dfrac{\pi}{2}\right)$;   (ii) $\cos\left(\theta+\dfrac{\pi}{2}\right)$;   (iii) $\tan\left(\theta+\dfrac{\pi}{2}\right)$.

9. Using only the statements

    (1) $\forall\,\theta\in R,\ \sin(\pi-\theta)=\sin\theta$,

    (2) $\forall\,\theta\in R,\ \sin\left(\dfrac{\pi}{2}-\theta\right)=\cos\theta$,

    (3) sine is an odd function,

    deduce that

    (ii) $\forall\,\theta\in R,\ \sin\left(\dfrac{\pi}{2}+\theta\right)=\cos\theta$;

    (ii) $\forall\,\theta\in R,\ \cos\left(\dfrac{\pi}{2}+\theta\right)=-\sin\theta$.

    [*Hint*: try replacing θ by $\dfrac{\pi}{2}\pm\theta$ in one of the given identities.]

10. Apply the angle bisector theorem of elementary geometry (to be found in most O-level text books) to an isosceles right-angled triangle to show that $\tan 22\frac{1}{2}^\circ = \dfrac{1}{1+\sqrt{2}}$, and rationalize the denominator.

11. (i) A turntable is rotating at 33 revolutions per minute. What is its angular speed in radians per second?
    (ii) Estimate the angular velocity in radians per second of
    (a) the earth about its axis of rotation;
    (b) the earth about the sun.

12. The two shorter sides of a right-angled triangle are of length 1 and 2 units. State the length of the hypotenuse. Given that $\tan\theta = \frac{1}{2}$ and θ is acute, state the values of $\sin\theta$ and $\cos\theta$.

13. By drawing suitable triangles as in question 12, find
    (i) $\tan\theta$ and $\sec\theta$ given that $\sin\theta = \frac{1}{3}$ and θ is acute;
    (ii) $\cot\theta$ and $\cos\theta$ given that $\operatorname{cosec}\theta = \frac{3}{2}$ and θ is obtuse.

14. Draw accurately on graph paper the curve $x^2 + y^2 = 1$. (Use compasses.) By drawing a suitable straight line graph on the same axes, find approximately the solution of

$$2\sin\theta + 3\cos\theta = 2\cdot 7$$

lying in the first quadrant. [*Hint*: $\sin\theta = y$, $\cos\theta = x$.]

15. Given that $\sin 3\theta = \frac{1}{2}$ and $\cos 3\theta = -\dfrac{\sqrt{3}}{2}$, show on a unit circle the only possible position of 3θ. Find and illustrate the three possible positions of θ.

16. (i) A triangle ABC has $AC = b$ and $BC = a$. Show, by dropping a perpendicular, that the area of the triangle is $\frac{1}{2}ab\sin C$. (Consider separately the cases in which C is acute and C is obtuse.)
    (ii) Hence find a formula for the area of the minor segment cut off by a chord subtending an angle $\theta\ (<\pi)$ at the centre of a circle of radius r.
    (iii) Find also the area of the major segment. Can this answer be obtained by substituting $(2\pi - \theta)$ for θ in the answer to (ii)?

17. In Chapter 13, we shall see that for small angles, $\theta \approx \sin\theta$. Calculate the percentage error in using θ as an approximation for $\sin\theta$ when
    (i) $\theta = 0\cdot 1$;   (ii) $\theta = 0\cdot 02$.
    Estimate by trial and error the range of positive values of θ for which the error would be less than 5%, converting your answer to degrees.

11.7 Simple trigonometric equations

In solving trigonometric equations, care must be taken to ensure that all the required solutions are found, and that no false solutions are introduced.

Example 3
Solve $\sin\theta = \frac{1}{2}$.
Solution
Clearly (or from tables) $\theta = 30^\circ$ is one solution, and of the three related

angles, only $150°$ (in the S quadrant) has the correct sign. To either of these solutions we may add or subtract any number of complete turns. The general solution is therefore

$$\theta = \begin{cases} 30° + 360n° \text{ for } n \in Z \\ \text{or } 150° + 360n° \text{ for } n \in Z. \end{cases}$$

Example 4
Solve $\operatorname{cosec} 2\theta = -1·5$.
Solution
(We are given a trigonometric ratio of 2θ; we must therefore start by finding the general solution for 2θ.) From tables,

$$\operatorname{cosec} 2\theta = +1·5 \Rightarrow 2\theta = 41·8°,$$

and from the CAST rule, cosec is negative in the 3rd and 4th quadrants. So

$$\operatorname{cosec} 2\theta = -1·5 \Rightarrow 2\theta = \left.\begin{array}{c} 221·8° \\ \text{or } 318·2° \end{array}\right\} + 360n°.$$

Dividing by 2 gives

$$\theta = \left.\begin{array}{c} 110·9° \\ \text{or } 159·1° \end{array}\right\} + 180n° \text{ for } n \in Z.$$

Qu.18 Draw a unit circle and illustrate the possible positions of θ in the solution to Example 4.

Notation
$\operatorname{Sin}^2 A$ means $(\sin A)^2$, and similarly $\sec^5 B$ means $(\sec B)^5$. This notation must be clearly distinguished from $\sin A^2$ which means $\sin(A^2)$ and which is encountered much less frequently. (See also Section 2.4.)

In the next example, only solutions between 0 and $360°$ are required. The best method remains, however, to find the general solution for θ and then to select the required values. Attempting to limit the range of values of θ at too early a stage can lead to errors.

Example 5
Solve

$$3 \sec^2(3\theta + 45°) = 4$$

for $0 \leqslant \theta < 360°$.
Solution

$$3 \sec^2(3\theta + 45°) = 4$$

$$\Rightarrow \quad \sec(3\theta + 45°) = \pm \frac{2}{\sqrt{3}}$$

$$\Rightarrow \quad \cos(3\theta + 45°) = \pm \frac{\sqrt{3}}{2}$$

$$\Rightarrow \quad 3\theta+45°= \left.\begin{array}{r} 30° \\ 150° \\ 210° \\ 330° \end{array}\right\} +360n° \text{ for } n\in Z.$$

More simply, this may be written

$$3\theta+45°= \left.\begin{array}{r} 30° \\ 150° \end{array}\right\} +180n°$$

$$\Rightarrow \quad 3\theta= \left.\begin{array}{r} -15° \\ 105° \end{array}\right\} +180n°$$

$$\Rightarrow \quad \theta= \left.\begin{array}{r} -5° \\ 35° \end{array}\right\} +60n°.$$

We now select the required values. They are

$35°, 55°, 95°, 115°, 155°, 175°, 215°, 235°, 275°, 295°, 335°, 355°.$

Qu.19 Find the general solutions of

(i) $\tan^2\theta=3$; (ii) $\sec\dfrac{\theta}{2}=-2$.

Qu.20 Solve

(i) $\cos(3\theta-90°)=\frac{1}{2}$ for $0\leqslant\theta<360°$;

(ii) $\cot^2(5\theta+20°)=1$ for $0\leqslant\theta\leqslant90°$.

The following question illustrates a number of useful techniques for dealing with some trigonometric equations.

Qu.21 Solve the following in the range $0\leqslant\theta<360°$. Hints are given in brackets.

(i) $2\sin\theta+\cos\theta=0$. (Divide through by $\cos\theta$, checking first whether $\cos\theta=0$ leads to a solution.)

(ii) $\sin^2\theta=3\sin\theta+1$. (A quadratic in $\sin\theta$.)

(iii) $2\sin^2\theta=\sin\theta$. (Factorize this quadratic. Do not simply divide through by $\sin\theta$; why not?)

(iv) $\sin\theta\cos\theta=0$.

(v) $\tan 2\theta=3\cot 2\theta$. $\left(\text{Write } \cot 2\theta \text{ as } \dfrac{1}{\tan 2\theta}.\right)$

(vi) $2\cot\theta=\cos\theta$. $\left(\text{Write } \cot\theta \text{ as} \dfrac{\cos\theta}{\sin\theta}.\right)$

Example 6

Solve

(i) $\sin(\theta+30°)=\sin 4\theta$;

(ii) $\sin(2\phi-45°)=\cos(3\phi+60°)$.

Solution

(Another method for this type of equation is given in question 9(i) of Exercise 12b.)

(i) If the sines of two angles are equal, then either they are in the same position of the unit circle, or each is the reflection of the other in the y-axis (so that they add up to $180° + 360n°$). Thus

$$\sin(\theta + 30°) = \sin 4\theta$$

$$\Rightarrow \begin{cases} \theta + 30° = 4\theta + 360n° \\ \text{or} \quad \theta + 30° = 180° - 4\theta + 360n° \end{cases}$$

$$\Rightarrow \begin{cases} 3\theta = 30° - 360n° \\ \text{or} \quad 5\theta = 150° + 360n° \end{cases}$$

$$\Rightarrow \theta = \begin{cases} 10° - 120n° \\ \text{or} \quad 30° + 72n° \quad \text{for } n \in Z. \end{cases}$$

(Note that the term $-120n°$ may be replaced by $+120n°$.)

(ii) Using the fact that $\cos A \equiv \sin(90° - A)$ (see Qu.8 part (iv)), we have

$$\cos(3\phi + 60°) \equiv \sin\{90° - (3\phi + 60°)\}$$
$$\equiv \sin(30° - 3\phi).$$

The given equation becomes

$$\sin(2\phi - 45°) = \sin(30° - 3\phi).$$

This may be solved by the method of part (i). The reader is left to obtain the answer

$$\phi = \begin{cases} 15° + 72n° \\ \text{or} \quad -195° + 360n°. \end{cases}$$

(Alternatively, we may convert each side to a cosine function and use the fact that $\cos\alpha = \cos\beta \Rightarrow \alpha = \beta + 360n°$ or $\alpha = -\beta + 360n°$.)

11.8 Graphs of the trigonometric functions

The graph of $y = \sin x$ may be drawn for all real x. Part of the graph is shown in Fig. 11.10.

Figure 11.10

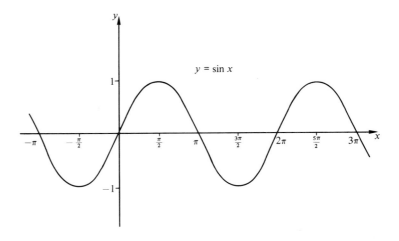

A number of important features of this graph follow from the definition of the sine function. For example, from the definition, $\sin x \equiv \sin(x+2\pi)$, and so the sine curve is periodic with period 2π.

Qu.22 (For discussion.) Check that each of these statements follows from the definition of the sine function and explain the significance of each for the graph of $\sin x$.

(i) $\sin(-x) \equiv -\sin x$;

(ii) $\sin\left(\dfrac{\pi}{2}+x\right) \equiv \sin\left(\dfrac{\pi}{2}-x\right)$;

(iii) $\sin(\pi-x) \equiv -\sin(\pi+x)$.

The general shape of the graph of $y=\operatorname{cosec} x$ may be deduced from the fact that it is the reciprocal of the sine curve (Fig. 11.11). Note that the curve has asymptotes† at the points where $\sin x=0$, i.e. at $x=n\pi$ for $n\in Z$. The cosecant curve shares some of the properties of the sine curve; for example it is an odd function, and it is periodic with period 2π.

Figure 11.11

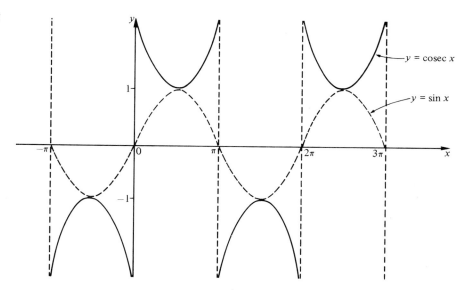

Qu.23 Sketch on the same axes the graphs of $y=\cos x$ and $y=\sec x$ for $-\pi \leqslant x \leqslant 3\pi$.

It can be shown—see, for example, Exercise 11a, question 8—that $\cos x \equiv \sin\left(x+\dfrac{\pi}{2}\right)$, and it follows that the graph of $y=\cos x$ is the same as the graph of $y=\sin x$ translated a distance $\dfrac{\pi}{2}$ to the left.

---

†An *asymptote* is a line which the curve gets 'very close to' at a great distance from the origin.

TRIGONOMETRIC FUNCTIONS   241

(Alternatively we may say that the y-axis has been moved $\dfrac{\pi}{2}$ to the right.)

Qu.24 (i) What curve is obtained if the cosine curve is moved $\dfrac{\pi}{2}$ to the left?

What curves are obtained if the curve is moved $\dfrac{\pi}{2}$ to the left

(ii) once more;   (iii) twice more?

Figure 11.12 shows the graphs of $y = \tan x$ and $y = \cot x$. They are both periodic with period π.

Qu.25 Solve the equation $\tan x = \cot x$ in the range $0 \leqslant x < \pi$. Hence find the coordinates of two of the points of intersection of the graphs in Fig. 11.12.

Figure 11.12

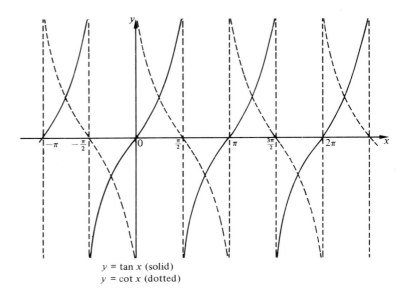

$y = \tan x$ (solid)
$y = \cot x$ (dotted)

Qu.26 Sketch on the same axes (and check by plotting a few points) the graphs of

   (i) $y = \sin x$;   (ii) $y = \sin 2x$;   (iii) $y = 2 \sin x$.

When sketching the graph of $y = A \sin nx$, the number A is the greatest value of y (and $-A$ is the least value of y). A is called the *amplitude* of the graph.

11.9 The inverse trigonometric functions

We have seen that the domain of the sine function is R and that its range is $\{y: -1 \leqslant y \leqslant 1\}$. The inverse function is called the *arcsine* function, and is defined only on the domain $\{x: -1 \leqslant x \leqslant 1\}$ (i.e. the range of sine).

The immediate problem that arises is that arcsine is not well-defined. For example, we have $\sin\dfrac{\pi}{6}=\dfrac{1}{2}$ and $\sin\dfrac{5\pi}{6}=\dfrac{1}{2}$; so do we take $\arcsin\dfrac{1}{2}$ to be $\dfrac{\pi}{6}$ or $\dfrac{5\pi}{6}$? The usual convention is to take the range of arcsine as $\left\{y\!:\!-\dfrac{\pi}{2}\leqslant y\leqslant\dfrac{\pi}{2}\right\}$. Then $\arcsin\dfrac{1}{2}$ would be $\dfrac{\pi}{6}$, and the value $\dfrac{\pi}{6}$ is thus called the *principal value* of $\arcsin\dfrac{1}{2}$.

In Fig. 11.13, the graph of arcsin x is shown, the part corresponding to the principal values being indicated by a solid line.

Figure 11.13

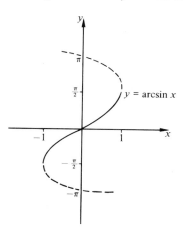

The other inverse trigonometric functions (arccosec, arctan etc.) may be dealt with similarly, but care must be taken with the ranges of principal values. For example, the graph of $y=\arccos x$ shows that $\left\{y\!:\,-\dfrac{\pi}{2}\leqslant y\leqslant\dfrac{\pi}{2}\right\}$ cannot be taken as the set of principal values. Instead, we take $\{y\!:\!0\leqslant y\leqslant\pi\}$, as shown in Fig. 11.14.

Figure 11.14

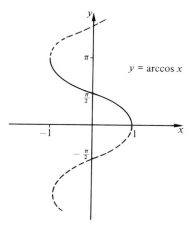

Qu.27 Find the principal values of

(i) $\arcsin\dfrac{\sqrt{3}}{2}$; (ii) $\arcsin-\dfrac{1}{2}$; (iii) $\arccos\dfrac{1}{\sqrt{2}}$; (iv) $\arccos-\dfrac{1}{\sqrt{2}}$.

Qu.28 Sketch the graphs of

(i) arctan x; (ii) arccot x;

(iii) arcsec x; (iv) arccosec x.

Suggest a range of principal values for each. (Choose, where possible, a range corresponding to a continuous part of the curve.)

Note

The notation $\sin^{-1} x$ is sometimes used for arcsin x, and so $\sin^{-1} x$ is not the same as $(\sin x)^{-1}$, which would normally be written as cosec x. Similarly, $\tan^{-1} x$ means arctan x etc.

Exercise 11b

1. Solve these equations, giving the general solutions in degrees. Do not use tables or a calculator.
   (i) $2 \sin \theta = 1$;
   (ii) $2 \cos \theta = -1$;
   (iii) $4 \sin^2 \theta = 1$;
   (iv) $\cot^2 \theta = 3$;
   (v) $3 \sec^2 \theta = 4$;
   (vi) $3 \tan^2 \theta = -1$.

2. Solve these equations, giving the general solution in radians. Do not use tables or a calculator.
   (i) $\sin 2\theta = \frac{1}{2}$;
   (ii) $\cos 3\theta = -\frac{\sqrt{3}}{2}$;
   (iii) $\operatorname{cosec} 5\theta = 2$;
   (iv) $\cos \frac{\theta}{2} = -1$;
   (v) $\tan 2\theta = 1$;
   (vi) $\cot \frac{\theta}{4} = \sqrt{3}$.

3. Use tables or a calculator to solve the following, giving all solutions in the range $0 \leqslant \theta < 360°$:
   (i) $\sin (\theta + 10°) = 0.4$;
   (ii) $\sec (\theta - 50°) = 3$;
   (iii) $\tan (2\theta + 180°) = \frac{1}{2}$;
   (iv) $\operatorname{cosec} (3\theta - 30°) = 1$.

4. Solve the following in the range $-\pi < \theta \leqslant \pi$:

   (i) $4 \sin^2 2\theta = 3$; (ii) $3 \tan \left(3\theta + \frac{\pi}{2} \right) = \sqrt{3}$;

   (iii) $3 \cot^2 \left(\frac{3\theta}{2} - \frac{\pi}{6} \right) = 1$;

   (iv) $2 \operatorname{cosec}^2 (3\theta - \pi) = -1$.

5. Solve the following in the range $0 \leqslant \theta < 2\pi$:
   (i) $\sin \theta - \sqrt{3} \cos \theta = 0$;
   (ii) $2 \sin^2 \theta + \sin \theta - 1 = 0$;
   (iii) $\tan^3 \theta = 3 \tan \theta$;
   (iv) $\tan \theta = 2 \sin \theta$;
   (v) $\operatorname{cosec} \theta \sec \theta = 0$.

6. On the same axes, sketch the graphs of $y = x^2$ and $y = \sin x$. Find the number of roots of $\sin x = x^2$.

7. By sketching two suitable graphs as in question 6, show that the equation $x = \tan x$ has infinitely many roots, and that for large integers n, there is a root near $(n + \frac{1}{2})\pi$.

TRIGONOMETRIC FUNCTIONS

8. Sketch the graphs of
   (i) $y = \sin 3x$;
   (ii) $y = \cos \frac{1}{2}x$;
   (iii) $y = 2 \operatorname{cosec} x$;
   (iv) $y = \sin(x^2)$.
   State the period of each where appropriate.

9. Sketch on the same axes $y = \sin x$ and $y = \sin\left(x + \frac{\pi}{3}\right)$, showing clearly the graphical significance of the $\frac{\pi}{3}$. (In this context, the $\frac{\pi}{3}$ is referred to as a *phase difference* between the two graphs.)

10. How many solutions of $0 = \sin\frac{1}{x}$ are there in the range $0 < x \leqslant 1$? Sketch a graph to show how this function behaves in the given range.

11. Solve for $0 \leqslant \theta < 2\pi$:
    (i) $\sin\theta \leqslant \frac{1}{2}$;   (ii) $\sin^2\theta \leqslant \frac{1}{4}$.

12. Write down the principal values of
    (i) arcsin $\frac{1}{2}$;
    (ii) arccosec 1;
    (iii) arccot $\sqrt{3}$;
    (iv) arcsec -2;
    (v) arctan -1.

13. Find the principal values of
    (i) arcsin $(\cos 20°)$;   (ii) arccos $(\sin 100°)$;
    (iii) arcsin $(\cos\theta)$ where θ is an acute angle.

14. Sketch on the same axes $y = \arcsin x$ and $y = \arcsin 2x$. On new axes, sketch $y = \arcsin x$ and $y = 2 \arcsin x$.

15. Solve for $0 \leqslant \theta \leqslant 2\pi$ or $0° \leqslant \theta \leqslant 360°$:
    (i) $\sin 2\theta = \sin 3\theta$;
    (ii) $\sin(\theta + 10°) = \sin\theta$;
    (iii) $\cos\left(2\theta - \frac{3\pi}{5}\right) = \cos\left(2\theta + \frac{\pi}{5}\right)$;
    (iv) $\cos 2\theta = \sin(\theta - 15°)$;
    (v) $\tan\theta = \tan\left(2\theta - \frac{\pi}{3}\right)$.

11.10 Some trigono-metric identities

In Section 11.2, we saw that it follows from the definitions of $\sin\theta$ and $\cos\theta$ and from Pythagoras's Theorem that

$$\boxed{\sin^2\theta + \cos^2\theta \equiv 1.} \tag{1}$$

Dividing through by $\cos^2\theta$ gives

$$\boxed{\tan^2\theta + 1 \equiv \sec^2\theta,}$$

and dividing (1) through by $\sin^2\theta$ gives

$$\boxed{1 + \cot^2\theta \equiv \operatorname{cosec}^2\theta.}$$

These three identities are important. Because of their derivation, they are sometimes called the Pythagorean identities.

Qu.29 Solve the equation

$$2 \sec^2 \theta - 3 \tan \theta = 1$$

for $0 \leqslant \theta \leqslant \dfrac{\pi}{2}$ by first writing it as a quadratic in $\tan \theta$.

In proving trigonometric identities, the following techniques may be helpful.

(1) Start with the more complicated side of the identity: it is easier to 'simplify' than to 'complicate'. It may be necessary to start with each side separately.

(2) Converting everything into sines and cosines is a useful, if at times inelegant, technique.

(3) Make use of the Pythagorean identities.

Example 7

Prove that

$$\tan A + \cot A \equiv \frac{\sec A + \operatorname{cosec} A}{\sin A + \cos A}.$$

Solution

$$\text{R.H.S.} \equiv \frac{\dfrac{1}{\cos A} + \dfrac{1}{\sin A}}{\sin A + \cos A}.$$

Multiplying top and bottom by $\cos A \sin A$ gives

$$\frac{\sin A + \cos A}{\cos A \sin A (\sin A + \cos A)}$$

$$\equiv \frac{1}{\cos A \sin A}.$$

And

$$\text{L.H.S.} \equiv \frac{\sin A}{\cos A} + \frac{\cos A}{\sin A}$$

$$\equiv \frac{\sin^2 A + \cos^2 A}{\cos A \sin A}$$

$$\equiv \frac{1}{\cos A \sin A}.$$

Hence L.H.S. $\equiv$ R.H.S.

Qu.30 Prove that $\dfrac{\sin A}{1 + \cos A} \equiv \operatorname{cosec} A - \cot A$. [*Hint*: one neat method is to start with the L.H.S. and multiply top and bottom by $1 - \cos A$.]

With the aid of the Pythagorean identities, it is now possible to express any of the six trigonometric ratios in terms of any other.

Example 8

If $\sin P = \frac{1}{3}$, find

(i) $\cos P$; (ii) $\cot P$.

First solution

(i) $\cos^2 P = 1 - \sin^2 P = \frac{8}{9}$.

So
$$\cos P = \pm\frac{\sqrt{8}}{3}.$$

(Note that the sign of $\cos P$ cannot be determined from the given information, since P could be in either the first or the second quadrant.)

(ii) $\cot P = \dfrac{\cos P}{\sin P}$

$$= \pm\frac{\sqrt{8}}{3} \div \frac{1}{3}$$

$$= \pm\sqrt{8}.$$

Second solution

(i) We draw a triangle (Fig. 11.15) in which $\sin P = \frac{1}{3}$. By Pythagoras, the 'adjacent' is $\sqrt{8}$. The numerical value of $\cos P$ is thus $\dfrac{\sqrt{8}}{3}$. The sign must be found by considering in which quadrants P may lie (see note above). In this case, $\cos P = \pm\dfrac{\sqrt{8}}{3}$.

(ii) Cot P = 'adjacent over opposite' = $\sqrt{8}$. Considering quadrants gives $\cot P = \pm\sqrt{8}$.

Figure 11.15

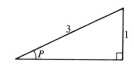

Example 9

Express $\tan B$ in terms of $\operatorname{cosec} B$ only.

Solution

Cot $B = \pm\sqrt{(\operatorname{cosec}^2 B - 1)}$.

So
$$\tan B = \pm\frac{1}{\sqrt{(\operatorname{cosec}^2 B - 1)}}.$$

Qu.31 If $\sec\theta = \lambda$, express
(i) $\sin\theta$, (ii) $\cot\theta$,
in terms of λ. (Use each of the methods of Example 8. In the second method, we have 'sec = hypotenuse over adjacent' in a right-angled triangle.)

Example 10

Eliminate θ between the equations

(i) $x = a + \cos\theta$
  $y = b + 2\sin\theta$;

(ii) $x = p\cos\theta$
  $y = q\cot\theta$.

Solution

(i) $\cos \theta = x - a$ and $\sin \theta = \dfrac{y-b}{2}$. Squaring and adding,

$$(x-a)^2 + \left(\frac{y-b}{2}\right)^2 = 1.$$

(ii) $\cot^2 \theta = \dfrac{\cos^2 \theta}{\sin^2 \theta} = \dfrac{\cos^2 \theta}{1-\cos^2 \theta}.$

So
$$\left(\frac{y}{q}\right)^2 = \frac{(x/p)^2}{1-(x/p)^2}$$

$$\Rightarrow \quad y^2(p^2 - x^2) = q^2 x^2.$$

11.11 More trigonometric equations

The methods used in proving the identities above can be applied to certain types of equation.

Example 11
Solve

$$\sec x + \tan x = 2 \cos x.$$

First solution
Multiplying each side by $(\sec x - \tan x)$ gives

$$\sec^2 x - \tan^2 x = 2 \cos x (\sec x - \tan x)$$
$$\Rightarrow 1 = 2 - 2 \sin x$$
$$\Rightarrow \sin x = \tfrac{1}{2}$$
$$\Rightarrow x = \left.\begin{array}{c} \dfrac{\pi}{6} \\[2mm] \text{or } \dfrac{5\pi}{6} \end{array}\right\} + 2\pi n \text{ for } n \in Z.$$

Second solution
Multiplying the given equation through by $\cos x$,

$$1 + \sin x = 2 \cos^2 x$$
$$\Rightarrow 1 + s = 2(1 - s^2) \quad \text{where } s \text{ stands for } \sin x$$
$$\Rightarrow 2s^2 + s - 1 = 0$$
$$\Rightarrow (2s - 1)(s + 1) = 0$$

which leads to the same solution as the first method above, together with the solution $x = \dfrac{3\pi}{2} + 2\pi n$. This apparent contradiction is discussed in Qu.32.

Qu.32 Show, by substitution, that $x = \dfrac{3\pi}{2}$ is not a solution of the equation given in Example 11. Explain how this false solution arises in the second method above.

Qu.33 Criticize the following solution to the equation

$$\sin \theta + 2 \cos \theta = 1.$$

Determine which are the false solutions and explain where they are introduced.

$$2\cos\theta = 1 - \sin\theta$$
$$\Rightarrow 4\cos^2\theta = (1 - \sin\theta)^2$$
$$\Rightarrow 4(1 - s^2) = 1 - 2s + s^2$$
$$\Rightarrow 5s^2 - 2s - 3 = 0$$
$$\Rightarrow (5s + 3)(s - 1) = 0$$
$$\Rightarrow \sin\theta = -0{\cdot}6 \text{ or } 1$$
$$\Rightarrow \theta = 323{\cdot}1° \left.\begin{array}{l} \\ \text{or } 216{\cdot}9° \\ \text{or } 90° \end{array}\right\} + 360n°.$$

(Other methods of solving this type of equation are given in Sections 12.6 and 12.7.)

1. Simplify

(i) $\dfrac{\tan A}{\sec A}$; (ii) $\dfrac{\operatorname{cosec} A}{\sqrt{(\operatorname{cosec}^2 A - 1)}}$;

(iii) $\dfrac{(\operatorname{cosec} A - 1)(\sin A + 1)}{\cos A}$;

(iv) $\dfrac{\cos A}{1 + \sin A} + \dfrac{1 + \sin A}{\cos A}$;

(v) $\dfrac{\tan A + \tan B}{\cot A + \cot B}$;

(vi) $(1 + \sec A + \tan A)(1 - \sec A + \tan A)$.

2. Prove the identities

(i) $\tan A + \cot A \equiv \sec A \operatorname{cosec} A$;

(ii) $\sec A - \cos A \equiv \sin A \tan A$;

(iii) $\dfrac{\cos A}{1 + \cot A} \equiv \dfrac{\sin A}{1 + \tan A}$;

(iv) $\dfrac{\operatorname{cosec} A + \sec A}{\operatorname{cosec} A \sec A} \equiv \pm \sqrt{(1 + 2\sin A \cos A)}$;

(v) $\pm \sqrt{\left(\dfrac{1 + \cos A}{1 - \cos A}\right)} \equiv \dfrac{\tan A}{\sec A - 1}$;

(vi) $\dfrac{\sin A}{1 + \cos A} \equiv \operatorname{cosec} A - \operatorname{cosec} A \sec A + \tan A$.

3. Solve the equations for $0 \leqslant \theta < 360°$:
   (i) $6\cos^2\theta - \sin\theta = 4$;
   (ii) $2 - \cos^2\theta = 3\sin\theta$;
   (iii) $4\sin\theta = 3\operatorname{cosec}\theta - 1$.
4. Find
   (i) $\cos x$ and $\cot x$ given that $\sin x = \tfrac{1}{10}$;

TRIGONOMETRIC FUNCTIONS

(ii) sec x and cos x given that $\tan x = \frac{3}{2}$;

(iii) cosec x and cos x given that $\cot x = \frac{1}{4}$;

(iv) cosec x given that $\cos x = -\frac{2}{3}$;

(v) sin x and cot x given that $\sec x = -3$.

5. Simplify

(i) $\sqrt{(9-u^2)}$ where $u = 3 \sin 2\theta$;

(ii) $\dfrac{1}{\sqrt{(4+9p^2)}}$ where $p = \frac{2}{3} \tan \dfrac{\theta}{4}$;

(iii) $\dfrac{5g}{\sqrt{\{(25g^2/4)-9\}}}$ where $g = \frac{6}{5} \operatorname{cosec} 3\theta$.

6. If $t = \tan \theta$, express $\cos \theta$, $\sec \theta$ and $\operatorname{cosec} \theta$ in terms of t.

7. If $z = 2 \operatorname{cosec} \theta$, express $\cos \theta$, $\tan \theta$ and $\sec \theta$ in terms of z.

8. Make θ the subject of

(i) $y = 2 \sin \theta$;   (ii) $y = \sin 2\theta$;   (iii) $y = 4 \sin^2 3\theta$.

9. Eliminate θ between the following pairs of equations:

(i) $x = 1 + \cos \theta$
$\ y = \sin \theta$;

(ii) $x = 2 \tan \theta - 3$
$\ y = \sec \theta + 1$;

(iii) $y = a \cos 3\theta$
$\ x = b \sin 3\theta$;

(iv) $x = a \operatorname{cosec} \theta$
$\ y = b + c \cot \theta$;

(v) $x = \sin 5\theta$
$\ y = \tan 5\theta$;

(vi) $x = p \operatorname{cosec} 3\theta + q$
$\ y = q \sec 3\theta + p$.

10. (In this question, inverse trigonometric functions are assumed to take their principal values.)

(i) If $\theta = \arcsin p$, in which quadrants may θ lie?

(ii) What can be said about the sign of $\cos(\arcsin p)$?

(iii) Simplify $\cos(\arcsin p)$. [*Hint*: we seek $\cos \theta$, where $\sin \theta = p$.]

(iv) Find similarly $\sin(\arccos p)$, taking care over the sign(s).

12 Trigonometric formulae

12.1 Compound angle formulae

It is certainly the case that, in general,

$$\sin(A+B) \neq \sin A + \sin B.$$

We shall therefore derive the correct expansions for $\sin(A+B)$ and $\cos(A+B)$. The method used assumes an elementary knowledge of vectors.

Consider the effect of an (anticlockwise) rotation about the origin O through an angle B. From the definition of the sine and cosine functions (see Fig. 12.1).

Figure 12.1

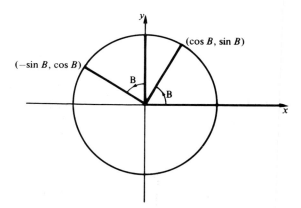

the vector $\begin{pmatrix} 1 \\ 0 \end{pmatrix}$ maps to $\begin{pmatrix} \cos B \\ \sin B \end{pmatrix}$

and (by congruent triangles)

$$\begin{pmatrix} 0 \\ 1 \end{pmatrix} \text{ maps to } \begin{pmatrix} -\sin B \\ \cos B \end{pmatrix}.$$

Therefore, if A is any angle,

$$\cos A \begin{pmatrix} 1 \\ 0 \end{pmatrix} \text{ maps to } \cos A \begin{pmatrix} \cos B \\ \sin B \end{pmatrix}$$

and $\sin A \begin{pmatrix} 0 \\ 1 \end{pmatrix}$ maps to $\sin A \begin{pmatrix} -\sin B \\ \cos B \end{pmatrix}$,

i.e. $\begin{pmatrix} \cos A \\ 0 \end{pmatrix}$ maps to $\begin{pmatrix} \cos A \cos B \\ \cos A \sin B \end{pmatrix}$ $\qquad$ (1)

and $\begin{pmatrix} 0 \\ \sin A \end{pmatrix}$ maps to $\begin{pmatrix} -\sin A \sin B \\ \sin A \cos B \end{pmatrix}$. (2)

Now let P be the point on the unit circle corresponding to angle A, and let Q be the foot of the perpendicular from P to the x-axis (Fig. 12.2). Then

$$OP = OQ + QP = \begin{pmatrix} \cos A \\ 0 \end{pmatrix} + \begin{pmatrix} 0 \\ \sin A \end{pmatrix}.$$

Figure 12.2

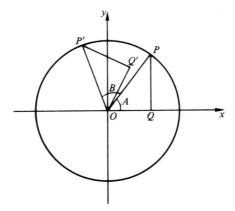

Rotating triangle OQP through angle B to triangle $OQ'P'$, we have

$$OP' = OQ' + Q'P'$$
$$= \begin{pmatrix} \cos A \cos B \\ \cos A \sin B \end{pmatrix} + \begin{pmatrix} -\sin A \sin B \\ \sin A \cos B \end{pmatrix} \quad \text{(by (1) and (2) above)}$$
$$= \begin{pmatrix} \cos A \cos B - \sin A \sin B \\ \cos A \sin B + \sin A \cos B \end{pmatrix}.$$

Hence the coordinates of P' are

$$(\cos A \cos B - \sin A \sin B, \ \sin A \cos B + \sin B \cos A).$$

But P' is the point on the unit circle corresponding to the angle $(A + B)$, and so its coordinates are

$$(\cos (A + B), \ \sin (A + B)).$$

Equating the two sets of coordinates for P' gives the identities

$$\boxed{\begin{aligned} \cos (A + B) &\equiv \cos A \cos B - \sin A \sin B \\ \sin (A + B) &\equiv \sin A \cos B + \cos A \sin B. \end{aligned}}$$
(1)
(2)

Qu.1 The reader who is familiar with matrix transformations may prefer† to set out the above argument thus:

(i) Verify that the matrix $\begin{pmatrix} \cos B & -\sin B \\ \sin B & \cos B \end{pmatrix}$ gives rise to an anti-

clockwise rotation centre O, through an angle B.

† This matrix method is not *genuinely* simpler, since to *prove* (i) involves an argument along the lines of the method above.

(ii) Let P be $(\cos A, \sin A)$ and find the point P' by

evaluating $\begin{pmatrix} \cos B & -\sin B \\ \sin B & \cos B \end{pmatrix}\begin{pmatrix} \cos A \\ \sin A \end{pmatrix}$. Then proceed as in the

given proof.

Replacing B by $-B$ in the formulae above, and remembering that $\cos(-B) = \cos B$ and $\sin(-B) = -\sin B$, we get

$$\cos(A - B) \equiv \cos A \cos(-B) - \sin A \sin(-B),$$

so
$$\boxed{\cos(A - B) \equiv \cos A \cos B + \sin A \sin B,} \tag{3}$$

and
$$\sin(A - B) \equiv \sin A \cos(-B) + \cos A \sin(-B),$$

so
$$\boxed{\sin(A - B) \equiv \sin A \cos B - \sin B \cos A.} \tag{4}$$

Qu.2 Use either the method given in the text for deriving the formulae (1) and (2), or the method of Qu.1, to derive formulae (3) and (4) directly. (Consider a clockwise rotation through angle B.)

Example 1
Find $\sin 15°$ and $\cos 15°$ in surd form.
Solution

$$\begin{aligned}
\text{Sin } 15° &= \sin(45° - 30°) \\
&= \sin 45° \cos 30° - \sin 30° \cos 45° \\
&= \frac{1}{\sqrt{2}} \cdot \frac{\sqrt{3}}{2} - \frac{1}{2} \cdot \frac{1}{\sqrt{2}} \\
&= \frac{\sqrt{3} - 1}{2\sqrt{2}}.
\end{aligned}$$

Similarly,

$$\begin{aligned}
\cos 15° &= \cos 45° \cos 30° + \sin 45° \sin 30° \\
&= \frac{1}{\sqrt{2}} \cdot \frac{\sqrt{3}}{2} + \frac{1}{\sqrt{2}} \cdot \frac{1}{2} \\
&= \frac{\sqrt{3} + 1}{2\sqrt{2}}.
\end{aligned}$$

Qu.3 Find in surd form:
(i) $\sin 75°$; (ii) $\cos 75°$.
Qu.4 Check the answers to Example 1 by using them to evaluate $\sin^2 15° + \cos^2 15°$.

Dividing formula (2) by (1) gives

$$\tan(A + B) \equiv \frac{\sin A \cos B + \sin B \cos A}{\cos A \cos B - \sin A \sin B}$$

and dividing top and bottom of the R.H.S. by $\cos A \cos B$

gives

$$\tan (A+B) \equiv \frac{\tan A + \tan B}{1 - \tan A \tan B}. \tag{5}$$

Similarly, dividing (4) by (3),

$$\tan (A-B) \equiv \frac{\tan A - \tan B}{1 + \tan A \tan B}. \tag{6}$$

Qu.5 Use formula (6) to find $\tan 15°$ in surd form.

The formulae (1) to (6) are known as the *compound angle formulae*.

12.2 The double angle formulae

Replacing B by A in each of formulae (1), (2) and (5), we obtain the double angle formulae:

$$\sin 2A \equiv 2 \sin A \cos A; \tag{7}$$

$$\cos 2A \equiv \cos^2 A - \sin^2 A; \tag{8}$$

$$\tan 2A \equiv \frac{2 \tan A}{1 - \tan^2 A}. \tag{9}$$

The expansion of $\cos 2A$ may be written in two other forms; since $\cos^2 A = 1 - \sin^2 A$,

$$\cos 2A \equiv (1 - \sin^2 A) - \sin^2 A$$

so

$$\boxed{\cos 2A \equiv 1 - 2 \sin^2 A} \tag{8a}$$

or, replacing $\sin^2 A$ in (8) by $1 - \cos^2 A$,

$$\boxed{\cos 2A \equiv 2 \cos^2 A - 1.} \tag{8b}$$

Example 2

If $\sin P = 0 \cdot 6$, find the possible values of $\sin \dfrac{P}{2}$.

Solution

If $\qquad\qquad\qquad \sin P = 0 \cdot 6,$

then $\qquad\qquad\qquad \cos^2 P = 1 - (0 \cdot 6)^2 = 0 \cdot 64.$

So $\qquad\qquad\qquad \cos P = \pm 0 \cdot 8.$

Now putting $A = \dfrac{P}{2}$ into formula (8a) gives the 'half angle' formula

$$\cos P \equiv 1 - 2 \sin^2 \frac{P}{2}.$$

So
$$\pm 0{\cdot}8 = 1 - 2\sin^2\frac{P}{2}$$

$$\Rightarrow \sin^2\frac{P}{2} = \frac{1}{2}(1 \pm 0{\cdot}8)$$

$$\Rightarrow \sin^2\frac{P}{2} = 0{\cdot}1 \text{ or } 0{\cdot}9$$

$$\Rightarrow \sin\frac{P}{2} = \pm\sqrt{0{\cdot}1} \text{ or } \pm\sqrt{0{\cdot}9}.$$

There appear to be four solutions. That all four are indeed possible values of $\sin\frac{P}{2}$ may be checked by observing that $\sin P = 0{\cdot}6$ allows two possible positions for P (see Fig. 12.3), each of which gives two positions (180° apart) for $\frac{P}{2}$ (Fig. 12.4). Thus $\frac{P}{2}$ may have four possible positions, each with a different sine.

Figure 12.3

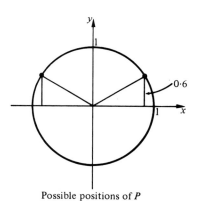

Possible positions of P

Figure 12.4

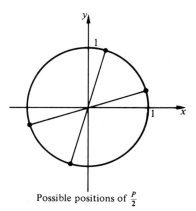

Possible positions of $\frac{P}{2}$

Qu.6 Use formulae (8a) and (8b) to find $\sin 22\frac{1}{2}°$ and $\cos 22\frac{1}{2}°$.
Qu.7 If $t = \tan 22\frac{1}{2}°$, use formula (9) to show that $1 - t^2 = 2t$. Hence find t. Explain the significance of the other root of the quadratic equation.

Qu.8 Show from formula (7) that

$$\sin P = 2 \sin \frac{P}{2} \cos \frac{P}{2}$$

and deduce that if $\sin P > 0$, then $\sin \frac{P}{2}$ and $\cos \frac{P}{2}$ have the same sign. In which quadrants may $\frac{P}{2}$ then lie? (Compare with the answer to Example 2.)

12.3 The angles 90° ± θ etc.

Trigonometric ratios such as $\sin(90° + \theta)$, $\sec(\theta - 270°)$ and $\cot(180° - \theta)$ may be simplified using the compound angle formulae. Thus, for example,

$$\sec(\theta - 270°) = \frac{1}{\cos(\theta - 270°)}$$

$$= \frac{1}{\cos\theta \cos 270° + \sin\theta \sin 270°}$$

$$= \frac{1}{-\sin\theta}$$

$$= -\operatorname{cosec}\theta.$$

Alternatively, we may consider the mapping

$$\theta \to \theta - 270°.$$

Under this mapping, (a clockwise rotation through 270°), the point (λ, μ) corresponding to angle θ, maps to $(-\mu, \lambda)$ corresponding to angle $\theta - 270°$. Then

$$\sec(\theta - 270°) = \frac{1}{x\text{-coordinate}}$$

$$= \frac{1}{-\mu}$$

$$= -\operatorname{cosec}\theta.$$

This latter method leads to the following rule for simplifying such expressions:
(i) If an angle of 90° or 270° (or 450° etc.) is involved, remove or add the prefix 'co'. (So sin changes to cos, cot changes to tan etc.) If an angle of 180° or 360° etc. is involved, keep the same function.
(ii) Determine the sign by assuming that θ is in the first quadrant.

Thus if we wish to simplify $\operatorname{cosec}(90° + \theta)$, we change the cosec to sec; and since $(90° + \theta)$ is in the second quadrant (for θ acute) where cosec is positive, we have

$$\operatorname{cosec}(90° + \theta) = +\sec\theta.$$

Qu.9 Use this method to simplify
(i) $\sin(\theta - 90°)$;   (ii) $\operatorname{cosec}(180° - \theta)$;

(iii) $\operatorname{cosec}(180° + \theta)$;

(iv) $\tan(270° - \theta)$;   (v) $\cot(270° + \theta)$.

Qu.10 (For discussion.) Explain precisely the reason for part (i) of the rule above, in terms of the corresponding mappings.

Exercise 12a

1. Evaluate in surd form:
    (i) $\sec 15°$;   (ii) $\operatorname{cosec} 75°$;   (iii) $\tan 105°$.

2. By first writing $3A$ as $2A + A$, express
    (i) $\sin 3A$ in terms of $\sin A$ only;
    (ii) $\cos 3A$ in terms of $\cos A$ only.

3. Expand
    (i) $\sin(A + B + C)$;   (ii) $\cos(A + B + C)$;
    (iii) $\tan(A + B + C)$.

4. Simplify

    (i) $\dfrac{\sqrt{3}}{2}\cos x + \dfrac{1}{2}\sin x$;

    (ii) $\sin 2x \cos x + \sin x \cos 2x$;

    (iii) $\dfrac{2\tan\frac{1}{2}A}{1 - \tan^2 \frac{1}{2}A}$.

5. Show that, for all A,

    $$\tan A \sin 2A \geqslant 0.$$

6. Expand $\sin\left(\dfrac{\pi}{4} - \theta\right)$ and $\cos\left(\dfrac{\pi}{4} + \theta\right)$. Verify that these expressions are equal.

7. If P is an acute angle whose cosine is $\frac{3}{5}$, and Q is an obtuse angle whose sine is $\frac{12}{13}$, find without using tables or a calculator:
    (i) $\sin(P + Q)$;   (ii) $\cos(Q - P)$;
    (iii) $\tan(P + Q)$;   (iv) $\sec 2P$.

8. Use the identity $1 - 2\sin^2 A \equiv \cos 2A$ to evaluate in surd form:
    (i) $\sin 15°$;   (ii) $\sin 22\frac{1}{2}°$;   (iii) $\sin 75°$.

9. Show that

    $$\operatorname{cosec}(A + B) \equiv \frac{\operatorname{cosec} A \operatorname{cosec} B}{\cot A + \cot B}$$

    $$\equiv \frac{\sec A \sec B}{\tan A + \tan B},$$

    and derive corresponding expressions for $\sec(A + B)$.

10. Show that

    $$\cot(A + B) \equiv \frac{\cot A \cot B - 1}{\cot A + \cot B}.$$

11. (i) Without using tables, find the tangent of the acute angle between the lines $y = 2x + 3$ and $3y = 4x - 7$.
    (ii) Find an expression for the tangent of the *acute* angle between the lines $y = m_1 x + c_1$ and $y = m_2 x + c_2$.

12. If $\sin A = s$, express in terms of s:
    (i) $\sin 2A$;   (ii) $\tan 2A$.
13. Let $\sin \theta = s$ and $\cos \theta = c$. Express
    (i) $\sin 4\theta$ in terms of c only;
    (ii) $\cos 4\theta$ in terms of s only.
14. Let $t = \tan \theta$. Simplify

(i) $\dfrac{2t}{1+t^2}$;   (ii) $\dfrac{1-t^2}{1+t^2}$;   (iii) $\dfrac{2t}{1-t^2}$.

15. Express as simply as possible in terms of $\dfrac{\theta}{2}$:

(i) $\sin \theta$;   (ii) $\sqrt{(1+\cos \theta)}$;   (iii) $\cot \theta$;   (iv) $\dfrac{\sin \theta}{1-\cos \theta}$;

(v) $\sqrt{(1+\sin \theta)}$;   (vi) $\sqrt{(1-\sin \theta)}$.

16. If $\cos B = -\frac{2}{3}$, find all possible values of
    (i) $\cos 2B$;   (ii) $\cos \frac{1}{2}B$.
    Prove the identities, questions 17 to 23.
17. $\dfrac{\sin^2 A}{1+\cos A} \equiv 2 \sin^2 \dfrac{A}{2}$.

18. $\sin x + \cos x \equiv \pm \sqrt{(1+\sin 2x)}$.
19. $\operatorname{cosec} 2\theta + \cot 2\theta \equiv \cot \theta$.
20. $4 \sin A \cos^3 A - 4 \cos A \sin^3 A \equiv \sin 4A$.
21. $\cos^2 \theta (\cot 2\theta + \tan \theta) \equiv \frac{1}{2} \cot \theta$.
22. $\tan (45° + x) - \tan (45° - x) \equiv 2 \tan 2x$.
23. $\tan (45° + A) \equiv \dfrac{\cos A + \sin A}{\cos A - \sin A} \equiv \dfrac{1 + \sin 2A}{\cos 2A}$.

24. By considering the sign of each bracket separately, show that, for A and B lying between 0 and π,

$$\left(\sin \frac{A}{2} - \sin \frac{B}{2} \right)\left(\cos \frac{A}{2} - \cos \frac{B}{2} \right) \leqslant 0.$$

Deduce that

$$\frac{\sin A + \sin B}{2} \leqslant \sin \left(\frac{A+B}{2} \right).$$

Illustrate this result concerning averages on a sine curve.
25. Show that if $\sin \theta = p$ and $\sin 2\theta = q$, then

$$q^2 + 4p^4 = 4p^2.$$

26. Eliminate θ between the following pairs of equations:
    (i) $a = \sin \theta$ (ii) $m = \sin 2\theta$
                                          $n = \tan \theta$;

    $b = \sin \left(\theta + \dfrac{\pi}{3} \right)$;

    (iii) $p = 2 \sin \theta$ (iv) $x = a \tan \theta$

    $q = 3 \cos 2\theta$; $y = b \tan \dfrac{\theta}{2}$.

27. (i) Show that $\arctan p + \arctan q = \arctan \dfrac{p+q}{1-pq}$. [*Hint*: start
with the expansion of $\tan(\theta + \phi)$ where $\tan \theta = p$ and $\tan \phi = q$.]
(ii) If $\arcsin x + \arcsin y = \arcsin z$, express z in terms of x and y
(without any trigonometric functions).

28. Simplify
    (i) $\operatorname{cosec}(270° - \theta)$;   (ii) $\sec(\theta + 270°)$;
    (iii) $\tan(180° - \theta)$;   (iv) $\sin(180° + \theta)$;
    (v) $\cos(270° - \theta)$;   (vi) $\cot(90° + \theta)$.

29. (i) Without using a calculator, show that $\sin 10°$, $\sin 130°$, $\sin 250°$
are all roots of $8x^3 - 6x + 1 = 0$. [*Hint*: consider the expansion of
$\sin 3A$.]
(ii) Hence *write down* the value of $\sin 10° + \sin 130° + \sin 250°$.

30. $ABCDE$ is a regular pentagon of side a, and AX, BY and DZ are
perpendiculars dropped on to CE.
(i) Find angles EAX and CED.
(ii) By calculating CE in two different ways, show that
$2\cos 36° = 1 + 2\sin 18°$.
(iii) Deduce that $\sin 18°$ is a root of $4x^2 + 2x - 1 = 0$, and find
$\sin 18°$ in surd form.
(iv) If all five diagonals are drawn and the 'inner' pentagon
labelled $A'B'C'D'E'$ so that A' is opposite A etc., show that

$EA' = a$ and that $A'C = \dfrac{a}{2}\sec 36°$. Deduce that

$$a + \frac{a}{2}\sec 36° = 2a\cos 36°$$

and hence show that $2\cos 36°$ satisfies $x^2 = 1 + x$.
(v) Find out what you can about the *golden number* (or golden
ratio) and its relationship with this question and with the
Fibonacci Sequence.

12.4 Converting a product to a sum or difference

It is often necessary to convert a product of two trigonometric
functions into the sum or difference of two such functions. This can be
done using the compound angle formulae.

Example 3
Express $\sin 40° \sin 70°$ as a sum or difference.
Solution
We notice that the term $\sin 40° \sin 70°$ occurs in the expansions of
$\cos(70° - 40°)$ and $\cos(70° + 40°)$:

$$\cos 30° = \cos(70° - 40°)$$
$$= \cos 70° \cos 40° + \sin 70° \sin 40°,$$
and $\qquad \cos 110° = \cos(70° + 40°)$
$$= \cos 70° \cos 40° - \sin 70° \sin 40°.$$

Subtracting removes the unwanted terms $\cos 70° \cos 40°$, giving

$$\cos 30° - \cos 110° = 2\sin 70° \sin 40°.$$
So $\qquad \sin 70° \sin 40° = \tfrac{1}{2}(\cos 30° - \cos 110°).$

Qu.11 By adding instead of subtracting in the example above, find an expression for $\cos 70° \cos 40°$.

Qu.12 Write out the expansions of $\sin(70° + 40°)$ and $\sin(70° - 40°)$, and deduce expressions for
(i) $\sin 70° \cos 40°$;   (ii) $\sin 40° \cos 70°$.

The method of Example 3, Qu.11 and Qu.12 may be used to derive the following formulae. (The actual derivation is left to the reader.)

$$\cos A \cos B \equiv \tfrac{1}{2}\{\cos(A - B) + \cos(A + B)\};$$
$$\sin A \sin B \equiv \tfrac{1}{2}\{\cos(A - B) - \cos(A + B)\};$$
$$\sin A \cos B \equiv \tfrac{1}{2}\{\sin(A + B) + \sin(A - B)\}.$$

12.5 Converting a sum or difference to a product

The process of factorizing the sum or difference of two sines or cosines is slightly more awkward than the reverse process discussed in the previous section, and, for this reason, the formulae derived below are usually memorized. But we begin by illustrating the derivation with a numerical example.

Example 4
Factorize $\cos 33° + \cos 19°$.
Solution
We first find numbers P and Q such that $P + Q = 33$ and $P - Q = 19$. Solving gives $P = 26$, $Q = 7$. Then

$$\cos 33° = \cos(26° + 7°)$$
$$= \cos 26° \cos 7° - \sin 26° \sin 7°$$
and
$$\cos 19° = \cos(26° - 7°)$$
$$= \cos 26° \cos 7° + \sin 26° \sin 7°.$$

Adding,

$$\cos 33° + \cos 19° = 2 \cos 26° \cos 7°.$$

Qu.13 By subtracting in Example 4, factorize

$$\cos 19° - \cos 33°.$$

Qu.14 Use the method of Example 4 to factorize
(i) $\cos 41° + \cos 59°$;   (ii) $\cos 41° - \cos 59°$;
(iii) $\sin 41° + \sin 59°$.

We can use the method of Example 4 to derive the *factor formulae*. Suppose that we wish to factorize $\sin A + \sin B$. We first find P and Q such that $P + Q = A$ and $P - Q = B$. Solving gives $P = \dfrac{A + B}{2}$ and $Q = \dfrac{A - B}{2}$. Then

$$\sin A = \sin(P + Q) = \sin P \cos Q + \sin Q \cos P,$$
and
$$\sin B = \sin(P - Q) = \sin P \cos Q - \sin Q \cos P.$$

Adding, $$\sin A + \sin B = 2 \sin P \cos Q$$

$$\Rightarrow \qquad \boxed{\sin A + \sin B \equiv 2 \sin \frac{A+B}{2} \cos \frac{A-B}{2}.}$$

Similarly, we have

$$\boxed{\begin{aligned} &\sin A - \sin B \equiv 2 \sin \frac{A-B}{2} \cos \frac{A+B}{2}; \\[2mm] &\cos A + \cos B \equiv 2 \cos \frac{A+B}{2} \cos \frac{A-B}{2}; \\[2mm] &\cos A - \cos B \equiv -2 \sin \frac{A+B}{2} \sin \frac{A-B}{2}. \end{aligned}}$$

These formulae are usually remembered in the form
'sin plus sin equals two sin semi-sum cos semi-difference',
etc.

Qu.15 Use the formulae above to factorize
(i) $\sin 20° + \sin 30°$;
(ii) $\sin(x+20°) + \sin(x-20°)$;
(iii) $\cos 2A - \cos 6A$;
(iv) $\sin \theta - \sin(\pi/2 - \theta)$.

An expression of the form $\sin A + \cos B$ may be factorized by writing it first as $\sin A + \sin\left(\dfrac{\pi}{2} - B\right)$ or as $\sin A + \sin\left(\dfrac{\pi}{2} + B\right)$. (Alternatively, we may write it first as the sum of two cosines.)

Example 5
Factorize $\sin 20° + \cos 40°$.
Solution

$$\begin{aligned} \sin 20° + \cos 40° &= \sin 20° + \sin 50° \\ &= 2 \sin 35° \cos 15°. \end{aligned}$$

Qu.16 Solve Example 5 by writing $\sin 20° + \cos 40°$ as $\sin 20° + \sin 130°$. Verify that the answer is the same.
Qu.17 Factorize
(i) $\sin 68° - \cos 52°$; (ii) $\cos 73° - \sin 55°$.

12.6 Further trigonometric equations
The formulae established in this chapter can be useful for some trigonometric equations.

Example 6
Solve

$$3 \sin \theta = \cos \frac{\theta}{2}$$

for $-180° < \theta \leqslant 180°$.

Solution

Writing $\sin \theta$ as $2 \sin \dfrac{\theta}{2} \cos \dfrac{\theta}{2}$, we have

$$6 \sin \frac{\theta}{2} \cos \frac{\theta}{2} = \cos \frac{\theta}{2}$$

$$\Rightarrow \cos \frac{\theta}{2}\left(6 \sin \frac{\theta}{2} - 1\right) = 0$$

$$\Rightarrow \cos \frac{\theta}{2} = 0 \text{ or } \sin \frac{\theta}{2} = \frac{1}{6}$$

$$\Rightarrow \frac{\theta}{2} = \begin{cases} 90° + 180n° \\ \text{or } 9\cdot59° + 360n° \\ \text{or } 170\cdot41° + 360n° \end{cases}$$

$$\Rightarrow \theta = \begin{cases} 180° + 360n° \\ \text{or } 19\cdot2° + 720n° \\ \text{or } 340\cdot8° + 720n°. \end{cases}$$

Selecting values in the given range,

$$\theta = 19\cdot2° \text{ or } 180°.$$

Example 7

Solve

$$\cos 3\theta - \cos 5\theta + \cos 7\theta = 0$$

for $0 \leqslant \theta \leqslant 90°$.

Solution

Combining the first and third terms,

$$(\cos 3\theta + \cos 7\theta) - \cos 5\theta = 0$$
$$\Rightarrow 2 \cos 5\theta \cos 2\theta - \cos 5\theta = 0$$
$$\Rightarrow \cos 5\theta (2 \cos 2\theta - 1) = 0$$
$$\Rightarrow \cos 5\theta = 0 \text{ or } \cos 2\theta = \tfrac{1}{2}$$
$$\Rightarrow 5\theta = 90° + 180n° \text{ or } 2\theta = \left.\begin{matrix} 60° \\ 300° \end{matrix}\right\} + 360n°$$
$$\Rightarrow \theta = 18° \text{ or } 30° \text{ or } 54° \text{ or } 90°.$$

Example 8 illustrates an important technique, one application of which is given in Example 9.

Example 8

Express $2 \cos \theta - 3 \sin \theta$ in the form $r \sin (\theta + \alpha)$, where r and α are constants, and $r > 0$.

Solution

Suppose $\quad r \sin (\theta + \alpha) \equiv 2 \cos \theta - 3 \sin \theta$.

Then $\quad r \sin \theta \cos \alpha + r \cos \theta \sin \alpha \equiv 2 \cos \theta - 3 \sin \theta$.

Equating coefficients of $\sin \theta$ and of $\cos \theta$ (see Qu.19),

$$\begin{cases} r \cos \alpha = -3, \\ r \sin \alpha = 2. \end{cases}$$

Squaring and adding these equations gives

$$r^2(\cos^2 \alpha + \sin^2 \alpha) = 9 + 4$$
$$\Rightarrow r = \sqrt{13};$$

and dividing them,

$$\tan \alpha = -\frac{2}{3}.$$

(Note that α must now be taken in the 2nd quadrant rather than the 4th, since $\cos \alpha$ is negative and $\sin \alpha$ is positive.)

Thus
$$2 \cos \theta - 3 \sin \theta \equiv \sqrt{13} \sin (\theta + 146 \cdot 3°).$$

Qu.18 (i) Express $3 \sin \theta - 5 \cos \theta$ in the form $r \sin (\theta + \alpha)$.
(ii) Express $2 \sin \theta + 5 \cos \theta$ in the form $r \cos (\theta + \beta)$.

Note that, in general, the amplitude (r) of $a \sin \theta \pm b \cos \theta$ is $\sqrt{(a^2 + b^2)}$.

Qu.19 Show that if
$$a \sin \theta + b \cos \theta \equiv p \sin \theta + q \cos \theta$$

then $a = p$ and $b = q$.

Example 9
Solve the equation

$$3 \cos \theta + \sin \theta = 2$$

for $0 \leqslant \theta < 360°$.
Solution
(Another solution is discussed in Section 12.7.) Using the method of Example 8, we find that $3 \cos \theta + \sin \theta$ may be written $\sqrt{10} \sin (\theta + 71 \cdot 57°)$. The equation thus becomes

$$\sqrt{10} \sin (\theta + 71 \cdot 57°) = 2$$
$$\Rightarrow \sin (\theta + 71 \cdot 57°) = 0 \cdot 6325$$
$$\Rightarrow \theta + 71 \cdot 57° = \left. \begin{array}{c} 39 \cdot 23° \\ 140 \cdot 77° \end{array} \right\} + 360n°.$$

So the required solutions are

$$\theta = 69 \cdot 2° \text{ or } 327 \cdot 7°.$$

Qu.20 Solve Example 9 by first writing the L.H.S. in the form $\sqrt{10} \cos (\theta + \beta)$.

12.7 The substitution $t = \tan \frac{1}{2}\theta$

A trigonometric equation or expression in θ can often be written more conveniently in terms of $\dfrac{\theta}{2}$. For example, the equation in Example 9,

$$3 \cos \theta + \sin \theta = 2,$$

becomes

$$3\left(\cos^2\frac{\theta}{2}-\sin^2\frac{\theta}{2}\right)+2\sin\frac{\theta}{2}\cos\frac{\theta}{2}=2.$$

Dividing through by $\cos^2\frac{\theta}{2}$,

$$3\left(1-\tan^2\frac{\theta}{2}\right)+2\tan\frac{\theta}{2}=2\sec^2\frac{\theta}{2}$$

$$\Rightarrow 3\left(1-\tan^2\frac{\theta}{2}\right)+2\tan\frac{\theta}{2}=2\left(1+\tan^2\frac{\theta}{2}\right),$$

which may be solved as a quadratic in $\tan\frac{\theta}{2}$.

Qu.21 Complete the solution of Example 9 by this method.

A useful short cut in this method is to write t for $\tan\frac{\theta}{2}$, and to know in advance the expressions for $\sin\theta$, $\cos\theta$ and $\tan\theta$ in terms of t. We have

$$\sin\theta=2\sin\frac{\theta}{2}\cos\frac{\theta}{2}$$

$$=2\tan\frac{\theta}{2}\cos^2\frac{\theta}{2}$$

$$=\frac{2\tan\frac{1}{2}\theta}{\sec^2\frac{1}{2}\theta}$$

$$=\frac{2t}{1+t^2};$$

$$\cos\theta=\cos^2\frac{\theta}{2}-\sin^2\frac{\theta}{2}$$

$$=\cos^2\frac{\theta}{2}\left(1-\tan^2\frac{\theta}{2}\right)$$

$$=\frac{1-t^2}{1+t^2};$$

and $\tan\theta=\dfrac{2t}{1-t^2}$ by the double angle formula.

Summarizing,

$$\boxed{\begin{aligned}\sin\theta&\equiv\frac{2t}{1+t^2};\\[2mm]\cos\theta&\equiv\frac{1-t^2}{1+t^2};\\[2mm]\tan\theta&\equiv\frac{2t}{1-t^2}.\end{aligned}}$$

Qu.22   Check that, in terms of t, $\dfrac{\sin\theta}{\cos\theta} = \tan\theta$.

The solution to Example 9 now begins

$$3\cos\theta + \sin\theta = 2$$

$$\Rightarrow 3\left(\frac{1-t^2}{1+t^2}\right) + \left(\frac{2t}{1+t^2}\right) = 2$$

$$\Rightarrow 3(1-t^2) + 2t = 2(1+t^2)$$

and so on.

Qu.23   Use this method to solve $4\cot\theta + 2\operatorname{cosec}\theta = 3$ for $0 \leqslant \theta < 360°$.

Qu.24   If a t-substitution is used to solve $3\sin 6\theta + \sin 6\theta = 2$, what does t stand for?

Qu.25   (For discussion.) Verify that $\theta = \pi$ is a solution of $3\sin\theta + 2\cos\theta = -2$, and show that the substitution $t = \tan\dfrac{\theta}{2}$ leads to the omission of this solution. Explain why.

Exercise 12b
1. Express the following as sums or differences:
      (i) $\sin 30° \cos 40°$;
      (ii) $\cos 20° \cos 50°$;
      (iii) $\cos 80° \sin 70°$;
      (iv) $\sin 15° \sin 25°$;
      (v) $\sin A \cos 3A$;
      (vi) $\sin (P+Q) \sin (P-Q)$;
      (vii) $\sin A \cos A$;
      (viii) $\sin (\theta + 20°) \cos (\theta - 30°)$.
2. Without using tables or a calculator, evaluate
      (i) $\cos 15° \cos 75°$;
      (ii) $\sin 45° \sin 15°$;
      (iii) $\sin 22\frac{1}{2}° \cos 67\frac{1}{2}°$.
3. Factorize
      (i) $\sin 50° + \sin 20°$;
      (ii) $\sin 80° - \sin 42°$;
      (iii) $\cos 32° - \cos 64°$;
      (iv) $\cos 18° + \cos 20°$;

      (v) $\cos (A + 30°) - \cos A$;
      (vi) $\sin\left(x + \dfrac{\pi}{3}\right) - \sin\left(x - \dfrac{\pi}{3}\right)$;

      (vii) $\sin 40° - \cos 70°$;
      (viii) $\cos A + \sin B$;

      (ix) $\sin\left(\dfrac{\pi}{4} - x\right) - \cos\left(\dfrac{\pi}{4} + x\right)$;
      (x) $\sin A - \cos A$;

      (xi) $1 - \cos 2A$.
4. Prove the following identities:

      (i) $\dfrac{\sin 2A + \sin 2B}{\cos 2A + \cos 2B} \equiv \tan (A + B)$;

      (ii) $\dfrac{\sin A - \sin B}{\cos A - \cos B} \equiv -\cot\left(\dfrac{A+B}{2}\right)$;

      (iii) $\dfrac{\sin A + \sin 3A + \sin 5A}{\cos A + \cos 3A + \cos 5A} \equiv \tan 3A$;

      (iv) $\cos 2A + 2\cos 4A + \cos 6A \equiv 4\cos^2 A \cos 4A$;

(v) $\dfrac{\sin 3A + \sin 5A + \sin 7A + \sin 9A}{\sin A + \sin 3A + \sin 5A + \sin 7A} \equiv \dfrac{\sin 6A}{\sin 4A}$.

5. Without using tables or calculator, evaluate
   (i) $\sin 75° + \sin 15°$;   (ii) $\cos 15° - \cos 75°$.

6. Without using tables or calculator, show that
   (i) $\cos 40° + \cos 80° = \cos 20°$;
   (ii) $1 + \sin 50° = 2 \sin^2 70°$;

   (iii) $\sin \theta + \cos \theta \equiv \sqrt{2} \cos \left(\theta - \dfrac{\pi}{4}\right)$.

7. Show that $\sec A \cos 3A \equiv 2 \cos 2A - 1$.
   Deduce that $\sec 15° = \sqrt{6} - \sqrt{2}$.

8. (i) Show that

   $$\tan A - \tan B \equiv (1 + \tan A \tan B) \tan(A - B).$$

   (ii) Find the corresponding factorization of $\tan A + \tan B$.
   (iii) Show that

   $$\sec A + \sec B \equiv 2 \sec A \sec B \cos \dfrac{A + B}{2} \cos \dfrac{A - B}{2}.$$

   (iv) Find the corresponding factorization of $\operatorname{cosec} A - \operatorname{cosec} B$.

9. Solve the following equations for $-180° < \theta \leqslant 180°$. (Remember to find the general solution first.)
   (i) $\sin \theta = \sin 5\theta$ (by first writing the equation as $\sin 5\theta - \sin \theta = 0$ and factorizing; compare with the method used in Example 6 of Chapter 11).
   (ii) $\sin \theta = 3 \sin 2\theta$;
   (iii) $\cos 2\theta - 3 \sin \theta = 1$;
   (iv) $\cos 2\theta - 3 \sin \theta = -1$;
   (v) $\tan 2\theta = 2 \tan \theta$.

10. Apply the method of question 9(i) to parts (i) to (iv) of question 15 of Exercise 11b.

11. Express the following in the form $r \sin(\theta \pm \alpha)$:
    (i) $\sin \theta + \cos \theta$;   (ii) $\sqrt{3} \sin \theta - \cos \theta$;
    (iii) $3 \cos \theta - 5 \sin \theta$.

12. Express in the form $r \cos(\theta \pm \alpha)$:
    (i) $3 \sin \theta + 4 \cos \theta$;   (ii) $\cos \theta - \sin \theta$;
    (iii) $2 \sin \theta - \cos \theta$.

13. *Write down* the maximum value of $\sin x + 2 \cos x$.

14. Solve the following equations for $0 \leqslant \theta < 360°$. (Each of the methods of Example 9 and Section 12.7 should be practised.)
    (i) $\sin \theta - \cos \theta = \frac{1}{2}$;
    (ii) $12 \cos \theta + 5 \sin \theta = 6·5$;
    (iii) $2 \cos \theta - 3 \sin \theta + 1·5 = 0$;
    (iv) $3 \sin \theta - \cos \theta = 3$.

15. With triangle OPQ as in Fig. 12.5, show that

    (i) $\mathbf{OP} = \sqrt{(a^2 + b^2)} \begin{pmatrix} \cos(\theta + \alpha) \\ \sin(\theta + \alpha) \end{pmatrix}$;   (ii) $\mathbf{OQ} = a \begin{pmatrix} \cos \theta \\ \sin \theta \end{pmatrix}$;

    (iii) $\mathbf{QP} = b \begin{pmatrix} -\sin \theta \\ \cos \theta \end{pmatrix}$.

Figure 12.5

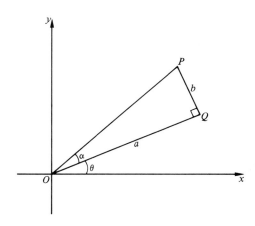

Deduce that $a \sin \theta + b \cos \theta = \sqrt{(a^2 + b^2)} \sin (\theta + \alpha)$ where $\tan \alpha = \dfrac{b}{a}$.

16. In this question, we show that $a \sin \theta + b \sin (\theta + \varepsilon)$ may be expressed in the form $r \sin (\theta + \alpha)$. This means that when two sine curves of the same period are added, they give a simple sine curve, even if they have different amplitudes (a and b) and are out of phase (by an angle ε).

(i) Find, in terms of a, b and ε, values of c and d such that

$$a \sin \theta + b \sin (\theta + \varepsilon) \equiv c \sin \theta + d \cos \theta.$$

(ii) Deduce that

$$a \sin \theta + b \sin (\theta + \varepsilon) \equiv r \sin (\theta + \alpha)$$

where $r = \sqrt{(a^2 + b^2 + 2ab \cos \varepsilon)}$ and $\tan \alpha = \dfrac{b \sin \varepsilon}{a + b \cos \varepsilon}$.

(iii) Show how this solution may be illustrated by Fig. 12.6 by finding in terms of a, b, ε and θ:
(a) **OQ** and **QP**.
Hence find
(b) **OP**;

Figure 12.6

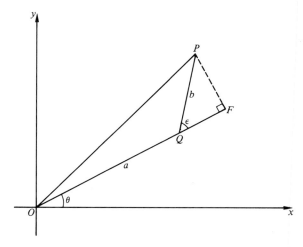

(c) the lengths PF and OF;
(d) the length of OP;
(e) $\tan \alpha$, where $\alpha = \angle POQ$.
Complete the deduction of (ii).

17. (Hard.) Without using tables or calculator, show that

$$\cos 70° = \frac{\cos 30°}{2 \cos 40° + 1}.$$

Deduce that $x = \cos 70°$ is a solution of

$$3x - 4x^3 = \frac{\sqrt{3}}{2}.$$

18. Solve
    (i) $\cos 2\theta + 3 \cos 5\theta + \cos 8\theta = 0$;
    (ii) $\sin \theta + \sin 5\theta + \sin 9\theta = 0$;
    (iii) $\cos \theta + \cos 4\theta + \cos 6\theta = 0$.
19. Solve
    (i) $7 \cos \theta + 8 \cos (\theta - 60°) = 13$;
    (ii) $\cos \theta + 2 \cos (\theta - 30°) = 1·5$.
20. Use the t-substitution formulae to solve

$$7 \cos 4\theta - 6 \sin 4\theta = 2.$$

21. Show that
    (i) $\cos^2 A - \cos^2 B \equiv \sin (B + A) \sin (B - A)$;

    (ii) $4 \cos \theta \cos \left(\theta + \frac{2\pi}{3} \right) \cos \left(\theta + \frac{4\pi}{3} \right) \equiv \cos 3\theta$.

22. Factorize
    (i) $\sin A + \sin B - \sin C - \sin (A + B - C)$;
    (ii) $\cos^2 A + \cos^2 B - 1$. [*Hint*: convert to double angles first.]
23. By using the factor formulae or the sum and difference formulae, find the maximum values of the following, and the corresponding values of x.
    (i) $\cos x \sin (x + 30°)$;   (ii) $\sin (x + 20°) \sin (x - 30°)$;
    (iii) $\cos x + \cos (x + 40°)$;   (iv) $\sin (x - 23°) - \cos (x + 7°)$.

12.8 Triangles: the sine and cosine rules

In proving an identity involving the three angles A, B and C of a triangle, we know that $A + B + C = \pi$, and so

$$\sin C = \sin \{\pi - (A + B)\} = \sin (A + B)$$
and $$\cos C = \cos \{\pi - (A + B)\} = -\cos (A + B).$$

The neatest method of proving many identities involves the use of the factor formulae.

Example 10
If A, B and C are the three angles of a triangle, show that

$$\sin 2A + \sin 2B + \sin 2C = 4 \sin A \sin B \sin C.$$

Solution

$$(\sin 2A + \sin 2B) + \sin 2C$$
$$= 2 \sin (A+B) \cos (A-B) + 2 \sin C \cos C$$
$$= 2 \sin (A+B) \cos (A-B) - 2 \sin (A+B) \cos (A+B)$$
$$= 2 \sin (A+B) \{\cos (A-B) - \cos (A+B)\}$$
$$= 2 \sin (A+B) \times 2 \sin A \sin B$$
$$= 4 \sin A \sin B \sin C.$$

The sine rule and cosine rule may already be familiar to the reader. Their proofs are essentially elementary and are outlined in Qus. 27 and 28. The rules are as follows:

Let a, b, c be the lengths of the sides of triangle ABC opposite angles A, B, C respectively (so $a = BC$ etc.). Then the *sine rule* states that

$$\frac{a}{\sin A} = \frac{b}{\sin B} = \frac{c}{\sin C};$$

and the *cosine rule* states that

$$a^2 = b^2 + c^2 - 2bc \cos A.$$

Note

(1) In using the sine rule, only one equality is needed at a time. The reciprocal form $\dfrac{\sin A}{a} = \dfrac{\sin B}{b}$ is more convenient for finding angles.

(2) When using the sine rule to find an angle, the possibility of an *obtuse* angle should be considered. (See Qu. 26.)

(3) The cosine rule may be written with b^2 or c^2 as the subject, for example $b^2 = a^2 + c^2 - 2ac \cos B$. It is also commonly used with $\cos A$, $\cos B$ or $\cos C$ as the subject, for example

$$\cos A = \frac{b^2 + c^2 - a^2}{2bc}.$$

(4) The cosine rule may be regarded as a generalization of Pythagoras's Theorem. (Putting $\cos A = 0$ gives Pythagoras.)

(5) When deciding whether to use the sine rule or the cosine rule, remember that the sine rule involves two sides and two angles, whereas the cosine rule involves three sides and one angle.

Qu.26 (i) In triangle ABC, $AB = 7$ cm, $AC = 5$ cm and $\angle B = 32°$. By means of a suitable sketch, show that there are two possible triangles, and state the relationship between the two values of $\angle C$.

(ii) Calculate $\angle C$ by the sine rule, showing at which stage the two answers appear.

Qu.27 (Two proofs of the sine rule.)

(i) Draw a triangle ABC and the perpendicular CD from C to AB. Show that $CD = a \sin B$ and $CD = b \sin A$, and deduce that

$\dfrac{a}{\sin A}=\dfrac{b}{\sin B}$. (Similarly $\dfrac{a}{\sin A}=\dfrac{c}{\sin C}$ and the full result follows.) Check that this argument is valid if any of the angles is obtuse.

(ii) Fig. 12.7 shows triangle ABC and its circumcircle, and AD is the diameter through A. Explain why $\angle ACD=90°$ and why $\angle B=\angle D$.

Use triangle ADC to show that $\dfrac{b}{\sin B}=2R$ where R is the radius.

$\left(\text{Similarly } \dfrac{a}{\sin A}=2R \text{ etc., and so } \dfrac{a}{\sin A}=\dfrac{b}{\sin B}=\dfrac{c}{\sin C}=2R.\right)$

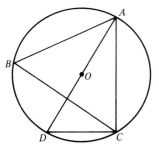

Figure 12.7

Qu.28 (Proof of cosine rule.)

(i) In Fig. 12.8, write down Pythagoras's Theorem for triangles ADC and BDC. Eliminate h between these equations and show that $c^2-2xc=a^2-b^2$. Deduce that $a^2=b^2+c^2-2bc\cos A$.

(ii) Modify this proof for the case when $\angle A$ or $\angle B$ is obtuse.

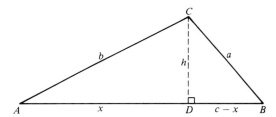

Figure 12.8

Example 11

With the usual notation, show that

$$a\cos A-b\cos B=\cos C\,(b\cos A-a\cos B).$$

Solution

Let $\dfrac{a}{\sin A}=2R$; then, by the sine rule, $\dfrac{b}{\sin B}=2R$ and $\dfrac{c}{\sin C}=2R$.

So

$$\begin{aligned}
\cos C(b&\cos A-a\cos B)\\
&=-\cos(A+B)\{2R\sin B\cos A-2R\sin A\cos B\}\\
&=2R\cos(A+B)\sin(A-B)\\
&=R(\sin 2A-\sin 2B)\\
&=2R(\sin A\cos A-\sin B\cos B)\\
&=a\cos A-b\cos B.
\end{aligned}$$

TRIGONOMETRIC FORMULAE

Qu.29 Use the cosine rule to show that

$$\frac{\cos A}{a} + \frac{\cos B}{b} + \frac{\cos C}{c} = \frac{a^2 + b^2 + c^2}{2abc}.$$

12.9 Two- and three-dimensional trigonometric problems

The following points should be borne in mind when solving two- and three-dimensional problems:

(1) A clear diagram is essential. In the case of a three-dimensional problem, additional two-dimensional diagrams (looking 'head-on' at particular faces) may be helpful.

(2) The angle between a line and a plane is the angle between the line and its perpendicular (or orthogonal) projection onto the plane. (In Fig. 12.9, FB is the projection of AB on the plane π; so α is the angle between AB and π.)

Figure 12.9

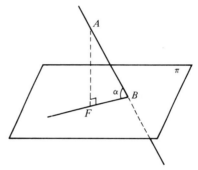

(3) The angle between two planes (Fig. 12.10) is the angle between two lines, one in each plane, which are perpendicular to the line of intersection of the two planes. (If one of the planes is horizontal, then the required angle is the angle made with the horizontal by a 'line of greatest slope' in the other plane.)

Figure 12.10

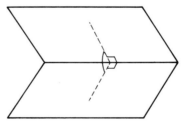

(4) The sine and cosine rules may be useful, but not usually when right-angled triangles are available.

(5) Many problems involve proving an identity. A common method of approaching these is to 'chase round' the diagram in two different ways to obtain two expressions for a particular length or angle, and then to equate these expressions.

Example 12

Q and B are the tops of two vertical towers QP and BA which stand on a slope so that the angles of elevation of A and B from P are α and β respectively. From Q, the angle of depression of B is θ. If AB is of height h, show that the height of PQ is

$$\frac{h\ \sin(\beta+\theta)\ \cos\alpha}{\cos\theta\ \sin(\beta-\alpha)}.$$

Solution

Figure 12.11

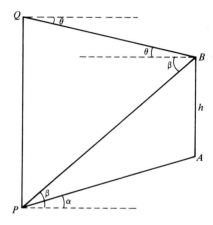

By the sine rule in triangle APB,

$$\frac{PB}{\sin A}=\frac{h}{\sin(\beta-\alpha)}.$$

But the angle at A is $90°+\alpha$, so $\sin A=\cos\alpha$

$$\Rightarrow PB=\frac{h\ \cos\alpha}{\sin(\beta-\alpha)}. \tag{1}$$

By the sine rule in triangle PQB,

$$\frac{PQ}{\sin(\beta+\theta)}=\frac{PB}{\sin(90°-\theta)}$$

$$\Rightarrow PQ=\frac{PB\ \sin(\beta+\theta)}{\cos\theta}.$$

Substituting for PB from (1),

$$PQ=\frac{h\ \cos\alpha\ \sin(\beta+\theta)}{\cos\theta\ \sin(\beta-\alpha)}.$$

Example 13
A hill-side slopes at an angle θ to the horizontal, and a path up the hill makes an angle ϕ with a line of greatest slope. Find an expression for the angle that the path makes with the horizontal. (Assume that the hill-side is a plane.)
Solution
Fig. 12.12 shows a slice of the hill with $DEBC$ a vertical rectangle. AE and AB are horizontal, AC is a line of greatest slope, and AD is the

Figure 12.12

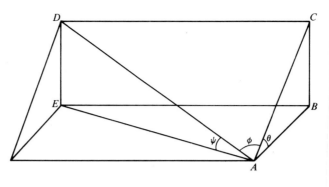

Figure 12.13

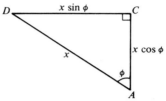

path. Let the required angle be ψ, and let $AD = x$. Then in triangle ADC

$$DC = x \sin \phi$$

and
$$AC = x \cos \phi.$$

And so, in triangle ACB,

$$CB = AC \sin \theta$$
$$= x \cos \phi \sin \theta.$$

So
$$DE = x \cos \phi \sin \theta.$$

Figure 12.14

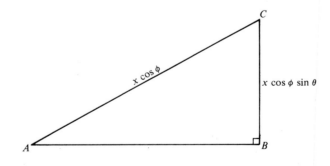

Figure 12.15

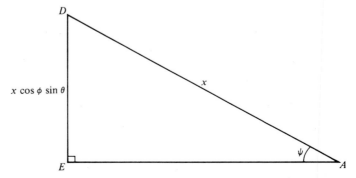

Then, in triangle DEA,

$$\sin \psi = \frac{x \cos \phi \, \sin \theta}{x}$$

$$\Rightarrow \quad \psi = \arcsin (\cos \phi \, \sin \theta).$$

Note

(1) In this solution, we took DA to be x; but we could have taken any of the lengths involved in the question as our basic length. Since it is easier to work with sines and cosines than with secants and cosecants, it is usually best to call the hypotenuse of some triangle by a suitable letter.

(2) As a (partial) check, we note that the answer, $\psi = \arcsin (\cos \phi \, \sin \theta)$, gives (i) $\psi \leqslant \theta$; (ii) $\phi = 0 \Rightarrow \psi = \theta$; (iii) $\phi = 90° \Rightarrow \psi = 0$; all of which are as we should expect from the diagram.

Qu.30 (See Example 13.) If the lines of greatest slope on the hill-side run due North, and a second path runs up the hill on a bearing of ϕ West of North, find the slope of this second path. (In Fig. 12.12, it is now angle EAB which is ϕ.)

Exercise 12c (In this exercise, A, B and C are the angles of a triangle, and a, b, c are the respective lengths of the opposite sides.)

1. Prove the following:

(i) $\sin A + \sin B + \sin C = 4 \cos \dfrac{A}{2} \cos \dfrac{B}{2} \cos \dfrac{C}{2}$;

(ii) $\cos A + \cos B + \cos C = 1 + 4 \sin \dfrac{A}{2} \sin \dfrac{B}{2} \sin \dfrac{C}{2}$;

(iii) $\cos 2A + \cos 2B + \cos 2C = -1 - 4 \cos A \cos B \cos C$;

(iv) $\cos^2 \dfrac{A}{2} + \cos^2 \dfrac{B}{2} + \cos^2 \dfrac{C}{2} = 2 + 2 \sin \dfrac{A}{2} \sin \dfrac{B}{2} \sin \dfrac{C}{2}$.

2. (i) If $A = 42°$, $B = 69°$, $a = 7 \cdot 3$ cm, find b.

(ii) If $c = 3 \cdot 5$ m, $C = 71 \cdot 2°$, $a = 2 \cdot 9$ m, find B. [*Hint*: find A first.]

(iii) If $a = 23 \cdot 5$ m, $c = 14 \cdot 8$ m, $B = 127°$, find b.

(iv) If $b = 9 \cdot 8$ cm, $a = 6 \cdot 5$ cm, $C = 51°15'$, find c.

(v) If $a = 2 \cdot 3$ cm, $b = 3 \cdot 4$ cm, $c = 5 \cdot 0$ cm, find the largest angle.

3. (For discussion.) A triangle has sides 8 cm, 9 cm, 12 cm. Determine as efficiently as possible whether it has an obtuse angle or not.

4. (i) (For discussion.) A standard method of finding the angles of a triangle given all three sides is to find one angle by the cosine rule, a second by the sine rule, and the third by the angle sum of a triangle. Explain why the largest angle should not be the second to be found (i.e. by the sine rule).

(ii) Find all the angles of a triangle with sides 4·87 cm, 5·36 cm, 7·25 cm.

5. Solve the following triangles (i.e. find all unknown sides and angles). Where there is an ambiguity, give both sets of solutions. Find also the radius of the circumcircle in each case.

(i) $A = 29°$, $c = 7·9$ cm, $a = 5·8$ cm;

(ii) $A = 29°$, $c = 5·8$ cm, $a = 7·9$ cm.

6. (i) Use the sine rule and the small angle approximation $\sin \theta \approx \theta$, to show that if A and B are small, then

$$A \approx \frac{a}{a+b}(A+B).$$

(ii) Use this result to estimate the two other angles of a triangle with sides 2 cm and 3 cm and included angle 165°. Check your answer by exact calculation.

(iii) (For discussion.) In what circumstances will this method give an exact answer?

7. Prove the following:

(i) $a \cos A + b \cos B + c \cos C = 2c \sin A \sin B$. (Use the sine rule.)

(ii) $a \sin (B - C) + b \sin (C - A) + c \sin (A - B) = 0$. (Use the sine rule.)

(iii) $\dfrac{\sin (A - B)}{\sin (A + B)} = \dfrac{a^2 - b^2}{c^2}$.

8. Use the sine rule to prove the *tangent rule*

$$\frac{a-b}{a+b} = \tan \frac{A-B}{2} \ \tan \frac{C}{2}.$$

9. (*Hero's formula* for the area of a triangle.)

(i) Starting with the formula $\triangle = \frac{1}{2}bc \sin A$ for the area of a triangle (see Exercise 11a, question 16), show that

$$\triangle^2 = \tfrac{1}{4}b^2c^2(1 - \cos^2 A)$$

and deduce that

$$16\triangle^2 = 4b^2c^2 - (b^2 + c^2 - a^2)^2.$$

(ii) Factorize the R.H.S. and simplify each factor to obtain

$$16\triangle^2 = \{a^2 - (b-c)^2\}\{(b+c)^2 - a^2\}.$$

(iii) Factorize each bracket and deduce Hero's formula

$$\triangle = \sqrt{\{s(s-a)(s-b)(s-c)\}}$$

where $s = \frac{1}{2}(a+b+c)$.

Questions 10 to 13 are two-dimensional.

10. A vertical tower stands on horizontal ground. The angle of elevation of the top of the tower from a point on the ground is α, and the corresponding angle measured from a point a distance x nearer to the tower is β. Show that the height of the tower is

$$\frac{x \sin \alpha \ \sin \beta}{\sin (\beta - \alpha)}.$$

11. Use the sine rule to show that, in Fig. 12.16, $\dfrac{x}{a} = \dfrac{y}{b}$. (This is one case of the angle bisector theorem.)

12. Show that in Fig. 12.17

$$\tan \theta = \frac{2 \sin \alpha \sin \beta}{\sin (\beta - \alpha)}.$$

Figure 12.16

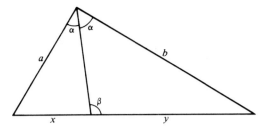

Figure 12.17

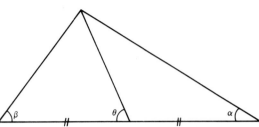

13. A is the foot of a straight tower AC which tilts to the North, making an angle θ with the horizontal ground. B is a point on the tower between A and C. From a point P a distance d due North of A, the angles of elevation of B and C and β and γ respectively. Show that the distance BC is

$$\frac{d \sin \theta \, \sin (\gamma - \beta)}{\sin (\theta + \beta) \, \sin (\theta + \gamma)}.$$

Questions 14 to 17 are three-dimensional.

14. A, B and C are three points on horizontal ground, with A due South of C and B due West of C. A vertical post at C subtends angles of α and 2α at A and B respectively. Show that the bearing of B from A is θ West of North, where

$$2 \tan \theta = 1 - \tan^2 \alpha.$$

15. (i) From a point A on level ground, the bearings of two vertical towers, each of height h, are due North and θ East of North. If the distance between the towers is d, and their angles of elevation from A are α and β, show that

$$d^2 = h^2(\cot^2 \alpha + \cot^2 \beta - 2 \cot \alpha \cot \beta \cos \theta).$$

(ii) If wires are joined from the tops of the towers to A, show that the angle between the wires is

$$\arccos (\sin \alpha \, \sin \beta + \cos \alpha \, \cos \beta \, \cos \theta).$$

16. Two paths cross at right angles on a hill which slopes at θ to the horizontal. If the paths slope at angles α and β to the horizontal, show that

$$\sin^2 \theta = \sin^2 \alpha + \sin^2 \beta.$$

17. Three points A, B and C are on level ground, with $AB = BC$. Vertical towers are built at B and C subtending angles at A of α and 2α respectively. The angle of elevation of the top of the tower at C from the top of the tower at B is also α. If $\angle ABC = 4\theta$, show that

$$\sec \alpha = 2 \sin(45° - \theta).$$

1. Prove that $\cos 3\theta = 4 \cos^3 \theta - 3 \cos \theta$.
   Hence, find solutions of $4x^3 - 27x = -27$, by substituting $x = 3 \cos \theta$.

[SUJB (O)]

2. (a) Prove the identity

$$\frac{2 \sin A + \sin 2A}{1 - \cos 2A} = \frac{\sin A}{1 - \cos A}.$$

(b) Find all the values of x between $0°$ and $360°$ which satisfy the equation

$$\cos 2x = \sin x.$$

(c) Prove the identity

$$\frac{\sin 2A + \sin 2B}{\cos 2A + \cos 2B} = \tan(A + B).$$

[SUJB (O)]

3. (a) The obtuse angle β is such that $\tan \beta = -\frac{3}{4}$.
   (i) Find, *without using tables*, the values of $\sin \beta$, $\cos \beta$, $\sin 2\beta$ and $\cos 2\beta$.
   (ii) Show that $5 \sin(\theta + \beta) = 3 \cos \theta - 4 \sin \theta$.
   (b) Solve the equation $\tan x = 2 \sin x$ for values of x between $-180°$ and $+180°$.

   (c) Prove that $\dfrac{\cos 10° - \cos 110°}{\sin 110° - \sin 10°} = \sqrt{3}$.

[W (O)]

4. Express $\cos 2\theta$ in terms of $\sin \theta$ and express $\tan 2\theta$ in terms of $\tan \theta$.
   A pyramid stands on a horizontal square base $ABCD$ of side 20 metres. The vertex P is 20 metres vertically above O, the centre of the base and Q is the mid-point of BC. Without using tables or calculators,
   (i) calculate $\sin \angle BPQ$ in surd form and show that $\cos \angle BPC = \frac{2}{3}$,
   (ii) state the value of $\tan \angle OPQ$ and show that two opposite slant faces of the pyramid are inclined to each other at an angle of $\tan^{-1}(\frac{4}{3})$,
   (iii) calculate the cosine of the angle between two opposite slant edges.

[JMB (O)]

5. (a) Prove that $\sin \theta + \sin(\theta + 120°) = \sin(\theta + 60°)$.
   (b) Solve the equation $\cos \theta + \sin 2\theta - \cos 3\theta = 0$, giving all possible values of θ in the range $0° \leqslant \theta \leqslant 360°$.
   (c) Express $5 \sin \theta - 6 \cos \theta$ in the form $R \sin(\theta - \alpha)$ where R is a positive constant and $0° < \alpha < 90°$.

Hence or otherwise solve the equations
(i) $5 \sin \theta - 6 \cos \theta = 4$,
(ii) $5 \sin \theta - 6 \cos \theta = 0$,
giving all possible values of θ in the range $0° < \theta < 360°$.

[JMB (O)]

6. In this question all angles are measured in degrees.
   (i) Prove the identity

   $$\cos(60 + x) + \sin(30 + x) = \cos x.$$

   (ii) The angle y is acute. In addition it is known that $\cos 2y$ is negative and that $\sin 3y$ is positive. Find the range of values within which y must lie.
   (iii) Solve the equation $\sin 2z = \frac{1}{2}$ giving values of z such that $0 \leqslant z \leqslant 360$.

[Ox (O)]

7. A surveyor takes readings of the angles of elevation of T, the top of a church spire, from two points P and Q which are on the same level. The point T is h metre above the level of P and Q. The point P is due south of the spire and the angle of elevation of T from P is $45°$. The point Q is due west of the spire and the angle of elevation of T from Q is θ, such that $\tan \theta = 0.5$. If the distance PQ is 120 m, calculate the value of h.

   Find, also, the greatest angle of elevation of T from any point in the line PQ.

[O & C (O)]

8. (a) Express $\cos x + 2 \sin x$ in the form $R \cos(x - \alpha)$ where R is positive. Hence or otherwise solve the equation $\cos x + 2 \sin x = 1.52$ for $0° \leqslant x \leqslant 360°$.
   (b) An equilateral triangle ABC is drawn on a plane inclined at $5°$ to the horizontal. AB is horizontal at a level lower than that of the point C and $AB = 10$ cm. Point P is the foot of the perpendicular from C to the horizontal plane through A. The midpoint of AB is M. Find
   (i) the length of CM,
   (ii) the inclination of AC to the horizontal.

[AEB '79]

9. (a) Prove that, if all the angles are positive and acute,
   (i) $\sin^{-1}\left(\frac{3}{5}\right) - \sin^{-1}\left(\frac{5}{13}\right) = \sin^{-1}\left(\frac{16}{65}\right)$,
   (ii) $\cos^{-1}\sqrt{(2x - x^2)} = \sin^{-1}(1 - x)$, where $0 < x < 1$.
   (b) Show that the equation

   $$3 \sec x + 4 \operatorname{cosec} x = 10$$

   can be put in the form

   $$\sin 2x = \sin(x + \alpha),$$

   where α is a constant. Hence find all the angles between $0°$ and $360°$ which satisfy the equation

   $$3 \sec x + 4 \operatorname{cosec} x = 10.$$

[C (S)]

10. (i) Find the general solution of the equation $\sin 2\theta = \cos \theta$.

(ii) The points A, B, C lie on a circle with centre O. Given that $AB = 11$ m, $BC = 13$ m, $CA = 20$ m, find the angles AOB, BOC, COA to the nearest tenth of a degree and the radius of the circle to the nearest tenth of a metre.

[L]

11. Triangle ABC is inscribed in a circle centre O and radius R. The sides of the triangle are of lengths $BC = a$, $CA = b$ and $AB = c$.

State the size of angle BOC and show that $\dfrac{a}{\sin A} = 2R$. Hence prove the Sine Rule.

A line is now drawn through the vertex A of the triangle to meet BC at D. Show that the ratio of the radii of the circumcircles of the triangles ABD and ACD is $\dfrac{c}{b}$. Hence, by letting D approach C, prove that the radius of a circle through A and touching BC at C is $\dfrac{b}{2 \sin C}$.

[SUJB (O)]

12. (a) Show that if $x + y = \frac{1}{4}\pi$ then

$$(1 + \tan x)(1 + \tan y) = 2.$$

(b) Without reference to tables find the value of

$$\tan^{-1}\left(\tfrac{1}{2}\right) + \tan^{-1}\left(\tfrac{1}{3}\right).$$

(c) Show that if $\theta = 72°$ then $\cos 3\theta = \cos 2\theta$. Hence prove that $\cos 72° = \frac{1}{4}(\sqrt{5} - 1)$, and deduce that $\cos 36° = \frac{1}{4}(\sqrt{5} + 1)$.

[W]

13. (a) Solve the equation

$$5 \cot \theta = 3 + 2 \operatorname{cosec} \theta$$

giving the general solution.

(b) Prove, for a triangle ABC, that

$$\tan\left(\frac{B - C}{2}\right) = \frac{b - c}{b + c} \cot \frac{A}{2}.$$

Given that $b = 7 \cdot 5$, $c = 2 \cdot 5$ and $A = 60°$, calculate B, C and a.

[C]

14. Using the formula for $\tan(A + B)$ deduce, or prove otherwise, that

$$\tan 3\theta = \frac{3 \tan \theta - \tan^3 \theta}{1 - 3 \tan^2 \theta}.$$

Deduce that the roots of the cubic equation

$$t^3 - 3t^2 - 3t + 1 = 0$$

are $\tan(\pi/12)$, $\tan(5\pi/12)$ and $\tan(3\pi/4)$. Find also a cubic equation with integer coefficients whose roots are $\sec^2(\pi/12)$, $\sec^2(5\pi/12)$ and $\sec^2(3\pi/4)$.

[L (S)]

15. A and B are two points on level ground, and B is a metres due

East of A; a tower, h metres high, is also on the same level ground.

From A, the tower is in a direction $N\theta E$ and from B it is $N\phi W$.

From the top of the tower, the angle of depression of A is α and of B is β.

Prove the following (in any order):

(i) $h \sin(\theta + \phi) = a \cos\phi \tan\alpha$

(ii) $\cos\phi \tan\alpha = \cos\theta \tan\beta$

(iii) $h^2(\cot^2\alpha - \cot^2\beta) - 2ha \cot\alpha \sin\theta + a^2 = 0$.

[SUJB]

16. A vertical flagpole AB of height h stands at a point A on an inclined plane which makes an angle α with the horizontal. From a point P on the plane, a distance a from A down the line of greatest slope through A, the angle of elevation of B, the top of the pole, is θ. From a point Q on the plane, a distance a from A up the line of greatest slope, the angle of elevation of B is ϕ. Prove that

(i) $\tan\alpha = \frac{1}{2}(\tan\theta - \tan\phi)$,

(ii) $h = a \sin(\theta + \phi)[4 \cos^2\theta \cos^2\phi + \sin^2(\theta - \phi)]^{-\frac{1}{2}}$.

[C]

13 The differentiation of trigonometric functions

13.1 An important limit

In differentiating the sine and cosine functions from first principles, we shall need to know the value of $\lim\limits_{\theta \to 0} \dfrac{\sin \theta}{\theta}$. The value of this limit is not immediately obvious, since, as $\theta \to 0$, both the numerator and the denominator tend to zero.

Theorem 13.1

$$\lim_{\theta \to 0} \frac{\sin \theta}{\theta} = 1.$$

(Note that θ is in radians.)

Proof

In Fig. 13.1 (in which PT is a tangent to the circle centre O),

$$\text{area of } \triangle\, OPQ < \text{ area of sector } OPQ < \text{area of } \triangle OPT.$$

Figure 13.1

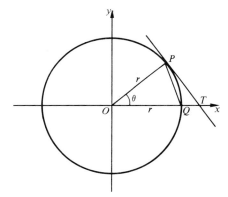

So, for $0 < \theta < \dfrac{\pi}{2}$,

$$\tfrac{1}{2} r^2 \sin \theta < \tfrac{1}{2} r^2\, \theta < \tfrac{1}{2} r^2\, \tan \theta$$
$$\Rightarrow \sin \theta < \theta < \tan \theta$$
$$\Rightarrow 1 < \frac{\theta}{\sin \theta} < \frac{1}{\cos \theta}.$$

Now these inequalities remain true when θ is very small, and, as $\theta \to 0$, $\dfrac{1}{\cos \theta} \to 1$. So the value of $\dfrac{\theta}{\sin \theta}$ is 'squashed' between 1 and a number which is approaching 1. Therefore $\lim\limits_{\theta \to 0} \dfrac{\theta}{\sin \theta} = 1$, and hence $\lim\limits_{\theta \to 0} \dfrac{\sin \theta}{\theta} = 1$.

It follows from this result that if θ is small, then $\sin\theta\approx\theta$. A more precise relationship is derived in Chapter 28. Be careful to avoid the nonsense of writing $\sin\theta\to\theta$ (see Section 2.14).

Qu.1 Using a calculator, find (and compare) the values of $\sin\theta$ and θ when θ is

(i) $\dfrac{\pi}{10}$;   (ii) $\dfrac{\pi}{20}$;   (iii) $\dfrac{\pi}{30}$;   (iv) $\dfrac{\pi}{40}$;   (v) $\dfrac{\pi}{50}$.

Qu.2 For what values of θ in the first quadrant is $\dfrac{\sin\theta}{\theta}>0{\cdot}95$?

(Use trial and error.)

Qu.3 Use the inequalities $\sin\theta<\theta<\tan\theta$ derived in the proof above to show that

$$\lim_{\theta\to0}\frac{\tan\theta}{\theta}=1.$$

13.2 The derivatives of sin x and cos x etc.

Starting from the definition of a derived function, we have

$$\frac{d}{dx}(\sin x)=\lim_{h\to0}\frac{\sin(x+h)-\sin x}{h}$$

$$=\lim_{h\to0}\frac{1}{h}\left\{2\sin\frac{h}{2}\cos\left(x+\frac{h}{2}\right)\right\}\ \text{(by the factor formula)}$$

$$=\lim_{h\to0}\frac{\sin h/2}{h/2}\cos\left(x+\frac{h}{2}\right).$$

Now as $h\to0$, then also $\dfrac{h}{2}\to0$; so by Theorem 13.1 with $\theta=\dfrac{h}{2}$, $\dfrac{\sin h/2}{h/2}\to1$. Further, $\cos\left(x+\dfrac{h}{2}\right)\to\cos x$. Thus

$$\frac{d}{dx}(\sin x)=1.\ \cos x=\cos x.$$

Qu.4 Before reading on, try differentiating $\cos x$ from first principles.

Similarly, we have

$$\frac{d}{dx}(\cos x)=\lim_{h\to0}\frac{\cos(x+h)-\cos x}{h}$$

$$=\lim_{h\to0}\frac{-2\sin(x+h/2)\sin h/2}{h}$$

$$=\lim_{h\to0}-\frac{\sin h/2}{h/2}\sin\left(x+\frac{h}{2}\right)$$

$$=-\sin x.$$

The derivatives of the other trigonometric functions can now be found using the quotient rule or the function of a function rule.

Thus

$$\frac{d}{dx}(\tan x) = \frac{d}{dx}\left(\frac{\sin x}{\cos x}\right)$$

$$= \frac{\cos^2 x + \sin^2 x}{\cos^2 x}$$

$$= \sec^2 x,$$

and

$$\frac{d}{dx}(\sec x) = \frac{d}{dx}(\cos x)^{-1}$$

$$= \sin x\,(\cos x)^{-2}$$

$$= \sec x \tan x.$$

Qu.5 Show that

$$\frac{d}{dx}(\cot x) = -\mathrm{cosec}^2 x$$

and

$$\frac{d}{dx}(\mathrm{cosec}\, x) = -\mathrm{cosec}\, x \cot x.$$

Summarizing,

$$\frac{d}{dx}(\sin x) = \cos x; \qquad \frac{d}{dx}(\cos x) = -\sin x;$$

$$\frac{d}{dx}(\tan x) = \sec^2 x; \qquad \frac{d}{dx}(\cot x) = -\mathrm{cosec}^2 x;$$

$$\frac{d}{dx}(\sec x) = \sec x \tan x; \qquad \frac{d}{dx}(\mathrm{cosec}\, x) = -\mathrm{cosec}\, x \cot x.$$

Qu.6 By considering their graphs, find
(i) which of the six trigonometric functions are increasing in the first quadrant;
(ii) the range of values of x for which $\tan x$ has a negative gradient.
Check that your answers are confirmed by the derivatives found above.

Qu.7 Use the derivatives above to find
(i) the gradient of the sine curve at the origin;
(ii) the maximum value of the gradient of the cotangent curve.

Qu.8 When writing $\dfrac{d}{dx}(\sin x) = \cos x$, we are assuming that x is measured in radians. At what point in this chapter was this assumption needed?

13.3 Further differentiation involving trigonometric functions

Example 1
Differentiate

(i) $x^2 \cos x$;   (ii) $\sin x \cos x$;   (iii) $\dfrac{2\,\mathrm{cosec}\, x}{1 + x^2}$.

Solution
By the product rule,

(i) $\dfrac{d}{dx}(x^2 \cos x) = 2x \cos x - x^2 \sin x;$

(ii) $\dfrac{d}{dx}(\sin x \cos x) = \cos^2 x - \sin^2 x;$

and by the quotient rule,

(iii) $\dfrac{d}{dx}\left(\dfrac{2 \operatorname{cosec} x}{1+x^2}\right) = \dfrac{-2 \operatorname{cosec} x \cot x(1+x^2) - 4x \operatorname{cosec} x}{(1+x^2)^2}.$

Example 2
Differentiate
  (i) $\sec(3x+1);$ (ii) $\cot^4 x;$ (iii) $\tan^3(x^2 - 3x - 1).$
Solution
(i) Let $y = \sec(3x+1).$ (Note that the sec is the 'outer' function, i.e. we think of y as 'sec something'.) Then

$$\frac{dy}{dx} = \sec(3x+1)\tan(3x+1).\frac{d}{dx}(3x+1)$$

$$= 3\sec(3x+1)\tan(3x+1).$$

(ii) Let $y = \cot^4 x.$ (Here, the 'fourth power' is the outer function, i.e. we think of y as 'something to the power four'.) Then

$$\frac{dy}{dx} = 4\cot^3 x.\frac{d}{dx}(\cot x)$$

$$= -4\operatorname{cosec}^2 x \cot^3 x.$$

(iii) Let $y = \tan^3(x^2 - 3x - 1).$ (This is a triple composite function. The outermost function is the 'cube' function, but the inner function is itself composite, and must be differentiated as such.) Then

$$\frac{dy}{dx} = 3\tan^2(x^2 - 3x - 1).\frac{d}{dx}\tan(x^2 - 3x - 1)$$

$$= 3\tan^2(x^2 - 3x - 1)\sec^2(x^2 - 3x - 1) \times \frac{d}{dx}(x^2 - 3x - 1)$$

$$= 3(2x - 3)\tan^2(x^2 - 3x - 1)\sec^2(x^2 - 3x - 1).$$

Exercise 13   Differentiate questions 1 to 28 with respect to x.

  1. $\sin 2x.$ **2.** $\sec(3x+1).$

  3. $\cot x^2.$ **4.** $\operatorname{cosec}\left(\dfrac{3}{2x^2}\right).$

  5. $x \tan x.$ **6.** $3x^2 \cos\left(x + \dfrac{\pi}{4}\right).$

  7. $\dfrac{\sin x}{x}.$ **8.** $\sin x \tan x.$

  9. $\cot 2x \sec 3x.$ **10.** $x \cos \tfrac{1}{2}x \operatorname{cosec} \tfrac{3}{2}x.$

11. $\dfrac{2x \sec x}{1 + x^2}$.

12. $x \cot (3x - 2)$.

13. $2x^3 \operatorname{cosec} 3x^2$.

14. $\dfrac{2 \tan (x^2 + x)}{x^2 + x}$.

15. $\dfrac{1 + x \cos x}{1 + x \sin x}$.

16. $\sec \sqrt{(x^2 + 1)}$.

17. $x\sqrt{x \cot 5x}$. ·

18. $\sin^2 x$.

19. $\tan^3 x$.

20. $\sqrt{\operatorname{cosec} (x + 3)}$.

21. $x \sec^2 x$.

22. $\dfrac{\cos^3 x}{x^3}$.

23. $\sqrt[3]{(\cot^2 x)}$.

24. $\tan^2 \left(3x - \dfrac{3\pi}{4} \right)$.

25. $\sin^3 4x^2$.

26. $\cos^4 \left(\dfrac{2}{x^2} \right)$.

27. $3\sqrt{\sin 2x}$.

28. $5\sqrt{(\operatorname{cosec}^3 4x^2)}$.

29. Verify that $y = A(\sec \theta + \tan \theta)$ is a solution of the differential equation $\dfrac{dy}{d\theta} \cos \theta = y$, for any constant A. Find a similar solution to the differential equation $\dfrac{dy}{d\theta} \sin \theta = -y$.

30. Use a calculus method to find the maximum and minimum values of $y = \sin x + \cos x$ in the region $0 \leqslant x < 2\pi$.

31. Find the maximum and minimum values of $\sin^2 x + \cos^2 x$ in the region $0 \leqslant x < 2\pi$.

32. Verify that $y = A \sin 2x + B \cos 2x$ (where A and B are constants) is a solution of $\dfrac{d^2 y}{dx^2} + 4y = 0$.

33. Find the equation of the tangent to the curve $y = \sin x$ at the point $(\arcsin \frac{3}{5}, \frac{3}{5})$.

34. Differentiate $y = \sin x°$.

35. Find the minimum value of $y = \tan \theta + 3 \cot \theta$ in the range $0 \leqslant \theta \leqslant \dfrac{\pi}{2}$.

36. Differentiate
   (i) $\tan^2 x$;   (ii) $\sec^2 x$.
   Explain the connection between your answers.

37. Differentiate $\tan x$ from first principles.

38. Find

   (i) $\dfrac{d^{10}}{d\theta^{10}} (\sin \theta)$;   (ii) $\dfrac{d^{20}}{d\theta^{20}} (\cos \theta)$.

39. Evaluate $\displaystyle\sum_{n=1}^{\infty} \dfrac{d^n}{dx^n} (\cos \tfrac{1}{2} x)$ when $x = \dfrac{\pi}{3}$.

40. A particle is moving in one dimension, its position at time t being given by $x = A \sin nt$, where A and n are constants.

(i) Obtain an expression for the acceleration of the particle at time t, and show that the acceleration is always directed towards the origin (rather than away from it).

(ii) If v is the velocity of the particle, show that $v^2 = n^2(A^2 - x^2)$ at all times.

(iii) Show that $\dfrac{dv}{dx} = -n \tan nt$. [*Hint*: use the chain rule.]

41. Find the values of θ for which $y = \sin^3 \theta$ has points of inflexion. Sketch the graph of this function.

42. (i) In Fig. 13.2, show that $PQ = \tan 2\theta + \cot \theta$.

(ii) Show that the minimum length of PQ as θ varies is about 3·33 units.

Figure 13.2

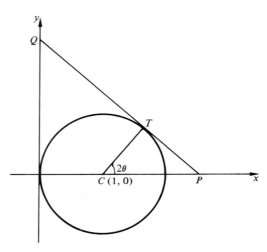

Miscellaneous exercise 13   **1.** Sketch the graph of $y = \cos x$ for $0 \leqslant x \leqslant 4\pi$. (An accurately plotted curve on graph paper is *not* necessary.) Use your sketch to find the *number* of positive roots of the equation $\cos x = \frac{1}{10}x$.

Prove that the equation of the tangent to the curve $y = \cos x$ at $(\frac{1}{2}\pi, 0)$ is $y = \frac{1}{2}\pi - x$.

By using $\frac{1}{2}\pi - x$ as an approximation for $\cos x$ (valid for x near $\frac{1}{2}\pi$), calculate an estimate of the smallest positive root of the equation $\cos x = \frac{1}{10}x$, giving your answer to 2 decimal places.

Show on a new sketch the part of the curve $y = \cos x$ near $x = \frac{1}{2}\pi$ and the tangent at $x = \frac{1}{2}\pi$. Hence determine whether the estimate of the root calculated above is larger or smaller than the correct value.

[C]

2. *ABCD* is a kite which has *AC* as its axis of symmetry and $\angle A = 90°$. The lengths *BC* and *DC* are each 20 cm and $\angle BCD = 2\theta$. Find the lengths of the diagonals of the kite in terms of θ and hence write down an expression for S cm^2, the area of the kite.

Verify that $\theta = 67\frac{1}{2}°$ makes S a maximum, and find this maximum value.

[Ox (O)]

THE DIFFERENTIATION OF TRIGONOMETRIC FUNCTIONS

3. Differentiate with respect to x

   (i) $\sin(2x+5)$,   (ii) $3x\cos 3x$,   (iii) $\dfrac{\tan x}{x^2}$.

[AEB (O) '80]

4. (a) Differentiate with respect to x
   (i) $\tan 2x$;   (ii) $\sin^3 3x$;   (iii) $(1+\sin x)^5$.

   (b) If $y=\dfrac{\sin x}{x}$, prove that

$$x\frac{d^2y}{dx^2}+2\frac{dy}{dx}+xy=0.$$

[SUJB (O)]

5. A right circular cone has a semi-vertical angle θ and slant edge of length 10 cm. Find an expression for its volume, V, in terms of θ. Hence show that the maximum volume of the cone as θ varies occurs when $\tan\theta=\sqrt{2}$, and find the maximum volume, rounding your answer to three significant figures.
   (*Hint*: When $\tan\theta=\sqrt{2}$, $\sin\theta=\sqrt{2}/\sqrt{3}$ and $\cos\theta=1/\sqrt{3}$.)

[Ox (O)]

6. (a) The function f is given by

$$f:x\to x^2+\frac{1}{x}\,(x\in R,\ x\neq 0).$$

   Prove that the graph of f has one turning point, and determine whether it is a maximum or a minimum.
      Sketch the graph of f.
   (b) An isosceles triangle ABC, with $AB=AC$, is inscribed in a fixed circle, centre O, whose radius is 1 unit. Angle $BOC=2\theta$, where θ is acute. Express the area of the triangle ABC in terms of θ, and hence show that, as θ varies, the area is a maximum when the triangle is equilateral.

[C]

7. Two tangents are drawn from a point at a fixed distance a from the centre of a circle of variable radius. The angle subtended at the centre of this circle by the minor arc joining the points of contact of the tangents is 2θ. Prove that the area of the region enclosed by the tangents and the minor arc is

$$a^2(\tfrac{1}{2}\sin 2\theta-\theta\cos^2\theta).$$

Show that this area is a maximum when the length of each tangent is equal to the length of the minor arc.

[MEI]

8. Given that $y=\sin 2x+2\cos x$, find the values of x in $[0,2\pi]$ for which $\dfrac{dy}{dx}=0$. For each such value of x determine whether the corresponding value of y is a maximum, a minimum or neither. Sketch the graph of y against x for $0\leqslant x\leqslant 2\pi$.
      If $k>0$, determine the set of values of k for which the equation $\sin 2x+2\cos x=kx$ has
   (i) one real root in $[0,2\pi]$,   (ii) two real roots in $[0,2\pi]$.

[C]

14 Further differentiation

In Chapter 6, we used differentiation to solve simple kinematics problems. We now consider some other problems in which time is the independent variable.

Example 1
Water is leaking from a hole in a tank in such a way that, after t seconds, the volume (q litres) of water in the tank is given by

$$q = 2 + \frac{3}{t} - \frac{1}{t^2}, \qquad \text{for } t > 1.$$

Find the rate at which water leaves the hole when $t = 3$.
Solution
We seek the rate at which q is changing.

$$\frac{dq}{dt} = -3t^{-2} + 2t^{-3},$$

so when $t = 3$,

$$\frac{dq}{dt} = -\frac{3}{9} + \frac{2}{27}$$

$$= -\frac{7}{27} \text{ litre/second.}$$

The volume of water in the tank is decreasing at a rate of $\frac{7}{27}$ litre/second, and so this is the rate at which the water is leaking at this instant.

Qu.1 (For discussion.) Explain why the above expression for q is unlikely to be valid for $t = \frac{1}{2}$ say, by evaluating $\frac{dq}{dt}$ at that time.
What does this mean would be happening?
Qu.2 What happens to q and $\frac{dq}{dt}$ as $t \to \infty$? Suggest a reason for this.
Qu.3 If, in Example 1, the value of q in the time interval $0 \leqslant t \leqslant 1$ is given by $q = 4$, determine whether q is
(i) continuous, (ii) differentiable,
at $t = 1$. Discuss the physical significance of your answers.

The next two examples use the chain rule. The general method should be noted: begin by expressing both the data and the 'unknown' in calculus notation.

Example 2

A spherical balloon is inflated so that its volume V increases at a constant rate of $3\,\text{cm}^3$ per second. Find the rate at which the radius r is changing when $r = 2\frac{1}{2}\,\text{cm}$.

Solution

We are given that $\dfrac{dV}{dt} = 3$, and we wish to find $\dfrac{dr}{dt}$ when $r = 2\frac{1}{2}$. By the chain rule,

$$\frac{dr}{dt} = \frac{dr}{dV} \times \frac{dV}{dt}. \tag{1}$$

Now $\dfrac{dr}{dV}$ is a purely *geometrical* measure; we have

$$V = \frac{4}{3}\pi r^3$$

$$\Rightarrow \quad \frac{dV}{dr} = 4\pi r^2$$

$$\Rightarrow \quad \frac{dr}{dV} = \frac{1}{4\pi r^2} \quad \text{(using the result of Section 7.4).}$$

Substituting into (1),

$$\frac{dr}{dt} = \frac{1}{4\pi r^2} \times 3.$$

Putting $r = 2\frac{1}{2}$ gives

$$\frac{dr}{dt} = \frac{3}{25\pi}.$$

So r is increasing at a rate of $\dfrac{3}{25\pi}\,\text{cm s}^{-1}$ when $r = 2\frac{1}{2}$.

Qu.4 In the example above, what would you expect to happen to the rate of change of r when r gets very big, assuming that the balloon does not burst? Check your answer by finding $\dfrac{dr}{dt}$ when $r = 500$ say.

Example 3

A hollow cone of semi-vertical angle $\arctan \frac{1}{2}$ is fixed vertex-downwards with its axis vertical (Fig. 14.1), and is filled with water at a constant rate k. A spider which can run at speed u is asleep

Figure 14.1

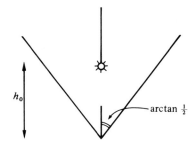

at the end of a web hanging at a height h_0 vertically above the vertex. Find, in terms of k and h_0, the minimum value of u that will enable the spider to escape a soaking, assuming that he starts to run upwards as soon as the water touches his feet.

Solution

Let the depth of the water be $h(t)$. We have $\dfrac{dV}{dt}=k$, and we wish to investigate $\dfrac{dh}{dt}$ when $h \geqslant h_0$. Now

$$\frac{dh}{dt}=\frac{dh}{dV}\times\frac{dV}{dt}. \qquad (2)$$

To find $\dfrac{dh}{dV}$, we must express V in terms of h only. We have

$$V=\frac{1}{3}\pi r^2 h,$$

where r is the radius of the surface of the water. But $r=\frac{1}{2}h$ (Fig. 14.2).

Figure 14.2

So

$$V=\frac{1}{3}\pi\left(\frac{1}{2}h\right)^2 h$$

$$=\frac{1}{12}\pi h^3$$

$$\Rightarrow \frac{dV}{dh}=\frac{1}{4}\pi h^2$$

$$\Rightarrow \frac{dh}{dV}=\frac{4}{\pi h^2}.$$

So (2) becomes

$$\frac{dh}{dt}=\frac{4}{\pi h^2}\times k.$$

The R.H.S. clearly decreases as h increases, so the spider needs to run fastest when the water first reaches him, i.e. when $h=h_0$. So, to avoid a soaking,

$$u \geqslant \frac{4k}{\pi h_0^2}.$$

Notation

A dot above a letter indicates differentiation w.r.t. time; thus $\dot{h}$ means $\dfrac{dh}{dt}$, and $\ddot{x}$ means $\dfrac{d^2x}{dt^2}$.

Exercise 14a

1. A capacitor is allowed to discharge in such a way that, at time t, the charge on it is given by

$$q = a + \frac{b}{1+t}$$

where a and b are constants. Find, in terms of a and b, the rate of discharge when $t=2$. If the initial charge is 3 coulombs and the initial rate of discharge is 2 amps, find the values of a and b.

2. A ship is rolling in a rough sea in such a way that the angle ($\theta°$) that its mast makes with the vertical at time t seconds is given by

$$\theta = 15 \sin \frac{\pi t}{6}.$$

Find
   (i) the angular velocity ($\dot\theta$) of the mast when $t=2$;
   (ii) the maximum angular velocity.

3. Two variables p and q are related by $p = \sqrt{(5+q^2)}$, and q is increasing at a constant rate of 6 units per second. At what rate is p increasing when $q=2$? Show that 'after a long time', p is increasing at about the same rate as q.

4. The radius of a sphere is increasing at 2 cm s^{-1}.
   (i) At what rate is the volume increasing when $r=4 \text{ cm}$?
   (ii) At what rate is the surface area increasing when $r=6 \text{ cm}$?

5. A vessel is shaped in such a way that when the depth of water in it is h cm, the volume of water in cm^3 is given by $V = h^2 + 2h$. If water is leaking at a rate of $5 \text{ cm}^3 \text{ s}^{-1}$, find the rate at which the water level is falling when $h=10$.

6. The amount of water in a leaking bucket at time t is given by $V = (10-t)^{3/2}$ for $0 \leqslant t \leqslant 10$. Show that the rate at which the water is leaking is proportional to the cube root of the volume of water in the bucket.

7. A colony of bacteria forms a circular disc whose area increases at a constant rate G. Find an expression for the rate at which the radius r is changing when $r = r_0$. Explain why it is to be expected that this rate is inversely proportional to the circumference.

8. The angle made by a swinging pendulum with the vertical approximately satisfies an equation of the form

$$\theta = A \sin nt + B \cos nt$$

where A, B and n are constants. Show that the 'angular acceleration' of the pendulum is proportional to θ.

9. A man stands 40 m from a railway line on which a train is travelling at 30 m s^{-1}. At time $t=0$, the train is at O, the nearest point to the man. Find

(i) the change in the distance of the train from the man between the times $t=0$ and $t=1$ second;

(ii) the rate at which this distance is changing when $t=2$ second;

(iii) at what rate this distance is changing after a long time.

10. Show that the volume of water in a cone (held as in Example 3) varies as the cube of the height of the water (h). If the rate at which water is leaking is proportional to h, show that h decreases at a rate inversely proportional to itself.

11. A circular coin of radius 1 cm rests on the floor, 2 m vertically below a point source of light. It is raised vertically so that the area of the shadow of the coin on the floor increases at a constant rate of π cm^2 s^{-1}. Find

(i) how fast the coin is moving when it is 1 m from the floor;

(ii) how long it took to reach that point;

(iii) how much longer it takes to reach the light.

12. A particle starts at the origin, and moves along the curve $y=x^2$ so that its x-coordinate increases at a rate which is inversely proportional to the x-coordinate itself. Show that the y-coordinate increases at a constant rate.

14.2 Small changes Suppose we wish to find approximately the value of $\sin 31°$ without using trigonometric tables (or the trigonometric functions on a calculator). As a first approximation, we might argue that since the sine function is continuous (i.e. has no sudden jumps),

$$\sin 31° \approx \sin 30°$$

i.e.,
$$\sin 31° \approx \tfrac{1}{2}.$$

The question now is whether we can find a better approximation. What increase in the value of the sine does the extra 1° make? We shall return to this problem shortly.

Qu.5 Use four-figure tables or a calculator to find the change in the value of $\sin x$ corresponding to a 1° increase in x, when the initial value of x is

(i) 0°;   (ii) 20°;   (iii) 50°;

(iv) 70°;   (v) 89°.

Explain the significance of the variation in these answers in terms of the graph of $\sin x°$.

Let y be a function of x, and let δy be the change in y resulting from a small change δx in x. The problem that we are concerned with here is that of finding δy assuming that we know the values of x and δx. In Chapter 5, we defined the derivative $\dfrac{dy}{dx}$ by

$$\frac{dy}{dx} = \lim_{\delta x \to 0} \frac{\delta y}{\delta x},$$

and we noted that for sufficiently small δx

$$\frac{dy}{dx} \approx \frac{\delta y}{\delta x}. \tag{3}$$

By evaluating $\dfrac{dy}{dx}$ exactly, we can use it to give an approximate value to $\dfrac{\delta y}{\delta x}$. $\left(\text{This is the reverse of what was done in Chapter 5, where we}\right.$ found values of $\dfrac{\delta y}{\delta x}$ to estimate $\left.\dfrac{dy}{dx}.\right)$ Then δy may be estimated by writing (3) as

$$\boxed{\delta y \approx \dfrac{dy}{dx}\delta x.}$$
(4)

Qu.6 Starting with the definition of the derivative in the form given in Section 5.4, show that, for small h,

$$f(x+h)-f(x) \approx h f'(x)$$

and verify that this says the same as (4) above.

The result of Qu.6 may be written

$$\boxed{f(x+h) \approx f(x) + h f'(x)}$$
(5)

and is sufficiently important to be worth learning in this form.

Example 4
Estimate $\sin 31°$ without using trigonometric tables or the trigonometric functions on a calculator.
Solution
(Note that we must work in radians since we are using calculus.) Let $f(x) = \sin x$, then $f'(x) = \cos x$. Writing

$$\sin 31° = \sin(30° + 1°)$$

$$= \sin\left(\dfrac{\pi}{6} + \dfrac{\pi}{180}\right)$$

and using (5) above, we have

$$\sin\left(\dfrac{\pi}{6} + \dfrac{\pi}{180}\right) \approx \sin\dfrac{\pi}{6} + \dfrac{\pi}{180}\cos\dfrac{\pi}{6}$$

$$= \dfrac{1}{2} + \dfrac{\pi}{180}\cdot\dfrac{\sqrt{3}}{2}$$

$$\approx 0.5 + 0.01511.$$

So $\qquad\qquad\qquad\qquad \sin 31° \approx 0.51511.$

(This compares with 0.51504 which is the correct value to 5 d.p.)

Qu.7 Write out a solution to Example 4 using the delta notation and (4) above.

A geometrical interpretation sheds further light on the approximation $\delta y \approx \dfrac{dy}{dx} \delta x$. In Fig. 14.3 (in which, for clarity, δx is drawn as being 'large') the true value of δy is the length AQ, whereas $\dfrac{dy}{dx}\delta x$ is the length AB, as can be shown thus:

$$\frac{AB}{AP} = \text{gradient at } P = \frac{dy}{dx};$$

so

$$AB = \frac{dy}{dx} \cdot AP = \frac{dy}{dx} \delta x.$$

Figure 14.3

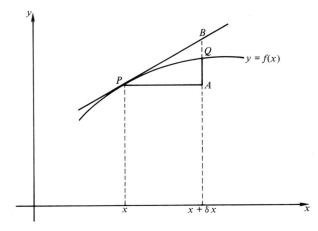

In using this approximation for δy therefore, we are in effect sliding along the tangent at P rather than moving along the chord PQ.

The remaining error (corresponding to BQ in Fig. 14.3) arises only because the gradient of the curve is changing, and so it seems likely that the error depends in some way on the size of $\dfrac{d^2 y}{dx^2}$ (evaluated at the point P say). This idea will be clarified by Taylor's Theorem in Chapter 28.

Qu.8 Let ε be the error in the approximation $f(x+h) \approx f(x) + hf'(x)$, so that we may write $f(x+h) = f(x) + hf'(x) + \varepsilon$. Verify that in Fig. 14.3, h is positive and $f''(x)$ and ε are both negative. Investigate, by drawing suitable diagrams, the sign of ε for other combinations of sign of h and $f''(x)$.

Example 5
Estimate the change in the total surface area of a solid cylinder of height 10 cm when the radius decreases from 2 cm to 1·97 cm.
Solution
Let the total surface area be S, and the radius r; we shall use the approximation $\delta S \approx \dfrac{dS}{dr} \delta r$.

$$S = 2\pi r^2 + 2\pi r h$$
$$= 2\pi r^2 + 20\pi r \text{ (since } h \text{ is constant at 10 cm)}$$

$$\Rightarrow \frac{dS}{dr} = 4\pi r + 20\pi$$

$$= 28\pi \text{ when } r = 2.$$

Also, $\delta r = -0\cdot03.$

So $\delta S \approx \dfrac{dS}{dr}\delta r$

$$= 28\pi \times (-0\cdot03)$$
$$= -0\cdot84\pi.$$

So the total surface area decreases by approximately $0\cdot84\pi$ cm².

The next example illustrates a method of dealing with percentage changes. Note that the percentage change in a variable x is given by $\dfrac{\delta x}{x} \times 100.$

Example 6

Two variables, r and ω, are related by the equation $Mr^2\omega = c$, where M and c are constants. If ω decreases by 3%, estimate the percentage change in r.

First solution

$$Mr^2\omega = c \Rightarrow \omega = \frac{c}{M}r^{-2}$$

$$\Rightarrow \frac{d\omega}{dr} = -\frac{2c}{M}r^{-3}$$

$$\Rightarrow \delta\omega \approx -\frac{2c}{M}r^{-3}\delta r$$

$$\Rightarrow \frac{\delta\omega}{\omega} \approx -\frac{2c}{M\omega}\frac{1}{r^2}\frac{\delta r}{r}.$$

Putting $M\omega r^2 = c$ on the R.H.S.,

$$\frac{\delta\omega}{\omega} \approx -\frac{2c}{c}\frac{\delta r}{r}$$

$$\Rightarrow \frac{\delta\omega}{\omega} \times 100 \approx -2\left(\frac{\delta r}{r} \times 100\right)$$

$$\Rightarrow -3 \approx -2\left(\frac{\delta r}{r} \times 100\right)$$

$$\Rightarrow \frac{\delta r}{r} \times 100 \approx \frac{3}{2}.$$

So r increases by $1\frac{1}{2}\%$.
(Note that we could instead have started by making r the subject and differentiating w.r.t. ω.)
Second solution

$$Mr^2\omega = c,$$

and so also

$$M(r+\delta r)^2(\omega+\delta\omega)=c$$
$$\Rightarrow \quad M(r^2\omega+2r\omega\delta r+r^2\delta\omega)=c$$

(where terms of second order smallness have been neglected)

$$\Rightarrow \quad M(2r\omega\delta r+r^2\delta\omega)=0.$$

Dividing by $Mr^2\omega$ gives

$$2\frac{\delta r}{r}+\frac{\delta\omega}{\omega}=0$$

etc. as in the first solution.

Exercise 14b

1. Let $y=\sqrt{x}$. Estimate the change in the value of y when x changes from 9 to 9·02. Deduce an approximate value for $\sqrt{9\cdot02}$.

2. Find an approximation for $\sqrt{4\cdot003}$.

3. Estimate the change in the value of $y=\sqrt{(1+x)}$ when x changes from 3 to $3\frac{1}{4}$. Hence estimate $\sqrt{4\cdot25}$ and deduce that $\sqrt{17}\approx4\frac{1}{8}$.

4. Estimate the change in $\sqrt{x}$ when x changes from $20\frac{1}{4}$ to 20. Deduce an approximation to $\sqrt{5}$ to 4 significant figures. Compare with the value given in tables.

5. Estimate the change in the value of $\sqrt[3]{x}$ when x changes from 343 to 344. Hence estimate $\sqrt[3]{344}$ and deduce $\sqrt[3]{43}$.

6. Estimate the change in $\sqrt[3]{x}$ when x changes from 125 to 128. Deduce a value for $\sqrt[3]{2}$.

7. Find an approximation to the value of $\dfrac{3}{1+4x^3}$ when $x=2\cdot0003$.

   Find approximate values of questions 8 to 12. Take $1°=0\cdot017453$ radian. Compare your answers with those given by the trigonometric functions on a calculator.

8. $\sin 29°$.   9. $\sin 61°$.

10. $\tan 47°$.   11. $\cos 44°$.

12. $\sec 134°$.

13. By considering the function $f(x)=\dfrac{1}{x^2}$, estimate the value of $\dfrac{1}{(2\cdot99)^2}$.

14. Use the method of small changes to estimate

   (i) $\dfrac{1}{2\cdot005}$;   (ii) $\sqrt{8\cdot98}$;   (iii) $\sqrt[3]{8\cdot03}$.

15. Find an approximation to the value of $y=x^3-3x^2+7x-2$ when $x=2\cdot03$.

16. Find the increase in the surface area of a sphere when the radius increases from 2 cm to 2·001 cm.

17. The gravitational force exerted by the earth on a body of mass m, on or above the earth's surface, is $\dfrac{GMm}{r^2}$ where G and M are constants and r is the distance of the body from the centre of the earth. Estimate the change in the size of this force if r changes from 6400 to 6450. Express this change as a fraction of the original value.

18. Estimate the percentage change in the volume of a cube if the length of each side increases by 2%.

19. Estimate the percentage change in the radius of a sphere if its surface area decreases by $1\frac{1}{2}\%$.

20. Estimate the percentage change in Ax^n if x changes by $k\%$ (where k is small).

21. The radius of a cone of fixed height h increases by $p\%$. Show that the percentage increase in the curved surface area depends on the ratio of r to h. Estimate this percentage if initially $r=h$. Show that if $r \gg h$ (r is much bigger than h) then this percentage is approximately $2p\%$. What shape does the curved surface of the cone then approximate to?

22. In economics, the arc elasticity of demand for a product is a measure of the sensitivity of consumer demand to price changes. It is defined by

$$e = \frac{\% \text{ change in quantity purchased } (q)}{\% \text{ change in price } (p)}.$$

(i) Explain why for most goods e is negative. (In practice, the negative sign is often omitted.)

(ii) Show that for a sufficiently small change, $e \approx \dfrac{p}{q}\dfrac{dq}{dp}$. (This quantity is called the point elasticity.)

(iii) For a certain commodity, it is known that p and q (when measured in suitable units) are related by $q=(1+p^2)^{-1}$. Find the point elasticity when
(a) $p=1$;   (b) $p=2$;
and the arc elasticity when
(c) p changes from 1 to 2;
(d) p changes from 2 to 1.

23. (See question 22.) Suppose that the total amount of money spent by consumers on a product is a constant c, regardless of the price of that product.
(i) Write down a relationship connecting p, q and c.
(ii) Show that the point elasticity is -1 for all values of p.

14.3 Implicit functions

If y is a function of x, and is written in the form $y=f(x)$ (i.e. with y as the 'subject of the formula'), then we say that y is an *explicit function* of x. Consider, on the other hand, the relationship

$$xy^3 + 2x^2y = 6 + x.$$

Here, y is not expressed explicitly in terms of x; we say that y is an *implicit function* of x. (Note that many implicit functions are not functions in the strict sense, since they are not well-defined; a simple example is $y^2=x$.)

If we wish to find $\dfrac{dy}{dx}$ where y and x are related implicitly, we might begin by trying to obtain y explicitly in terms of x. In general, this is either difficult or impossible. A better method is simply to differentiate the function *as it stands*. To do this, we need to be able to differentiate terms like those in the following example.

Example 7

Differentiate the following with respect to x, where y is a function of x:

(i) y;   (ii) y^2;   (iii) xy;

(iv) $\dfrac{2x}{3y}$;   (v) $\dfrac{dy}{dx}$;   (vi) $2xy\dfrac{dy}{dx}$.

Solution

(i) $\dfrac{dy}{dx}$.

(ii) By the chain rule,

$$\frac{d}{dx}(y^2) = \frac{d}{dy}(y^2) \times \frac{dy}{dx}$$

$$= 2y\frac{dy}{dx}.$$

$\Bigg($ Note: this can be done mentally by thinking 'differentiate with respect to y and multiply by $\dfrac{dy}{dx}$'. $\Bigg)$

(iii) By the product rule,

$$\frac{d}{dx}(xy) = y + x\frac{dy}{dx}.$$

(iv) By the quotient rule,

$$\frac{d}{dx}\left(\frac{2x}{3y}\right) = \frac{2.3y - 3\dfrac{dy}{dx}.2x}{(3y)^2}$$

$$= \frac{2\left(y - x\dfrac{dy}{dx}\right)}{3y^2}.$$

(v) $\dfrac{d^2y}{dx^2}$.

(vi) By the extended product rule (see Exercise 7a, question 5),

$$\frac{d}{dx}\left(2xy\frac{dy}{dx}\right) = 2y\frac{dy}{dx} + 2x\left(\frac{dy}{dx}\right)^2 + 2xy\frac{d^2y}{dx^2}.$$

Example 8

Find the gradient of the curve

$\quad y^2 - 2xy + 3x^2 y^3 = 9$ at the point (2, 1).

Solution

Differentiating each side as it stands,

$$2y\frac{dy}{dx} - 2y - 2x\frac{dy}{dx} + 6xy^3 + 9x^2y^2\frac{dy}{dx} = 0$$

$$\Rightarrow \frac{dy}{dx}(2y - 2x + 9x^2y^2) = 2y - 6xy^3$$

$$\Rightarrow \frac{dy}{dx} = \frac{2y - 6xy^3}{2y - 2x + 9x^2y^2}.$$

So when $x = 2$ and $y = 1$,

$$\frac{dy}{dx} = -\frac{10}{34} = -\frac{5}{17}.$$

So the gradient of the curve at (2, 1) is $-\frac{5}{17}$.

Example 9

Let r and θ be functions of time, and let $x = r\cos\theta$. Find an expression for $\ddot{x}$.

Solution

$$x = r\cos\theta$$

$$\Rightarrow \dot{x} = \dot{r}\cos\theta - r\dot{\theta}\sin\theta$$

(using the chain rule to differentiate $\cos\theta$ with respect to time)

$$\Rightarrow \ddot{x} = \ddot{r}\cos\theta - \dot{r}\dot{\theta}\sin\theta - \dot{r}\dot{\theta}\sin\theta - r\ddot{\theta}\sin\theta - r\dot{\theta}^2\cos\theta$$

$$= (\ddot{r} - r\dot{\theta}^2)\cos\theta - (2\dot{r}\dot{\theta} + r\ddot{\theta})\sin\theta.$$

Qu.9   Let $y^2 = 1 + x^3$. Differentiate this twice, and show that $\frac{d^2y}{dx^2} = \frac{1}{y}\left\{3x - \left(\frac{dy}{dx}\right)^2\right\}$. By expressing $\frac{dy}{dx}$ in terms of x and y, find an expression for $\frac{d^2y}{dx^2}$ in terms of x and y only.

14.4 Parametric differentiation

Parametric form was introduced in Chapter 3, where we remarked that it is often inconvenient or impossible to eliminate the parameter to give a single equation connecting x and y. To find $\frac{dy}{dx}$ and $\frac{d^2y}{dx^2}$ (and higher order derivatives) therefore, we use the chain rule:

$$\frac{dy}{dx} = \frac{dy}{dt} \times \frac{dt}{dx}$$

$$\Rightarrow \boxed{\frac{dy}{dx} = \frac{dy}{dt} \div \frac{dx}{dt},}$$

and

$$\frac{d^2y}{dx^2} = \frac{d}{dx}\left(\frac{dy}{dx}\right)$$

$$= \frac{d}{dt}\left(\frac{dy}{dx}\right) \times \frac{dt}{dx}$$

$$\Rightarrow \boxed{\frac{d^2 y}{dx^2} = \frac{d}{dt}\left(\frac{dy}{dx}\right) \div \frac{dx}{dt}.}$$

Example 10
If $x = t^2 - 3$, $y = t^3 + t$, find

(i) $\dfrac{dy}{dx}$;   (ii) $\dfrac{d^2 y}{dx^2}$;

(iii) the equation of the tangent to this curve at the point $t = 3$.
Solution

(i) $\dfrac{dy}{dt} = 3t^2 + 1$ and $\dfrac{dx}{dt} = 2t$.

So
$$\frac{dy}{dx} = (3t^2 + 1) \div 2t$$

$$= \frac{3}{2}t + \frac{1}{2t}.$$

(ii) $\dfrac{d}{dt}\left(\dfrac{dy}{dx}\right) = \dfrac{3}{2} - \dfrac{1}{2t^2}$;

so
$$\frac{d}{dx}\left(\frac{dy}{dx}\right) = \left(\frac{3}{2} - \frac{1}{2t^2}\right) \div 2t$$

$$\Rightarrow \frac{d^2 y}{dx^2} = \frac{3}{4t} - \frac{1}{4t^3}.$$

(iii) Putting $t = 3$ in the answer to (i),
$$\frac{dy}{dx} = \frac{3}{2} \times 3 + \frac{1}{6}$$

$$= \frac{14}{3}.$$

And when $t = 3$, $x = 6$ and $y = 30$. So we seek the line passing through $(6, 30)$ with gradient $\frac{14}{3}$. This is

$$y - 30 = \frac{14}{3}(x - 6)$$

$$\Rightarrow 3y - 14x = 6.$$

Qu.10   Let $x = \sin t$ and $y = \sec t$.

(i) Find $\dfrac{dy}{dx}$ in terms of t. Plot the graph of y against x for values of t

from 0 to $\dfrac{\pi}{3}$. (Take 10° intervals for t.) Draw and measure the gradient of the tangent at the point $t = \dfrac{\pi}{6}$. Compare this with the value obtained by calculation of the gradient.

(ii) Show that $x^2 y^2 = y^2 - 1$. By differentiating this implicitly with respect to x, find $\dfrac{dy}{dx}$ in terms of x and y. Verify that this expression for $\dfrac{dy}{dx}$ is the same as that obtained in the first part of (i).

14.5 Differentiation of inverse functions

Suppose that $y = f^{-1}(x)$ and we wish to find $\dfrac{dy}{dx}$. It may be that we know sufficient about f^{-1} simply to differentiate it. For example, if f is the 'square' function, then f^{-1} is the 'square root' function, and differentiation presents no problem.

It often happens, however, that a function f is much more manageable than its inverse, and that we can easily differentiate f but not f^{-1}. We may then proceed as follows:

$$y = f^{-1}(x) \implies f(y) = x. \tag{7}$$

Differentiating w.r.t. x,

$$f'(y)\frac{dy}{dx} = 1$$

$$\implies \frac{dy}{dx} = \frac{1}{f'(y)}.$$

It is often possible to rewrite this in terms of x.

Qu.11 Obtain the same result by differentiating (7) w.r.t. y and using $\dfrac{dy}{dx} = 1 \bigg/ \dfrac{dx}{dy}$.

We can use this method to differentiate the inverse trigonometric functions.

Example 11
Differentiate

(i) $\arcsin x$;   (ii) $\arctan(x^2 + 3)$.

Solution
(i) Let $y = \arcsin x$; then $\sin y = x$. So

$$\cos y \frac{dy}{dx} = 1$$

$$\implies \frac{dy}{dx} = \frac{1}{\cos y}.$$

Now

$$\cos y = \pm \sqrt{(1 - \sin^2 y)}$$
$$= \pm \sqrt{(1 - x^2)}.$$

But in the range under consideration, $-\dfrac{\pi}{2} \leqslant y \leqslant \dfrac{\pi}{2}$ (the principal values of arcsin), $\cos y$ is always positive; so we take the positive square root. Thus

$$\frac{d}{dx}(\arcsin x) = \frac{1}{\sqrt{(1-x^2)}}.$$

(ii) Let $y = \arctan(x^2 + 3)$.

Then
$$\tan y = x^2 + 3$$

$$\Rightarrow \sec^2 y \frac{dy}{dx} = 2x$$

$$\Rightarrow \frac{dy}{dx} = \frac{2x}{\sec^2 y}.$$

But
$$\sec^2 y = 1 + \tan^2 y$$
$$= 1 + (x^2 + 3)^2.$$

So
$$\frac{dy}{dx} = \frac{2x}{x^4 + 6x^2 + 10}.$$

Qu.12 Differentiate
(i) $\arccos x$; (ii) $\arctan x$; (iii) $\operatorname{arccot} x$.
In (i), remember to take care with the sign of the square root.

Note
A slight problem occurs in the differentiation of $\operatorname{arcsec} x$ and $\operatorname{arccosec} x$, and the reader should work through the differentiation of $y = \operatorname{arcsec} x$ while reading this note. We reach the stage

$$\frac{dy}{dx} = \frac{1}{\sec y \tan y}.$$

Now $\sec y = x$, and so $\tan y = \pm \sqrt{(x^2 - 1)}$. But in the range of principal values of arcsec $(0 \leqslant y \leqslant \pi)$, $\tan y$ takes both positive and negative values. As it happens, in this region $\tan y$ has the same sign as $\sec y$, so that $\sec y \tan y$ is non-negative. Then

$$\frac{dy}{dx} = \left| \frac{1}{x\sqrt{(x^2 - 1)}} \right|$$

or
$$\frac{d}{dx}(\operatorname{arcsec} x) = \frac{1}{|x|\sqrt{(x^2 - 1)}}.$$

(Remember that the domain of the arcsec function is $|x| \geqslant 1$.)

Qu.13 Sketch the graph of the principal values of $y = \operatorname{arcsec} x$, and verify that the final result above is correct as far as signs are concerned.

Qu.14 Use the method of the note above to differentiate $\operatorname{arccosec} x$.

Exercise 14c

In questions 1 to 7, find $\frac{dy}{dx}$ by (i) immediate (implicit) differentiation, and (ii) first making y the subject. Verify that the two answers are the same.

1. $y^2 = 1 + x$.

2. $xy = 3 + 2x^2$.

3. $y^3 = \sin 2x$.

4. $y \sin x = \cos x$.

5. $x^2 y - \tan x = y \sin x$.

6. $\dfrac{1}{1+y^2} = 2 + x$.

7. $\dfrac{3x}{y^2} = 2 \sec^2 x$.

In questions 8 to 13, find $\dfrac{dy}{dx}$.

8. $x^2 y + y^2 x = 1$.

9. $x \sin y = 2y^2$.

10. $\sin xy = 1 + x$.

11. $x^2 + y^2 = 1 - \dfrac{2x}{y}$.

12. $\dfrac{x}{1+xy} = \tan y^3$.

13. $2x + 3y + 3y^2 + 2xy = \sin^2 y$.

In questions 14 to 21, find $\dfrac{dy}{dx}$ and $\dfrac{d^2 y}{dx^2}$ in terms of t.

14. $x = 2t,\ y = t^2$.

15. $x = 1 + t,\ y = 1 - t^2$.

16. $x = t^2,\ y = t^3$.

17. $x = a \cos t,\ y = b \sin t$.

18. $x = \sec t,\ y = \tan t$.

19. $x = \dfrac{1}{1+t^2},\ y = \dfrac{t}{1+t^2}$.

20. $x = \dfrac{1}{1+t^3},\ y = \dfrac{t}{1+t^3}$.

21. $x = ct,\ y = \dfrac{c}{t}$.

22. Find $\dfrac{d^3 y}{dx^3}$ in question 16 above.

23. The position at time t of a projectile fired from the origin with speed $V\,\mathrm{ms}^{-1}$ at an angle α to the horizontal is given approximately by

$$x = Vt \cos \alpha$$
$$y = Vt \sin \alpha - 5t^2.$$

Plot the path of the projectile in the case where $V = 50$ and $\alpha = \arcsin \frac{3}{5}$. Calculate the gradient of this curve at time t.

24. Show that

$$\frac{d^2}{dx^2}(y^2 x) = 2y_1(2y + xy_1) + 2xyy_2,$$

where y_n means $\dfrac{d^n y}{dx^n}$.

25. If $y^2 = 2xy + 3$, obtain an expression for $\dfrac{d^2 y}{dx^2}$ in terms of x and y only.

26. P is a point on the circumference of a disc of radius a which is rolling along a horizontal straight line. Initially, P is in contact with O, the origin of the line. Show that the path of P is given parametrically by

$$x = a\theta - a\sin\theta, \quad y = a - a\cos\theta.$$

Choosing any convenient value for a, plot the graph of y against x for two complete periods. (The curve is called a *cycloid*.) Calculate the gradient at the general point θ, expressing it in terms of $\dfrac{\theta}{2}$. Hence find the direction in which P begins to move.

27. A particle of mass m is moving in one dimension. Show that (with the usual notation)

(i) $a = v\dfrac{dv}{dx}$;   (ii) $\dfrac{d}{dt}(\tfrac{1}{2}mv^2) = mav$;

(iii) $\dfrac{d}{dx}(\tfrac{1}{2}mv^2) = ma$.

28. Differentiate the following with respect to x:

(i) $\arcsin 2x$;   (ii) $\arccos x^2$;

(iii) $\arctan\dfrac{1}{x}$;   (iv) $\operatorname{arccosec}(1-x^3)$.

29. If $y = x\arcsin\dfrac{1}{x}$, find $\dfrac{dy}{dx}$ in terms of x only.

30. If, in question 12 of Exercise 14a, the particle moves so that its x-coordinate at time t is given by $x = 1 - \cos t$, write down an expression for the y-coordinate in terms of t. Verify that
$$\dfrac{dy}{dx} = \dfrac{dy}{dt} \div \dfrac{dx}{dt}.$$

31. Two variables, p and v, are related by $pv = 50$. Show that $\dfrac{dv}{dp} = -\dfrac{v}{p}$. Write down the values of v and $\dfrac{dv}{dp}$ when $p = 2$, and hence estimate v when $p = 2\cdot003$.

32. When light from an object passes through a lens of focal length f, the object distance u and the image distance v, both measured from the lens, are related by $\dfrac{1}{v} + \dfrac{1}{u} = \dfrac{1}{f}$. An object is moved at $5\,\mathrm{cm\,s^{-1}}$ towards a lens of focal length $20\,\mathrm{cm}$. Find the speed at which the image is moving when

(i) $u = 60\,\mathrm{cm}$;   (ii) $u = 30\,\mathrm{cm}$.

33. A spaceship is falling directly towards a planet. Its distance h from the centre of the planet and its velocity v are related by

$$\tfrac{1}{2}mv^2 = \dfrac{c}{h} - k,$$

where m, c and k are constants. By differentiating this equation w.r.t. time, derive an expression for the acceleration of the spaceship in terms of h and the constants only.

34. Find the equation of the tangent to the ellipse $x^2 + xy + 2y^2 = 23$ at the point $(3, 2)$.

35. If u and v are functions of time, and are related by $uv=1$, prove that

$$\frac{\ddot{u}}{u}+\frac{\ddot{v}}{v}+2\dot{u}\dot{v}=0.$$

[*Hint*: it may help to remember that $uv=1$.]

36. Let $y^2=x^3$. Find $\dfrac{dy}{dx}$ by

    (i) differentiating implicitly;
    (ii) making y the subject and differentiating w.r.t. x;
    (iii) making x the subject, differentiating w.r.t. y and using

$$\frac{dy}{dx}=1\bigg/\frac{dx}{dy}.$$

Verify that all three answers are the same.

37. Let $x=\dfrac{1}{1+t},\ y=\dfrac{t}{1+t}.$

    (i) Find $\dfrac{dy}{dx}.$

    (ii) Eliminate t and hence check your answer to (i).

38. Show that the curve $xy^2+2yx^2=8$ has zero gradient when $y(4x+y)=0$. By considering each factor separately, show that there is only one point on the curve with zero gradient, and find its coordinates.

39. Find the equations of the tangents and normals to the curves

    (i) $x=1+t^3,\ y=t(1+t^3)$ at the point $t=2$;

    (ii) $x=2\sin t,\ y=\cos 2t$ at the point $t=t_0$.

40. In question 9 of Exercise 14a, let M be the man, and T the train, and let $\angle OMT=\theta$. Express θ in terms of t. Find $\dfrac{d\theta}{dt}$ and verify that after a long time, the man scarcely needs to move his head (or eyes) to watch the train. What is the greatest rate (in radians per second) at which the man must turn his head while watching the train?

41. Differentiate the following w.r.t. ϕ as they stand, and hence find expressions for $\dfrac{d\theta}{d\phi}$:

    (i) $\sin\theta=2\sin\phi$;
    (ii) $2\cos\theta\tan\phi=3$.

Miscellaneous exercise 14

1. A viscous liquid is poured on to a flat surface. It forms a circular patch which grows at a steady rate of $5\,\text{cm}^2/\text{s}$. Find, in terms of π,

(i) the radius of the patch 20 seconds after pouring has commenced.

(ii) the rate of increase of the radius at this instant.

[C (O)]

2. A curve is represented parametrically by the equations

$$x = t + \frac{1}{t},$$

$$y = t^3 - 3t.$$

Find $\frac{dy}{dx}$ in terms of t and reduce it to its simplest form.

Given that $y = 12x + 28$ is a tangent to the curve, find the co-ordinates of the point of contact.

[C (O)]

3. (i) The time taken to travel 150 km at v km/h is t hours. Find v in terms of t.

When t is decreased by a small amount δt, find an approximation for δv, the corresponding increase in v. If t was originally 125 minutes, find the approximate increase in v necessary to reduce t by 5 minutes.

(ii) The volume, V cm^3, of water in a vessel is given by $V = 3x^2 + 5x$ where x cm is the depth of water. Water is being poured into the vessel at a constant rate of 650 cm^3/s; find the rate at which the depth of water is increasing when $x = 10$, stating the units in your answer.

[Ox (O)]

4. At the instant when the depth of water in a reservoir is h metres, the volume V, in m^3, of the water is given by

$$V = \frac{\pi h^3}{12}.$$

If the volume of water in the reservoir is increasing at the *constant rate* of 2 m^3 per minute, calculate, in terms of π, the rate at which the depth is increasing at the instant when the depth is 4 m.

Calculate the time taken for the depth to increase from 6 m to 12 m, leaving your answer in terms of π.

[AEB (O) '80]

5. The height of a cone remains constant at 4 cm. The radius of the base increases by 0·1 cm s^{-1}. When the radius is 10 cm, calculate the rate at which

(i) the volume of the cone is increasing;

(ii) the curved surface area is increasing.

Leave your answers in terms of π. Surds need not be evaluated.

[SUJB (O)]

6. If $x = \cot \theta$ and $y = \sin^2 \theta$, find $\frac{dy}{dx}$ in terms of θ and show that

$$\frac{d^2 y}{dx^2} = 2 \sin^3 \theta \sin 3\theta.$$

[JMB]

7. The points A and B are on the same horizontal level, and at a distance b apart. A particle P falls vertically from rest at B, so that, at time t, its depth below B is kt^2, where k is a constant. At this time, the angle of depression of P from A is θ. Prove that

$$\frac{d\theta}{dt} = \frac{2bkt}{b^2 + k^2 t^4}.$$

Show that $\dfrac{d\theta}{dt}$ is greatest (and not least) when $\theta = \frac{1}{6}\pi$.

[C]

8. A street lamp is 5 m high and a boy, of height 1·5 m, starts at the lamp and runs from it in a straight line on level ground at a uniform speed of 3 m/s. Find the length of his shadow when he is x m from the lamp, and also the rate at which this length is increasing.

   After running for 3 s his speed begins to decrease uniformly so that after 5 more seconds it is 2 m/s. Find the rate at which the shadow is growing t seconds from the moment he begins to slow down. Find also the length of his shadow when he stops moving.

[Ox (O)]

9. A curve is given parametrically by the equations

$$x = a(2 + t^2), \quad y = 2at.$$

Find the values of the parameter t at the points P and Q in which this curve is cut by the circle with centre $(3a, 0)$ and radius $5a$. Show that the tangents to the curve at P and Q meet on the circle, and that the normals to the curve at P and Q also meet on the circle.

[L]

15 The exponential and logarithmic functions †

15.1 The number e
In Chapter 4, we met the family of functions of the form $y = a^x$ $(a > 0)$; these were called exponential or power functions, and their inverses were called logarithmic functions. We begin this chapter by attempting to differentiate $y = a^x$.

Let $f(x) = a^x$ where $a > 0$. Then

$$f'(x) = \lim_{h \to 0} \frac{a^{x+h} - a^x}{h}$$

$$= \lim_{h \to 0} a^x \left(\frac{a^h - 1}{h} \right)$$

$$= a^x \lim_{h \to 0} \left(\frac{a^h - 1}{h} \right),$$

where the factor a^x may be taken outside the limit, since it does not depend on h. Now $\lim_{h \to 0} \dfrac{a^h - 1}{h}$ is a number whose value depends only on the value of a; let us denote it by $L(a)$. We then have

$$\frac{d}{dx}(a^x) = L(a) \cdot a^x.$$

Qu.1 (i) Evaluate $\dfrac{2^h - 1}{h}$ for $h = 1$, $h = 0 \cdot 1$, $h = 0 \cdot 01$, $h = 0 \cdot 002$, $h = 0 \cdot 0005$. Hence estimate $L(2)$ to 3 d.p.
(ii) Repeat for $L(3)$ and $L(4)$.
(iii) Write down the value of $L(1)$.

Later in this chapter, the precise nature of the function $L(a)$ will emerge; for the present, we shall assume that there is a unique value of a satisfying $L(a) = 1$, and we shall call this number e. (A proof of this assumption is outlined in questions 16, 17 and 18 of Exercise 15a.) Thus we have:

Definition
The number e is the number such that

$$\lim_{h \to 0} \left(\frac{e^h - 1}{h} \right) = 1.$$

† The introduction to the appendix on page 497 should be read before this chapter.

The values obtained in Qu.1 suggest that e lies between 2 and 3, and we shall see in Chapter 28 that its value is approximately 2·7182818. It is an irrational number.

We now have that

$$\frac{d}{dx}(e^x) = e^x . L(e)$$

$$\Rightarrow \boxed{\frac{d}{dx}(e^x) = e^x.}$$

Thus the function $y = e^x$ satisfies the differential equation

$$\frac{dy}{dx} = y.$$

(The question naturally arises as to whether there are any other functions which are equal to their own derivatives. This is answered in Exercise 15a, question 7.)

15.2 The exponential function

Definition
The function $y = e^x$ is known as *the* exponential function. It is sometimes written as exp x, and it possesses all the properties of the other power functions established in Chapter 4.

Qu.2 Plot the graph of $y = e^x$ for $|x| \leqslant 3$. Draw by eye the tangents at several points and verify that at each point the gradient and the height of the curve are equal.

Example 1
Differentiate

(i) $\exp x^2$;   (ii) $\exp(\sin x)$;

(iii) $\sin(e^{3x})$;   (iv) $x^2 e^x$.

Solution
The exponential function is its own derivative, and so
(i) by the chain rule,

$$\frac{d}{dx}(\exp x^2) = \exp x^2 . \frac{d}{dx}(x^2)$$

$$= 2x \exp x.$$

(ii) Similarly,

$$\frac{d}{dx}\{\exp(\sin x)\} = \cos x \exp(\sin x).$$

(iii) Here, the exponential function is the inner function:

$$\frac{d}{dx}\{\sin(e^{3x})\} = \cos e^{3x} . 3e^{3x}$$

$$= 3e^{3x} \cos e^{3x}.$$

(iv) By the product rule,

$$\frac{d}{dx}(x^2 e^x) = 2xe^x + x^2 e^x$$

$$= xe^x(x+2).$$

Qu.3 Differentiate

(i) e^{4x}; (ii) e^{-x}; (iii) $\exp(\sec x)$; (iv) $\exp(\exp x)$;

(v) $\tan(2e^x)$; (vi) xe^x; (vii) $\dfrac{e^x}{x}$.

Qu.4 Differentiate e^{5x}
 (i) directly;
 (ii) by writing it as the product $e^{2x}e^{3x}$.

Qu.5 Differentiate

(i) $(\exp x)^3$; (ii) $\exp 3x$.

Explain why your answers should be equal, and show that they are.

Qu.6 If $y=e^{2x}$, find $\dfrac{d^2 y}{dx^2}, \dfrac{d^3 y}{dx^3}$ and $\dfrac{d^n y}{dx^n}$.

Some differentiations are made simpler by writing the question in an alternative form, as in Qu.7.

Qu.7 Differentiate $y = \sqrt{(\exp x^2)}$ by first writing it as $\exp(\tfrac{1}{2}x^2)$.

Exercise 15a   **1.** Simplify

(i) $e^x e^x$; (ii) $\exp x \exp(-x)$; (iii) $\dfrac{\exp(x+1)}{\exp(x-1)}$;

(iv) $\left(\exp\dfrac{1}{n}\right)^n$; (v) $\sqrt{(\exp 4x^2)}$.

2. Differentiate

(i) e^{-4x}; (ii) $4e^x$; (iii) e^{-x^2}; (iv) e^2; (v) $e^x \sin x$;

(vi) $e^x(\sin x + \cos x)$; (vii) $e^x\left(1-x+\dfrac{1}{2}x^2-\dfrac{1}{6}x^3\right)$;

(viii) $\sin(e^x)$; (ix) $xe^x(\sec x + \tan x)$.

3. Differentiate the following by first writing them in a simpler form:
 (i) $(\exp 3x^2)^3$; (ii) $\sqrt{(e^{\sin x})}$.

4. (i) Show that $y=Ae^{nx}$ satisfies the differential equation
 $\dfrac{dy}{dx}-ny=0.$
 (ii) What can be said about the real numbers p and q if $y=pe^{qx}$
 and $\dfrac{d^2 y}{dx^2}=y$?

5. Prove by induction that $\dfrac{d^n}{dx^n}(xe^x)=(x+n)e^x.$

THE EXPONENTIAL AND LOGARITHMIC FUNCTIONS

6. Let $S=e^x \sin x$ and $C=e^x \cos x$. Show that the derivative of $aS+bC$ is $(a-b)S+(a+b)C$. Hence find a function whose derivative is $e^x(3 \sin x+7 \cos x)$.

7. Suppose that $y_1(x)$ and $y_2(x)$ are two functions which are both solutions of $\dfrac{dy}{dx}=y$. Suppose also that y_2 is not identically zero.

Use the quotient rule to find $\dfrac{d}{dx}\left(\dfrac{y_1}{y_2}\right)$ and deduce that for some constant k, $y_1=ky_2$. Hence show that any solution of this differential equation is of the form $y=Ae^x$ where A is a constant.

8. Show that $y=e^x-x-1$ has a minimum at the origin and no other stationary points. Deduce that for all values of x, $e^x \geqslant 1+x$, with equality only when $x=0$. Show further that, for any positive n,
$$e>\left(1+\frac{1}{n}\right)^n.$$

9. Find
   (i) the equation of the tangent to the curve $y=e^x$ at the point $x=a$;
   (ii) the point where this tangent cuts the x-axis.

10. (i) Let D be an operator meaning $\dfrac{d}{dx}$. $\left(\text{So, for example, } D^2 \right.$
    means DD or $\dfrac{d^2}{dx^2}$, and $(1+D)y=y+\dfrac{dy}{dx}.\Big)$
    Show by induction that
    $$D^n(ye^x)=e^x(1+D)^n y.$$

    (ii) Find the corresponding results for
       (a) $D^n(ye^{2x})$;   (b) $D^n(ye^{-x})$.

    (iii) Use the result of (i) to write down $\dfrac{d^{10}}{dx^{10}}(x^2 e^x)$.

11. Prove by induction that
    $$\frac{d^n}{dx^n}(e^x \sin x)=2^{n/2}e^x \sin\left(x+\frac{n\pi}{4}\right).$$

12. (i) Show that the only stationary point of $y=e^x x^{-n}$ (where $n>0$) is a minimum.

    (ii) Show also that $\dfrac{dy}{dx}=y\left(1-\dfrac{n}{x}\right)$ and deduce that as $x\to\infty$,

    $\dfrac{e^x}{x^n}\to\infty$. (You will need to explain your argument carefully.)
    This means that given any $n>0$, e^x will 'eventually' be much bigger than x^n.

13. If $e^{xy}=x+y$, show that $\dfrac{dy}{dx}=-\dfrac{y^2+xy-1}{x^2+xy-1}$.

14. Find the x-coordinates of the points of inflexion of
    (i) $y=\exp\left(-\frac{1}{2}x^2\right)$;   (ii) $y=xe^{-x}$.

15. Particles P and Q move in one dimension, their positions at time t

being given by $x_P = A \sin nt$ and $x_Q = Ae^{nt} \sin nt$, where A and n are constants. Show that

(i) the particles start at the same point with the same velocity;

(ii) they repeatedly pass through the origin together;

(iii) the particles repeatedly come to rest, Q doing so at time $\dfrac{\pi}{4n}$ after P;

(iv) the acceleration of Q is zero when the velocity of P is zero.

16. Let $L(a) = \lim\limits_{h \to 0} \left(\dfrac{a^h - 1}{h} \right)$.

    (i) Show that $L(a^n) = \lim\limits_{h \to 0} \left(\dfrac{a^k - 1}{k/n} \right)$ where $k = nh$.

    (ii) Explain why $\lim\limits_{h \to 0}$ and $\lim\limits_{k \to 0}$ are equivalent here.

    (iii) Deduce that $L(a^n) = nL(a)$.

17. (This question derives the same result as question 16.) We have seen in the chapter that $\dfrac{d}{dx}(a^x) = a^x L(a)$.

    (i) Show by the chain rule that $\dfrac{d}{dx}(a^{nx}) = na^{nx} L(a)$.

    (Write $u = nx$ if you wish to do this formally.)

    (ii) Substitute $b = a^n$ in $\dfrac{d}{dx}(b^x) = b^x L(b)$ to obtain another expression for $\dfrac{d}{dx}(a^{nx})$.

    (iii) Deduce from (i) and (ii) that $L(a^n) = nL(a)$.

18. In Qu.1 in the chapter, we saw that $L(3)$ seemed to be about 1·099. In this question, its actual value does not matter, so long as it is not zero.

    (i) Explain why $L(3)$ cannot be zero, by considering
$$\frac{d}{dx}(3^x) = 3^x L(3).$$

    (ii) Use the result of questions 16 and 17 to find the value of $L(3^{1/L(3)})$. (This establishes the existence of e.)

    (iii) (We now need to show that e is the only number satisfying $L(a) = 1$.) Suppose that $L(e) = 1$ and $L(f) = 1$. Let $f = e^\lambda$. Use the result of questions 16 and 17 to show that $\lambda = 1$, so that $e = f$.

19. The proportion (p) of a certain strain of bacteria surviving a poison solution of strength s is given by
$$p = \exp(-s^2 - 2s).$$

    (i) Verify that the values of p when $s = 0$ and when $s \to \infty$ are what might be expected.

    (ii) Use the method of small changes to estimate the proportion of bacteria killed by a strength of 0·03.

20. A function f has the property that
$$f(x + p) = f(x).f(p) \tag{1}$$

for all values of x and p.

(i) Show that either $f(t) \equiv 0$ or $f(0) = 1$.

(ii) By differentiating each side of (1) w.r.t. p, treating x as a constant, show that

$$f'(x+p) = f(x) . f'(p).$$

(iii) If f has the further property that $f'(0) = 1$, deduce from (i) that $f(0) = 1$ and from (ii) that $f'(x) = f(x)$.

(iv) Hence show, using the result of question 7, that $f(x) = \exp x$.

15.3 The logarithmic function

Since we have now given a special status to one particular exponential function, and called it *the* exponential function, it follows that its inverse function, $\log_e x$, acquires a special status. It is known as *the* logarithmic function, or the *natural* logarithmic function, and is usually denoted by $\ln x$ rather than $\log_e x$. Thus we have

$$p = \exp q \Leftrightarrow q = \ln p$$

where $p > 0$, since the range of the exponential function (and hence the domain of the logarithmic function) is the positive reals. The logarithmic function possesses all the properties of other logarithmic functions established in Chapter 4. From this point on, any reference to logarithms is understood to mean natural logarithms unless otherwise indicated.

Qu.8 Sketch the graph of $y = \ln x$. State the values of

(i) $\lim\limits_{x \to \infty} \ln x$; (ii) $\lim\limits_{x \to 0} \ln x$.

15.4 The derivative of ln x

We may obtain the derivative of $\ln x$ by the general method for differentiating inverse functions (Section 14.5).

$$y = \ln x \Rightarrow \exp y = x.$$

Differentiating w.r.t. x,

$$\exp y . \frac{dy}{dx} = 1$$

$$\Rightarrow \frac{dy}{dx} = \frac{1}{\exp y} = \frac{1}{x}.$$

So

$$\boxed{\frac{d}{dx}(\ln x) = \frac{1}{x}.}$$

Thus the derivative of the logarithmic function is the reciprocal function.

Qu.9 Differentiate the following w.r.t. x:

(i) $x \ln x$; (ii) $\dfrac{\ln x}{x^2 + 1}$; (iii) $\dfrac{\ln x}{1 - 3 \ln x}$.

Example 2

Differentiate w.r.t. x:

(i) $\ln(3x^4 + 2x)$;   (ii) $\ln(\cos x)$;   (iii) $(\ln x)^3$.

Solution

(i) By the chain rule,

$$\frac{d}{dx}\{\ln(3x^4+2x)\} = \frac{1}{3x^4+2x}\cdot\frac{d}{dx}(3x^4+2x)$$

$$= \frac{12x^3+2}{3x^4+2x}.$$

(ii) Similarly,

$$\frac{d}{dx}\{\ln(\cos x)\} = \frac{1}{\cos x}\cdot(-\sin x)$$

$$= -\tan x.$$

(iii)

$$\frac{d}{dx}\{(\ln x)^3\} = 3(\ln x)^2\cdot\frac{1}{x}$$

$$= \frac{3}{x}(\ln x)^2.$$

Qu.10   Differentiate
(i) $\ln(2x^3 - x^2 + 7x + 3)$;   (ii) $\ln(\sec x)$;   (iii) $\sec(\ln x)$;   (iv) $\ln 2x$.

It may at first seem surprising that the derivatives of $\ln 2x$ and $\ln x$ are equal (see Qu.10 (iv)); but the reason becomes clear when it is remembered that

$$\ln 2x = \ln x + \ln 2$$

and $\ln 2$ is a constant.

Qu.11   Differentiate $\ln x^3$
(i) directly;
(ii) by first writing it as $3 \ln x$.

Qu.12   Differentiate $\ln\dfrac{(2x+3)^2}{(5x+1)^3}$ by writing it as

$2\ln(2x+3) - 3\ln(5x+1)$.

Qu.13   *Explain* the relationship between the derivatives of $\ln(\cos x)$ and $\ln(\sec x)$. (See Example 2 (ii) and Qu.10 (ii).)

15.5 The derivative of a^x

At the beginning of this chapter, we attempted to differentiate $y = a^x$ from first principles and found that

$$\frac{d}{dx}(a^x) = a^x \lim_{h \to 0}\left(\frac{a^h - 1}{h}\right).$$

We were unable to evaluate this limit, and denoted it by $L(a)$.

We can now differentiate a^x by an alternative method, and the value of $L(a)$ will emerge, Two methods are given.

Method 1.
$$y = a^x$$
$$\Rightarrow \ln y = \ln (a^x)$$
$$\Rightarrow \ln y = x \ln a.$$

Differentiating each side w.r.t. x,

$$\frac{1}{y}\frac{dy}{dx} = \ln a$$

$$\Rightarrow \frac{dy}{dx} = y \ln a$$

$$\Rightarrow \frac{dy}{dx} = a^x \ln a.$$

Method 2.
$$y = a^x.$$
But
$$a = e^{\ln a};$$
so
$$y = (e^{\ln a})^x$$
$$= e^{x \ln a}$$

$$\Rightarrow \frac{dy}{dx} = \ln a) \cdot e^{x \ln a}$$

$$= a^x \ln a.$$

Thus

$$\boxed{\frac{d}{dx}(a^x) = a^x \ln a,}$$

and it follows that the value of $L(a) = \lim\limits_{h \to 0} \left(\dfrac{a^h - 1}{h} \right)$ is $\ln a$.

Qu.14   Calculate the gradient of the curve $y = 3^x$ at the point $(2, 9)$.

15.6 Logarithmic differentiation

In the first method of differentiating a^x above, we used the technique of taking logarithms before differentiating. We now give some further examples.

Example 3
Differentiate
(i) $x^{1/x}$;   (ii) $(\sin 2x)^x$.
Solution

(i)
$$y = x^{1/x} \Rightarrow \ln y = \frac{1}{x} \ln x$$

$$\Rightarrow \frac{1}{y}\frac{dy}{dx} = -\frac{1}{x^2} \ln x + \frac{1}{x^2}$$

$$\Rightarrow \frac{dy}{dx} = \frac{y}{x^2}(1 - \ln x)$$

$$\Rightarrow \frac{dy}{dx} = x^{1/x - 2}(1 - \ln x).$$

(ii)
$$y = (\sin 2x)^x \implies \ln y = x \ln (\sin 2x)$$

$$\implies \frac{1}{y}\frac{dy}{dx} = \ln (\sin 2x) + 2x \cot 2x$$

$$\implies \frac{dy}{dx} = (\sin 2x)^x \{\ln (\sin 2x) + 2x \cot 2x\}.$$

Qu.15 Differentiate x^x.
Qu.16 Differentiate $y = 5^{x^2}$ by
(i) taking logarithms of each side;
(ii) using the result of Section 15.5 and the chain rule.

Taking logarithms of each side can sometimes simplify a complicated differentiation which would otherwise involve the product and quotient rules.

Example 4
Differentiate

$$y = \frac{(x-3)^2(4x+3)^3}{(2x^2+1)^7}$$

w.r.t. x.
Solution
Taking logarithms of each side,

$$\ln y = \ln \frac{(x-3)^2(4x+3)^3}{(2x^2+1)^7}$$

$$\implies \ln y = 2 \ln(x-3) + 3 \ln(4x+3) - 7 \ln(2x^2+1).$$

Differentiating each side w.r.t. x,

$$\frac{1}{y}\frac{dy}{dx} = \frac{2}{x-3} + \frac{12}{4x+3} - \frac{28x}{2x^2+1}$$

$$\implies \frac{dy}{dx} = \left(\frac{2}{x-3} + \frac{12}{4x+3} - \frac{28x}{2x^2+1}\right)\frac{(x-3)^2(4x+3)^3}{(2x^2+1)^7}.$$

Exercise 15b   **1.** Simplify

(i) $\ln e$;   (ii) $\exp \ln 2$;   (iii) $\ln (\exp 2)$;

(iv) $\ln \dfrac{1}{e}$;   (v) $\ln 2x - \ln x$;   (vi) $\ln \sqrt{e}$;

(vii) $e^{2 \ln x}$;   (viii) $\ln (x^2 - 4) - \ln (x - 2)$.

2. Differentiate the following. Rewrite first where appropriate.

(i) $\ln 3x$;   (ii) $\ln x^4$;

(iii) $\ln 2x^4$;   (iv) $\ln \dfrac{1}{x}$;

(v) $\ln (1 + x)$;   (vi) $\ln (1 + x^2)$;

(vii) $\ln(\sin x)$;   (viii) $\ln(\sec x + \tan x)$;

(ix) $3 \ln x$;   (x) $x \ln x$;

(xi) $\ln(\exp x)$;   (xii) $\dfrac{3}{4 \ln x}$;

(xiii) $x^2 \ln x$;   (xiv) $\ln\left(\dfrac{x^2}{x^3 - 3}\right)$;

(xv) $(\ln x)^2$;   (xvi) $e^x \ln x$;

(xvii) $\exp(x \ln x)$;   (xviii) $\ln\dfrac{(3x-7)^2}{(x+5)^5}$;

(xix) $\ln\left(\dfrac{x^2+1}{x^2}\right)$;   (xx) $\ln(\operatorname{cosec} x)$;

(xxi) $\ln(\operatorname{cosec}^2 x)$.

3. For what values of x do the following exist?
(i) $\ln \ln x$;   (ii) $\ln \ln \ln x$;
(iii) $\ln \ln \sin x$.

4. Sketch the graph of $y = \ln |x|$.

5. Write down the derivative of $\ln\{f(x)\}$. Hence find functions whose derivatives are

(i) $\dfrac{2x}{x^2+1}$;   (ii) $\dfrac{3x^2+4}{x^3+4x+1}$;   (iii) $\dfrac{1+\tan^2 x}{\tan x}$;

(iv) $\tan x$;   (v) $\dfrac{1}{(1+x^2)\arctan x}$.

6. Investigate, for all values of n, the stationary points of $y = x^n \ln x$. (Remember that $\ln x$ is defined only for $x > 0$.)

7. Find the equations of any tangents to $y = \ln x$ which pass through the origin. Hence *write down* the equations of any tangents to $y = e^x$ which pass through the origin.

8. The position of a particle at time $t > 0$ is given by $s = \ln(1+t)$. Verify that
(i) the particle is always moving forwards;
(ii) it is always slowing down;
(iii) there is no limit to its eventual distance from the origin.
Which of these three statements would be true if the position were given by $s = 1 - \dfrac{1}{t}$?

9. Differentiate

(i) 4^x;   (ii) $x \cdot 2^x$;   (iii) 2^{2x};
(iv) $\ln 2^x$;   (v) $2x^2$;   (vi) $x^{\sin x}$;
(vii) $x^{\ln x}$;   (viii) $(\tan 2x)^{\exp 3x}$.

10. Find $\displaystyle\sum_{n=1}^{\infty} \frac{d^n}{dx^n}(2^x)$ when $x = 0$.

11. Find the maximum value of x^{1/x^2} and show that it is a maximum. (It should not be necessary to find $\frac{d^2y}{dx^2}$.)

12. Use the method of Example 4 to differentiate

   (i) $\dfrac{(x-1)^2}{(x+2)^3(2x-3)^3}$; (ii) $\sqrt{\left/\left\{\dfrac{x-2}{(x+1)(x+2)^3}\right\}\right.}$.

13. Simplify

   (i) $a^{1/\ln a}$; (ii) $a^{\ln b/\ln a}$.

14. Find any points of inflexion of $y=\dfrac{1}{x}\ln x$.

15. Solve (giving answers to 3 s.f.);

   (i) $e^x=3$; (ii) $e^{2x}=3e^x+5$;
   (iii) $e^{2x}+2e^{-2x}=3$.

16. A continuous function F is defined on the set of positive reals (i.e. R^+ is its domain) and possesses the properties that

   (1) for all x, $y\in R^+$, $F(xy)=F(x)+F(y)$;

   and (2) F is not identically zero.

   Show that

   (i) $F(1)=0$;
   (ii) $F^{-1}(p+q)=F^{-1}(p).F^{-1}(q)$ [*Hint*: let $F(x)=p$ and $F(y)=q$];
   (iii) $F(1/b)=-F(b)$;
   (iv) $F(a/b)=F(a)-F(b)$;
   (v) $F(x^n)=n\,F(x)$ [*Hint*: prove for $n\in Z$ then for $n\in Q$, and argue that it must be true for $n\in R$ by continuity];
   (vi) $F(x)=\log_a x$ for some a (You will need first to establish carefully the existence of an a such that $F(a)=1$.)

Miscellaneous exercise 15

1. Find the gradient of the curve $y=\log_e x$ at the point $(e, 1)$. Show that the tangent to the curve at $(e, 1)$ passes through the origin.

   By means of a sketch, find the range of values of the constant k such that the equation $\log_e x=kx$ has two real roots.

   Draw the graph of $y=\log_e x$ in the range $1\leqslant x\leqslant 2$, using values of x at intervals of 0·1. Take 2 cm to represent 0·2 on the x-axis and 0·1 on the y-axis. By drawing a suitable straight line, estimate from your graph the smaller root of the equation $\log_e x=0·3x$.

   [AEB (O) '80]

2. Show that the curve

   $$y=(1+x)e^{-2x}$$

   has just one maximum and one point of inflexion.

   Sketch the curve, marking the coordinates of the maximum point, the point of inflexion and the point where the curve crosses the x-axis.

   [You may assume that $(1+x)e^{-2x}\to0$ as $x\to+\infty$.]

   [L]

3. (i) Differentiate with respect to x

   (a) $\sin^2 3x$, (b) $\dfrac{\ln x}{x^2}$.

(ii) Show that the curve $y = x e^{-x^2/2}$ has two stationary points and investigate their nature.
Sketch the curve. (You may assume that $x e^{-x^2/2} \to 0$ as $x \to \infty$.)

[L]

4. A curve has the equation $y = \dfrac{a}{b + e^{-cx}}$ $(a \neq 0,\ b > 0,\ c > 0)$. Show that it has one point of inflexion, and that the value of y at the point of inflexion is half the limiting value of y as $x \to \infty$.

[Ox]

5. The function f is defined by

$$f(x) = \frac{e^{ax}}{1 + x^2}, \text{ for all real } x.$$

Find conditions upon a for f to have both a maximum and a minimum. *Sketch* (do *not* plot) the graphs of f for the following values of a:
(a) $-\frac{1}{3}$ (b) 1, (c) 3.

[O & C]

16 Curve sketching and inequalities

16.1 The aims of curve sketching

There are many reasons why we may wish to sketch a graph: to illustrate an inequality; to clarify a problem in calculus; to investigate the roots of an equation; and so on. But in all cases, the immediate aim is the same: to illustrate the important features of the graph without becoming involved in too much numerical detail. Thus, for example, it is important to show any turning points that a curve may have, and they must be in the correct position in relation to other features of the curve; but it is not normally necessary to calculate the coordinates of the turning points. Accuracy of scale is not particularly important, since any numerical values that are significant (such as where the curve crosses the axes) can be marked on the graph.

16.2 The techniques of curve sketching

Another important feature of curve sketching is that it should be carried out as *efficiently* as possible, i.e. with the minimum amount of work. This usually involves selecting, from the many techniques available, those which are appropriate to the particular graph being sketched.

The following list gives some of the more important aids in curve sketching, but it must be emphasized that (a) the list is not intended to be complete, and (b) the list should not be used in its entirety for every graph: some methods are not appropriate for some graphs.

(i) *A knowledge of standard curves.* Familiarity with the graphs of $y = \exp x$, $y = \ln x$, $y = a^x$, the trigonometric graphs, the graph of the modulus function, and the work on polynomial graphs in Chapters 6 and 10 is assumed.

(ii) *Points where the graph meets the axes.* It will meet the x-axis where $y = 0$, and the y-axis where $x = 0$.

(iii) *Form at infinity.* This may simply be a question of noticing what happens to x as $y \to \pm \infty$ and to y as $x \to \pm \infty$. But see also Examples 2 and 3 and Section 16.8.

(iv) *Form at the origin.* See Section 16.6.

(v) *Calculus methods.* The values of $\dfrac{dy}{dx}$ and $\dfrac{d^2 y}{dx^2}$ at particular points may help. Stationary points may be found.

(vi) *Symmetry.* See Section 16.4.

(vii) *Restrictions on x or y.* It may be clear that the range of values of x or y is restricted. For example, if $y^2 = 1 + x^2$, then $|y| \geq 1$, giving a part of the plane in which the graph cannot lie (namely the strip $|y| < 1$).

16.3 The graph of a rational function

A function of the form $y = \dfrac{P(x)}{Q(x)}$ where P and Q are polynomials is called a *rational function*. (Compare with the term 'rational number'.) It is not defined for values of x for which $Q(x) = 0$.

Example 1

Sketch the graph of $y = \dfrac{1}{1+x}$.

Solution

(In studying this solution, the reader should draw a pair of axes, and illustrate each piece of information as it is found, ending up with Fig. 16.1.)

Figure 16.1

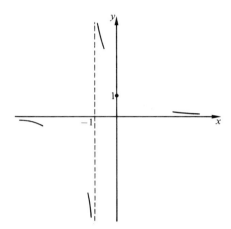

We observe that
(i) when $x = 0$, $y = 1$, so the curve meets the y-axis at $(0, 1)$ only;
(ii) no value of x makes y zero, so the curve does not meet the x-axis;
(iii) when x is 'near' -1, y has a numerically large value. More precisely, as $x \to -1$ from below, $y \to -\infty$. (Think of x as $-1 \cdot 01$ say.) And as $x \to -1$ from above ($x = -0 \cdot 99$ say), $y \to +\infty$. Thus there are two distinct 'tails' approaching the line $x = -1$. When a curve approaches a line in this way, the line is called an *asymptote*. Asymptotes are usually indicated by dotted lines.

Figure 16.2

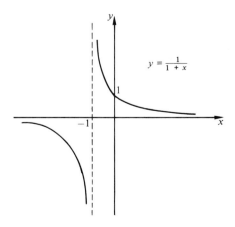

$y = \frac{1}{1+x}$

(iv) As $x \to \infty$, $y \to 0$ from above; and as $x \to -\infty$, $y \to 0$ from below. (So the x-axis is also an asymptote.)

This information is shown in Fig. 16.1. It is reasonable to assume that the curve is continuous except at $x = -1$, and so it may be completed as in Fig. 16.2. Note that there are only two significant numerical values, and both are marked on the graph.

Example 2
Sketch the graph of

$$y = \frac{(x+2)(2x-1)(x-3)}{(x+1)(x-4)(x-5)}.$$

Solution
(i) $y = 0$ when $x = -2$ or $\frac{1}{2}$ or 3; and when $x = 0$, $y = \frac{6}{20} = \frac{3}{10}$.

(ii) There are vertical asymptotes at $x = -1$, $x = 4$ and $x = 5$ (the values of x at which the denominator is zero). The sign of y must be determined on each side of each asymptote. This may be done by finding the sign of each of the six brackets in the expression for y. For example, to the left of the line $x = -1$, when $x = -1 \cdot 01$ say,

$$y = \frac{(+)(-)(-)}{(-)(-)(-)}; \text{ so } y \text{ is negative.}$$

Similarly,

$$x = -0 \cdot 99 \Rightarrow y \text{ is positive;}$$
$$x = 3 \cdot 99 \Rightarrow y \text{ is positive;}$$
$$x = 4 \cdot 01 \Rightarrow y \text{ is negative;}$$
$$x = 4 \cdot 99 \Rightarrow y \text{ is negative;}$$
$$x = 5 \cdot 01 \Rightarrow y \text{ is positive.}$$

(iii) To investigate the values of y as $x \to \infty$ and as $x \to -\infty$, we note that the dominant terms in the numerator and denominator (when the expression for y is multiplied out) are $2x^3$ and x^3 respectively. So, by the method of Example 11 in Chapter 2,

$$\lim_{x \to \infty} y = \lim_{x \to -\infty} y = 2.$$

To find *how* the curve approaches the asymptote $y = 2$ (i.e. whether from above or below), we must examine the next most important terms (in x^2) in the numerator and denominator. Multiplying out gives

$$y \approx \frac{2x^3 - 3x^2}{x^3 - 8x^2}$$

$$\Rightarrow y \approx \frac{2(x^3 - 8x^2) + 13x^2}{x^3 - 8x^2}$$

$$\Rightarrow y \approx 2 + \frac{13x^2}{x^3 - 8x^2}.$$

Now as $x \to \infty$, the term $\dfrac{13x^2}{x^3 - 8x^2}$ is positive, and as $x \to -\infty$, it is negative. So

as $x \to \infty$, $y \to 2$ from above,

and as $x \to -\infty$, $y \to 2$ from below.

Marking all this information on our axes gives Fig. 16.3, and 'joining up' gives the required graph, Fig. 16.4. Note that we have not found the coordinates of the two turning points on the graph; to do so would be very tedious. However, some light can be shed on the position of these turning points by the method of Qu.1.

Figure 16.3

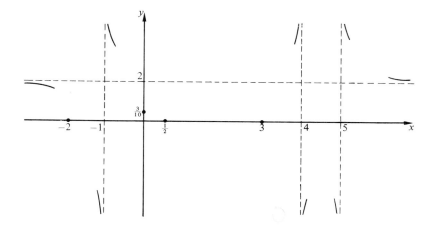

Figure 16.4

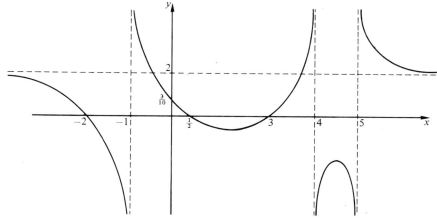

Qu.1 (i) By starting to solve their equations simultaneously, show that the straight line $y = mx + c$ will meet the curve in Example 2 at the points whose x-coordinates are roots of

$$(mx + c)(x + 1)(x - 4)(x - 5) = (x + 2)(2x - 1)(x - 3).$$

(ii) Deduce that no straight line can cut the curve at more than four points.

(iii) Similarly show that no straight line parallel to the x-axis can cut the curve more than three times.

(iv) What can you deduce about the y-coordinates at the two turning points? (A simple result only is required.)

The method of Qu.1 is important; it can provide a useful check of a sketch.

Qu.2 Let $P(x)$ and $Q(x)$ be polynomials of degree m and n respectively. Find an expression for the maximum number of intersections of the curve $y = \dfrac{P(x)}{Q(x)}$ with

(i) a horizontal straight line;

(ii) any straight line.

Qu.3 In Fig. 16.4, it will be seen that the sign of y changes at each asymptote (and at each root). Sketch the graph of $y = \dfrac{1}{(x-1)^2 (x-2)^3}$ and hence suggest a rule connecting sign change at an asymptote with the 'multiplicity' of the asymptote. Compare with the corresponding result for multiple roots (Section 10.7).

Qu.4 (i) Let $y = \dfrac{1}{x^2 + 3x + 4}$. Show that there is no value of x for which $y \to \infty$.

(ii) By completing the square in the denominator, find the coordinates of the maximum point and sketch the curve.

In the next example, we shall concentrate on investigating what happens as $x \to \pm \infty$.

Example 3

Investigate the nature of the following curves as $x \to \pm \infty$;

(i) $y = \dfrac{(2x-1)(x+4)}{(x+1)}$;   (ii) $y = \dfrac{(x+3)(2x-1)^2}{(x-2)}$.

Solution

(i) Certainly, as $x \to \infty$, $y \to \infty$. Examining dominant terms in the numerator and denominator suggests that y eventually increases at about the same rate as $\dfrac{2x^2}{x} = 2x$, i.e. that the curve has a gradient of approximately 2 a long way from the origin. In some cases this may be all that is required; but we can go further. By algebraic long division, we have

$$y = \frac{2x^2 + 7x - 4}{x + 1}$$

$$= 2x + 5 - \frac{9}{x + 1}.$$

So, for large x,

$$y \approx 2x + 5.$$

The line $y = 2x + 5$ is an *oblique asymptote*. The 'error' term, $-\dfrac{9}{x+1}$, shows that

as $x \to \infty$, $y < 2x + 5$,

and   as $x \to -\infty$, $y > 2x + 5$.

The relevant part of the graph is therefore as shown in Fig. 16.5.

Figure 16.5

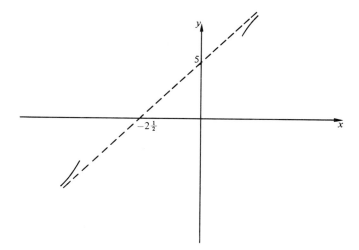

(ii) Examining dominant terms only, y behaves like $4x^2$ for large x, i.e. there is no limit to the gradient of the curve as $x \to \pm \infty$. More precisely,

$$y = \frac{(x+3)(2x-1)^2}{x-2}$$

$$= \frac{4x^3 + 8x^2 - 11x + 3}{x-2}$$

$$= 4x^2 + 16x + 21 + \frac{45}{x-2}.$$

Thus the curve approaches the *parabolic asymptote* $y = 4x^2 + 16x + 21$. In practice, we rarely need to find parabolic asymptotes.

Qu.5 (i) Complete the sketch of the curve in Example 3(i).
(ii) Sketch the parabola $y = 4x^2 + 16x + 21$, and show how the curve in Example 3(ii) approaches it as $x \to \infty$ and as $x \to -\infty$. (Use the 'error' term $+\dfrac{45}{x-2}$.)
(iii) Sketch the curve in Example 3(ii) without showing the parabolic asymptote.

Qu.6 Find the oblique asymptote of the curve
$$y = \frac{x^2(x-1)}{(x+1)(x+2)}$$
and show how the curve approaches the asymptote.

Sketch the curve, and calculate the x-coordinate of the point where the curve cuts this asymptote.

One very useful technique in sketching a rational function is to find the range of values that y can take. The values of y at a local maximum or minimum can often be found in this way. This technique is discussed in Section 16.21 (Example 20 and Qu.51).

Exercise 16a   **1.**   Sketch the graphs of $y=f(x)$, where $f(x)$ is

(i) $\dfrac{1}{x}$;   (ii) $\dfrac{1}{x^2}$;   (iii) $\dfrac{2}{x-3}$;

(iv) $\dfrac{x-2}{(x+3)^2}$;   (v) $\dfrac{x}{(x-1)(x-2)}$;

(vi) $\dfrac{x-1}{x(x-2)}$;   (vii) $\dfrac{x(x-1)}{(x-2)(x-4)(x-5)}$;

(viii) $\dfrac{(x+1)(x-2)}{x(x-4)(x-5)}$;

(ix) $\dfrac{2x-3}{x(x-4)^2}$;   (x) $\dfrac{(x+1)^2}{x^2(x+2)}$.

2.   Sketch the following curves on the same axes, showing only the parts in the region $x>0$.

(i) $y=\dfrac{1}{x}$;   (ii) $y=\dfrac{1}{x^2}$;   (iii) $y=\dfrac{1}{x^3}$.

3.   Sketch the graphs of $y=f(x)$ where $f(x)$ is as given. State the equations of any line asymptotes. Do not find any parabolic asymptotes.

(i) $\dfrac{2x+5}{x-1}$;   (ii) $\dfrac{(x+1)(x+2)}{(x+3)}$;

(iii) $\dfrac{x(x^2-1)}{2x-1}$;   (iv) $\dfrac{3x-7}{2x+1}$;

(v) $\dfrac{x^2-x}{x^2-4}$;   (vi) $\dfrac{x^4}{x-1}$;

(vii) $\dfrac{x(x^2+5x-6)}{x^2-5x+6}$;   (viii) $\dfrac{(x+1)^3}{(x+2)^3}$.

4.   (i) Find the parabolic asymptote in question 3(iii).
(ii) Find the cubic asymptote in question 3(vi).

5.   Sketch the graphs of the following functions:

(i) $\dfrac{1}{x^2+2x+2}$, finding the coordinates of the turning point (use an algebraic method);

(ii) $\dfrac{x-1}{x^2+2x+2}$;   (iii) $\dfrac{2x^2-x+3}{x^2+1}$;

(iv) $\dfrac{x^3-x}{2x^2+1}$, finding the equation of the oblique asymptote;

(v) $\dfrac{x^2-4}{x^2+2}$.

6.   (Miscellaneous.) Sketch

(i) $y=\dfrac{(2x+1)^2}{2x^2-8}$;   (ii) $y=\dfrac{1}{1+x^3}$;

(iii) $xy = x^2 + 1$;　　　　　(iv) $y = \dfrac{x^4}{(1+x)^2}$;

(v) $x^2(y-1) = 1 - 2y$;　　　　　(vi) $y = \dfrac{(x-1)^2(x-2)^3}{(2x-1)^3}$;

(vii) $(y-1)(x-1) = x$;　　　　　(viii) $y + x = \dfrac{1}{x}$;

(ix) $y = \dfrac{(x-1)(x-3)^2(x-5)}{x(x-2)^2(x-4)}$.

7.　Let $f(x) = \dfrac{ax^2 + bx + c}{dx^2 + ex + f}$, where a and d are non-zero, and
no linear factor will cancel. By considering $f'(x)$, show that the
graph of f has at most two turning points.

8.　(i) With $f(x)$ as in question 7, show that the graph of f can cross
its horizontal asymptote at most once. [*Hint*: first find the
equation of this asymptote.]
(ii) Find the condition (on the coefficients) for the curve not to
cross this asymptote at all.
(iii) Sketch $y = \dfrac{(x-1)(x-2)}{x(x-4)}$. Where does the curve cross the
horizontal asymptote?

16.4 Symmetry　　**Qu.7** (For discussion.) Describe the geometrical effect of the mappings
from R^2 (the plane) to R^2 in which the point (x, y) maps to
　(i) $(-x, y)$;　(ii) $(x, -y)$;　(iii) $(-x, -y)$;
(iv) (y, x);　(v) $(-y, -x)$.

(i) *Line-symmetry in the y-axis*
If, in the equation $x^2 y - 2y^2 + 3x^4 - 2 = 0$, we replace x by $-x$, the
equation remains unchanged, since only even powers of x occur. So if
the coordinates (λ, μ) satisfy the equation, then so will the coordinates
$(-\lambda, \mu)$. Geometrically, this means that if a point lies on the curve,
then so does its image under reflection in the y-axis. Thus a sufficient
condition for a curve to have line-symmetry in the y-axis is that only
even powers of x appear in its equation (after multiplying out
brackets, removing fractions etc.). That this is not a necessary
condition can be seen by considering say $y = \cos x$.
(ii) *Line-symmetry in the x-axis*
Similarly, if only even powers of y appear, the curve will have line-
symmetry in the x-axis.
(iii) *S-symmetry about the origin*
If simultaneously replacing x by $-x$ and y by $-y$ has no effect on the
equation, then for each point (λ, μ) on the curve, the point $(-\lambda, -\mu)$
also lies on the curve. So the curve has S-symmetry (or rotational
symmetry of order 2) about the origin. If by the *degree* of a term in
the equation, we mean the sum of the powers of x and y (so that a
term $2x^2 y^3$ would be of degree 5), then a sufficient condition for S-
symmetry is that the terms are either all of even degree or all of odd

degree, after multiplying out brackets etc. The reader is left to verify this.

(iv) *Line-symmetry in* $y = \pm x$.

If interchanging x and y has no effect on the equation, then for each point (λ, μ) on the curve, (μ, λ) also lies on the curve, and the curve has line-symmetry in $y = x$. Similarly, the curve has symmetry in the line $y = -x$ if replacing y by $-x$ and x by $-y$ has no effect.

Qu.8 State what types of symmetry the following curves have:

(i) $y = x(x^2 - 1)$;

(ii) $y = x^2(x^2 - 1)$;

(iii) $x^3 y^2 + y^3 x^2 = 1$;

(iv) $\dfrac{x - y}{x^2 + y^2} = xy$;

(v) $\dfrac{x^2 - 2y^2}{x^2 + 1} = 1 + y^2$;

(vi) $x^2 - y^2 = 2x - 3$;

(vii) $x^3 + 2x^2 y + 2y^2 x - y^3 = 0$.

16.5 Graphs of the form $y^2 = f(x)$

Example 4

Sketch the graph of $y^2 = \dfrac{1 + x}{1 - x}$.

Solution

We note that

(i) the x-axis is an axis of symmetry;

(ii) the value of $\dfrac{1 + x}{1 - x}$ cannot be negative (since $y^2 \geqslant 0$). By inspection (or by the method of Section 16.20) this means that $-1 \leqslant x \leqslant 1$; so the whole graph lies in this vertical strip.

(iii) $x = 0 \quad \Rightarrow y^2 = 1 \Rightarrow y = \pm 1$;

(iv) $x = -1 \Rightarrow y^2 = 0 \Rightarrow y = 0$;

(v) $x \to 1 \quad \Rightarrow y^2 \to \infty \Rightarrow y \to \pm \infty$.

If we now attempt to sketch the curve, we find that there is some uncertainty about the nature of the curve near the point $(-1, 0)$. The gradient could be infinite, zero, or some finite non-zero value, as illustrated in Figs. 16.6a, 16.6b, and 16.6c and d respectively. The 'points' on the last three of the curves shown are called *cusps*. To decide between these possibilities, we can proceed thus:

$$y^2 = \frac{1 + x}{1 - x} \;\Rightarrow\; y^2 = -1 + \frac{2}{1 - x}.$$

(This step is not essential, but it makes the differentiation easier.)

$$\Rightarrow 2y \frac{dy}{dx} = 2(1 - x)^{-2}$$

$$\Rightarrow \frac{dy}{dx} = \frac{1}{(1 - x)^2 y}.$$

At the point $(-1, 0)$ the gradient is therefore infinite. The complete curve is shown in Fig. 16.7.

Figure 16.6   (a)

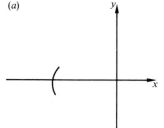

(b)

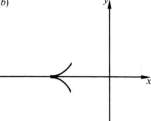

(c)

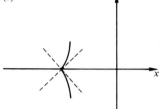

(d)

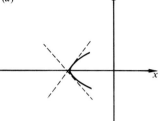

Figure 16.7

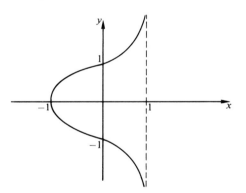

Qu.9 Fig. 16.7 shows that the curve has points of inflexion for some value of x. Show, by differentiating $2y \dfrac{dy}{dx} = 2(1-x)^{-2}$, that

$$y\frac{d^2y}{dx^2} + \left(\frac{dy}{dx}\right)^2 = 2(1-x)^{-3}$$

and hence find the coordinates of the points of inflexion.

    The techniques used in Example 4 are useful for many curves of the form $y^2 = f(x)$. The problem concerning the gradient at points where $y = 0$ (i.e. where the two 'halves' of the curve meet) often arises. We have

$$y^2 = f(x)$$

$$\Rightarrow 2y\frac{dy}{dx} = f'(x)$$

$$\Rightarrow \frac{dy}{dx} = \frac{f'(x)}{2y}.$$

Now $y=0$, and so, provided that $f'(x) \neq 0$ at the point, the gradient of the curve will be infinite. If $f'(x)=0$, then $\lim \dfrac{f'(x)}{2y}$ as the curve approaches the point in question must be found, the method depending on the particular function f. The whole problem can often be avoided, however, by using the method of finding the 'form at the origin' (Section 16.6). (If the point in question is not the origin, it can in effect be moved there by a 'change of origin'; see Section 16.9.)

Qu. 10 By differentiating, find the gradient of the curve $y^2=(x-1)^3$ at the point $(1,0)$. Try
(i) differentiating the expression as it stands;
(ii) first writing $y=(x-1)^{3/2}$.

Qu.11 Show that the difficulty described above arises when trying to find the gradient of $y^2=1-\cos x$ at the origin. Show also that, in this particular case, the problem of finding $\lim\limits_{x \to 0} \dfrac{\sin x}{2\sqrt{(1-\cos x)}}$ may be solved by multiplying top and bottom by $\sqrt{(1+\cos x)}$ and simplifying. What is the required gradient? $\left(\text{Another method here would be to write the given equation as } y^2=2\sin^2\dfrac{x}{2}.\right)$

Another useful technique in dealing with $y^2=f(x)$ is given in Qu.12. The method is to sketch $y=f(x)$ first.

Qu.12 It is required to sketch the graph of $y^2=(x-1)(x-2)(x-4)$.
(i) Sketch $y=(x-1)(x-2)(x-4)$.
(ii) Hence write down the range of values of x in which the required graph lies, and its gradient at $x=1$, $x=2$ and $x=4$ (see the discussion above).
(iii) Sketch the required graph.
Qu.13 Sketch the line pair $y^2=x^2$.

16.6 Form at the origin

Qu.14 What is the condition for the polynomial equation $a_0+a_1x+\ldots+a_nx^n=0$ to have
(i) a root at $x=0$;   (ii) a double root at $x=0$;
(iii) a root of multiplicity k at $x=0$?

Suppose† a curve passes through the origin. If we remove fractions and substitute $y=mx$, we shall obtain a polynomial in x, one of whose roots is zero. This corresponds to the fact that any straight line through the origin ($y=mx$) meets the curve at the origin. If the curve has two branches passing through the origin, then any line $y=mx$ will meet it twice at the origin, and similarly for n branches.
Now suppose that we choose m so that the coefficient of the lowest power of x vanishes; then we have chosen a line which meets the

† The argument that follows may be omitted, but the rule it leads to is important, and is stated just before Example 5.

curve at the origin one more time than 'any line'. The line we have chosen must therefore be a tangent to (one branch of) the curve. An example will make this clear.

Consider

$$2y = \frac{x^2}{x^2 + y}.$$

Removing fractions

$$2x^2y + 2y^2 = x^2,$$

and putting $y = mx$

$$2mx^3 + (2m^2 - 1)x^2 = 0.$$

This has a double root at $x = 0$ for any value of m (so the origin is a double point of the curve). If we take $m = \pm \sqrt{\frac{1}{2}}$, it has a triple root, so these values of m must correspond to the tangents to the two branches of the curve at the origin. (Note that this argument will fail for a curve such as $y^3 = x^2$, where the gradient at 0 is infinite. We can write $x = my$ instead. The rule given below remains valid.)

This method leads directly to the following rule:

The tangents to a curve at the origin may be found by equating to zero the terms of lowest degree in the equation.

Example 5
Find the tangents at the origin to the curve

$$y^2 - 3xy + x^3 = 0.$$

Solution
Equating to zero the terms of lowest degree (degree 2),

$$y^2 - 3xy = 0$$
$$\Rightarrow y(y - 3x) = 0$$
$$\Rightarrow y = 0 \text{ or } y = 3x.$$

These are the tangents.

It is often fairly easy to obtain a better approximation to the form at the origin than merely giving the tangents. Such a 'second approximation' will show how the curve meets the tangents. In Example 5, we may write the equation of the curve as

$$y(y - 3x) = -x^3$$
$$\Rightarrow y - 3x = \frac{-x^3}{y}. \tag{1}$$

For the branch whose tangent is $y = 3x$, we have $y \approx 3x$ for points on the curve. So (1) may be written

$$y - 3x \approx \frac{-x^3}{3x}$$

$$\Rightarrow y \approx 3x - \frac{x^2}{3}$$

$$\Rightarrow y \leqslant 3x \text{ as it passes through the origin.}$$

So this branch of the curve is as shown in Fig. 16.8.

Figure 16.8

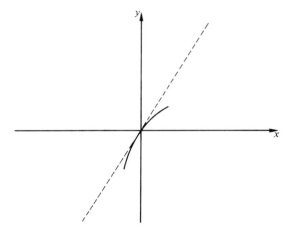

Qu.15 By writing $y(y - 3x) = -x^3$ as $y = \dfrac{-x^3}{y - 3x}$, show that, for

the other branch of the curve, $y \approx \dfrac{x^2}{3}$. Hence show how this branch

meets the tangent $y = 0$.

Qu.16 Sketch the complete graph of the curve in Example 5. Check your answer by the following methods.

(i) By considering the signs of y, $y - 3x$ and $-x^3$, show that $y(y - 3x) = -x^3$ cannot lie

(a) in the 4th quadrant (i.e. where $y < 0$ and $x > 0$);

(b) in the first quadrant where $y > 3x$;

(c) in the 3rd quadrant where $y > 3x$.

(ii) By regarding $y^2 - 3xy + x^3 = 0$ as a quadratic in y, show that (a) as $x \to -\infty$, there are two values for y, and $y \to \pm \infty$, (b) for $x > 0$, there is a maximum value of x to give a real value for y. Find this value of x.

Qu.17 (i) Find the tangents at the origin to $x^2 = \dfrac{y^2}{y + 1}$. Find also a

second approximation showing how the curve meets the tangent.

(ii) Show that $y + 1 > 0$ for all points on the curve. What happens to x as $y \to -1$?

(iii) Does the curve cross the y-axis other than at the origin?

(iv) Show that $x^2 = y - 1$ is a parabolic asymptote.

(v) Sketch the curve.

Qu.18 (i) Let $x^2 = \dfrac{y^2}{y - 1}$. Show that $y > 1$ for all points on the curve except the origin. The origin is therefore an example of an *isolated point* of the curve.

(ii) Try to find the tangents at the origin by the method of Example 5.

(iii) Show that $x^2 = \dfrac{y^4}{y-1}$ also has an isolated point at the origin, but that the method of Example 5 does not break down. (Be warned!)

16.7 Line pairs Consider the equation $3y^2 - 5xy - 2x^2 = 0$. Each term is of degree two, and we may factorize, giving

$$(3y + x)(y - 2x) = 0$$
$$\Rightarrow 3y + x = 0 \text{ or } y - 2x = 0$$
$$\Rightarrow y = -\tfrac{1}{3}x \text{ or } y = 2x.$$

Thus the graph is a pair of straight lines through the origin.

Any equation of the form $ax^2 + bxy + cy^2 = 0$ gives either a line pair (if the discriminant $b^2 - 4ac$ is positive) or a single line (if $b^2 - 4ac = 0$) or is a single point at the origin (if $b^2 - 4ac < 0$). This last possibility may be illustrated by the method of completing the square (see Qu.19).

Qu.19 By completing the square, show that

$$y^2 + 2xy + 2x^2 = 0 \Rightarrow y + x = 0 \text{ and } x = 0$$
$$\Rightarrow x = y = 0.$$

Qu.20 Describe the graph of $y^2 + 2xy + x^2 = 0$.

(Further work on line pairs is given in Chapter 30.)

16.8 Form at infinity† We have already met a method of finding the 'form at infinity' of the graph of a rational function; we now give a method of finding the asymptotes of other types of curve.

The gradient of a curve a long way from the origin (and hence the gradient of any asymptotes) may be found by equating to zero the terms of highest degree. (This may be seen by noting that we seek values of m such that $y = mx$ meets the curve at some large value of x. Thus the polynomial obtained by substituting must have a 'large' root. This will be the case when the leading coefficient vanishes, as may be seen by dividing through by the highest power of x to give a polynomial in $\dfrac{1}{x}$. As before, the case $m = \infty$ must be argued separately.)

Only the directions of the asymptotes are given by this method; 'better' approximations are needed to give the actual asymptotes and the way in which the curve approaches them. Also, the method gives only the eventual directions of the curve; there may not be a line

†This section may be postponed or omitted.

asymptote in each of the directions found, there may only be a parabolic (or higher degree) asymptote. (See part (iii) of Example 6.)

Example 6
Investigate the form at infinity of the following:

(i) $x^2 + 2xy - 3y^2 - x - 7y = 1$;
(ii) $x^2 + 2xy + 3y^2 - x - 7y = 1$;

(iii) $x^2 = \dfrac{1 + y^2}{2x - y}$.

Solution
(i) Equating to zero the terms of highest degree,

$$x^2 + 2xy - 3y^2 = 0$$
$$\Rightarrow (x - y)(x + 3y) = 0.$$

So any asymptotes are parallel to $x = y$ or $x = -3y$. Writing the original equation as

$$(x - y)(x + 3y) = 1 + x + 7y$$

gives, for the asymptote parallel to $x = y$,

$$x - y = \frac{1 + x + 7y}{x + 3y} \tag{1}$$

$$\approx \frac{1 + x + 7x}{x + 3x} \quad \text{(since, for large } x, \ y \approx x)\dagger$$

$$\approx \frac{8}{4} = 2 \text{ for large } x.$$

So the asymptote is $x - y = 2$.

We may now substitute the more accurate approximation $y \approx x - 2$ into (1) to find how the curve approaches this asymptote:

$$x - y \approx \frac{1 + x + 7(x - 2)}{x + 3(x - 2)}$$

$$\approx 2 - \frac{1}{4x - 6}.$$

$\dagger$ Strictly we should write $\dfrac{y}{x} \approx 1$ to give

$$x - y = \frac{1 + x + 7y}{x + 3y}$$

$$= \frac{1/x + 1 + 7y/x}{1 + 3y/x}$$

$$\approx \frac{1/x + 1 + 7}{1 + 3}$$

$$\approx 2.$$

The notation used here and in part (iii) is a convenient abbreviation.

CURVE SKETCHING AND INEQUALITIES

Thus
$$y \approx x - 2 + \frac{1}{4x - 6}$$

and so
$$y > x - 2 \text{ as } x \to \infty$$

and
$$y < x - 2 \text{ as } x \to -\infty.$$

For the asymptote parallel to $x + 3y = 0$, we write the original equation as

$$x + 3y = \frac{1 + x + 7y}{x - y}$$

and proceed as above, putting $y \approx -\frac{1}{3}x$. (See Qu.21.)

(ii) The direction at infinity is given by

$$x^2 + 2xy + 3y^2 = 0.$$

This has no real factors, so the curve does not approach infinity.

(iii) Asymptotic directions are given by $x^2(2x - y) = 0$, i.e. $x = 0$ or $2x - y = 0$.

For the direction parallel to $x = 0$,

$$x^2 = \frac{1 + y^2}{2x - y}$$

$$\approx \frac{1 + y^2}{-y}$$

$$\approx -y \text{ for large } y.$$

So part of the curve (a long way from the origin) resembles $y = -x^2$. (This is not the actual parabolic asymptote; a further approximation would be needed to find it.)

For the direction parallel to $2x - y = 0$,

$$2x - y = \frac{1 + y^2}{x^2} \qquad (2)$$

$$\approx \frac{1 + 4x^2}{x^2} \approx 4.$$

So the asymptote is $2x - y = 4$.

Putting $y = 2x - 4$ in (2),

$$2x - y \approx \frac{4x^2 - 16x + 17}{x^2}$$

$$\approx 4 - \frac{16}{x}.$$

So
$$y \approx 2x - 4 + \frac{16}{x}.$$

The full curve is shown in Fig. 16.9.

Qu.21 (i) Complete the investigation of the second asymptote in Example 6(i), giving your final approximation for y.
(ii) Show that the curve does not meet either asymptote.
(iii) Sketch the curve.

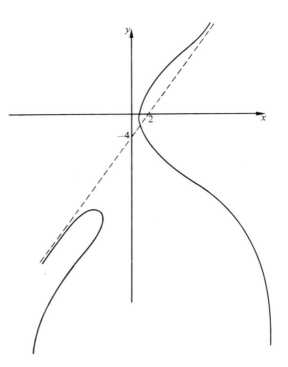

Figure 16.9

Qu.22 Investigate the form at infinity of $x^2 = \dfrac{1+y}{2x-y}$

(i) by the method of Example 6 (showing why there is no asymptote parallel to $x=0$);

(ii) by making y the subject, and using the method of Example 3.

Qu.23 Determine whether or not the following curves have asymptotes:

(i) $x^2 + 2y^2 = x - y$;

(ii) $x^2 - 2x = y^2 - 3y$;

(iii) $(x+y)^2 + (x-2y)^2 = 2$.

16.9 Change of origin

A mapping $f : R^2 \rightarrow R^2$ of the form

$$(x, y) \rightarrow (x+a, \; y+b)$$

where a and b are constants, is called a *translation*. Geometrically, each point in the plane moves a distance a to the right and b up. (Alternatively, we may say that we have kept each point still and moved the axes, a to the left and b down. We have thus 'changed the origin'.)

Given the equation of some curve C, it is a simple matter to find the equation of C' where C' is a translation of C. Consider, for example, the curve C

$$y = x^2 + 3x - 5 \tag{1}$$

and the curve C'

$$y = (x-1)^2 + 3(x-1) - 5. \tag{2}$$

CURVE SKETCHING AND INEQUALITIES

If (λ, μ) satisfies (1), then $(\lambda + 1, \mu)$ satisfies (2). So

$$(\lambda, \mu) \in C \Leftrightarrow (\lambda + 1, \mu) \in C'.$$

So C' is a translation of C by 1 unit in the direction of the positive x-axis (Fig. 16.10). The equation of C' can of course be simplified to $y = x^2 + x - 7$.

Figure 16.10

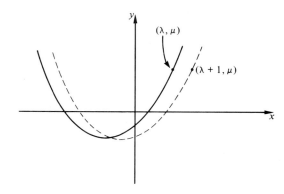

This idea may be generalized to give the rule: Replacing x by $(x - a)$ and y by $(y - b)$ in the equation of a curve gives rise to the translation

$$(x, y) \rightarrow (x + a, y + b)$$

of the curve.

Example 7
Show that the curve $y = x^3 + 6x^2 + 11x + 10$ has S-symmetry about the point $(-2, 4)$.
Solution
We wish to translate the curve so that the point $(-2, 4)$ on the curve moves to the origin; i.e. we wish to effect the translation

$$(x, y) \rightarrow (x + 2, y - 4).$$

So we replace x by $(x - 2)$ and y by $(y + 4)$ in the equation of the curve. This gives

$$y + 4 = (x - 2)^3 + 6(x - 2)^2 + 11(x - 2) + 10$$
$$\Rightarrow y = x^3 - x.$$

Now this curve has S-symmetry about the origin (it is an odd function), so the original curve has S-symmetry about $(-2, 4)$.

Qu.24 Sketch $y^2 = x^3$ and hence sketch $(y - 1)^2 = (x + 1)^3$.

Exercise 16b   Sketch the following graphs, numbers 1 to 17:

1. $y^2 = x^2 + 1$; 2. $y^2 = x^5$;
3. $y^2 = x^3 - 1$; 4. $xy^2 = x^3 - 1$;
5. $y^2 = x(x - 1)$; 6. $x^2 = y^3 - y$;
7. $x^4 - x^2 y = y^2$; 8. $x + y + x^2 y^3 = 0$;
9. $x^2 - y^2 = xy^3$; 10. $xy^2 - yx^2 = 1$;

11. $x^3 + y^3 = x^2 y^2$; **12.** $y^4 - y^2 + x^2 = 0$;

13. $y^4 - y^2 + x^4 = 0$; **14.** $yx^2 = (y-1)(y-2)^2$;

15. $y^2 = \dfrac{x^2 - 1}{x + 2}$; **16.** $y^3 = x(x+1)$;

17. $y^3 = x^2(x+1)$.

18. On the same axes, sketch
 (i) the line pair $x^2 + 3xy + 2y^2 = 0$;
 (ii) the hyperbola $x^2 + 3xy + 2y^2 = 1$.

19. Sketch the ellipse $x^2 + 2xy + 2y^2 = 1$. $\left(\text{It may help to find an expression for } \dfrac{dy}{dx}.\right)$

20. (i) Sketch the graph of $xy^3 - yx^2 = 1$.
 (ii) By replacing x by $-x$, write down the equation of its reflection in the y-axis.
 (iii) Similarly find the equation of
 (a) its reflection in the x-axis;
 (b) its reflection in the line $y = x$;
 (c) its reflection in the line $y = -x$;
 (d) the curve when it has been rotated through $180°$ about the origin.

21. (i) Show that

$$P \text{ lies on } y^2 - yx^2 = 0$$
$$\Leftrightarrow P \text{ lies on } y = x^2 \text{ or } P \text{ lies on the } x\text{-axis.}$$

 (ii) Sketch the following, each of which is in two or more parts as in (i):
 (a) $(x - y)(x - y^2) = 0$; (b) $y^2 = x^4$;
 (c) $xy = 0$; (d) $xy^3 = yx^2$.

22. Investigate the nature of the curve $y^2 x + (x-1)^2(x-2) = 0$ at the point $(1, 0)$ by transferring the origin to this point. Hence sketch the (original) curve.

Sketch the following curves, numbers 23 to 29, where a is a positive constant.

23. $y^2 = \dfrac{x^3}{2a - x}$ (Cissoid of Diocles).

24. $y^2 = \dfrac{x^2(a + x)}{a - x}$ (Right strophoid).

25. $x^3 + y^3 = 3axy$ (Folium of Descartes).

26. $x(x^2 + y^2) = a(y^2 - 3x^2)$ (Trisectrix of Maclaurin).

27. $(x^2 + y^2)^2 = a^2(x^2 - y^2)$ (Lemniscate of Bernoulli).

28. $y(x^2 + a^2) = a^3$ (Witch of Agnesi).

29. $x^{2/3} + y^{2/3} = a^{2/3}$ (Astroid).

30. Factorize and sketch the line pair

$$2x^2 + 3y^2 - 5xy + 5x - 8y - 3 = 0.$$

31. Sketch the curve $x^2 - 2xy + y^2 + x - 2y = 0$.

16.10 Graphs involving particular functions

Simple graphs involving the trigonometric, exponential, logarithmic or modulus functions have been met in earlier chapters. We now give some further examples.

Example 8
Sketch
(i) $y=x+e^{-x}$;   (ii) $y=e^x \sin x$.
Solution
(i) Considering the graphs of $y=x$ and $y=e^{-x}$ (both shown in Fig. 16.11) and mentally 'adding' them gives Fig. 16.12. Note that
(a) e^{-x} is small when x is large and positive, so $y \approx x$, i.e. $y=x$ is an asymptote;

Figure 16.11

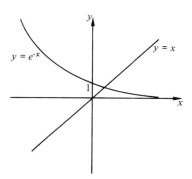

Figure 16.12

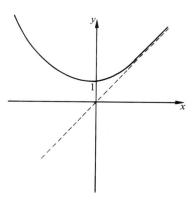

(b) when $x \to -\infty$, e^{-x} is the dominant term;
(c) differentiation shows that $(0, 1)$ is the minimum.
(ii) We may think of $e^x \sin x$ as a sine curve with an increasing amplitude. (Compare with $y=A \sin x$.) Thus we have Fig. 16.13. The dotted lines are $y= \pm e^x$. (Further properties of this function are established in question 15 of Exercise 15a.)

Example 9
Sketch
(i) $y=|x-1|+|2x+3|$;

(ii) $y= \left| \dfrac{(x-1)(x-2)}{x+1} \right|$.

Figure 16.13

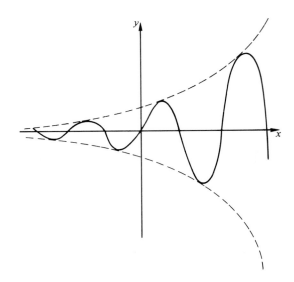

Solution

(i) For $x < -\frac{3}{2}$,

$$y = -(x-1)-(2x+3)$$
$$\Rightarrow\ y = -3x-2.$$

For $-\frac{3}{2} \leqslant x < 1$,

$$y = -(x-1)+(2x+3)$$
$$\Rightarrow\ y = x+4.$$

For $x \geqslant 1$,

$$y = (x-1)+(2x+3)$$
$$\Rightarrow\ y = 3x+2.$$

The curve may now be sketched in three parts. Its continuity provides a check (Fig. 16.14).

Figure 16.14

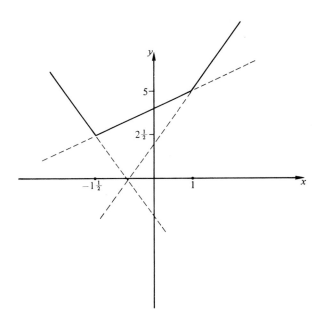

(ii) We first sketch $y = \dfrac{(x-1)(x-2)}{x+1}$ (Fig. 16.15).

Figure 16.15

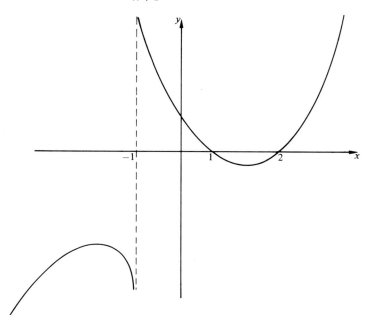

The effect of the modulus sign is to 'reflect' any part of the curve below the x-axis, so that the whole curve lies on or above the x-axis (Fig. 16.16).

Figure 16.16

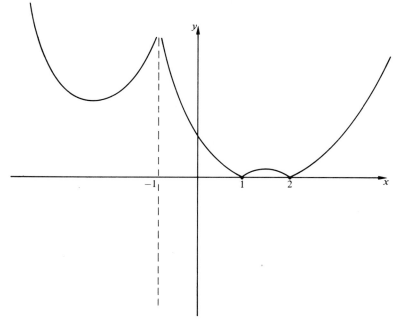

Exercise 16c   Sketch the following, numbers 1 to 17:

      1. $y = e^{-x^2}$;   **2.** $y = x + \sin x$;

3. $y = e^{-x} \cos x$; **4.** $y = \frac{1}{2}(e^x + e^{-x})$;

5. $y = \frac{1}{2}(e^x - e^{-x})$; **6.** $y^2 = \sin x$;

7. $y = \dfrac{\sin x}{x}$; **8.** $y = \ln(\sin x)$;

9. $y = |\sin x|$; **10.** $y = |x^2 - x|$;

11. $y = \left| \dfrac{x+2}{(x-1)(x+3)} \right|$;

12. $y = |x-1||x-2||x-4|$;

13. $y = |x-3| + |3x-6|$;

14. $y = |x-1| + |x+1|$;

15. $y = |2x-7| - |x+1|$;

16. $|y-1| = |x-2|$;

17. $y = \ln|x|$.

18. Which points satisfy $y^2 + |x^2 - 2x| = 0$?

16.11 Parametric graphs

Parametric graphs were introduced in Section 3.10. In sketching a parametric graph (with parameter t), it is useful to pay particular attention to any 'special' values of t; these usually include $t = 0$ and $t \to \pm \infty$.

Example 10
Sketch the graph of

$$x = \frac{t}{1+t^3}, \quad y = \frac{t^2}{1+t^3}.$$

Solution
Starting at $t = +\infty$ and working down to $t = -\infty$, we have:
(1) When t is 'near' $+\infty$, x and y are both positive, and, as $t \to \infty$, $(x,y) \to (0,0)$. Also, as $t \to \infty$, the ratio $\frac{y}{x} \to \infty$, (i.e. y is relatively much larger than x when they are both very small. This can also be seen by noting that $y = tx$.) So the y-axis is a tangent at the origin (Fig. 16.17a).
(2) When t decreases to zero, (x, y) remains in the 1st quadrant, and $t = 0 \Rightarrow x = y = 0$. But now, the ratio $\frac{y}{x} \to 0$, i.e. the x-axis is a tangent (Fig. 16.17b).
(3) For $-1 < t < 0$, the curve lies in the 2nd quadrant, and as $t \to -1$ from above, $x \to -\infty$, $y \to \infty$. Here, $y = tx$ suggests an asymptote parallel to $y = -x$ as $t \to -1$. (See (5) below.) This gives Fig. 16.17c.
(4) For $t < -1$, the curve lies in the 4th quadrant. As $t \to -1$ from below, $x \to \infty$ and $y \to -\infty$ (with $y = -x$ as a first approximation). As $t \to -\infty$, $(x, y) \to (0, 0)$ with $\frac{y}{x} \to -\infty$, so that the y-axis is again the tangent (Fig. 16.17d).

Figure 16.17   (a)

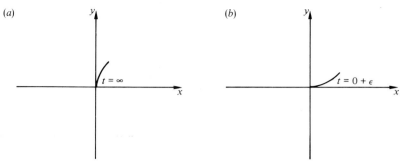

(b)

(c) $t = -1 + \epsilon$

(d)

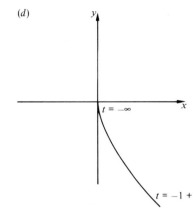

(5) As a second approximation to the asymptote corresponding to $t = -1$, we have

$$\lim_{t \to -1} (y + x) = \lim_{t \to -1} \frac{t^2 + t}{1 + t^3}$$

$$= \lim_{t \to -1} \frac{t}{1 - t + t^2}$$

$$= -\frac{1}{3}.$$

So the asymptote is $y + x + \frac{1}{3} = 0$.

(6) It is useful to know whether the curve meets this asymptote (other than at $t = -1$). Substituting into $y + x + \frac{1}{3} = 0$,

$$\frac{t^2}{1 + t^3} + \frac{t}{1 + t^3} + \frac{1}{3} = 0$$

$$\Rightarrow \frac{t^2 + t}{1 + t^3} = -\frac{1}{3}$$

$$\Rightarrow \frac{t}{1 - t + t^2} = -\frac{1}{3}$$

$$\Rightarrow 1 + 2t + t^2 = 0$$

$$\Rightarrow t = -1.$$

So the curve does not cross the asymptote.

The whole curve is now shown in Fig. 16.17e.

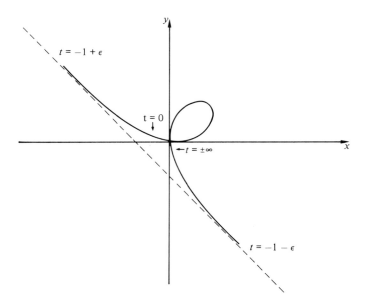

Figure 16.17 (e)

$t = -1 + \epsilon$

$t = 0$

$\leftarrow t = \pm\infty$

$t = -1 - \epsilon$

Fig. 16.17e leads us to ask whether the curve is symmetrical in the line $y = x$. Substituting $1/t$ for t in the original equations and simplifying, we find that the expressions for x and y are interchanged. (Try it.) So the curve is symmetrical in $y = x$. (See Qu.25.) Note that replacing t by $1/t$ happens to indicate symmetry in this particular case it is certainly not a general rule. See also Qu.27.

Qu.25 Find the coordinates of the points on the curve in Example 10 corresponding to
(i) $t = 2$ and $t = \frac{1}{2}$;   (ii) $t = -3$ and $t = -\frac{1}{3}$.
Qu.26 Find the value of t, and hence the values of x and y, where the loop of the curve in Example 10 meets the axis of symmetry $y = x$.

It is worth noting that, in Example 10, a small range of values of t corresponds to a 'long' piece of curve (e.g. $-1 < t < 0$), while an infinite range of values of t gives a finite piece of curve ($t > 0$). This is quite common with parametric graphs.

Qu.27 Sketch $x = \dfrac{\theta}{1-\theta^2}$, $y = \dfrac{\theta^2}{1-\theta^2}$. What symmetry is indicated when θ is replaced by $-\theta$?

Exercise 16d   **1.**   Sketch the following; a is a positive constant. (Use question 2 to check your answers.)

(i) $x = at^2$, $y = 2at$;   (ii) $x = a\cos\theta$, $y = a\sin\theta$;

(iii) $x = a\sec\theta$, $y = a\tan\theta$;   (iv) $x = t^2$, $y = t^3$;

(v) $x = at$, $y = \dfrac{a}{t}$;   (vi) $x = t + 1$, $y = t^2 + 1$;

344   CURVE SKETCHING AND INEQUALITIES

(vii) $x = \dfrac{t-1}{t}, \ y = \dfrac{t^2-1}{t};$ (viii) $x = \dfrac{t}{1+t}, \ y = \dfrac{t^2}{1+t};$

(ix) $x = \dfrac{1}{1+t^2}, \ y = \dfrac{t}{1+t^2}.$

2. Eliminate the parameters in each part of question 1, and use the new equations to sketch the curves again.

3. Sketch the following, where a is a positive constant:

(i) $x = \dfrac{t}{1-t^2}, \ y = \dfrac{t^2}{1-t^2};$ (ii) $x = \dfrac{t^2}{1+t}, \ y = \dfrac{t^3}{1+t};$

(iii) $x = 2\cos\theta, \ y = 3\sin\theta;$ (iv) $x = a\sin\theta, \ y = a\tan\theta;$

(v) $x = a\sin\theta, \ y = a\sin 2\theta;$ (vi) $x = \sin 3\theta, \ y = \sin 2\theta;$

(vii) $x = t^2, \ y = t^4;$ (viii) $x = t - \dfrac{1}{t}, \ y = 1 - t^2;$

(ix) $x = a\cos^3\theta, \ y = a\sin^3\theta$ (Astroid).

4. Sketch the cycloid $x = a(\phi - \sin\phi), \ y = a(1 - \cos\phi)$. (See also question 26 of Exercise 14c.)

16.12 Polar co-ordinates

So far in this book, we have used only rectangular Cartesian coordinates, i.e. the ordinary x-y coordinates measured from a pair of perpendicular axes. We now introduce polar coordinates, in which the position of a point P in the plane is specified by giving

(1) its distance (r) from a fixed point O called the *pole* (or origin);
(2) the angle (θ) that OP makes with some fixed line called the *initial line*. (The initial line is conventionally drawn in the same direction as the positive x-axis, and θ is measured anticlockwise.)

Thus, for example, the point $P(3, 120°)$ would be as shown (Fig. 16.18).

Figure 16.18

Note that if any multiple of 360° is added to θ, the point remains the same. In Fig. 16.18, the coordinates of P could have been given as $(3, 480°)$ or $(3, -240°)$ etc. A further convention governing polar coordinates is that negative values of r are permissible, the distance from the pole being then measured 'backwards'. For example, $Q(-3, 20°)$ is as shown in Fig. 16.19; and P in Fig. 16.18 could have been given as $(-3, -60°)$. So changing the sign of r is equivalent to adding 180° to θ.

Figure 16.19

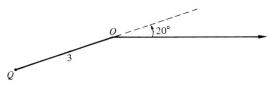

Qu.28 Plot accurately (on polar graph paper if available) $r = \sin 3\theta$, first writing out a table of values for r against θ at $10°$ intervals.

16.13 Converting between polar and Cartesian coordinates

Suppose a point P has Cartesian coordinates (x, y) and polar coordinates (r, θ) as in Fig. 16.20. From the definition of the sine and cosine functions,

$$x = r \cos \theta \quad \text{and} \quad y = r \sin \theta.$$

We can therefore find x and y given r and θ. Similarly, given x and y,

$$r^2 = x^2 + y^2 \quad \text{and} \quad \theta = \arctan \frac{y}{x}$$

(the correct quadrant for θ being found by examining the signs of x and y).

Figure 16.20

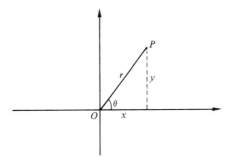

Qu.29 (i) Give the Cartesian coordinates of $(2, 120°)$ and of $(-\sqrt{2}, -45°)$.
(ii) Give the polar coordinates of $(1, -1)$ and of $(-\sqrt{3}, 1)$, choosing $r > 0$ and $0 \leqslant \theta < 360°$ in each case.

Example 11
Convert the equations
 (i) $r^3 = \operatorname{cosec} \theta \cot \theta$ to Cartesian form;
 (ii) $x^2 - y^2 = 2$ to polar form.
Solution

(i) $r^3 = \operatorname{cosec} \theta \cot \theta \;\Rightarrow\; r^3 = \dfrac{1}{\sin \theta} \dfrac{\cos \theta}{\sin \theta}$

$$\Rightarrow\; r^2 = \frac{r \cos \theta}{(r \sin \theta)^2}$$

$$\Rightarrow\; x^2 + y^2 = \frac{x}{y^2}.$$

(ii) $x^2 - y^2 = 2 \;\Rightarrow\; r^2 \cos^2 \theta - r^2 \sin^2 \theta = 2$

$$\Rightarrow\; r^2 \cos 2\theta = 2$$

$$\Rightarrow\; r^2 = 2 \sec 2\theta.$$

Qu.30 (i) Convert these equations to Cartesian form:
(a) $r^2 = \sin \theta$;   (b) $r = 2 \sec \theta$.

(ii) Convert to polar form:
(a) $y = x^2$;   (b) $x^2 + y^2 = 7$.

16.14 Sketching polar graphs

Occasionally, the best method of sketching a polar graph is to convert the equation to Cartesian form first; but in most cases, the following list of 'aids' should help. The reader should verify them.

(i) *Line symmetry.* The reflection of the point (r_0, θ_0) in the line $\theta = \alpha$ is $(r_0, 2\alpha - \theta_0)$. So the curve will be symmetrical in $\theta = \alpha$ if replacing θ by $(2\alpha - \theta)$ leaves the equation unchanged. In particular, if θ can be replaced by $-\theta$, the curve is symmetrical in the initial line, and if θ can be replaced by $\pi - \theta$, then there is symmetry in the line $\theta = \dfrac{\pi}{2}$ (the y-axis).

(ii) *Point symmetry.* The curve has S-symmetry about the pole if the equation is unchanged by replacing r by $-r$, or by replacing θ by $\theta + \pi$. More generally, the curve has rotational symmetry of order n if θ may be replaced by $\theta + \dfrac{2\pi}{n}$.

(iii) *Range of values for r and θ.* It may be clear that r or θ can take only a limited range of values.

(iv) *Asymptotes.* It is useful to note what happens to θ as $r \to \infty$ and to r as $\theta \to \infty$. If $r \to a$ as $\theta \to \infty$, then $r = a$ is an asymptotic circle.

(v) *Plotting.* It nearly always helps to plot (roughly) a few points at significant values of θ.

(v) *Form at the origin.* If $r = 0$ when $\theta = \alpha$, then the line $\theta = \alpha$ is a tangent at the origin (provided that the curve is suitably 'well-behaved', and the origin is not, for example, an isolated point of the curve).

(vii) *Spotting some geometrical 'trick'.* This is done in the solution to Example 12. There are a few tricks which are standard, and which are not difficult once they are familiar.

(viii) *The gradient of the curve.* This is not discussed until Chapter 40.

Example 12

Sketch the graphs of

(i) $r \cos(\theta - \alpha) = 2$, where α is a constant between 0 and $\dfrac{\pi}{2}$;

(ii) $r \sec\left(\theta + \dfrac{\pi}{4}\right) = 3$.

Solution

(i) In Fig. 16.21, P is the point (r, θ) and Q is the foot of the perpendicular from P to the line $\theta = \alpha$. The equation $r \cos(\theta - \alpha) = 2$ can now be expressed as $OQ = 2$. Thus the locus of P is as shown in Fig. 16.22.

(ii) We mark the general point $P(r, \theta)$; and the angle $\theta + \dfrac{\pi}{4}$ in the question suggests that we should draw the line $\theta = -\dfrac{\pi}{4}$. Writing the

Figure 16.21

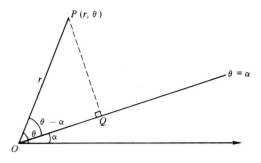

Figure 16.22

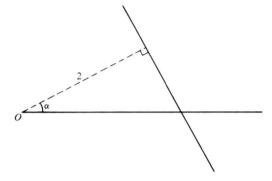

given equation as $r = 3\cos(\theta + \frac{\pi}{4})$ indicates that we need to make r the 'adjacent' of a triangle with hypotenuse 3. In Fig. 16.23, Q is the point $\left(3, -\frac{\pi}{4}\right)$. As P varies subject to the condition that $\angle OPQ = 90°$, it will trace out a circle with diameter OQ (by the angle in a semi-circle theorem).

Figure 16.23

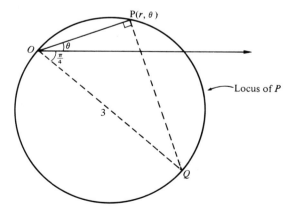

Locus of P

Qu.31 (1) Discuss the solution to Example 12(i) for values of θ such that

(a) $\theta < \alpha$; (b) $\theta > \frac{\pi}{2} + \alpha$.

(ii) If in the equation $r\cos(\theta - \alpha) = 2$ the value of θ is increased by π, what happens to r? Explain the geometrical significance of this.

Qu.32 Solve each part of Example 12 by converting the equations to Cartesian form. (Expand $\cos(\theta - \alpha)$ first in (i), and expand $\cos\left(\theta + \dfrac{\pi}{4}\right)$ in (ii).)

Example 13
Sketch $r = a\,(1 + \cos\theta)$.
Solution
(1) Replacing θ by $-\theta$ leaves r unchanged, so the curve is symmetrical in the initial line.
(2) For θ between 0 and π, r is decreasing. When $\theta = 0$, $r = 2a$; when $\theta = \dfrac{\pi}{2}$, $r = a$; when $\theta = \pi$, $r = 0$, and so the line $\theta = \pi$ is a tangent at the origin. This is sufficient to sketch the curve (Fig. 16.24). It is an example of a *cardioid* (meaning heart-shaped). The question of the gradient at $(2a, 0)$ is discussed in Qu.33.

Figure 16.24

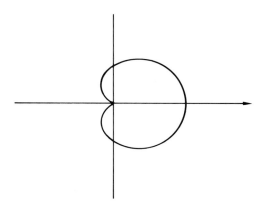

Qu.33 Find the value of $\dfrac{dr}{d\theta}$ at the point $(2a, 0)$. Explain how this shows that the curve is perpendicular to the radial line at this point.

Example 14
Sketch $(r - 1)\theta = 1$ for $\theta > 0$.
Solution
(1) As $\theta \to 0$, $r \to \infty$. So there is some sort of asymptote parallel to the initial line.
(2) As $\theta \to \infty$, $r \to 1$. So $r = 1$ is an asymptotic circle.
(3) $r = 1 + \dfrac{1}{\theta}$, so r decreases as θ increases, and $\forall\,\theta > 0$, $r > 1$, i.e. the whole curve lies outside the circle $r = 1$.
(4) To see exactly what happens as $\theta \to 0$ (from above), we have

$$r - 1 = \frac{1}{\theta} \quad \Rightarrow \quad r\sin\theta - \sin\theta = \frac{\sin\theta}{\theta}$$

$$\Rightarrow \quad y - \sin\theta = \frac{\sin\theta}{\theta}.$$

Now as $\theta \rightarrow 0$, $\dfrac{\sin \theta}{\theta} \rightarrow 1$ (see Section 13.1). So $y \rightarrow 1$, i.e. $y=1$ is an asymptote. More precisely,

$$y = \frac{\sin \theta}{\theta} + \sin \theta,$$

and this can be shown† to tend to 1 from above.

Fig. 16.25 gives the required curve. Note that the curve winds round infinitely many times as it approaches the circle.

Figure 16.25

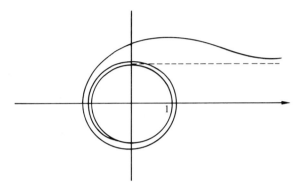

Qu.34 Find the three greatest values of y at which the curve in Fig. 16.25 cuts the y-axis.

Qu.35 (i) Sketch the spiral $r = 2\theta$ for
(a) $\theta \geqslant 0$; (b) $\theta \leqslant 0$.
(Remember that θ is in radians.)

(ii) Plot it accurately for $0 \leqslant \theta \leqslant \dfrac{\pi}{2}$.

Qu.36 (i) Sketch the petal curves $r = a \cos 2\theta$ and $r = a \cos 3\theta$ (where $a > 0$).
(ii) How many times is each curve traced out as θ moves from 0 to 2π?
(iii) Sketch $r = a \sin 2\theta$ and $r = a \sin 3\theta$.

Qu.37 Sketch $r = \sin \theta$. Check your answer by converting to Cartesian form.

16.15 The intersection of polar graphs

Finding the points of intersection of two polar graphs is essentially a matter of solving their equations simultaneously for r and θ. But a

---

† The best method uses the Maclaurin expansion of $\sin \theta$ (Chapter 28). Meanwhile, the following method is quite elegant.

From Section 13.1, $\dfrac{\sin \theta}{\theta} > \cos \theta$ (for small positive θ). So

$$\frac{\sin \theta}{\theta} + \sin \theta > \cos \theta + \sin \theta.$$

But from the definition of sine and cosine, $\cos \theta + \sin \theta > 1$ in the 1st quadrant, since two sides of a triangle are greater than the third side. So

$$\frac{\sin \theta}{\theta} + \sin \theta > 1.$$

CURVE SKETCHING AND INEQUALITIES

slight problem arises owing to the fact that each point in the plane has more than one 'name', so that, for example, $(2, 30°)$ and $(-2, 210°)$ are in fact the same point.

Consider the curves

$$r = 2 \cos 2\theta \tag{1}$$
$$r = 7 + 9 \cos \theta. \tag{2}$$

The coordinates $(2, 0)$ satisfy (1), and $(-2, 180°)$ satisfy (2); so the curves intersect at this point. But solving (1) and (2) simultaneously by writing

$$2 \cos 2\theta = 7 + 9 \cos \theta \tag{3}$$

does not give this point of intersection. Similarly, both curves pass through the pole, and this is not given by (3).

The safest rule, therefore, to overcome this difficulty is always to sketch the two curves concerned.

Qu.38 (i) Sketch the two curves (1) and (2) above.
(ii) Solve (3) above to find two of the points of intersection.
(iii) One method of searching for 'hidden' intersections is to rewrite one of the equations in an alternative form, replacing r by $-r$ and θ by $\theta + 180°$. Solve simultaneously

$$r = 2 \cos 2\theta$$
$$-r = 7 + 9 \cos(\theta + 180°).$$

The pole is particularly susceptible to being omitted by an algebraic solution, since it has so many 'names': it is the point $(0, \theta)$ for any value of θ.

Qu.39 Sketch $r = \sin \theta$ and $r = \cos \theta$ and find all points of intersection.

Exercise 16e

1. Convert to polar equations, simplifying where possible. Make r or r^2 the subject.

   (i) $y = 2x + 1$;   (ii) $y^3 = x^2$;
   (iii) $x^2 + y^2 = 4$;   (iv) $x^2 - y^2 = 4$;
   (v) $xy = 1$;   (vi) $(x^2 + y^2)^{3/2} = x - y$.

2. Convert to Cartesian equations:

   (i) $r = 3$;   (ii) $r = \sin \theta$;
   (iii) $r = 2 \sec \theta$;   (iv) $r \sec \theta = 2$;
   (v) $r^2 = \sin 2\theta$;   (vi) $\tan \theta = 1$;

   (vii) $r \sin \left(\theta + \dfrac{\pi}{3} \right) = 4$;   (viii) $r^3 = \sec 2\theta$.

   Sketch the following, numbers 3 to 18; a and α are constants.
   3. $r = 2 + \sin \theta$.   4. $r = 2 + \sin 2\theta$.
   5. $r = a(1 - \cos \theta)$.   6. $r = 1 + 2 \cos \theta$.
   7. $r = a \cos \theta$.   8. $r \sin(\theta - \alpha) = a$.
   9. $(r - 1) \sin(\theta - \alpha) = a$.

CURVE SKETCHING AND INEQUALITIES

10. $r \sin 2\theta = a$.

11. $r^2 = a^2 \sin \theta$.

12. $r^2 \sin 2\theta = 2a^2$.

13. $r^2 = a^2(1 + \cos \theta)$.

14. $r = a \cos 4\theta$.

15. $r = a \sin 5\theta$.

16. $(r - a) = a \cos 3\theta$.

17. $r \cos 3\theta = a$.

18. $r \sin 4\theta = a$.

19. (i) By writing the equation $r = \tan \theta$ in the form $r \cos \theta = \sin \theta$, find
   (a) limits for the range of values of x;
   (b) the Cartesian equations of the vertical asymptotes.
   (ii) Sketch $r = \tan \theta$.

20. (i) Use the method of question 19(i) to find the vertical asymptote(s) of $r^2 = \tan \theta$.
   (ii) In which quadrants does this curve lie?
   (iii) Sketch the curve.

21. (Hard.) Sketch $r = 1 + \tan \theta$.

Sketch the spiral curves, numbers 22 to 27.

22. $r = \theta$ (i) for $\theta \geqslant 0$;   (ii) for $\theta \leqslant 0$.

23. $r = e^\theta$.

24. $r = \dfrac{1}{\theta}$. Find what happens as θ tends to zero from above and from below.

25. $r(1 + \theta) = \theta$.

26. $r^2 = \theta$.

27. $r(1 + \theta^2) = a\theta^2$.

Sketch the following, numbers 28 to 31.

28. $r = 2a \tan \theta \sin \theta$ (Cissoid of Diocles).

29. $r = a(\sec \theta + \tan \theta)$ (Right strophoid).

30. $r^2 = a^2 \cos 2\theta$ (Lemniscate of Bernoulli).

31. $r = a + b \sec \theta$.   (Draw separately the cases $a > b$, $a = b$ and $a < b$. The Conchoid of Nicomedes.)

32. Sketch $r^2 = \cos \theta$, finding the greatest value of the 'y-coordinate'.

Find the points of intersection of the following pairs of curves, numbers 33 to 35.

33. $r = 1 + \cos \theta$ and $r = 1 - \cos \theta$.

34. $r = 3 + 2 \cos 2\theta$ and $r = 2 + 3 \cos 2\theta$.

35. $r = \theta$ (for $\theta \geqslant 0$) and $r = 2\theta$ (for $\theta \geqslant 0$).

16.16   The algebra of inequalities

The rules governing the manipulation of inequalities (i.e. statements involving the signs $>$, $<$, $\geqslant$, $\leqslant$) differ slightly from those for equations; for example,

$$a = b \quad \Rightarrow \quad a^2 = b^2$$

but

$$a > b \quad \not\Rightarrow \quad a^2 > b^2.$$

(Find a counter-example.)

Perhaps the most significant difference is that when multiplying or dividing each side of an inequality by a negative number, the sign must be reversed. So, for example,

$$a \geqslant b \quad \Rightarrow \quad -3a \leqslant -3b.$$

Qu.40 Which of the following are true for all real a, b, c, d? Give a counter-example for each false statement.

(i) $a > b \Rightarrow a + c > b + c$;

(ii) $a > b$ and $k \geqslant 0 \Rightarrow ak > bk$;

(iii) $a > b$ and $c > d \Rightarrow a + c > b + d$;

(iv) $a > b$ and $c > d \Rightarrow ac > bd$;

(v) $a > b \Leftrightarrow a^3 > b^3$;

(vi) $|a| > |b| \Leftrightarrow a^2 > b^2$;

(vii) $a^2 + b^2 + c^2 > 0$;

(viii) $a > b$ and $ab \neq 0 \Rightarrow \dfrac{1}{a} < \dfrac{1}{b}$;

(ix) $a > b$ and $ab > 0 \Rightarrow \dfrac{1}{a} < \dfrac{1}{b}$.

16.17 Solving inequalities
To solve an inequality in x means to find the set of values of x which satisfy the inequality. In this sense solving an inequality is similar to solving an equation. There are, however, two important differences:

(i) The different algebraic rules for inequalities means that the actual techniques are rather different from those for equations.

(ii) An equation usually has a finite number of 'point' solutions; an inequality is usually satisfied by whole regions of the real line.

16.18 Linear inequalities

Example 15
Solve $x + 1 \leqslant 2(3 - x) < 8$.
Solution
This is a pair of inequalities which must both be satisfied. We solve each separately, and then take the intersection of our answers.

$$x + 1 \leqslant 2(3 - x)$$
$$\Rightarrow x + 1 \leqslant 6 - 2x$$
$$\Rightarrow 3x \leqslant 5$$
$$\Rightarrow x \leqslant 5/3;$$

and
$$2(3 - x) < 8$$
$$\Rightarrow 3 - x < 4$$
$$\Rightarrow -x < 1$$
$$\Rightarrow x > -1.$$

Thus the required solution is $\{x \in R : -1 < x \leqslant 5/3\}$ or, more briefly, $-1 < x \leqslant 5/3$.

Qu.41 Solve
(i) $x + 6 < 3x - 4 < 2x - 1$;
(ii) $5x - 21 \leqslant 3x - 7 \leqslant 13 - x$.

In the next example, both methods are important.

Example 16
Solve $x^2 - 3x \geqslant 4$.
First solution

$$x^2 - 3x \geqslant 4 \implies x^2 - 3x - 4 \geqslant 0$$
$$\implies (x+1)(x-4) \geqslant 0.$$

We now sketch the graph of $y = (x-4)(x+1)$, Fig. 16.26, and see that the values of x satisfying $y \geqslant 0$ are

$$x \leqslant -1 \text{ or } x \geqslant 4.$$

(Note that this solution is the union of two disjoint subsets of R. There is no simpler way of writing the set. For example, $4 \leqslant x \leqslant -1$ is nonsense, since it would mean that x must satisfy *both* the conditions $x \geqslant 4$ and $x \leqslant -1$.)

Figure 16.26

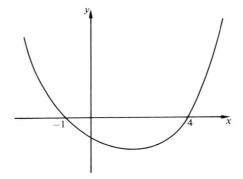

Second solution
As before, $(x+1)(x-4) \geqslant 0$. We consider the possible signs of the brackets if the inequality is to be satisfied. Either both brackets are positive, or both are negative (or one of them is zero). We can illustrate on a number line the region where each bracket is positive (Fig. 16.27).

Figure 16.27

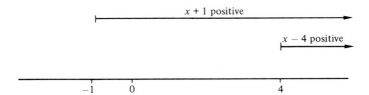

It is then clear that for $x < -1$, both brackets are negative, and for $x > 4$, both are positive. The two critical values (-1 and 4) can be considered separately. The full solution is therefore

$$x \leqslant -1 \text{ or } x \geqslant 4.$$

Qu.42 Solve by each of the methods of Example 16:
(i) $x^2 < 5x - 6$;   (ii) $3 + x - 2x^2 \leqslant 0$.

Qu.43 (i) What is meant by the terms
(a) discriminant, (b) positive definite,
as applied to the expression $ax^2 + bx + c$?
(ii) Solve $2x^2 - x + 1 \leqslant 0$.
(iii) What is the least value of p for the solution of $x^2 - 6x + p \geqslant 0$ to be the set R?

The second solution to Example 16 may be used for any polynomial inequality (as indeed may the first solution). A product of a number of brackets will be negative if an odd number of brackets are negative, and the product will be positive if an even number of brackets are negative. For example, the solution of $x(x+1)(x-1)^3(3-2x) < 0$ is illustrated by Fig. 16.28, the solution being $x < -1$ or $0 < x < 1$ or $x > 1\frac{1}{2}$.

Figure 16.28

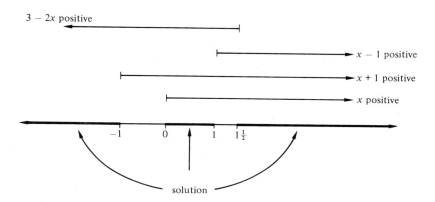

Qu.44 Draw a number line diagram to find the solution of $x(x+2)(x+1)^2(2-x)(2x-5) \leqslant 0$. (Note that the factor $(x+1)^2$ may be omitted from the diagram since it is non-negative everywhere. But it *does* contribute to the solution.)
Qu.45 (i) Show that the cubic $(x^2 - x + 5)(3x - 1)$ does not factorize further.
(ii) By considering the sign of each bracket, solve $(x^2 - x + 5)(3x - 1) \geqslant 0$.
Qu.46 Solve $x^2(x-1)^3(x+1) < 0$
(i) by sketching a graph;
(ii) by drawing a number line diagram.

Example 17
Find the range of values of k such that

$$kx^2 + (k+1)x + (2k-1) > 0 \quad \text{for all } x \in R.$$

Solution
There are two conditions (see Section 2.12) for this quadratic to be positive definite:

$$k > 0; \tag{1}$$

$$(k+1)^2 - 4k(2k-1) < 0. \tag{2}$$

Solving (2), we have

$$-7k^2 + 6k + 1 < 0$$
$$\Rightarrow -(7k+1)(k-1) < 0$$
$$\Rightarrow k > 1 \text{ or } k < -\tfrac{1}{7}.$$

Now *both* conditions (1) and (2) must be satisfied, i.e. we seek the intersection of the sets $\{k \in R : k > 0\}$ and $\{k \in R : k > 1 \text{ or } k < -\tfrac{1}{7}\}$. The required range of values is therefore $k > 1$.

Qu.47 (i) Show that for any value of k, $(k^2+1)x^2 - kx + 1$ is positive for all x.
(ii) What can you deduce from (i) about the graph of
$(y^2+1)x^2 - xy + 1 = 0$.
(iii) Show that $k^2 x^2 - kx + 1$ is positive for all choices of x and k. (Try to find a quick method.)

16.20 Inequalities involving rational functions

Example 18

Solve $\dfrac{(1-x)(x+3)}{(2x+1)(x+4)} \geqslant 0.$

Solution

The sign of the L.H.S. depends only on the total number of brackets (in the numerator and denominator) that are negative. Thus a number line diagram may be used as in Fig. 16.28. This is left to the reader. Care must be taken with the four 'critical' values: $x = 1$ and $x = -3$ are solutions, but $x = -\tfrac{1}{2}$ and $x = -4$ are not, since the L.H.S. is not defined for these values. The complete solution is

$$-4 < x \leqslant -3 \text{ or } -\tfrac{1}{2} < x \leqslant 1.$$

Qu.48 Solve $\dfrac{2x^2 + x + 2}{x^2 - 2x - 3} < 0.$

Example 19

Solve $\dfrac{x+1}{3x-5} \geqslant 3.$

Solution

If we multiply through by $3x - 5$, we must immediately split the solution into two cases: $3x - 5$ positive and $3x - 5$ negative. (In the latter case, the inequality must be reversed.) The following method is therefore preferable:

$$\frac{x+1}{3x-5} - 3 \geqslant 0$$
$$\Rightarrow \frac{x+1-3(3x-5)}{3x-5} \geqslant 0$$
$$\Rightarrow \frac{16-8x}{3x-5} \geqslant 0.$$

The method of Example 18 now gives $\tfrac{5}{3} < x \leqslant 2$.

Qu.49 Solve $\dfrac{5x+3}{x-x^2}<1$ by the method of Example 19.

Qu.50 Solve $\dfrac{1}{1-x}\geqslant\dfrac{x-2}{3x-4}$, by first taking both fractions to the L.H.S. and then writing them as a single fraction (i.e. with a common denominator).

16.21 Finding the range of values of a rational function

Example 20

Find the range of possible values of $\dfrac{x}{x^2+x+1}$ (as x takes all real values).

Solution

Suppose that we choose some number, say 4, and ask whether $\dfrac{x}{x^2+x+1}$ can take the value 4. This is the same as asking whether $\dfrac{x}{x^2+x-1}=4$ can be solved for x. In general, therefore, we wish to find the range of values of y for which the equation

$$y=\frac{x}{x^2+x+1} \tag{1}$$

(regarded as an equation in x) has a solution for x.
Now (1) may be written as

$$y(x^2+x+1)=x$$
$$\Rightarrow yx^2+(y-1)x+y=0,$$

and this quadratic in x has a root if and only if

$$(y-1)^2-4y^2\geqslant 0$$
$$\Leftrightarrow 3y^2+2y-1\leqslant 0$$
$$\Leftrightarrow (3y-1)(y+1)\leqslant 0$$
$$\Leftrightarrow -1\leqslant y\leqslant\frac{1}{3}.$$

So for all $x\in R$,

$$-1\leqslant\frac{x}{x^2+x+1}\leqslant\frac{1}{3}.$$

Finding the range of values of a rational function can be a valuable aid in sketching its graph (see Qu.51).

Qu.51 (See Example 20.) Without using calculus, find the values of x that give the maximum and minimum values of the function. Hence sketch $y=\dfrac{x}{x^2+x+1}$ indicating the coordinates of its turning points.

Qu.52 (i) Find the range of possible values of $\dfrac{x^3-x^2-1}{x^3+1}$ using the method of Example 20. Take care over each step of your argument.

(ii) What is the graphical significance of the 'missing' value?

16.22 Inequalities involving the modulus function

There is not always an easy method for dealing with inequalities involving moduli, although the fact that $|a|>|b| \Leftrightarrow a^2 > b^2$ means that squaring each side can sometimes be useful.

Example 21
Solve $|2x-3|<6$.
Solution
The given inequality may be written $-6<2x-3<6$, and this may be solved by the method of Example 15 to give $-\frac{3}{2}<x<\frac{9}{2}$.

Qu.53 Solve
(i) $|3x+1|>4$;   (ii) $|3x+1|<-1$.

Example 22
Solve $3|x-1| \geqslant |2x+3|$.
Solution
Each side of the inequality is certainly non-negative and so we may square each side:

$$9(x^2-2x+1) \geqslant 4x^2+12x+9$$
$$\Leftrightarrow 5x^2-30x \geqslant 0$$
$$\Leftrightarrow x(x-6) \geqslant 0$$
$$\Leftrightarrow x \geqslant 6 \text{ or } x \leqslant 0.$$

Example 23
Solve $|2x-1| \geqslant x+4$.
Solution
(The method of Example 24 may also be used.) Sketching the graphs of $y=|2x-1|$ and $y=x+4$ (Fig. 16.29), we see that the required region is $x \leqslant a$ or $x \geqslant b$ where a and b are as shown. Now b is the value of x

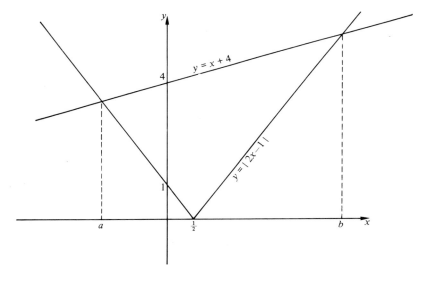

Figure 16.29

CURVE SKETCHING AND INEQUALITIES

where $2x-1=x+4$, so $b=5$; and a is the solution of $-(2x-1)=x+4$, so $a=-1$. So the solution is

$$x \leqslant -1 \text{ or } x \geqslant 5.$$

Example 24
Solve $|x-3|+2<|2x-1|$.
Solution
We examine separately the three regions $x \geqslant 3$, $\frac{1}{2} \leqslant x < 3$, $x < \frac{1}{2}$.
(1) In the region $x \geqslant 3$, the inequality is

$$(x-3)+2<(2x-1) \Rightarrow x>0.$$

So the whole region $x \geqslant 3$ is part of the solution. (We cannot say that $x>0$ is all part of the solution, since we are examining specifically the region $x \geqslant 3$.)
(2) In the region $\frac{1}{2} \leqslant x < 3$, the inequality is

$$(3-x)+2<(2x-1) \Rightarrow x>2.$$

So $2<x<3$ is part of the solution.
(3) In the region $x<\frac{1}{2}$, the inequality is

$$(3-x)+2<(1-2x) \Rightarrow x<-4.$$

So $x<-4$ is part of the solution.
   The whole solution is thus

$$x<-4 \text{ or } x>2.$$

Qu.54   Illustrate the solution to Example 24 by sketching on the same axes the graphs of $y=|x-3|+2$ and $y=|2x-1|$.
Qu.55   Solve Example 23 by the method of Example 24.

16.23   Some standard inequalities

So far, we have considered inequalities that, like equations, need to be solved. We now consider some inequalities that are always true; they correspond to algebraic identities. An important inequality forms the basis of all these standard inequalities; it expresses the fact that the square of a real number cannot be negative.
   Let $a_1, a_2, \ldots, a_n$ be real numbers, then $\sum a_i^2 \geqslant 0$. Moreover, equality (with zero) will occur only in the case $a_1=a_2=\ldots=a_n=0$.
   Proving an inequality is often a case of spotting what to 'consider', as we see in the next example.

Example 25
Prove that, for any real numbers a, b, c.

$$a^2+b^2+c^2 \geqslant ab+bc+ca.$$

When will equality occur?
Solution

$$(a-b)^2+(b-c)^2+(c-a)^2 \geqslant 0$$
$$\Rightarrow 2a^2+2b^2+2c^2-2ab-2bc-2ca \geqslant 0$$
$$\Rightarrow a^2+b^2+c^2 \geqslant ab+bc+ca.$$

From the first line, we see that equality occurs only when each bracket is zero, i.e. when $a=b=c$.

Definitions

(1) The *arithmetic mean* of the n numbers $a_1, a_2, \ldots, a_n$ is $\dfrac{1}{n}\sum a_i$.

(2) If $a_1, a_2, \ldots, a_n$ are positive numbers, then their *geometric mean* is $(a_1 a_2 a_3 \ldots a_n)^{1/n}$.

Qu.56 Find
(i) the arithmetic mean of 6 and 54;
(ii) their geometric mean.

Theorem 16.1
(Inequality of the means. See also Theorem 16.2.) The arithmetic mean (A.M.) of two positive numbers is not less than their geometric mean (G.M.).

Proof
Let the two positive numbers be a and b. Then

$$(\sqrt{a}-\sqrt{b})^2 \geqslant 0$$
$$\Rightarrow a+b-2\sqrt{(ab)} \geqslant 0$$
$$\Rightarrow \frac{a+b}{2} \geqslant \sqrt{(ab)} \quad \text{as required.}$$

We can see also that equality occurs only when $a=b$.

Theorem 16.2
(The general inequality of the means.) Let S be a set of n positive real numbers, and let A and G be the arithmetic and geometric means of S. Then $A \geqslant G$.

Proof
If all the members of S are equal, then $A=G=$ their common value. If not, then S certainly contains an $a_i > G$ and an $a_j < G$. Then

$$(a_i - G)(a_j - G) < 0$$
$$\Rightarrow a_i a_j + G^2 < (a_i + a_j)G$$
$$\Rightarrow \frac{a_i a_j}{G} + G < (a_i + a_j).$$

Thus if these two members of S are replaced by $\dfrac{a_i a_j}{G}$ and G, to give a new set S_1, the product of the members is unchanged, and the sum of the members is reduced; i.e.

arithmetic mean (A_1) of $S_1 < A$
geometric mean of $S_1 = G$.

Repeating this argument until all the members are equal to G gives

$$G = A_k < A_{k-1} < \ldots < A_2 < A_1 < A$$

where k will not be greater than $n-1$. So $A \geqslant G$ as required. (We have also shown that equality occurs only when the n numbers are all equal.)

Qu.57 (Theorem 16.2 is not to be used in this question.) Let $w, x, y, z \in R$.

(i) Use Theorem 16.1 to show that $\dfrac{w^4 + x^4}{2} \geqslant w^2 x^2$.

(ii) Write down the corresponding statement using y and z, and deduce that

$$\frac{w^4 + x^4}{4} + \frac{y^4 + z^4}{4} \geqslant \frac{w^2 x^2 + y^2 z^2}{2}.$$

(iii) Hence show (using Theorem 16.1 again) that

$$\frac{w^4 + x^4 + y^4 + z^4}{4} \geqslant wxyz.$$

The following theorem (Cauchy's inequality) occurs in various forms; in some of them it is also associated with the name of Schwarz. (Perhaps the best way of remembering the inequality is in terms of the scalar product of two vectors; see Qu.23 of Chapter 20.)

Theorem 16.3 (Cauchy's inequality.)
Given $2n$ real numbers $a_1, a_2, \ldots, a_n$ and $b_1, b_2, \ldots, b_n$, $b_2, \ldots, b_n$,

$$\left(\sum_{i=1}^{n} a_i^2 \right) \left(\sum_{i=1}^{n} b_i^2 \right) \geqslant \left(\sum_{i=1}^{n} a_i b_i \right)^2.$$

Proof
(An alternative proof is given in Exercise 16f, question 19.)
For any such set of $2n$ numbers,

$$\sum_{i=1, \; j=1}^{i=n, \; j=n} (a_i b_j - a_j b_i)^2 \geqslant 0,$$

where the summations over i and j are independent (i.e. the summation is over n^2 terms, in n of which i and j will take the same value). So

$$\Sigma(a_i^2 b_j^2 + a_j^2 b_i^2) \geqslant \Sigma 2 a_i b_j a_j b_i$$
$$\Rightarrow \; 2\Sigma a_i^2 b_j^2 \geqslant 2\Sigma(a_i b_i)(a_j b_j)$$
$$\Rightarrow \; \Sigma a_i^2 b_j^2 \geqslant \Sigma(a_i b_i)(a_j b_j)$$
$$\Rightarrow \; \Sigma a_i^2 \, \Sigma b_i^2 \geqslant (\Sigma a_i b_i)^2.$$

(This last step is clearer if the expressions are written out in full, i.e. without the Σ notation.)

Qu.58 Write out the statement of Cauchy's inequality in full in the cases $n = 2, 3, 4$. Write out the proof in full in the cases $n = 2$ and 3.

Exercise 16f   **1.** Solve

(i) $x - 4 < 3$;   (ii) $2x + 5 \geqslant 9$;

(iii) $5x - 2 > 2x - 5$;

(iv) $4(x - 1) \leqslant 2(3x + 2) - (x + 1)$;

(v) $x(x + 1) \geqslant (x - 2)(x + 3)$;

(vi) $(x + 2)(x - 6) < (x - 3)(x - 5)$.

2. For each part of question 1, test, by direct substitution, whether $x = 1$ is part of the solution; then check whether you have included it in your answer.

3. Solve the following. (Practise each of the methods of Example 16.)

(i) $x^2 - 5x - 6 \leqslant 0$; (ii) $2x^2 + 3 > 7x$;

(iii) $(x + 3)(6 - x) < 8$;

(iv) $x^2 > 4$;   (v) $x(x - 1)(x - 2) \leqslant 0$;

(vi) $(x - 1)^2 (x - 2) \geqslant 0$;

(vii) $(2x + 1)(x - 1)(3 - x) < 0$;

(viii) $x^3 \geqslant 4x$;

(ix) $(x - 1)(x - 2)(x - 5)(2x + 3) \geqslant 0$;

(x) $(x^2 + 1)(x^2 + 2) > 0$;

(xi) $(2x + 1)^3 (x + 4) \geqslant 0$;

(xii) $x^4 > 16$;

(xiii) $x^4 > 8x$;

(xiv) $(x^2 + 2x + 3)(x^2 + 2x - 3) < 0$;

(xv) $(2x - 5)^3 (x + 1)(x - 7)^2 \geqslant 0$.

4. For what values of p are the following positive definite (i.e. positive for all x)?

(i) $x^2 - 2x + p$; (ii) $2x^2 + px + 18$;

(iii) $x^2 + (p + 3)x + 3p$;

(iv) $(p + 2)x^2 + 2px - (p - 3)$.

5. For what values of p is question 4(iv) negative definite?

6. Solve

(i) $\dfrac{x - 1}{x^2 - 3x - 10} \leqslant 0$;

(ii) $\dfrac{15 - x - 2x^2}{x^2 + 2x - 8} \geqslant 0$;

(iii) $\dfrac{3 - x}{2x^2 + 2x + 3} < 0$;

(iv) $\dfrac{x + 1}{2x - 1} < 2$;

(v) $\dfrac{4x + 14}{x - 4} \geqslant x + 1$;

(vi) $\dfrac{x^2 + 4x - 6}{(x - 5)(1 - x)} > 2$;

(vii) $\dfrac{3x^2 + x + 1}{(x - 2)(x - 3)} < 3$;

(viii) $\dfrac{1}{x} \geqslant \dfrac{x}{x + 1}$;

(ix) $\dfrac{2}{(x - 1)(x - 2)} \leqslant \dfrac{1}{(x - 2)(x - 3)}$; (x) $\dfrac{x + 1}{x - 1} < \dfrac{x - 1}{x + 1}$.

7. (i) Illustrate the solution to question 6(v) by sketching the graphs of $y = \dfrac{4x + 14}{x - 4}$ and $y = x + 1$ on the same axes.

(ii) Illustrate similarly the solutions to 6(viii), (ix) and (x).

8. Find the range of values of $f(x)$ and hence sketch $y = f(x)$ when $f(x)$ is

(i) $\dfrac{x^2 + 4x + 1}{x + 4}$; (ii) $\dfrac{x^2 - 17}{x^2 + 3x - 4}$; (iii) $\dfrac{x^2 + 2x}{4x^2 + 2}$.

9. Solve the following:

(i) $|2x+3|<5$;

(ii) $|x+1|\geqslant|2x-1|$;

(iii) $|3x+4|>2|5-x|$;

(iv) $|x+1|<|x+2|$;

(v) $|x-3|\geqslant 2x+5$;

(vi) $|3x+1|\geqslant x$;

(vii) $|2x-4|<x+3$;

(viii) $2|x-4|>x-1$;

(ix) $|x-1|+|x-2|<5$;

(x) $|2x+3|<|x-1|+4$;

(xi) $3|x+1|\geqslant|2x-1|+10$.

10. (i) Solve $1<|x-3|<2$.

(ii) Solve in terms of a, b and c (where $0\leqslant a<b$)

$$a<|x-c|<b.$$

11. Solve the following, illustrating the solutions with suitable graphs:

(i) $0\leqslant\dfrac{2x-1}{x-2}\leqslant 1$;   (ii) $0<\dfrac{x(x-3)}{2x-4}<1$;   (iii) $\dfrac{1}{5}\leqslant\dfrac{x}{x^2-6}\leqslant 1$.

12. Solve

(i) $|x^2-9|<|x^2-4|$;

(ii) $|x^2-16|-|x^2-9|<1$.

13. (i) Let x, y and z be three positive numbers. Show that, using the notation introduced in Section 10.11,

$$2\sum x^3-6xyz=\left(\sum x\right)\left(\sum(x-y)^2\right).$$

(ii) Deduce that $x^3+y^3+z^3\geqslant 3xyz$, and state the condition for equality.

(iii) Obtain the same result directly from the (general) inequality of the means.

14. (i) Show that, for $a>0$, $a+\dfrac{1}{a}\geqslant 2$.

(ii) Show that, for a, $b>0$,

$$\left(a+\frac{1}{b}\right)\left(b+\frac{1}{a}\right)\leqslant\left(a+\frac{1}{a}\right)\left(b+\frac{1}{b}\right).$$

15. Using the general inequality of the means or otherwise, prove that, for a, $b>0$,

$$ab^2+ba^2+1\geqslant 3ab.$$

When will equality occur?

16. Use Cauchy's inequality to show that

$$1.2+2.3+3.4+\ldots+(n-1).n+n.1\leqslant\sum_1^n r^2.$$

17. The *Harmonic Mean* of a set of positive numbers $a_1, a_2, \ldots, a_n$ is defined to be the reciprocal of the average of the reciprocals, i.e.

$$H^{-1}=\frac{1}{n}\left(\frac{1}{a_1}+\frac{1}{a_2}+\ldots+\frac{1}{a_n}\right).$$

(i) Show that, for two positive numbers, $G^2 = AH$, where G and A are the G.M. and the A.M.

(ii) Use Cauchy's inequality to show that, for n positive numbers $a_1, a_2, \ldots, a_n$,

$$\sum a_r \sum \frac{1}{a_r} \geqslant n^2$$

and deduce that the harmonic mean of n positive numbers cannot exceed their arithmetic mean.

(iii) A sequence $\{u_r\}$ begins $2, 1, \ldots$ and satisfies

u_i is the harmonic mean of u_{i-1} and u_{i+1}.

Find the next four terms in the sequence. What do you notice about the sequence $\left\{\dfrac{1}{u_r}\right\}$?

18. The *Root Mean Square* of a set of n numbers is the positive square root of the average of the squares, i.e.

$$R^2 = \frac{1}{n}(a_1^2 + a_2^2 + \ldots + a_n^2).$$

Prove that $R \geqslant A$ (where A is the A.M.) for any set of numbers. (It may help to use Cauchy.)

19. (Alternative proof of Cauchy's inequality.)

The quadratic in x, $\sum\limits_{1}^{n} (a_i x - b_i)^2$ cannot be negative. Prove

Cauchy's inequality by expressing the fact that its discriminant ('$b^2 - 4ac$') cannot be positive.

20. Obtain the condition for equality in Cauchy's inequality from
(i) the text proof;
(ii) the proof in question 19.

16.24 Inequalities in the plane

Example 26

Shade the region of the plane satisfying

$$y^2 > x^2 + 1.$$

Solution

Imagine that y^2 and $x^2 + 1$ are evaluated at some point $P(x, y)$ which moves around in the plane. Both functions are 'well-behaved' (they do not have any sudden 'jumps' in value) and so it is reasonable to assume that $y^2 - (x^2 + 1)$ will only change sign where $y^2 = x^2 + 1$. We therefore sketch the curve $y^2 = x^2 + 1$, and find that it divides the plane into three regions. Trying three 'test points', one in each region:

$$
\begin{array}{lll}
\text{at } (0, 0), & y^2 < x^2 + 1; \\
\text{at } (0, 10), & y^2 > x^2 + 1; \\
\text{at } (0, -10), & y^2 > x^2 + 1.
\end{array}
$$

The solution is therefore the two regions shown (Fig. 16.30). (Boundary lines which are not part of the solution are usually dotted.)

Figure 16.30

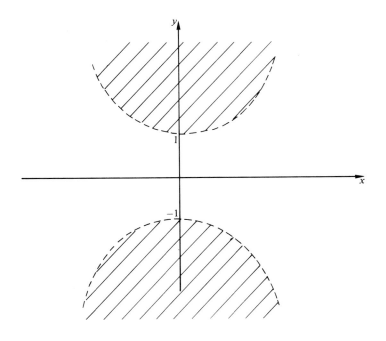

Qu.59   Shade the regions

(i) $y^2 < 49$;   (ii) $(y - x - 1)^2 > 0$.

Example 27

Shade the region

$$(x^2 + y^2 - 4)(y - x^2 - 1) \geqslant 0.$$

Solution

We seek the region where either both brackets are positive, or both are negative (or one of them is zero). The regions $x^2 + y^2 - 4 \geqslant 0$ and $y - x^2 - 1 \geqslant 0$ are shown in Figs. 16.31 and 16.32. The solution is shaded in Fig. 16.33.

Figure 16.31

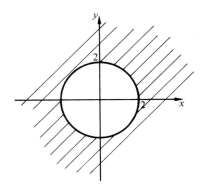

Figure 16.32

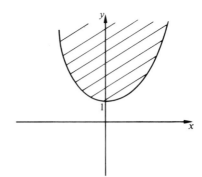

Figure 16.33

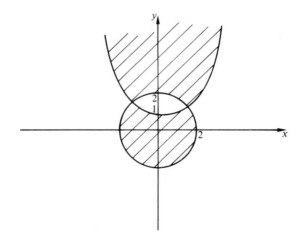

Exercise 16g   Indicate the following subsets of the plane either by shading them, or by shading their complement (i.e. leaving the required region unshaded). State which convention you are using.

1. $y < x^2$.

2. $\dfrac{y}{x} < x$. (Contrast with questions 1 and 3.)

3. $\dfrac{x^2}{y} > 1$. (Contrast with questions 1 and 2.)

4. $y^2 \leqslant x^2$.   5. $y^2 < x^2 - 1$.

6. $(y - x + 1)(y + x - 1) \leqslant 0$.   7. $(x^2 + y^2 - 9)(x - 2) \geqslant 0$.

8. $(y - x^2)(y - x - 1) \geqslant 0$.

9. $x\sqrt{y} \geqslant \sqrt{x}$. (Note that the positive y-axis is part of the solution.)

10. $y \leqslant \sin x$.   11. $|y| \leqslant \sin x$.

12. $\log y + \log x \geqslant 0$.

13. $\log xy \geqslant 0$. (Compare with question 12.)

14. $e^x e^{y^2} \leqslant e^y e^{x^2}$.   15. $x(y - e^x) > 0$.

16. $\sin x \leqslant \sin y$.

CURVE SKETCHING AND INEQUALITIES

1. The points of a curve are given parametrically in the form $x = a(t^2 + 1)$, $y = a(t + 1)$. Eliminate t to find the equation connecting x and y. Explain why the curve is symmetrical about $y = a$ and has no points for which $x < a$.

   Find $\dfrac{dx}{dt}$, $\dfrac{dy}{dt}$ and hence $\dfrac{dy}{dx}$.

   State which values of t make $\dfrac{dy}{dx}$ positive and which values make $\dfrac{dy}{dx}$ negative. Hence sketch the curve, indicating which values of t correspond to which parts of the curve.

   [SUJB (O)]

2. By considering the equation $yx^2 + (y-1)x + (y-1) = 0$, where $y \neq 0$, as a quadratic equation in x, find the range of values of y for which this equation has real roots. State the value of x which satisfies this equation when $y = 0$.

   Using these results, or otherwise, determine the range of the function $f : x \to \dfrac{x+1}{x^2 + x + 1}$, where x is real.

   Sketch the graph of $f(x)$ giving the coordinates of any turning points.

   Explain why f has no inverse function.

   [JMB]

3. Find the coordinates of the stationary point and of the points of inflexion of the curve

   $$y = \frac{x^2}{1 + x^2}.$$

   Show that the curve lies entirely in the region $0 \leqslant y < 1$, and that $\dfrac{dy}{dx} > 0$ for $x > 0$.

   Sketch the curve.

   Find the ranges of values of k for which the equation

   $$\frac{x^2}{1 + x^2} = kx$$

   has three real and distinct roots.

   [JMB]

4. If $y = \dfrac{3x - 5}{x^2 - 1}$, prove that for any real value of x, the value of y cannot lie between $\frac{1}{2}$ and $\frac{9}{2}$. Find the coordinates of the points on the curve $y = \dfrac{3x - 5}{x^2 - 1}$ at which $\dfrac{dy}{dx} = 0$, and the equations of the three asymptotes of the curve. Sketch the curve.

   [C]

5. Find the equations of the asymptotes of the curve

   $$y = (2x^2 - 2x + 3)/(x^2 - 4x).$$

Using the fact that x is real, show that y cannot take any value between -1 and $\frac{5}{4}$.

Sketch the curve, showing its stationary points, its asymptotes and the way in which the curve approaches its asymptotes.

[L]

6. Sketch the curve $r = a(1 + \cos\theta)$ for $0 \leqslant \theta \leqslant \pi$ where $a > 0$. Sketch also the line $r = 2a\sec\theta$ for $-\pi/2 < \theta < \pi/2$ on the same diagram. The half-line $\theta = \alpha$, $0 < \alpha < \pi/2$, meets the curve at A and the line $r = 2a\sec\theta$ at B. If O is the pole, find the value of $\cos\alpha$ for which $OB = 2OA$.

[L]

7. For the curve with equation

$$y = \frac{x}{(x-2)},$$

find
(i) the equation of each of the asymptotes,
(ii) the equation of the tangent at the origin.
Sketch the curve, paying particular attention to its behaviour at the origin and as it approaches its asymptotes.
On a separate diagram, sketch the curve with equation

$$y = \left| \frac{x}{x-2} \right|.$$

[L]

8. The curve $y = \dfrac{ax^2 + bx + c}{x^2 + px + q}$ cuts the y-axis at the point $(0,1)$ and has asymptotes $x = 2$, $x = 4$, $y = 3$. The line $y = 1$ is a tangent to the curve. Find the values of a, b, c, p and q and sketch the curve.

[AEB '77]

9. (a) Solve the inequality $\dfrac{3}{3x-2} > \dfrac{4}{4x-3}$.

(b) If $y = \dfrac{1}{(x-1)(x-2)}$ and x is real, show that y cannot lie between 0 and -4.
Sketch the graph of

$$y = \frac{1}{(x-1)(x-2)},$$

showing the main features clearly.

[SUJB]

10. Find the stationary points of the function f given by
$f(x) = \dfrac{3x+2}{x(x-2)}$ and sketch its graph.
From consideration of your graph, or otherwise, find
(i) the domain and range of f;
(ii) the set $f^{-1}(B)$ where B is the closed interval $[-5, -4]$.
Mark on your graph the approximate position of the point of inflexion.

[W]

11. (i) Show that, for all real values of λ, the roots of the equation

$$x^2 + (3\lambda - 7)x + \lambda^2 - 3\lambda = 0$$

are real and cannot differ by less than 2.

(ii) Find the set of values of k for which the expression

$$kx^2 - 2x + 3k - 2$$

is always negative.

[L (S)]

12. Given that $y = (2x+1)/(x^2+2)$, where x is real, show algebraically that y can only take values in the range $-\frac{1}{2} \leqslant y \leqslant 1$. Hence find the turning points on the curve with equation $y = (2x+1)/(x^2+2)$ and sketch the curve.

Explain how the intersection of this curve with a suitable straight line can show that the cubic equation $2x^3 + 2x - 1 = 0$ has only one real root. Using any method, find this root correct to two decimal places.

[O & C]

17 The theory of integration

17.1 An intuitive approach to integration

For a suitably 'smooth' curve, $y = f(x)$, we have an intuitive idea of what is meant by the area under the curve between two values of x. We therefore define *the definite integral of* f *from* a *to* b (where $a \leqslant b$) to be the area shown in Fig. 17.1. It is written $\int_a^b f(x)\,dx$, the reason for this notation appearing later. The numbers a and b are called the *lower* and *upper limits* respectively.

Figure 17.1

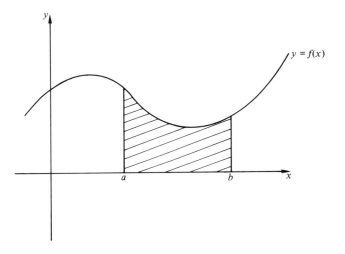

From this intuitive definition, it seems reasonable to deduce the following:

(1) If the curve $y = f(x)$ lies below the x-axis, it contributes a negative quantity of area. Thus in Fig. 17.2,

Figure 17.2

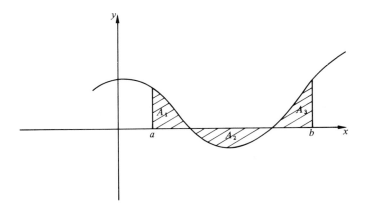

THE THEORY OF INTEGRATION

$$\int_a^b f(x)\,dx = A_1 - A_2 + A_3.$$

(2) $\displaystyle\int_a^b f(x)\,dx + \int_b^c f(x)\,dx = \int_a^c f(x)\,dx.$

(3) Putting $c=a$ in (2) suggests that

$$\int_a^b f(x)\,dx + \int_b^a f(x)\,dx = \int_a^a f(x)\,dx.$$

But clearly $\displaystyle\int_a^a f(x)\,dx = 0.$ So

$$\int_a^b f(x)\,dx = -\int_b^a f(x)\,dx.$$

We can thus give a meaning to a definite integral in which the upper limit is smaller than the lower limit.

For example, $\displaystyle\int_3^1 f(x)\,dx = -\int_1^3 f(x)\,dx.$

Qu.1 Use the standard formulae for areas of triangles and trapezia to find $\displaystyle\int_1^4 f(x)\,dx$ in the following cases. Draw a diagram for each.

(i) $f(x)=5$;
(iii) $f(x)=2x+1$;
(v) $f(x)=3x-4$;
(ii) $f(x)=x-1$;
(iv) $f(x)=-2$;
(vi) $f(x)=1-2x$.

Qu.2 Find

(i) $\displaystyle\int_0^5 (2x+7)\,dx$;

(ii) $\displaystyle\int_{-1}^1 (3x+2)\,dx$;

(iii) $\displaystyle\int_{-4}^{-3} (6-x)\,dx$;

(iv) $\displaystyle\int_5^2 (x-1)\,dx$;

(v) $\displaystyle\int_4^5 (2x-15)\,dx$;

(vi) $\displaystyle\int_1^{-2} (2x-3)\,dx$.

The calculations in Qu.1 and Qu.2 were simple enough because each of the 'curves' was a straight line. Now consider, as an example, the problem of finding the area under the curve $y=\sqrt{x}$ between $x=1$ and $x=2$. Figure 17.3 shows the required area divided into five strips of width $1/5$. If the top of each strip is replaced by a horizontal line lying below the curve (as shown), then the total area of the five rectangles provides an approximation to the area under the curve. We have

total area of 'lower' rectangles

$$= \tfrac{1}{5}\sqrt{1} + \tfrac{1}{5}\sqrt{1\cdot2} + \tfrac{1}{5}\sqrt{1\cdot4} + \tfrac{1}{5}\sqrt{1\cdot6} + \tfrac{1}{5}\sqrt{1\cdot8}$$

$$\approx 1\cdot177.$$

Figure 17.3

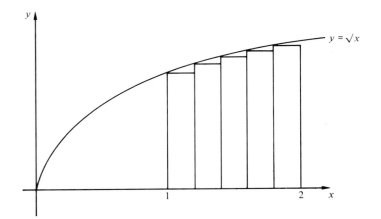

Since the rectangles lie entirely inside the shaded area

$$\int_1^2 \sqrt{x}\, dt > 1{\cdot}177.$$

Similarly, if each strip is replaced by a rectangle whose top lies just above the curve (Fig. 17.4), then

total area of 'upper' rectangles

$$= \tfrac{1}{5}\sqrt{1{\cdot}2} + \tfrac{1}{5}\sqrt{1{\cdot}4} + \tfrac{1}{5}\sqrt{1{\cdot}6} + \tfrac{1}{5}\sqrt{1{\cdot}8} + \tfrac{1}{5}\sqrt{2}$$
$$\approx 1{\cdot}260,$$

Figure 17.4

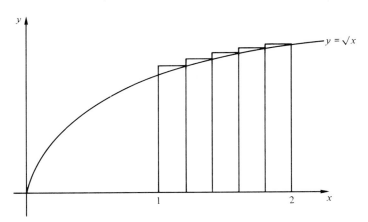

so that

$$\int_1^2 \sqrt{x}\, dx < 1{\cdot}260.$$

Therefore the required area lies between 1·177 and 1·260.

We have only succeeded in *estimating* the value of $\int_1^2 \sqrt{x}\, dx$. Better estimates can be found by dividing the area into thinner strips (see Qu.3), but a method of calculating the exact value does not appear until later in this chapter. (Numerical methods for estimating integrals are discussed more fully in Chapter 29.)

Qu.3 Draw a diagram showing this area divided into ten strips of width 0·1. Verify from your diagram that the total area of the ten 'lower' rectangles will give a better estimate of the area under the curve than that obtained with five 'lower' rectangles. Show that

$$1\cdot1981 < \int_1^2 \sqrt{x}\, dx < 1\cdot2396.$$

17.2 A formal approach to integration

There are two basic questions raised by the previous section:
(1) What precisely do we mean by the area under a curve?
(2) How can we calculate such areas?
In this section, we shall answer both questions, although we shall find that they are really the same question; thus we shall define 'area' by showing how to calculate it. (It is interesting to compare with the corresponding questions in differentiation, namely (1) what do we mean by the gradient at a point? and (2) how can we calculate it? Both questions are answered by saying that 'it is the limit of the gradient of a chord . . .'.)

Consider the curve $y = f(x)$ in the interval $a \leqslant x \leqslant b$. For simplicity we shall assume that $f(x) \geqslant 0$ throughout the interval; only minor modifications to the argument are needed if $f(x) < 0$. We split the interval up into n strips of width h, so that $nh = b - a$ (Fig. 17.5). Then, for each strip, we calculate the area of the largest rectangle that will fit

Figure 17.5

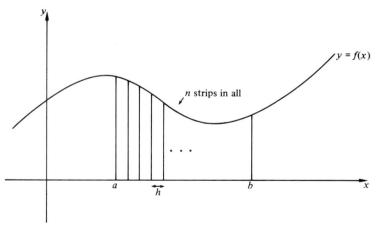

'inside' the strip. The width of this rectangle is, of course, h, and its height is equal to the least value of $f(x)$ in the strip. For a continuous function f, some of the possible geometrical relationships between the tops of the rectangles and the tops of the strips are shown in Fig. 17.6.

The tops of the rectangles define a new function which is an example of a *step function* (Fig. 17.7). It will normally be discontinuous at the ends of each step. (See also Fig. 17.8.)

We now define the *lower sum* to be the total area of these n rectangles, and denote it by L_n, where the suffix indicates the number

Figure 17.6

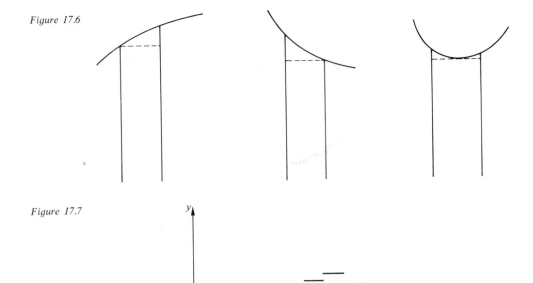

Figure 17.7

of strips. Now since the rectangles lie 'under' the curve, it follows that $L_n \leqslant A$, where A is our intuitive idea of the area under the curve. Also, it seems likely that, as $n \to \infty$ (and $h \to 0$), $L_n \to A$. But this is an intuitive argument, and so will not do. We therefore define U_n, the *upper sum*, found by taking the total area of the smallest rectangles that will fit 'above' the curve. Now certainly

$$L_n \leqslant A \leqslant U_n \text{ for all } n \in Z^+,$$

so that if L_n and U_n approach the same limit as $n \to \infty$, the value of A will be 'squashed' between them, and must also take this value. We therefore make the following definitions:

Definitions
(1) Let f be a function defined for $a \leqslant x \leqslant b$. Let L_n and U_n be defined as explained above. Then if $\lim_{n \to \infty} L_n = \lim_{n \to \infty} U_n$, we say that f is *integrable* between a and b.
(2) If f is integrable between a and b, then the common value of $\lim L_n$ and $\lim U_n$ is called the *area under the curve between* a *and* b. It is also called the *definite integral of* f *from* a *to* b, and is denoted by
$$\int_a^b f(x) \, dx.$$

THE THEORY OF INTEGRATION

Note

(1) It can be shown (but it is beyond the scope of this book to do so) that any continuous function is integrable. (But continuity is not a necessary condition: many discontinuous functions are integrable.)

(2) As an example of a non-integrable function, consider

$$f(x) = \begin{cases} 0 \text{ for } x \in Q \\ 1 \text{ for } x \text{ irrational.} \end{cases}$$

However narrow we make the strips, they will always include both rationals and irrationals, so $f(x)$ will take the values 0 and 1 in each strip. The lower rectangles will all have height zero, and the upper rectangles will have height 1. So, for all n, $L_n = 0$ and $U_n = b - a$, and so $\lim L_n \neq \lim U_n$.

(3) The reason for the notation $\int_a^b f(x)\,dx$ is that if the heights of the lower rectangles are $y_1, y_2, \ldots y_n$, then

$$L_n = \sum y_i h = \sum y_i \delta x$$

if δx is used in place of h. In the limit, the letter Σ is replaced by $\int$, an elongated S, δx is replaced by dx, and the suffices are dropped from the y's. This gives $\int y\,dx$ or $\int f(x)\,dx$.

(4) When we write $\int_a^b f(x)\,dx$, the function $f(x)$ is called the *integrand*.

(5) The value of $\int_a^b f(x)\,dx$ depends only on the function f and the

two limits. The letter x acts merely as a 'dummy' to illustrate the effect

of f; any letter will do just as well. Thus $\int_a^b f(x)\,dx = \int_a^b f(t)\,dt$.

Graphically, we have merely relabelled the horizontal axis in Fig. 17.1; the shaded area is unchanged.

Qu.4 (i) Sketch the graph of $y = 1 + x^2$, and divide the interval $1 \leqslant x \leqslant 5$ into four strips. Calculate L_4 and U_4.
(ii) Calculate also L_8 and U_8, and verify that

$$L_4 < L_8 < U_8 < U_4.$$

Properties of integrals that may seem 'obvious' in terms of areas can be proved in terms of our formal definition. As an example, we prove a result which will be discussed further in Chapter 23.

Theorem 17.1

If f and g are integrable functions in the interval $[a, b]$, and for each $x \in [a, b]$, $f(x) \geqslant g(x)$, then

$$\int_a^b f(x)\,dx \geqslant \int_a^b g(x)\,dx.$$

Proof

Suppose that the interval $[a, b]$ is cut into n strips, and that the corresponding lower sums for f and g are $L_n(f)$ and $L_n(g)$. Since

$f(x) \geqslant g(x)$ at each point, each lower rectangle for f cannot be smaller than the corresponding lower rectangle for g. Summing over all strips gives

$$L_n(f) \geqslant L_n(g).$$

Now this remains true for very large n, and so

$$\lim_{n \to \infty} L_n(f) \geqslant \lim_{n \to \infty} L_n(g)$$

i.e.

$$\int_a^b f(x)\, dx \geqslant \int_a^b g(x)\, dx.$$

17.3 Integration from first principles

Example 1

Find $\displaystyle\int_0^a x^2\, dx$ working from first principles.

Solution

We divide the required area into n strips of width h (so that $nh = a$), and evaluate the lower sum, L_n (Fig. 17.8). If we denote the heights of the lower rectangles by $y_0, y_1, y_2, \ldots, y_{n-1}$, then

Figure 17.8

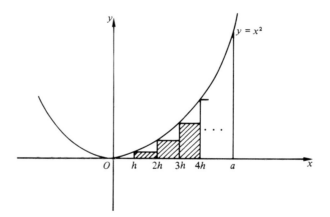

$$y_0 = 0$$
$$y_1 = h^2$$
$$y_2 = (2h)^2 = 4h^2$$
$$y_3 = (3h)^2 = 9h^2 \text{ etc.}$$

and so

$$\begin{aligned}
L_n &= hy_0 + hy_1 + hy_2 + \ldots + hy_{n-1} \\
&= h^3 \{0 + 1 + 4 + 9 + \ldots + (n-1)^2\} \\
&= h^3 \tfrac{1}{6}(n-1)(n)(2n-1)
\end{aligned}$$

(using the result of Exercise 8f, question 1(iv) with $n-1$ in place of n. Note that this, and other similar results will be met more formally in Chapter 28.)

Now we seek the limit of L_n as $n \to \infty$ and $h \to 0$. Since n and h

THE THEORY OF INTEGRATION

remain connected by $nh = a$, we can replace h by $\dfrac{a}{n}$. Then

$$L_n = \frac{a^3}{n^3} \frac{1}{6}(n-1)(n)(2n-1)$$

$$= a^3 \left(\frac{1}{3} - \frac{1}{2n} + \frac{1}{6n^2} \right).$$

So
$$\lim_{n \to \infty} L_n = \frac{1}{3}a^3.$$

Strictly, we should now check that $\lim\limits_{n \to \infty} U_n$ is the same. This is left for the reader in Qu.5.

Thus the required area is $\frac{1}{3}a^3$.

Qu.5   Repeat Example 1 using the upper sum U_n.

Qu.6   Find $\displaystyle\int_0^a x^3 \, dx$ from first principles, using

(i) the lower sum;   (ii) the upper sum.

$$\left(\text{You will need the result that } \sum_1^n r^3 = \tfrac{1}{4}n^2(n+1)^2. \right)$$

The result of Example 1 enables us to write down the value of $\displaystyle\int_a^b x^2 \, dx$. Regarding $x = 0$ as a 'reference point', we have

$$\int_a^b x^2 \, dx = \int_0^b x^2 \, dx - \int_0^a x^2 \, dx$$

$$= \frac{1}{3}b^3 - \frac{1}{3}a^3 \quad \text{by Example 1.}$$

Qu.7   Use the result of Qu.6 to evaluate $\displaystyle\int_1^4 x^3 \, dx$.

We saw in Example 1 that integration from first principles involves summing a series. In general, the series will be either very difficult or impossible to sum, so that integrating from first principles is not a very useful way of dealing with most functions. Fortunately there is another way of dealing with the practical problem of integrating; it is based upon the Fundamental Theorem of Calculus (Section 17.5).

17.4 Area-measuring functions

Suppose that f is a function which is integrable everywhere (i.e. it is integrable between any two points on the real line). Let

$$G(x) = \int_0^x f(t) \, dt.$$

(Note that we use t in the integrand to avoid confusion with x which is now our upper limit. It is this upper limit which is now to be regarded as the variable.) Then $G(x)$ is an area-measuring function, since it measures the area up to the point x from some fixed 'reference' point (in this case, $t=0$). (See Fig. 17.9.) We have, moreover,

Figure 17.9

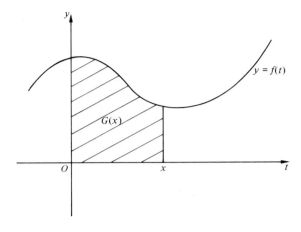

$$\int_a^b f(t)\,dt = \int_0^b f(t)\,dt - \int_0^a f(t)\,dt$$

$$= G(b) - G(a).$$

So the area between $t=a$ and $t=b$ is given by $G(b)-G(a)$.

Qu.8 Illustrate this last sentence graphically. (Take $0<a<b$.)

Now there is no particular reason for taking $t=0$ as our reference point; we could equally well measure from $t=5$ say. This would give the area-measuring function

$$H(x) = \int_5^x f(t)\,dt$$

which, like $G(x)$, satisfies

$$\int_a^b f(t)\,dt = H(b) - H(a).$$

In fact, the values of $G(x)$ and $H(x)$ will always differ by a constant, since

$$G(x) - H(x) = \int_0^x f(t)\,dt - \int_5^x f(t)\,dt$$

$$= \int_0^5 f(t)\,dt$$

which is constant.

To clarify these ideas, it may help to draw an analogy with the measurement of height in a map. The usual reference point is 'sea level', and the figures marked indicate height above sea level. To find

THE THEORY OF INTEGRATION

the difference in height between two points, we simply subtract their heights above sea level. Now suppose that another map breaks with convention, and marks the heights above a fixed landmark which is itself 50 m above sea level. The difference in height between any two points can still be found by subtracting the figures given on the map; but all the figures on this second map will be 50 m less than on the first map.

Definition

Given an integrable function f, the *indefinite integral* of f, written $\int f(x)\,dx$, is the family of area-measuring functions of the form

$$F(x) = \int_k^x f(t)\,dt$$

where k is some constant. Different values of k will give different members of the family; but, as we have seen above, the various members of the family differ from one another only by a constant. We have also seen that, for any member of the family,

$$\int_a^b f(t)\,dt = F(b) - F(a).$$

As an example, let us consider the indefinite integral of $f(t) = t^2$. From Example 1, we have

$$\int_0^x t^2\,dt = \frac{1}{3}x^3$$

and it follows that, for example,

$$\int_1^x t^2\,dt = \frac{1}{3}x^3 - \frac{1}{3}$$

and

$$\int_2^x t^2\,dt = \frac{1}{3}x^3 - \frac{8}{3}.$$

In general, this family of functions may be written as $\frac{1}{3}x^3 + c$ where c is an arbitrary constant. Thus we could say that the indefinite integral of the function x^2 is the function $\frac{1}{3}x^3 + c$, and we would write

$$\int x^2\,dx = \frac{1}{3}x^3 + c.$$

(The instruction 'integrate' is normally taken to mean 'find the indefinite integral of'.)

17.5 The Fundamental Theorem of Calculus

The theorem which we now introduce links the two main processes of calculus (differentiation and integration) by stating that the derivative of the indefinite integral of a function is the original function; i.e. integration and differentiation are, in a sense, mutually inverse processes. The theorem has one important practical consequence: it reduces the problem of integrating a function f to one of finding a function F such that $\dfrac{dF}{dx}=f$. This idea is used in Example 2 below; but first we prove the theorem.

Theorem 17.2 (The Fundamental Theorem of Calculus.)

Let f be a continuous function, and let $F(x)=\displaystyle\int_{k}^{x} f(t)\,dt$ where k is any constant. Then

$$\frac{d}{dx}F(x)=f(x).$$

Proof

From the definition of a derived function, we have

$$\frac{d}{dx}F(x)=\lim_{h\to 0}\frac{1}{h}\{F(x+h)-F(x)\}$$

$$=\lim_{h\to 0}\frac{1}{h}\left\{\int_{k}^{x+h} f(t)\,dt-\int_{k}^{x} f(t)\,dt\right\}$$

$$=\lim_{h\to 0}\frac{1}{h}\int_{x}^{x+h} f(t)\,dt.$$

Now, in Fig. 17.10, $\dfrac{1}{h}\displaystyle\int_{x}^{x-h} f(t)\,dt$ is the area of the shaded strip divided by the width of the strip; and so it is the average height of the strip. Since the function f is continuous, as $h\to 0$ this average height tends to $f(x)$, and so the result is proved. (More precisely, if $y_{\max}$ and $y_{\min}$ are the greatest and least values of $f(t)$ between x and $x+h$, then certainly

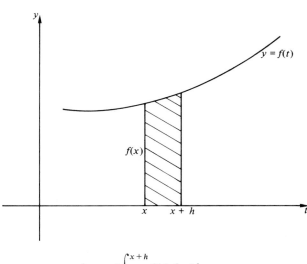

Figure 17.10

$$hy_{\min}\leqslant\int_{x}^{x+h} f(t)\,dt\leqslant hy_{\max}.$$

So

$$y_{\min}\leqslant\frac{1}{h}\int_{x}^{x+h} f(t)\,dt\leqslant y_{\max}.$$

THE THEORY OF INTEGRATION

But the curve $y=f(t)$ is continuous, and so as $h\to 0$, y_{max} and y_{min} both tend to $f(x)$. So $\dfrac{1}{h}\displaystyle\int_x^{x+h} f(t)\, dt$ is squashed between two values tending to $f(x)$, and so it too tends to $f(x)$.)

One way of 'visualizing' this theorem is by noting that in Fig. 17.11 the rate of change of A with respect to x is the rate at which the right hand boundary is sweeping out new area as x increases. This is clearly equal to the *height* of this boundary, i.e. $f(x)$.

Figure 17.11

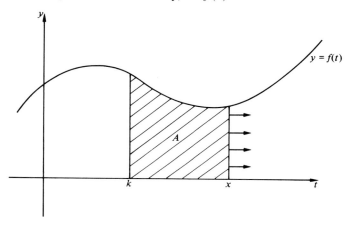

Example 2

(i) Integrate $2x^4+4x$.

(ii) Evaluate $\displaystyle\int_2^4 3x^5\, dx$.

Solution

(i) By the Fundamental Theorem, we seek a function which gives $2x^4+4x$ when differentiated. The answer can be spotted: $\frac{2}{5}x^5+2x^2+c$ where c is an arbitrary constant.

(ii) (The standard way of setting out this sort of problem is given in Example 3.)

We have seen that $\displaystyle\int_a^b f(x)\, dx=F(b)-F(a)$ where F is the indefinite integral of f. Now the indefinite integral of $3x^5$ is $F(x)=\frac{1}{2}x^6+c$. Then

$$F(4)-F(2)=\tfrac{1}{2}.4^6-\tfrac{1}{2}.2^6.$$

So
$$\int_2^4 3x^5\, dx=2048-32=2016.$$

Many functions can now be integrated by trial and error. In particular,

$$\int x^n\, dx=\frac{x^{n+1}}{n+1}+c \quad \text{for } n\neq -1;$$

$$\int x^{-1}\, dx=\ln x+c;$$

$$\int \sin x \, dx = -\cos x + c;$$

$$\int \cos x \, dx = \sin x + c;$$

$$\int \exp x \, dx = \exp x + c.$$

Qu.9 Write down the indefinite integrals of
(i) $\sec^2 x$;   (ii) $\operatorname{cosec} x \cot x$;
(iii) x^7;   (iv) $3x^3$;
(v) $4x^2 - x + 1$;   (vi) $\sin 2x$.

It is important always to check, by differentiating, integrals that have been 'written down'; mistakes can easily be made. For example,

$$\int \exp 2x \, dx = \frac{1}{2} \exp 2x + c$$

but

$$\int \exp x^2 \, dx \neq \frac{1}{2x} \exp x^2 + c.$$

(Try differentiating the R.H.S. in each case.)

The next example shows the standard way of setting out the calculation of a definite integral. Note how, after the integration has been performed, the limits are conventionally written beside the right hand square bracket; note also that the arbitrary '$+c$' is omitted.

Example 3

Evaluate
$$\int_{-\pi/4}^{\pi/6} \sin 3x \, dx.$$

Solution

$$\int_{-\pi/4}^{\pi/6} \sin 3x \, dx = \left[-\frac{1}{3} \cos 3x \right]_{-\pi/4}^{\pi/6}$$

$$= \left(-\frac{1}{3} \cos \frac{\pi}{2} \right) - \left(-\frac{1}{3} \cos \frac{-3\pi}{4} \right)$$

$$= 0 - \frac{1}{3\sqrt{2}}$$

$$= -\frac{1}{3\sqrt{2}}.$$

Qu.10 Sketch the area corresponding to the integral in Example 3, and hence check the sign of the answer.

Qu.11 Evaluate

(i) $\int_{-1}^{3} x^2 \, dx$;   (ii) $\int_{3}^{-1} x^2 \, dx$.

Finally in this chapter, we observe that where an integrand is a function which is defined by different 'rules' on different parts of the range of integration, it is necessary to split up this range and perform the various integrations separately, as in Qu.12.

Qu.12 A function is defined by

$$f(x) = \begin{cases} -x & \text{for } x < 0 \\ x^2 & \text{for } 0 \leqslant x < 1 \\ 1/x^2 & \text{for } x \geqslant 1. \end{cases}$$

Sketch the graph of $y = f(x)$ and find $\int_{-2}^{5} f(x)\, dx$ by evaluating

$$\int_{-2}^{0} -x\, dx + \int_{0}^{1} x^2\, dx + \int_{1}^{5} \frac{1}{x^2}\, dx.$$

Exercise 17

1. (i) Plot the graph of $y = 1 + 9x - 2x^2$ between $x = -1$ and $x = 4$. Divide the interval $0 \leqslant x \leqslant 3$ into 3 strips of width 1. Draw in the step functions corresponding to the lower and upper rectangles, and *calculate* the lower and upper sums L_3 and U_3.
(ii) Repeat the calculations with six strips of width $\frac{1}{2}$.
(iii) Write down and evaluate the integral which gives the area under the curve between $x = 0$ and $x = 3$.

2. Find the following integrals from first principles:

(i) $\int_{0}^{a} 3x\, dx$;   (ii) $\int_{2}^{5} x\, dx$;   (iii) $\int_{p}^{q} 1\, dx$.

3. Integrate
   (i) x^5;   (ii) $5x^{-2}$;   (iii) $2x^3 - 2x^{-3}$;   (iv) $x + 1$;
   (v) e^{2x};   (vi) $\sin(3x + 1)$;   (vii) $\dfrac{2}{x}$;   (viii) $\operatorname{cosec}^2(2x - 1)$.

4. Evaluate the following. Set out the working as in Example 3.

(i) $\int_{1}^{2} 4x^3\, dx$;   (ii) $\int_{0}^{1} x^4\, dx$;

(iii) $\int_{3}^{5} (2x + 3)\, dx$;   (iv) $\int_{1}^{3} \frac{1}{x^2}\, dx$;

(v) $\int_{2}^{4} \frac{1}{x}\, dx$;   (vi) $\int_{1}^{5} \left\{ 2(x+1) + \frac{2}{x+1} \right\} dx$;

(vii) $\int_{0}^{\pi/2} \sin x\, dx$;   (viii) $\int_{\pi/3}^{\pi/4} \sec^2 x\, dx$.

5. (i) Show that the area under the 'upper step function' obtained from the curve $y = \dfrac{1}{1+x}$ by splitting the interval $0 \leqslant x \leqslant 1$ into n strips is

$$\frac{1}{n} + \frac{1}{n+1} + \frac{1}{n+2} + \ldots + \frac{1}{2n-1}.$$

(ii) Evaluate $\int_0^1 \dfrac{1}{1+x}\,dx$ and deduce that

$$\lim_{n\to\infty}\left(\frac{1}{n}+\frac{1}{n+1}+\frac{1}{n+2}+\ldots+\frac{1}{2n-1}\right)=\ln 2.$$

(iii) Put $n=5$ to estimate $\ln 2$, and compare with the exact value of $\ln 2$. Repeat with $n=10$.

6. Evaluate $\int_1^2 x^{-2}\,dx$. Apply the method of question 5 to this integral to show that

$$\lim_{n\to\infty} n\left\{\frac{1}{n^2}+\frac{1}{(n+1)^2}+\frac{1}{(n+2)^2}+\ldots+\frac{1}{(2n-1)^2}\right\}=\frac{1}{2}.$$

7. Evaluate $\int_0^{2\pi} \sin x\,dx$. Explain the graphical significance of your

answer.

8. Let $P(x)=\int_0^x (1+t^2)\,dt$ and $Q(x)=\int_2^x (1+t^2)\,dt$. For each of the following, sketch the curve $y=1+t^2$, and shade the region corresponding to the given expression.
(i) $P(4)$; (ii) $Q(4)$; (iii) $P(7)-P(4)$;
(iv) $Q(7)-Q(4)$; (v) $P(6)-Q(6)$.

9. The function f is continuous, and $F(x)=\int_0^x f(t)\,dt$. If, for all $x,y\in R$,

$$x\geqslant y \iff F(x)\geqslant F(y)$$

what can you deduce about the function f?
10. With the aid of a suitable diagram, simplify

$$\frac{d}{dx}\int_x^k f(t)\,dt$$

where k is a constant.
11. Show diagrammatically that

$$\int_a^b \lambda f(x)\,dx=\lambda\int_a^b f(x)\,dx$$

where $\lambda\in R$. Give a specific counter-example to show that

$$\int_a^b x f(x)\,dx\neq x\int_a^b f(x)\,dx.$$

12. (i) Obtain the result corresponding to question 5(ii) using the lower step function obtained from $y=\dfrac{1}{1+x}$.
(ii) Combine the two results to show that, for sufficiently large n,

$$\frac{1}{2n}+\left(\frac{1}{n+1}+\frac{1}{n+2}+\ldots+\frac{1}{2n-1}\right)+\frac{1}{4n}\approx\ln 2.$$

THE THEORY OF INTEGRATION

(iii) By sketching on the same axes the graphs of $y = \dfrac{1}{1+x}$, and of the two step functions, determine whether this approximation will be too large or too small. [*Hint*: the average of the areas of an upper rectangle and of the corresponding lower rectangle is equal to the area of a suitably drawn trapezium.]

13. (i) Let $f(t) = \begin{cases} 5 \text{ for } t \leqslant 2 \\ 10 \text{ for } t > 2. \end{cases}$

Let $F(x)$ be the area under f from $t = 0$ to $t = x$ (for $x > 0$). Express $F(x)$ explicitly in terms of x. (Your answer will be in two parts: $x \leqslant 2$ and $x > 2$.)

(ii) Show that F is not differentiable at $x = 2$.

(iii) Repeat (i) for $f(t) = \begin{cases} 5 \text{ for } t \neq 2 \\ 10 \text{ for } t = 2. \end{cases}$

(iv) Show that in this case F is differentiable at $x = 2$, but that $\dfrac{dF}{dx} \neq f(x)$ at $x = 2$.

(Note that in the Fundamental Theorem, f was required to be continuous.)

14. (Suitable for discussion.) Show that the function $f(t)$ in question 13(i) is integrable between $t = 0$ and $t = 4$.

15. The function f is defined by the statements

(a) $f(x) = \begin{cases} x^3 \text{ for } 0 \leqslant x < 2 \\ 8 \text{ for } 2 \leqslant x < 3; \end{cases}$

(b) f is periodic with period 3.

(i) Evaluate $\displaystyle\int_0^3 f(x)\,dx.$

(ii) Hence write down $\displaystyle\int_0^6 f(x)\,dx.$

(iii) Find $\displaystyle\int_4^8 f(x)\,dx.$

18 Techniques of integration

18.1 Introduction

We saw in Chapter 17 that the problem of integrating a function f is identical to that of finding a function whose derivative is f. In this chapter, we introduce a number of techniques that are useful in integrating. It is important to realise, however, that whereas differentiation is almost always an easy process, integration is often either difficult—possibly requiring techniques beyond the scope of this book—or simply impossible.

Where a numerical answer to a definite integral is required, and no 'analytical' solution can be found, an approximate answer can be found using the methods of Chapter 29.

18.2 Integration of powers of x

$$\int x^n \, dx = \frac{x^{n+1}}{n+1} + c \text{ for } n \neq -1$$

and

$$\int x^{-1} \, dx = \ln x + c.$$

Example 1
Integrate

(i) $y = \dfrac{2x^2 - 3x - 4}{x}$; (ii) $y = \sqrt{x}(2 - x)$.

Solution
(i) $y = 2x - 3 - 4x^{-1}$, so

$$\int y \, dx = x^2 - 3x - 4\ln x + c.$$

(ii) $y = 2x^{1/2} - x^{3/2}$, so

$$\int y \, dx = \frac{4}{3}x^{3/2} - \frac{2}{5}x^{5/2} + c.$$

18.3 The basic method:guessing

By far the best method of integrating is to *write down* or *guess* the answer; it is a method that should always be tried. It does not matter if the 'guess' turns out to be wrong by a constant factor; this can be adjusted afterwards. (But simple adjustments will not cope with a non-constant factor.)

Example 2
Find

(i) $\int x \sin x^2\,dx$;   (ii) $\int x\sqrt{(x^2+1)}\,dx$;   (iii) $\int \cos x \sin^6 x\,dx$.

Solution
(In fact, each of these integrands is the result of 'function of a function' differentiation. This is something to look out for.)
(i) Try $\cos x^2$.

$$\frac{d}{dx}(\cos x^2) = -2x\sin x^2.$$

This is correct except for the unwanted factor of -2. So we try $-\frac{1}{2}\cos x^2$, and find that it works. So

$$\int x \sin x^2\,dx = -\frac{1}{2}\cos x^2 + c.$$

(ii) First rewrite as $\int x(x^2+1)^{1/2}\,dx$. The $(x^2+1)^{1/2}$ leads us to try $(x^2+1)^{3/2}$:

$$\frac{d}{dx}(x^2+1)^{3/2} = 3x(x^2+1)^{1/2}.$$

So
$$\int x\sqrt{(x^2+1)}\,dx = \frac{1}{3}(x^2+1)^{3/2} + c.$$

(iii) Try $\sin^7 x$. Differentiating would give $7\sin^6 x \cos x$. So

$$\int \sin^6 x \cos x\,dx = \frac{1}{7}\sin^7 x + c.$$

In each part of Example 2, the presence of the 'extra bit' that arises in function of a function differentiation is very convenient. For example, the integral of $\sqrt{(x^2+1)}$ is much more difficult than the integral in (ii) above.

Another important type of function which arises from function of a function differentiation is that in which the numerator is (essentially) the derivative of the denominator, as in parts (ii) and (iii) of Example 3.

Example 3
Find

(i) $\int \dfrac{2}{3x+5}\,dx$;   (ii) $\int \dfrac{x+1}{x^2+2x+3}\,dx$;   (iii) $\int \dfrac{2\sin x}{1+\cos x}\,dx$.

Solution
(The reader should check each part by differentiating.)

(i) $\int \dfrac{2}{3x+5}\,dx = \dfrac{2}{3}\ln(3x+5) + c$.

(ii) Here the top is, except for a factor of 2, the derivative of the bottom; so

$$\int \frac{x+1}{x^2+2x+3}\,dx = \frac{1}{2}\ln(x^2+2x+3)+c.$$

(iii) Similarly,

$$\int \frac{2\sin x}{1+\cos x}\,dx = -2\ln(1+\cos x)+c.$$

Exercise 18a   Integrate:

1. $\dfrac{3}{x^2}-\dfrac{2}{3x^4}$.

2. $\dfrac{1+2x}{x^3}$.

3. $\sqrt{x}+\sqrt[3]{x}$.

4. $\sqrt{x}(2+x+3x^2)$.

5. $\dfrac{1}{(x+3)^2}$.

6. $\dfrac{1}{(2x+3)^4}$.

7. $\sin 3x$.

8. $\cos(4x+1)$.

9. $\sec^2 3x$.

10. $\operatorname{cosec}\dfrac{x}{2}\cot\dfrac{x}{2}$.

11. 2^x.

12. $5\sin^4 x\cos x$.

13. $-4\sin x\cos^3 x$.

14. $\sin^6 x\cos x$.

15. $\sin x\cos^5 x$.

16. $\cos 2x\sin^4 2x$.

17. $\sin 3x\cos^2 3x$.

18. $\tan^4 x\sec^2 x$.

19. $\cot^3 x\operatorname{cosec}^2 x$.

20. $\tan^5 4x\sec^2 4x$.

21. $\cot^2(2x-1)\operatorname{cosec}^2(2x-1)$.

22. $5\sec^5 x\tan x$.

23. $-3\operatorname{cosec}^3 x\cot x$.

24. $\sec^4 5x\tan 5x$.

25. $\operatorname{cosec}^7(3x+4)\cot(3x+4)$.

26. $\sin x\cos x$. (Find two answers and explain the connection between them.)

27. $\tan x\sec^2 x$. (Find two answers.)

28. $2x(x^2+1)^5$.

29. $x^2(2x^3-3)^6$.

30. $(4x+1)^3$.

31. $x\sqrt{(x^2+1)}$.

32. $\sqrt{(3x-2)}$.

33. $3x^2(2x^3+1)^{-4}$.

34. $\dfrac{3x}{(x^2+4)^2}$.

35. $(x+1)\sqrt[3]{(x^2+2x)}$.

36. $\dfrac{x}{\sqrt{(2x^2-3)}}$.

37. $2xe^{x^2}$.

38. e^{2x-1}.

39. $x\operatorname{cosec}x^2\cot x^2$.

40. $x^2\sec^2 x^3$.

41. $\cos x\sqrt{(\sin x)}$.

42. $\dfrac{\cos x}{(\sin x-1)^4}$.

43. $\sec^2 x\,e^{\tan x}$.

TECHNIQUES OF INTEGRATION

44. $\dfrac{1}{2x}$.

45. $\dfrac{2x}{x^2+1}$.

46. $\dfrac{2}{4x+3}$.

47. $\dfrac{\cos 2x}{\sin 2x+1}$.

48. $\dfrac{4x-6}{x^2-3x+2}$.

49. $\dfrac{e^x}{1-e^x}$.

50. $\dfrac{\cos x}{\sin x}$.

51. $\dfrac{\sin x}{\cos x}$.

52. $\dfrac{\sec^2 (2x+1)}{\tan (2x+1)}$.

53. $\dfrac{6x^3}{\sqrt{(x^4+3)}}$.

54. $\left(\dfrac{\sec \frac{1}{2}x}{\tan \frac{1}{2}x+1} \right)^2$.

55. $\dfrac{\sin x}{\cos^4 x}$.

56. $\dfrac{x-2}{x^2-4x+5}$.

57. $(1+x^2)e^{x^2/2}$.

18.4 Rewriting the integrand

Many integrals cannot be 'spotted' immediately but become much simpler if the integrand is rewritten with the aid of some algebraic or trigonometric identity. Example 4 uses partial fractions and algebraic division.

Example 4
Find

(i) $\displaystyle\int \dfrac{x^2+x-3}{x^3-2x^2+x}\,dx$; (ii) $\displaystyle\int \dfrac{2x^2-3}{x^2+x-6}\,dx$.

Solution
(i) Factorizing the denominator, and expressing in partial fractions,

$$\int \left\{ -\dfrac{3}{x} + \dfrac{4}{x-1} - \dfrac{1}{(x-1)^2} \right\} dx$$

$$= -3 \ln x + 4 \ln (x-1) + \dfrac{1}{(x-1)} + c.$$

(ii) Here, the degree of the numerator is not smaller than that of the denominator, so we begin by dividing.

$$\int \dfrac{2x^2-3}{x^2+x-6}\,dx = \int \left\{ 2 - \dfrac{2x-9}{(x-2)(x+3)} \right\} dx$$

$$= \int \left\{ 2 + \dfrac{1}{x-2} - \dfrac{3}{x+3} \right\} dx$$

$$= 2x + \ln (x-2) - 3 \ln (x+3) + c$$

$$\text{or } 2x + \ln \dfrac{x-2}{(x+3)^3} + c.$$

The method of Example 4 will not work if the denominator cannot be factorized. Thus, for example, $\int \dfrac{x+1}{x^2+x+1} dx$ would require a substitution (see Section 18.7), although $\int \dfrac{2x+1}{x^2+x+1} dx$ could be written down as $\ln(x^2+x+1)+c$.

Qu.1 Find

(i) $\displaystyle\int \frac{1}{x^2-2} dx$; (ii) $\displaystyle\int \frac{2x+1}{(x-1)^2} dx$.

Qu.2 (i) *Write down* the integral of $\dfrac{2x+5}{x^2+5x+6}$.

(ii) Express this function in partial fractions and hence integrate it. Reconcile your two answers.

The next example uses the Pythagorean trigonometric identities.

Example 5
Find

(i) $\displaystyle\int \cos^5 \theta \sin^4 \theta \, d\theta$; (ii) $\displaystyle\int \sec^3 2\theta \tan^3 2\theta \, d\theta$.

Solution

(i) $\displaystyle\int \cos^5 \theta \sin^4 \theta \, d\theta$

$= \displaystyle\int \cos \theta (1 - \sin^2 \theta)^2 \sin^4 \theta \, d\theta$

$= \displaystyle\int (\cos \theta \sin^4 \theta - 2 \cos \theta \sin^6 \theta + \cos \theta \sin^8 \theta) \, d\theta$.

Each term can now be integrated to give

$$\frac{1}{5} \sin^5 \theta - \frac{2}{7} \sin^7 \theta + \frac{1}{9} \sin^9 \theta + c.$$

(ii) $\displaystyle\int \sec^3 2\theta \tan^3 2\theta \, d\theta$

$= \displaystyle\int \sec^3 2\theta (\sec^2 2\theta - 1) \tan 2\theta \, d\theta$

$= \displaystyle\int (\sec^5 2\theta \tan 2\theta - \sec^3 2\theta \tan 2\theta) \, d\theta$

$= \dfrac{1}{10} \sec^5 2\theta - \dfrac{1}{6} \sec^3 2\theta + c.$

In Example 5(i), it is the *odd* power of $\cos \theta$ which makes the method possible, since it allows a single factor of $\cos \theta$ to be left in each term. An odd power of $\sin \theta$ will work just as well.

Qu.3 Find

(i) $\int \cos^4 \theta \sin^3 \theta \, d\theta$; (ii) $\int \sin^5 \theta \, d\theta$.

Qu.4 (i) Integrate $\sin^3 \theta \cos^3 \theta$ in two different ways.
(ii) Explain the relationship between the two answers. (See question 26 of Exercise 18a).

This method will clearly not work for integrals of the form $\int \sin^m \theta \cos^n \theta \, d\theta$ where both m and n are even. Such integrals will be discussed in Chapter 40.

Qu.5 Integrate $\sec^4 \theta \tan^2 \theta$ by first rewriting it as $\sec^2 \theta (\tan^2 \theta + 1) \tan^2 \theta$.
Qu.6 (i) Integrate $\sec^4 \theta \tan^3 \theta$ in two different ways.
(ii) For what positive integral values of m and n will the methods of Example 5(ii) or Qu.5 not work for

$$\int \sec^m \theta \tan^n \theta \, d\theta.$$

Qu.7 Integrate $\cos 3x \cos 7x$ by first writing it as a sum or difference.

Example 6 illustrates a number of specific methods which should be memorized.

Example 6
Find

(i) $\int \sin^2 \theta \, d\theta$; (ii) $\int \tan \theta \, d\theta$;

(iii) $\int \sec \theta \, d\theta$.

Solution

(i) $\int \sin^2 \theta \, d\theta = \int \frac{1}{2} (1 - \cos 2\theta) \, d\theta$

$$= \frac{1}{2}\theta - \frac{1}{4} \sin 2\theta + c.$$

(ii) $\int \tan \theta \, d\theta = \int \frac{\sin \theta}{\cos \theta} \, d\theta$

$$= -\ln (\cos \theta) + c.$$

(See also Qu.9.)

(iii) $\int \sec \theta \, d\theta = \int \frac{\sec \theta (\sec \theta + \tan \theta)}{\sec \theta + \tan \theta} \, d\theta$

$$= \ln (\sec \theta + \tan \theta) + c.$$

(Another method for this integral is given in Qu.27.)

Qu.8 Integrate

(i) $\cos^2 3\theta$; (ii) $\cot \theta$; (iii) $\operatorname{cosec} \dfrac{\theta}{2}$.

Qu.9 Integrate $\tan \theta$ by writing it as $\dfrac{\sec \theta \tan \theta}{\sec \theta}$. Reconcile your answer with that of Example 6 (ii).

Exercise 18b Integrate

1. $\dfrac{2x-1}{(x-1)(x+2)}$.

2. $\dfrac{x-3}{(x+1)(2x+1)}$.

3. $\dfrac{3x-4}{(2x-1)(x-3)}$.

4. $\dfrac{2x+1}{2x^2+3x-2}$.

5. $\dfrac{3x+7}{3x^2-2x-1}$.

6. $\dfrac{x^2+2x-1}{(x-1)(x+1)(x+2)}$.

7. $\dfrac{1}{x(x-1)(x-2)}$.

8. $\dfrac{x+3}{(x-2)^2}$.

9. $\dfrac{4x-7}{(x-1)^2(x-4)}$.

10. $\dfrac{3}{x(x-2)^2}$.

11. $\dfrac{x^2+2x+5}{(x+1)(x-1)^2}$.

12. $\dfrac{4x-9}{x^2(2x-3)^2}$.

13. $\dfrac{x}{x+2}$.

14. $\dfrac{3x+1}{2x-4}$.

15. $\dfrac{x^2+2x+4}{x+1}$.

16. $\dfrac{x^2+3x-1}{x^2+5x+4}$.

17. $\dfrac{x^3+2}{x^2-1}$.

18. $\dfrac{x+2}{x^2+4x+5}$.

19. $\cos^2 2x \sin^3 2x$.

20. $\operatorname{cosec}^4 x \cot^2 x$.

21. $\operatorname{cosec}^3 (3x-1) \cot (3x-1)$.

22. $\tan^4 \tfrac{1}{2}x$.

23. $\sin^2 2x$.

24. $\sin^2 x \cos^2 x$.

25. $\sin^4 x$.

26. $\sin 3x \sin 5x$.

27. $\cos 2x \sin 3x$.

28. $\tan 2x$.

29. $\sec (1-x)$.

30. $x \cot x^2$.

18.5 Integration by substitution

The method of substitution will often work where the methods of the earlier parts of this chapter fail. A rigorous proof of the validity of the method is beyond the scope of this book, but an outline of one proof is given at the end of this section. First, however, we see the method in operation.

Example 7

Find
$$\int x \sqrt{(2x+1)} \, dx.$$

Solution

We try the substitution $u = 2x + 1$. The method is to rewrite the whole integral (including the dx) in terms of u. We have

$$u = 2x + 1 \;\Rightarrow\; \frac{du}{dx} = 2$$

$$\Rightarrow\; dx = \tfrac{1}{2} \, du.$$

So
$$\int x \sqrt{(2x+1)} \, dx = \int \tfrac{1}{2}(u-1) \sqrt{u} \cdot \tfrac{1}{2} \, du$$

$$= \frac{1}{4} \int (u^{3/2} - u^{1/2}) \, du$$

$$= \frac{1}{4} \left(\frac{2}{5} u^{5/2} - \frac{2}{3} u^{3/2} \right) + c$$

$$= \frac{1}{30} u^{3/2} \, (3u - 5) + c.$$

Rewriting in terms of x,

$$\frac{1}{30} (2x+1)^{3/2} \, (6x-2) + c$$

$$= \frac{1}{15} (2x+1)^{3/2} \, (3x-1) + c.$$

The reader might like to check this by differentiating.

For many integrals, it is necessary to select the 'right' substitution, and if one substitution does not work (i.e. if the new integrand cannot be integrated), another may be tried. (We may either start again with the original integral, or perform a second substitution on the integral which resulted from the first substitution. Sometimes, successive substitutions can lead to a whole string of integrals, none of which can be performed!) It may be that two different substitutions will work. As an example, we solve Example 7 by a different substitution:

Example 8

Integrate $x \sqrt{(2x+1)}$ using the substitution $u = \sqrt{(2x+1)}$.
Solution

$$u = \sqrt{(2x+1)} \;\Rightarrow\; u^2 = 2x + 1$$

$$\Rightarrow\; x = \frac{1}{2}(u^2 - 1)$$

$$\Rightarrow\; \frac{dx}{du} = u$$

$$\Rightarrow\; dx = u \, du.$$

So

$$\int x\sqrt{(2x+1)}dx = \int \frac{1}{2}(u^2-1)u \cdot u \, du$$

$$= \frac{1}{2}\int (u^4-u^2) \, du$$

$$= \frac{1}{2}\left(\frac{1}{5}u^5 - \frac{1}{3}u^3\right) + c$$

$$= \frac{1}{30}u^3(3u^2-5) + c$$

$$= \frac{1}{15}(2x+1)^{3/2}(3x-1) + c.$$

Integration by substitution can be used for the 'function and its derivative' type of integral that we met in Examples 2 and 3. (Of course, spotting the answer is a better method!) The following example shows how this works.

Example 9

Find $\int \dfrac{3x}{2\sqrt{(x^2-2)}} \, dx$ using the substitution $u = x^2 - 2$.

Solution

$$u = x^2 - 2 \implies \frac{du}{dx} = 2x$$

$$\implies dx = \frac{du}{2x}.$$

(This factor of x in the denominator will cancel the x in the integrand.) So

$$\int \frac{3x}{2\sqrt{(x^2-2)}} \, dx = \int \frac{3}{2\sqrt{u}} \frac{du}{2}$$

$$= \frac{3}{2}u^{1/2} + c$$

$$= \frac{3}{2}\sqrt{(x^2-2)} + c.$$

'Qu.10 Solve Example 9 using the substitution $u = \sqrt{(x^2-2)}$. (Note that, differentiating implicitly,

$$u^2 = x^2 - 2 \implies 2u \, du = 2x \, dx.)$$

Qu.11 Solve each part of Example 2 by a substitution.
(Take (i) $u = x^2$; (ii) $u = x^2 + 1$; (iii) $u = \sin x$.)

We now outline a proof of the method of integration by substitution. We saw in Chapter 17 that, provided f is integrable,

$$\int_a^b f(x) \, dx = \lim \sum f(x_i) \, \delta x, \tag{1}$$

where the summation is over n strips, the $f(x_i)$ are the minimum (or maximum) values in the strips, and $\delta x = (b-a)/n$. The limit is then as $n \to \infty$. Now if we make the substitution $x = g(u)$, then we know that $\delta x \approx \dfrac{dx}{du} \, \delta u$ (see Section 14.2), where the approximation becomes a better one as $\delta x \to 0$. So the R.H.S. of (1) may be written

$$\sum f(g(u_i)) \frac{dx}{du} \, \delta u,$$

giving, in the limit,

$$\int f(g(u)) \frac{dx}{du} \, du.$$

So the method is:
(i) substitute $g(u)$ for x in the integrand;
(ii) replace dx by $\dfrac{dx}{du} \, du$.

18.6 Trigonometric substitution

So far, we have substituted only algebraic functions when the integrand has been algebraic; the examples below show how trigonometric substitutions may be used. For some integrals, the substitution of a trigonometric function is the only reasonable method.

Example 10
Find

$$\int \frac{3}{x^2 + 4} \, dx.$$

Solution
We put $x = 2 \tan \theta$. (With practice, this will not seem such an obscure choice.) Then

$$dx = 2 \sec^2 \theta \, d\theta,$$

and so

$$\int \frac{3}{x^2 + 4} \, dx = \int \frac{3}{4 \tan^2 \theta + 4} \cdot 2 \sec^2 \theta \, d\theta$$

$$= \int \frac{3}{2} \, d\theta \ (\text{since } \tan^2 \theta + 1 = \sec^2 \theta)$$

$$= \frac{3}{2} \theta + c$$

$$= \frac{3}{2} \arctan \frac{x}{2} + c.$$

It may seem surprising that a purely algebraic integral should have a trigonometric answer, but it should be remembered that the basic trigonometric functions (sin and cos) satisfy the algebraic equation $x^2 + y^2 = 1$.

In an integral such as $\int \dfrac{2}{3x^2+7}\,dx$, it is simplest first to make the coefficient of x^2 unity, thus: $\frac{2}{3}\int \dfrac{1}{x^2+7/3}\,dx$. The substitution $x=\sqrt{\frac{7}{3}}\tan\theta$ may then be used.

Qu.12 (i) Complete the integration above.

(ii) Find $\int \dfrac{5}{4x^2+9}\,dx$.

This method is sufficiently important to be expressed in a standard formula. Consider $\int \dfrac{1}{x^2+a^2}\,dx$. Putting $x=a\tan\theta$,

$$\int \frac{a\sec^2\theta}{a^2\tan^2\theta+a^2}\,d\theta = \int \frac{1}{a}\,d\theta$$

$$= \frac{\theta}{a}+c$$

$$= \frac{1}{a}\arctan\frac{x}{a}+c.$$

So

$$\boxed{\int \frac{1}{x^2+a^2}\,dx = \frac{1}{a}\arctan\frac{x}{a}+c.}$$

Then, for example,

$$\int \frac{2}{3x^2+9}\,dx = \frac{2}{3}\int \frac{1}{x^2+3}\,dx$$

and putting $a=\sqrt{3}$ in the formula gives

$$\frac{2}{3\sqrt{3}}\arctan\frac{x}{\sqrt{3}}+c.$$

Qu.13 Use the formula to find

(i) $\int \dfrac{3}{2x^2+8}\,dx$; (ii) $\int \dfrac{5}{3x^2+1}\,dx$.

Now consider the following example.

Example 11

Find $\int \dfrac{1}{x^2+6x+13}\,dx$.

Solution
Completing the square in the denominator, the integral is

$$\int \frac{1}{(x+3)^2+4}\,dx.$$

We may now either use the substitution $x+3=2\tan\theta$ or use the formula with $a=2$, and $(x+3)$ replacing x. The latter method gives immediately $\frac{1}{2}\arctan\dfrac{x+3}{2}+c$.

Note

When using the formula, we may replace x by $x+b$, but not by an expression of the form $kx+b$ where $k\neq1$; the reader is left to explore why this is so. The safe rule when using the formula is always to start by making the coefficient of x^2 unity.

Qu.14 (i) Solve Example 11 using the substitution $x+3=2\tan\theta$.

(ii) Find $\displaystyle\int\frac{3}{(2x+1)^2+9}\,dx$ by putting $2x+1=3\tan\theta$.

(iii) Solve (ii) by rewriting as

$$\frac{3}{4}\int\frac{1}{(x+\frac{1}{2})^2+9/4}\,dx \text{ and using the formula.}$$

Remember that if the denominator is not positive definite (i.e. if it has roots) then partial fractions should be used; a tan substitution will not work.

We now consider integrals where a sine substitution is appropriate.

Example 12

Find $$\int\frac{1}{\sqrt{(4-x^2)}}\,dx.$$

Solution

Let $x=2\sin\theta$, then $dx=2\cos\theta\,d\theta$.

$$\int\frac{1}{\sqrt{(4-x^2)}}\,dx=\int\frac{2\cos\theta\,d\theta}{\sqrt{(4-4\sin^2\theta)}}$$

$$=\int 1\,d\theta$$

$$=\theta+c$$

$$=\arcsin\frac{x}{2}+c.$$

Note

(1) In this example, the substitution $x=2\cos\theta$ would have been equally effective. (Try it, and explain the relationship between the two answers.)

(2) The reader is left (in Qu.16) to verify the general formula

$$\boxed{\int\frac{1}{\sqrt{(a^2-x^2)}}\,dx=\arcsin\frac{x}{a}+c.}$$

(3) The formula may be used for more complicated examples, provided that the coefficient of x^2 has been reduced to -1. For example,

$$\int \frac{5}{\sqrt{(9+6x-3x^2)}}\,dx = \frac{5}{\sqrt{3}} \int \frac{1}{\sqrt{(3+2x-x^2)}}\,dx$$

$$= \frac{5}{\sqrt{3}} \int \frac{1}{\sqrt{\{4-(x-1)^2\}}}\,dx$$

$$= \frac{5}{\sqrt{3}} \arcsin \frac{x-1}{2} + c.$$

Qu.15 Find

(i) $\displaystyle\int \frac{1}{\sqrt{(18-2x^2)}}\,dx$; (ii) $\displaystyle\int \frac{3}{\sqrt{(-1-4x-2x^2)}}\,dx$.

Qu.16 Verify the formula in note (2) above by
(i) using a suitable substitution for the L.H.S.;
(ii) differentiating the R.H.S..

18.7 Further trigono-metric substitutions

Consider the integral $\displaystyle\int \frac{3x^2+x+13}{x^2+2x+5}\,dx$. By algebraic division it may be written as

$$\int \left(3 - \frac{5x+2}{x^2+2x+5}\right) dx.$$

Now since the denominator does not factorize, the problem is to eliminate the term in x in the numerator to leave an arctan integral. This may be done by splitting up the numerator of the fraction as $\frac{5}{2}(2x+2)-3$, so that the integral becomes

$$\int 3\,dx - \frac{5}{2} \int \frac{2x+2}{x^2+2x+5}\,dx + \int \frac{3}{x^2+2x+5}\,dx$$

$$= 3x - \frac{5}{2} \ln(x^2+2x+5) + \int \frac{3}{(x+1)^2+4}\,dx$$

$$= 3x - \frac{5}{2} \ln(x^2+2x+5) + \frac{3}{2} \arctan \frac{x+1}{2} + c.$$

In general, a linear term in the numerator can be dealt with by splitting the numerator into
(i) a suitable multiple of the derivative of the denominator (giving a ln integral), and
(ii) a constant term (giving an arctan integral). (See Exercise 18c, question 10(iv) for the corresponding procedure in the arcsin case.)

Qu.17 Find $\displaystyle\int \frac{2x^3-7x^2+2x+16}{x^2-4x+6}\,dx.$

If a substitution is performed on a *definite* integral, then the limits may also be changed to correspond to the new variable. This avoids the need to rewrite the integral in terms of the original variable after the integration has been performed. If the limits are *not* changed, then it must be made clear that they refer to the original variable. (Unless otherwise stated, it is assumed that limits refer to the variable of integration, i.e. the variable attached to the '*d*' as in '*du*' or '*dx*'.) For example, we may write either

$$\int_0^3 x \sqrt{(x+1)}\,dx$$

$$= \int_1^4 (u-1)\sqrt{u}\,du \quad (\text{putting } u = x+1)$$

$$= \left[\frac{2}{5}u^{5/2} - \frac{2}{3}u^{3/2}\right]_1^4$$

$$= \left(\frac{2}{5}\times 32 - \frac{2}{3}\times 8\right) - \left(\frac{2}{5}\times 1 - \frac{2}{3}\times 1\right)$$

$$= 7\tfrac{11}{15},$$

or

$$\int_0^3 x \sqrt{(x+1)}\,dx$$

$$= \int_{x=0}^{x=3} (u-1)\sqrt{u}\,du$$

$$= \left[\frac{2}{5}u^{5/2} - \frac{2}{3}u^{3/2}\right]_{x=0}^{x=3}$$

$$= \left[\frac{2}{5}(x+1)^{5/2} - \frac{2}{3}(x+1)^{3/2}\right]_0^3$$

$$= \text{etc. (as above).}$$

In general, the former (i.e. changing the limits) is simpler.

Difficulties can sometimes arise in using a substitution for a definite integral; these are discussed in Chapter 23.

Qu.18 (i) Complete the second solution above, and also find this integral by putting $v = \sqrt{(x+1)}$ and changing the limits.

(ii) Find $\displaystyle\int_{-2}^2 \frac{dx}{x^2+4}$ by

(a) using a suitable trigonometric substitution and changing the limits;

(b) using the formula for $\displaystyle\int \frac{dx}{x^2+a^2}$.

Exercise 18c

1. Using the given substitutions, find the following.

(i) $\displaystyle\int x\sqrt{(x-1)}\,dx$;　(a) $u = x - 1$,　(b) $u = \sqrt{(x-1)}$.

(ii) $\int (x+1)(x-4)^5 \, dx; \quad u=x-4.$

(iii) $\int \dfrac{x-1}{\sqrt{(3x+2)}} \, dx; \quad$ (a) $u=3x+2,$ (b) $u=\sqrt{(3x+2)}.$

(iv) $\int \dfrac{1}{(x-1)\sqrt{x}} \, dx; \quad u=\sqrt{x}. \left(\text{Also } try \ u=\dfrac{1}{x}. \right)$

(v) $\int \dfrac{x(x-2)}{(x-3)^5} \, dx; \quad u=x-3.$

2. Using suitable substitutions, find

(i) $\int (2x+1)\sqrt{(x+4)} \, dx;$ (ii) $\int x^2 \sqrt{(2x+3)} \, dx;$

(iii) $\int x(x+1)^7 \, dx;$ (iv) $\int 2x(3x-1)^{2/3} \, dx;$

(v) $\int (x^2-4)\sqrt{(x-2)} \, dx;$ (vi) $\int \dfrac{x-3}{(x-2)^4} \, dx;$

(vii) $\int \dfrac{x^3}{\sqrt{(x-5)}} \, dx;$ (viii) $\int \dfrac{\sqrt{x}}{x-4} \, dx.$

3. Solve each of the following
   (a) by inspection;
   (b) using the given substitution.

(i) $\int 2x\sqrt{(x^2+1)} \, dx; \quad u=x^2+1;$ (ii) $\int \dfrac{x}{\sqrt{(x^2-1)}} \, dx; \quad u=x^2-1;$

(iii) $\int \dfrac{3x}{x^2-4} \, dx; \quad u=x^2-4;$ (iv) $\int \dfrac{x}{(x^2+3)^2} \, dx; \quad u=x^2+3;$

(v) $\int \cos x \, \sin^4 x \, dx; \quad u=\sin x;$

(vi) $\int \sec^2 2x \, \tan^3 2x \, dx; \quad u=\tan 2x;$

(vii) $\int x^2 \sin x^3 \, dx; \quad u=x^3;$ (viii) $\int \dfrac{e^x+\cos x}{e^x+\sin x} \, dx; \quad u=e^x+\sin x.$

4. Use suitable substitutions to find

(i) $\int \dfrac{1}{\sqrt{(x-1)(x-2)}} \, dx;$

(ii) $\int \dfrac{\sqrt{(x+3)}}{x-6} \, dx;$

(iii) $\int \dfrac{1}{(x+1)\sqrt{(1-x^2)}} \, dx. \left(\text{Try } u=\dfrac{1}{x+1}. \right)$

5. Find $\int \dfrac{1}{x^3+x} \, dx$ by using

(i) the substitution $u=\dfrac{1}{x}$;

(ii) partial fractions.

Check that your two answers are the same.

6. Use substitutions for the following if inspection fails.

(i) $\displaystyle\int\frac{\sec^2 x}{\sqrt{(\tan x)}}\,dx$;

(ii) $\displaystyle\int\frac{(x-1)}{(x+1)^4(x-3)^4}\,dx$ (Put $u=x-1$);

(iii) $\displaystyle\int\tan x\sqrt{(\sec x)}\,dx$;

(iv) $\displaystyle\int\frac{\cot 2x}{\operatorname{cosec}^5 2x}\,dx$;

(v) $\displaystyle\int\frac{\cos x}{\sqrt{(1+\sin x)}}\,dx$;

(vi) $\displaystyle\int\frac{\sqrt{x}}{(2+\sqrt{x})^3}\,dx$;

(vii) $\displaystyle\int\frac{x}{2x^2+3}\,dx$;

(viii) $\displaystyle\int\frac{2x}{\sqrt{(1-x^2)}}\,dx$.

7. Integrate

(i) $\dfrac{1}{\sqrt{(4-x^2)}}$;

(ii) $\dfrac{1}{4+x^2}$;

(iii) $\dfrac{3}{\sqrt{(3-x^2)}}$;

(iv) $\dfrac{8}{x^2+2}$;

(v) $\dfrac{1}{\sqrt{(9-4x^2)}}$;

(vi) $\dfrac{1}{16+9x^2}$;

(vii) $\dfrac{7}{\sqrt{(3-75x^2)}}$;

(viii) $\dfrac{5}{8x^2+2}$;

(ix) $\dfrac{1}{\sqrt{\{12-3(x-1)^2\}}}$;

(x) $\dfrac{3}{18+2(x+2)^2}$;

(xi) $\dfrac{1}{\sqrt{\{1-(2x+3)^2\}}}$;

(xii) $\displaystyle\int\frac{1}{(3x-1)^2+5}$;

(xiii) $\displaystyle\int\frac{2}{\sqrt{(7+6x-x^2)}}$;

(xiv) $\displaystyle\int\frac{5}{x^2+8x+41}$;

(xv) $\displaystyle\int\frac{6}{\sqrt{(1-2x-3x^2)}}$;

(xvi) $\displaystyle\int\frac{2}{5x^2+4x+1}$.

8. (i) Find $\displaystyle\int\frac{1}{x^2+4}\,dx$ and $\displaystyle\int\frac{1}{x^2-4}\,dx$ and $\displaystyle\int\frac{x}{x^2+4}\,dx$.

(ii) Find $\displaystyle\int\frac{dx}{\sqrt{(9-x^2)}}$ and $\displaystyle\int\frac{x\,dx}{\sqrt{(9-x^2)}}$.

(iii) Find $\displaystyle\int\frac{dx}{x^2+x+2}$ and $\displaystyle\int\frac{dx}{x^2+x-2}$.

9. Find $\displaystyle\int\frac{dx}{x\sqrt{(x^2-4)}}$ by using

(i) the substitution $x=2\sec\theta$;

(ii) the substitution $t^2=x^2-4$, and then a second substitution.

Explain how the two methods are connected.

(iii) Solve also by putting $u = 1/x$ and then using a second substitution.

10. (Miscellaneous)

(i) $\displaystyle\int \frac{\sqrt{(x^2+1)}}{x}\,dx$;

(ii) $\displaystyle\int (x^2+4)^{-3/2}\,dx$;

(iii) $\displaystyle\int (x^2-4)^{-3/2}\,dx$;

(iv) $\displaystyle\int \frac{x\,dx}{\sqrt{(2x-x^2)}}$ [*Hint*: you will need to split the numerator into the form $A(-2x+2)+B$];

(v) $\displaystyle\int (4-x^2)^{-3/2}\,dx$;

(vi) $\displaystyle\int \sqrt{(1-x^2)}\,dx$;

(vii) $\displaystyle\int \frac{dx}{(x-3)\sqrt{(4x^2-24x+35)}}$, $\left(\text{Try } u = \dfrac{1}{x-3}\right)$;

(viii) $\displaystyle\int \frac{dx}{(x-1)\sqrt{(x-2)}}$;

(ix) $\displaystyle\int \frac{x^2\,dx}{\sqrt{(7-6x-x^2)}}$. (Try a sine substitution.)

11. Integrate

(i) $\dfrac{3x+8}{x^2+4x+13}$;

(ii) $\dfrac{x+1}{\sqrt{(2-x)}}$;

(iii) $\dfrac{x^2-3x+5}{2x^2+2x+1}$;

(iv) $\dfrac{x^3}{x^2+2x+2}$.

12. Evaluate the following, changing the limits at the same time as making the necessary substitution. (Definite integrals should be set out as in Example 3 of Chapter 17.)

(i) $\displaystyle\int_0^3 x(x+1)^{3/2}\,dx$;

(ii) $\displaystyle\int_{-3}^{-2} x\sqrt[3]{(2x+5)}\,dx$;

(iii) $\displaystyle\int_1^3 (3x+1)(3x-1)^{-3/2}\,dx$;

(iv) $\displaystyle\int_9^{16} \frac{\sqrt{x}}{x-4}\,dx$;

(v) $\displaystyle\int_0^2 \frac{dx}{\sqrt{(4-x^2)}}$ using

(a) a sine substitution;

(b) a cosine substitution.

(vi) $\displaystyle\int_{-2}^0 \frac{dx}{x^2+2x+2}$.

18.9 Integration by parts

The technique of integration by parts is derived from the method of differentiating a product; it is often the best method when the

TECHNIQUES OF INTEGRATION

integrand is itself a product of two functions. The method may be embodied in a formula, but first we see how it works without using the formula.

Consider the integral $\int x \cos x \, dx$. We might reasonably try the function $x \sin x$. We get

$$\frac{d}{dx}(x \sin x) = x \cos x + \sin x,$$

and integrating each side gives

$$x \sin x = \int x \cos x \, dx + \int \sin x \, dx$$

$$\Rightarrow \int x \cos x \, dx = x \sin x - \int \sin x \, dx$$

$$= x \sin x + \cos x + c.$$

Qu.19 Check this answer by differentiating.
Qu.20 Use this method to find

(i) $\int x \sin x \, dx$; (ii) $\int x \, e^x \, dx.$

Check each by differentiating.

We now derive a formula for this method. Let u and v be functions of x, then

$$\frac{d}{dx}(uv) = u\frac{dv}{dx} + v\frac{du}{dx}$$

$$\Rightarrow uv = \int u\frac{dv}{dx}dx + \int v\frac{du}{dx}dx$$

$$\Rightarrow \boxed{\int u\frac{dv}{dx}dx = uv - \int v\frac{du}{dx}dx.}$$

Let us use this formula to solve our original example.

Example 13

Find $\int x \cos x \, dx$ (by the formula above).

Solution

The formula gives the integral of $u\frac{dv}{dx}$. So let $u = x$ and $\frac{dv}{dx} = \cos x$, so that $\frac{du}{dx} = 1$ and $v = \sin x$. Substituting into the formula,

$$\int x \cos x \, dx = x \sin x - \int \sin x \, dx$$

$$= x \sin x + \cos x + c.$$

Note

In this example, we chose to put $u = x$ and $\dfrac{dv}{dx} = \cos x$. We might

have tried putting $u = \cos x$ and $\dfrac{dv}{dx} = x$. Unfortunately, this will fail

because the 'other integral' that arises is worse than the original one.
(Try it.) With practice, it is usually possible to tell which function to

take as u and which to take as $\dfrac{dv}{dx}$, but the reader should be

prepared to reverse the functions if the first attempt fails.

Qu.21   Solve Qu.20 by the formula.

Some integrals require the repeated use of 'parts'. For example, the
'other integral' that arises from $\int x^2 \sin x \, dx$ itself requires parts.

Qu.22   Using integration by parts (twice), find $\int x^2 \sin x \, dx$.

Repeated use of parts is also necessary in the following example
which illustrates an important method.

Example 14
Find

(i) $\displaystyle\int e^x \sin x \, dx$;   (ii) $\displaystyle\int \sec^3 x \, dx$.

Solution
(i) (Note that another method for this integral is outlined in question
6 of Exercise 15a.) Let $I = \int e^x \sin x \, dx$. Then by parts (taking $u = \sin x$

and $\dfrac{dv}{dx} = e^x$)

$$I = e^x \sin x - \int e^x \cos x \, dx. \tag{1}$$

Now, tackling this other integral also by parts (taking $u = \cos x$ and
$\dfrac{dv}{dx} = e^x$),

$$\int e^x \cos x \, dx = e^x \cos x + \int e^x \sin x \, dx$$

$$= e^x \cos x + I.$$

Substituting into (1),

$$I = e^x \sin x - e^x \cos x - I$$

$$\Rightarrow I = \tfrac{1}{2} e^x (\sin x - \cos x) + c.$$

(ii) $\displaystyle\int \sec^3 x \, dx$

$$= \int (\sec^2 x) \sec x \, dx$$

$$= \sec x \tan x - \int \sec x \tan^2 x \, dx$$

$$= \sec x \tan x - \int \sec x \, (\sec^2 x - 1) \, dx$$

$$\Rightarrow \int \sec^3 x \, dx = \sec x \tan x - \int \sec^3 x \, dx + \int \sec x \, dx$$

$$\Rightarrow \int \sec^3 x \, dx = \tfrac{1}{2} \sec x \tan x + \tfrac{1}{2} \ln (\sec x + \tan x) + c.$$

18.10 Integrating inverse functions

Integration by parts can be useful for integrating functions which are essentially inverse functions: $\ln x$, $\arcsin x$ etc. The trick is to insert a factor of 1.

Example 15

Find $\int \arcsin x \, dx$.

Solution

Write the integral as $\int 1 . \arcsin x \, dx$, and take $u = \arcsin x$ and $\dfrac{dv}{dx} = 1$, so that $\dfrac{du}{dx} = \dfrac{1}{\sqrt{(1-x^2)}}$ and $v = x$. Then

$$\int 1 . \arcsin x \, dx = x \arcsin x - \int x . \frac{1}{\sqrt{(1-x^2)}} dx$$

$$= x \arcsin x + (1-x^2)^{1/2} + c.$$

Qu.23 Find

(i) $\int \ln x \, dx$; (ii) $\int \log_{10} x \, dx$.

(In (ii), use the 'chain rule' for logarithms (Section 4.5) to convert to a ln function.)

Exercise 18d

Integrate numbers 1 to 18 by parts. (Note that in a few of these, 'parts' is not actually the best method.)

1. $x \cos (x+1)$.

2. $x e^{2x}$.

3. $x \sin 3x$.

4. $x \ln x$.

5. $x \sec^2 x$.

6. $(2x+3) \sec(3x-1) \tan (3x-1)$.

7. $x(1+x)^{-1/2}$.

8. $x^3 \sqrt{(1+x^2)}$.

9. $x^2 \cos x$.

10. $(x+3)^2 \exp(-x)$.

11. $e^x \sin 2x$.

12. $x \tan^2(2x+5)$.

13. $\sec x \tan^2 x$.

14. $x^2 \ln x$.

15. $x \, 2^x$.

16. $\arctan x$.

17. $\operatorname{arccot} 2x$.

18. $\ln(2x+1)$.

19. Show that $\dfrac{d}{dx} x \sqrt{(x^2+1)} = \dfrac{2x^2+1}{\sqrt{(x^2+1)}}$. Deduce that

$$\int \frac{x^2 \, dx}{\sqrt{(x^2+1)}} = \frac{1}{2} x \sqrt{(x^2+1)} - \frac{1}{2} \int \frac{dx}{\sqrt{(x^2+1)}}$$

and hence find

$$\int \frac{x^2 \, dx}{\sqrt{(x^2+1)}}.$$

20. (i) Let $I_n = \displaystyle\int \sec^n \theta \, d\theta$. Show that

$$I_n = \sec^{n-2} \theta \tan \theta - \int (n-2) \sec^{n-2} \theta \tan^2 \theta \, d\theta$$

and deduce that, for $n \geqslant 2$,

$$(n-1)I_n = (n-2)I_{n-2} + \sec^{n-2} \theta \tan \theta.$$

(This is an example of a *reduction formula*; see Chapter 40.)
(ii) Find I_2 and use the reduction formula above to find I_4 and I_6.

18.11 t-substitutions and other methods

Qu.24 Find $\displaystyle\int \frac{1}{1+\cos \theta} \, d\theta$ by showing that it is $\displaystyle\int \frac{1}{2}\sec^2 \frac{1}{2}\theta \, d\theta$.

If we write the integrand of $\displaystyle\int \frac{1}{5 \cos \theta + 4} \, d\theta$ in half angles, we get

$$\int \frac{1}{10 \cos^2 \frac{1}{2}\theta - 1} \, d\theta = \int \frac{\sec^2 \frac{1}{2}\theta}{10 - \sec^2 \frac{1}{2}\theta} \, d\theta$$

$$= \int \frac{\sec^2 \frac{1}{2}\theta}{9 - \tan^2 \frac{1}{2}\theta} \, d\theta.$$

This form suggests the substitution $t = \tan \frac{1}{2}\theta$ (so that $dt = \frac{1}{2}\sec^2 \frac{1}{2}\theta \, d\theta$). The integral becomes

$$\int \frac{2}{9-t^2} \, dt = \frac{1}{3} \ln \frac{3+t}{3-t} + c \text{ (using partial fractions)}$$

$$= \frac{1}{3} \ln \frac{3+\tan \frac{1}{2}\theta}{3-\tan \frac{1}{2}\theta} + c.$$

This substitution will work for any integral of the form
$\int \dfrac{1}{a\ \sin\theta + b\ \cos\theta + c}\,d\theta$. In practice, the substitution can be made immediately using the trigonometric identities $\sin\theta = \dfrac{2t}{1+t^2}$, $\cos\theta = \dfrac{1-t^2}{1+t^2}$ and $\tan\theta = \dfrac{2t}{1-t^2}$, where $t = \tan\frac{1}{2}\theta$. These were derived in Section 12.7.

Example 16

Find
$$\int \dfrac{1}{\sin\theta + 2\ \cos\theta + 3}\,d\theta.$$

Solution
Putting $t = \tan\frac{1}{2}\theta$, so that $dt = \frac{1}{2}\sec^2\frac{1}{2}\theta\ d\theta = \frac{1}{2}(1+t^2)\ d\theta$, the integral becomes

$$\int \dfrac{1}{\dfrac{2t}{1+t^2} + \dfrac{2-2t^2}{1+t^2} + 3} \cdot \dfrac{2\ dt}{(1+t^2)}$$

$$= \int \dfrac{2\ dt}{2t + 2 - 2t^2 + 3(1+t^2)}$$

$$= \int \dfrac{2\ dt}{t^2 + 2t + 5}$$

$$= \int \dfrac{2\ dt}{(t+1)^2 + 4}$$

$$= \arctan\left(\dfrac{1+\tan\frac{1}{2}\theta}{\cdot 2 \cdot}\right) + c.$$

Qu.25 Find $\displaystyle\int \dfrac{1}{\sin\theta - 3\ \cos\theta - 1}\,d\theta$.

Qu.26 Find $\displaystyle\int \dfrac{1}{1 - 5\ \sin^2\theta}\,d\theta$ using the substitution $t = \tan\theta$.

Qu.27 Use the $t = \tan\frac{1}{2}\theta$ substitution to show that
$$\int \sec\theta\ d\theta = \ln\left(\dfrac{1+\tan\frac{1}{2}\theta}{1-\tan\frac{1}{2}\theta}\right) + c$$
and verify that this may be written as
$$\int \sec\theta\ d\theta = \ln\left\{\tan\left(\dfrac{1}{2}\theta + \dfrac{\pi}{4}\right)\right\} + c.$$

Our next example uses again the technique of splitting the numerator.

Example 17

Find $\displaystyle\int \frac{4 \sin\theta + 3 \cos\theta}{\sin\theta + 2\cos\theta}\, d\theta$.

Solution

We write the numerator in the form

$$A(\sin\theta + 2\cos\theta) + B(\cos\theta - 2\sin\theta),$$

the second bracket being the derivative of the denominator. Equating coefficients gives $A = 2$ and $B = -1$. So

$$\int \frac{4\sin\theta + 3\cos\theta}{\sin\theta + 2\cos\theta}\, d\theta = \int 2\, d\theta - \int \frac{\cos\theta - 2\sin\theta}{\sin\theta + 2\cos\theta}\, d\theta$$

$$= 2\theta - \ln(\sin\theta + 2\cos\theta) + c.$$

The final technique in this chapter is that of rationalizing the numerator (Qu.28).

Qu.28 Integrate $\displaystyle\sqrt{\frac{x-1}{2-x}}$ by first writing it as $\displaystyle\frac{x-1}{\sqrt{\{(x-1)(2-x)\}}}$.

Exercise 18e Integrate

1. $\displaystyle\sqrt{\left(\frac{x-1}{3-x}\right)}$.

2. $\displaystyle x\sqrt{\left(\frac{x-1}{3-x}\right)}$.

3. $\displaystyle\frac{\sin\theta}{\sin\theta + \cos\theta}$.

4. $\displaystyle\frac{2\cos\theta - \sin\theta}{2\cos\theta + \sin\theta}$.

5. $\displaystyle\frac{1}{\sin\theta + \cos\theta}$.

6. $\displaystyle\frac{1}{1 + \sin\theta}$.

7. $\displaystyle\frac{\sin\theta}{3 + 2\cos\theta}$.

8. $\displaystyle\frac{1 + \sin\theta - 3\cos\theta}{2 - \sin\theta + \cos\theta}$.

9. $\displaystyle\frac{1}{2 + 3\cos^2\theta}$.

10. $\displaystyle\frac{\sin^2\theta}{2 - 5\cos^2\theta}$.

Exercise 18f (This exercise is intended to provide practice at selecting the correct method for an integral; it should not be attempted until each individual technique in the chapter has been thoroughly mastered. If desired, the limits may be ignored and the questions treated as indefinite integrals; both answers are provided. Definite integrals should be set out as in Example 3 of Chapter 17.)

1. $\displaystyle\int_0^{\pi/8} \cos 5x \cos 11x \, dx$.

2. $\displaystyle\int_1^{10} \log_{10} x \, dx$.

3. $\displaystyle\int_0^2 x e^{\frac{1}{2}x^2} \, dx$.

4. $\displaystyle\int_2^4 \frac{5x-9}{(x+1)(3x-4)} \, dx$.

5. $\displaystyle\int_{-2}^2 \sqrt{(4-x^2)} \, dx$.

6. $\displaystyle\int_0^{\pi/6} \frac{dx}{(\sin x - \cos x)^2}$.

7. $\displaystyle\int_0^1 (x+1)\sqrt{x}\,dx.$

8. $\displaystyle\int_0^2 \frac{2x+1}{x+2}\,dx.$

9. $\displaystyle\int_1^2 \frac{dx}{x\sqrt{(x^2-1)}}.$

10. $\displaystyle\int_0^{2/3} \frac{2x-3}{9x^2-12x+8}\,dx.$

11. $\displaystyle\int_{\sqrt{\pi/2}}^{\sqrt{(\pi/2)}} x\cot x^2\,dx.$

12. $\displaystyle\int_1^5 x^2\sqrt{(2x-1)}\,dx.$

13. $\displaystyle\int_{\pi/18}^{\pi/6} \operatorname{cosec}^4 3x\cot 3x\,dx.$

14. $\displaystyle\int_1^{\sqrt{3}} \frac{x\,dx}{\sqrt{(3x^2-1)}}.$

15. $\displaystyle\int_0^1 x^2 e^{-2x}\,dx.$

16. $\displaystyle\int_{-1/2}^{1/2} \frac{dx}{\sqrt{(1-2x^2)}}.$

17. $\displaystyle\int_{\pi/16}^{\pi/8} \operatorname{cosec} 4x\,dx.$

18. $\displaystyle\int_0^1 \frac{x\,dx}{1+x^3}.$

19. $\displaystyle\int_1^2 \operatorname{arcsec} x\,dx.$

20. $\displaystyle\int_3^7 \frac{4x^3}{2x^2-3x-5}\,dx.$

Miscellaneous exercise 18

1. By means of the substitution $x=\sin^2\theta$, or otherwise, evaluate

$$\int_0^{1/2}\left(\frac{x}{1-x}\right)^{1/2}dx,$$

giving your answer in terms of π.

[JMB]

2. (a) Find

(i) $\displaystyle\int \frac{1}{(2x+1)^2}\,dx,$

(ii) $\displaystyle\int \frac{x}{(2x+1)^3}\,dx.$

(b) Show that

$$\int_0^{\frac{1}{4}\pi} (1-\tan x)^2\,dx = 1-\ln 2.$$

[C]

3. (a) (i) Evaluate $\displaystyle\int_0^4 \frac{1}{\sqrt{(3x+4)}}\,dx.$

(ii) Find $\displaystyle\int x\ln x\,dx.$

(b) Using the substitution $u=e^x$, or otherwise,

find $\displaystyle\int \frac{1}{e^x-e^{-x}}\,dx.$

[C]

4. (a) Using the substitution $y = x + 1$, or otherwise, evaluate

$$\int_{-1}^{2} \frac{3}{x^2 + 2x + 10} \, dx.$$

(b) Find $\displaystyle\int 2xe^{2x} \, dx.$

(c) Find $\displaystyle\int \frac{x - 4}{(x-1)^2 (2x+1)} \, dx.$

[W]

5. (a) Evaluate $\displaystyle\int_0^{\pi/2} \frac{\cos^3 \theta}{2 - \cos^2 \theta} \, d\theta.$

(b) Evaluate the integrals

(i) $\displaystyle\int xe^{-ax^2} \, dx,$ (ii) $\displaystyle\int_0^R x^3 e^{-ax^2} \, dx.$

[JMB (S)]

6. (i) Use the substitution $t = \tan(x/2)$ to show that

$$\int_0^{\pi/2} \frac{1}{1 + \sin x + \cos x} \, dx = \ln 2.$$

(ii) Obtain $\displaystyle\int x \sec^2 x \, dx.$

[L]

7. (a) Evaluate $\displaystyle\int_0^{\pi/3} \frac{d\theta}{1 + \sin \theta}.$

(b) Given that $y = \tan^{-1} 3x$, prove that $\dfrac{dy}{dx} = \dfrac{3}{1 + 9x^2}$. Hence, or otherwise, evaluate

$$\int_0^1 x^2 \tan^{-1} 3x \, dx.$$

[AEB '78]

8. (a) Find the indefinite integrals of:
   (i) $\tan x$ (ii) $\sin^2 x \cos^3 x$ (iii) xe^{2x}

   (iv) $\dfrac{1}{1 + \cos 2x}.$

(b) Evaluate $\displaystyle\int_0^{\pi/2} \sin^2 x \, dx.$

(c) By means of a suitable substitution, or otherwise, find the indefinite integral of $\dfrac{1}{x^2 \sqrt{(9 - x^2)}}.$

[SUJB]

9. (i) Determine $\displaystyle\int \frac{dx}{x^{1/2} + x^{1/3}}.$

(ii) Determine $\displaystyle\int e^{-x}\sin x\,dx$.

Prove that $\displaystyle\int_{n\pi}^{(n+1)\pi} e^{-x}|\sin x|\,dx = \frac{1}{2}e^{-n\pi}(1+e^{-\pi})$,

where n is a positive integer, and hence find

$$\int_0^{(n+1)\pi} e^{-x}|\sin x|\,dx.$$

Show that this integral has limit

$$\frac{e^{\pi}+1}{2(e^{\pi}-1)}$$

as $n \to \infty$.

[O & C (S)]

10. Work out the following indefinite integrals:

(i) $\displaystyle\int \frac{x}{1+x^4}\,dx$ 　　　　　　　(ii) $\displaystyle\int x\tan^{-1}x\,dx$

(iii) $\displaystyle\int \frac{\sin 2x}{1+\cos^2 x}\,dx$ 　　　　　(iv) $\displaystyle\int \frac{1}{e^{-x}+1}\,dx$

(v) $\displaystyle\int \frac{1}{1-\cos x}\,dx$.

[SUJB (S)]

19 Simple applications of integration

19.1   Areas   In Chapter 17, we saw that the area under a curve between $x=a$ and $x=b$ may be divided into 'thin' strips of height y and width δx to give

$$A \approx \sum_{x=a}^{b} y\,\delta x,$$

and in the limit

$$A = \int_{a}^{b} y\,dx.$$

Now suppose that we wish to calculate the area bounded, as in Fig. 19.1, by two curves $f(x)$ and $g(x)$ between $x=a$ and $x=b$. The height of a typical strip is now the difference in the y-coordinates of the two curves, i.e. $f(x)-g(x)$; so the area is given by

Figure 19.1

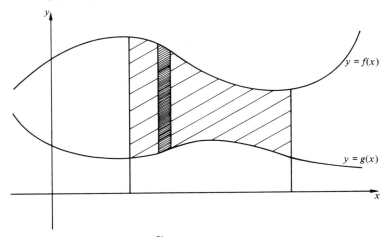

$$A = \int_{a}^{b} \{f(x)-g(x)\}\ dx.$$

This may also be written as

$$\int_{a}^{b} f(x)\,dx - \int_{a}^{b} g(x)\,dx,$$

in which form it may be regarded as the area under $g(x)$ subtracted from the area under $f(x)$.

If, for some part of the interval $a \leqslant x \leqslant b$, $f(x) < g(x)$, then there will be a negative contribution to this area.

Qu.1   Sketch the curves $y=x^2$ and $y=x^2+1$, and shade the area enclosed by them and the lines $x=1$ and $x=3$. Find this area.

Qu.2 Sketch the graphs of $y = \sin 2\theta$ and $y = \sin \theta$ for $0 \leqslant \theta \leqslant \pi$, and find where they cross. Hence evaluate $\int_0^\pi |\sin 2\theta - \sin \theta| \, d\theta$. Shade the region corresponding to this integral.

For some areas, it is not easy to calculate the heights of vertical strips, and it is easier to divide the region into horizontal strips, so that a small element of area is given by $\delta A \approx x \, \delta y$, and

$$A = \lim \sum x \, \delta y = \int x \, dy,$$

where the limits of integration are the appropriate y-coordinates. (Other ways of dividing areas are discussed in Chapter 25.)

Example 1
Find the area enclosed between the curve $x = 4y^2$ and the line $x - 4y = 8$.

Solution
Fig. 19.2 shows that it is simpler to calculate the lengths of horizontal rather than vertical strips, the length of a typical horizontal strip being

Figure 19.2

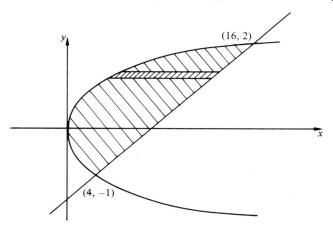

$(8 + 4y) - 4y^2$. Eliminating x between the two equations and solving for y gives the y-coordinates of the points of intersection as $y = -1$ and $y = 2$. So the area is

$$A = \int_{-1}^2 (8 + 4y - 4y^2) \, dy$$

$$= \left[8y + 2y^2 - \frac{4}{3} y^3 \right]_{-1}^2$$

$$= 18.$$

19.2 The mean value and root mean square value of a function

The mean value of a function $f(x)$ between two points $x = a$ and $x = b$ is the 'average value' of the function between these two points. If the graph of f is drawn, then the mean value may be interpreted as the average height of the curve. More formally, we define

$$\text{mean value} = \frac{1}{b-a} \int_a^b f(x)\,dx.$$

Thus the mean value is the area under the curve divided by the 'base', and so it is the height of a rectangle that would have the same area as is under the curve (Fig. 19.3).

Figure 19.3

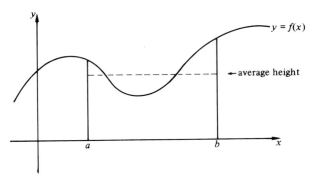

Qu.3 Find the mean value of $y = x^2$ between $x = 1$ and $x = 4$.

Qu.4 Find the mean value of $y = \cos\theta$ between $\theta = 0$ and $\theta = \pi$.

It is important to appreciate that the mean value of a function between two limits depends on the function and the limits (and only on these). If we are asked to find the average velocity of a particle for a certain 'journey', it is understood that we mean the *time-average*, i.e. the mean value of the velocity expressed as a function of time. The *distance-average* would be different, since if v is expressed as a function of distance, the function is different. (This is clarified by question 33 of Exercise 19.)

The *root mean square* (R.M.S.) value of a function f between two limits is, as the name indicates, the square root of the mean value of the square of f, i.e.

$$\text{R.M.S.} = \left[\frac{1}{b-a} \int_a^b \{f(x)\}^2\,dx \right]^{1/2}$$

Clearly, a R.M.S. value will always be non-negative.

Qu.5 Find the root mean square value of $y = \sin x$ between $x = 0$ and $x = 2\pi$.

19.3 Solids of revolution

Imagine that the area under part of the curve $y = f(x)$ is rotated (in three dimensions) about the x-axis. The resulting solid is said to be a *solid of revolution*. Thus a solid of revolution is one which has 'circular symmetry' about an axis (see Fig. 19.4a, b).

SIMPLE APPLICATIONS OF INTEGRATION

Figure 19.4 (a)

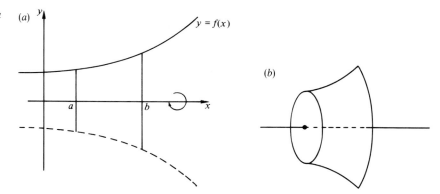

(b)

Suppose we wish to find the volume of such a solid. We cut it, perpendicular to the x-axis, into many thin discs, each of which is approximately a cylinder (Fig. 19.5). A typical disc is shown in Fig. 19.6.

Figure 19.5

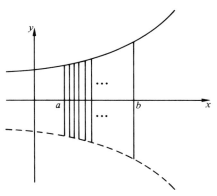

Figure 19.6

If the ordinate (y-value) of the curve $y = f(x)$ at the ith strip is y_i, then this is the radius of the ith cylinder. (Strictly, this ordinate varies over the width of the strip, but this variation becomes negligible in the limit.) Taking the width of each strip to be δx, we have:

$$\text{volume of } i\text{th cylinder} \approx \pi y_i^2 \delta x;$$

so
$$\text{total volume} \approx \Sigma \pi y_i^2 \delta x.$$

In the limit (as $\delta x \to 0$, and the number of strips $\to \infty$) we get:

$$\boxed{\text{volume of solid of revolution} = \int_a^b \pi y^2 \, dx}$$

where a and b are the limits for x on the original curve.

Example 2

A spool is in the shape formed by rotating the area under the curve $y=1+x^2$ between $x=-1$ and $x=1$ about the x-axis. Find its volume.

Solution

Figure 19.7

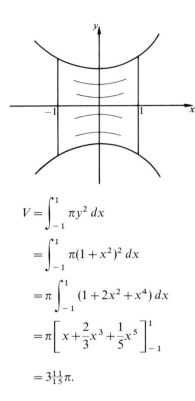

$$V = \int_{-1}^{1} \pi y^2 \, dx$$

$$= \int_{-1}^{1} \pi (1+x^2)^2 \, dx$$

$$= \pi \int_{-1}^{1} (1+2x^2+x^4) \, dx$$

$$= \pi \left[x + \frac{2}{3}x^3 + \frac{1}{5}x^5 \right]_{-1}^{1}$$

$$= 3\tfrac{11}{15}\pi.$$

Qu.6 Sketch the curve $y=\sqrt{(x-1)}$ and find the volume of the bowl obtained by rotating the area enclosed by the curve and the lines $y=0$ and $x=2$ about the x-axis.

We may also rotate areas about the y-axis, the corresponding formula being

$$V = \int_{y_1}^{y_2} \pi x^2 \, dy.$$

Qu.7 The area enclosed by the two positive axes, the line $y=1$ and the curve $y=\ln x$ is rotated about the y-axis. Find the volume swept out.

Qu.8 The area enclosed by the curve $y=\sec x$, the line $y=1$ and the line $x=\dfrac{\pi}{4}$ is rotated about the line $y=1$. Show that the volume of revolution is $\pi \displaystyle\int_0^{\pi/4} (\sec x - 1)^2 \, dx$ and evaluate this integral.

Qu.9 The area enclosed by the curve $y=5x-x^2$ and the line $y=4$ is rotated about the x-axis. Find the volume swept out. [*Hint*: subtract the cylindrical hole at the end of the problem.]

Suppose† that a bowl is formed by rotating the part of the curve $y = f(x)$ between $x = 0$ and $x = a$ about the y-axis. If water is poured into the bowl at a constant rate, then the rate at which the depth increases will not in general be constant. In fact, if the depth is $h(t)$ and the volume $V(t)$, then

$$\frac{dh}{dt} = \frac{dV}{dt} \div \frac{dV}{dh}, \tag{1}$$

where, in the problem we are considering, $\dfrac{dV}{dt}$ is constant.

One method of finding $\dfrac{dV}{dh}$ would be to obtain an explicit expression for V in terms of h by evaluating

$$V = \int_{f(0)}^{f(0) + h} \pi x^2 \, dy,$$

and then to differentiate. A more efficient method is available, however. We note that if the area of the surface of the water is $S(t)$, then, to a first approximation (see Fig. 19.8),

$$\delta V \approx S \, \delta h,$$

so that

$$\frac{\delta V}{\delta h} \approx S.$$

Figure 19.8

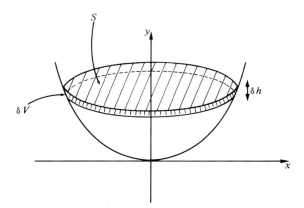

In the limit (as $\delta h \to 0$),

$$\frac{dV}{dh} = S.$$

Then (1) above becomes

$$\frac{dh}{dt} = \frac{dV}{dt} \div S.$$

†The rest of this section may be postponed.

If $\dfrac{dV}{dt}$ is constant, then $\dfrac{dh}{dt}$ varies inversely as S.

This argument applies to any bowl, and not merely to one having circular symmetry.

Qu.10 A bowl is formed by rotating the part of the curve $y = x \sin x$ between $x = 0$ and $x = \dfrac{\pi}{2}$ about the y-axis. It is filled with water at a constant rate of 3 cubic units per second, and it is required to find the rate at which the depth is increasing when the depth is $\dfrac{\pi}{12}$.

(i) State the value of x which gives a depth of $\dfrac{\pi}{12}$.

(ii) Find the surface area of the water when x has this value.

(iii) Hence solve the problem.

19.4 Kinematics   In Chapter 6 we saw that, for a particle moving in one dimension,
$$v = \frac{dx}{dt} \text{ and } a = \frac{dv}{dt}. \text{ It follows that } x = \int v \, dt \text{ and } v = \int a \, dt.$$

If we know the position function, $x(t)$, of a particle, then we can find its velocity function, $v(t)$, by differentiating; but we cannot find $x(t)$ from $v(t)$ alone, because of the arbitrary constant that arises from the integration. An example will provide an interpretation of this constant.

A railway engine of length 8 m moves so that the position (measured in metres) of the front of the engine at time t seconds is given by
$$x_1 = 2t^2 + t + 10.$$

Then the back of the engine, being 8 m behind, is at
$$x_2 = 2t^2 + t + 2.$$

Differentiating either x_1 or x_2 gives $v = 4t + 1$. So the front and the back of the engine have the same velocity at any instant (which is hardly surprising). Any other part of the engine also has the same velocity function. Clearly therefore, specifying $v(t)$ is not sufficient to enable us to find the distance function; further information is needed to distinguish the particle in question from its fellow travellers.

Example 3
A particle P has velocity at time t given by $v = 2t + 3$. Find its position at time t, given that when $t = 3$ it is at the point $x = 20$.
Solution
$$x = \int (2t + 3) \, dt$$
$$= t^2 + 3t + c.$$

When $t=3$, $x=20$, so

$$9+9+c=20$$
$$\Rightarrow c=2.$$
So
$$x=t^2+3t+2.$$

Example 4
A particle starts from rest at the origin, and moves with acceleration $a=3\sin 3t$. Find where the particle is at time t.
Solution

$$a=3\sin 3t$$

$$\Rightarrow v= \int 3\sin 3t\, dt$$

$$= -\cos 3t+c.$$

But when $t=0$, $v=0$ (since the particle starts from rest). Substituting gives $c=1$. So

$$v=-\cos 3t+1.$$

Then
$$x= \int (-\cos 3t+1)\, dt$$

$$= -\frac{1}{3}\sin 3t+t+k.$$

But when $t=0$, $x=0$ (since the particle starts at the origin), and substituting gives $k=0$. So

$$x=-\frac{1}{3}\sin 3t+t.$$

In the next example, the arbitrary constant cannot be found; but it is not needed, since every particle with the given velocity will travel the same distance. Put another way, the constant is not needed since the integral is 'definite'.

Example 5
A particle moves with velocity $v=t^2$. How far does it travel between the times $t=4$ and $t=10$?
Solution

$$\text{Distance travelled}= \int_4^{10} t^2\, dt$$

$$= \left[\frac{1}{3}t^3 \right]_4^{10}$$

$$= \frac{1}{3}(1000-64)$$

$$= 312 \text{ m}.$$

Qu.11 Find the distance travelled by this particle in the 3rd second of its motion.

(Where appropriate, units are metres and seconds.)

1. Find the area enclosed by the curve $y = 8x - x^2$ and the lines $x = 2$, $x = 5$ and $y = 2x + 1$.

2. Sketch the curves $y = x^2 - 3x + 2$ and $y = 2x - x^2$. Find the area enclosed between them.

3. Find the areas enclosed between
   (i) $y = x^2$ and $x = y^2$;   (ii) $y = x^{2n}$ and $x = y^{2n}$ (for $n \in Z^+$).
   What is the limit of the answer to (ii) as $n \to \infty$? Illustrate this by sketching the curves for large n.

4. Sketch on the same axes the curves $y = \sin x$ and $y = \cos x$. Find the area enclosed by the curves between two consecutive intersections of the curves.

5. (i) (Hard.) Use integration to find the area enclosed between two circles of radius a if the centre of each lies on the circumference of the other. [*Hint*: the top half of the circle $x^2 + y^2 = a^2$ is given by $y = \sqrt{(a^2 - x^2)}$ and the lower half by $y = -\sqrt{(a^2 - x^2)}$.]
   (ii) Solve (i) by a non-calculus method.

6. Find the mean value of
   (i) $y = x^2$ between $x = 0$ and $x = 1$;
   (ii) $y = x^3$ between $x = 3$ and $x = 5$;
   (iii) $y = \ln(1 + x)$ between $x = 0$ and $x = 10$;
   (iv) $y = \sin x$ between $x = 0$ and $x = \pi$.

7. Find the R.M.S. values of the following functions over the given ranges:
   (i) $y = x$; between $x = 2$ and $x = 4$;
   (ii) $y = 2 \sin x$; a complete period;
   (iii) $y = \sin 2x$; a complete period;
   (iv) $y = 3 \sec x$; $\frac{\pi}{6} \leqslant x \leqslant \frac{\pi}{3}$.

8. A function f is defined by the following properties:
   $$f(x) = x \text{ for } 0 \leqslant x \leqslant 1$$
   $$f(x) = 1 \text{ for } 1 < x < 2$$
   f is periodic with period 2.
   (i) Find the mean value of $f(x)$ for a complete period.
   (ii) Find the R.M.S. value of $f(x)$ for a complete period.

   In questions 9 to 15, find the volume swept out when the area bounded by the given curves is rotated about the given line.

9. $y = 4 - x^2$, $x = 1$, $x = -1$ and the x-axis; about the x-axis.

10. $y = \sec x$, $x = 0$, $y = 0$, $x = \dfrac{\pi}{4}$; about the x-axis.

11. $yx^2 = 4$, $x = 2$, $x = 1$, $y = 0$; about the x-axis.

12. $yx^2 = 4$, $y = 1$, $y = 4$, $x = 0$; about the y-axis.

13. $xy = 12$, $x = 1$, $x = 2$, $y = 1$; about the x-axis.

14. The area in question 13; about the line $y = 1$.

15. $y = x^2$, $y = x$; about the x-axis.

16. Find by integration the volume generated when the area enclosed by $y = mx$, the x-axis and $x = h$ is rotated about the x-axis. Show that the result agrees with the formula for the volume of a cone.

17. The part of the curve $y = x^3$ between $(0, 0)$ and (k, k^3) is rotated about each of the axes. If the two 'bowls' formed have equal volumes, find k.

SIMPLE APPLICATIONS OF INTEGRATION

18. Show that the area under the curve $y = \dfrac{1}{x}$ in the region $x > 1$ is infinite, but that if this area is rotated about the x-axis, the volume generated is finite. (Integrals in which one of the limits is infinite are discussed in Chapter 23.) Discuss the paradox that a 'thin' sheet of metal in the shape of the (infinite) area can be painted by inserting it into the volume of revolution filled with a finite amount of paint.

19. Obtain the formula for the volume of a sphere.

20. Let C be a closed curve lying entirely above the x-axis, such that no ordinate (vertical line) cuts C more than twice. Let A and B be as shown in Fig. 19.9, and let the upper and lower parts of the curve between A and B be given by $y_1 = f(x)$ and $y_2 = g(x)$ respectively. If C is rotated about the x-axis, show that the volume enclosed is given by

Figure 19.9

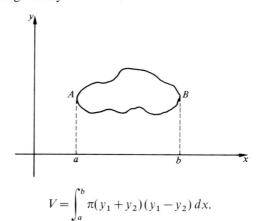

$$V = \int_a^b \pi(y_1 + y_2)(y_1 - y_2)\, dx.$$

Show further that if C is a circle of radius r with centre at a height c $(>r)$, then

$$V = 2\pi c \int_a^b (y_1 - y_2)\, dx.$$

Interpret this integral geometrically, and hence *write down* the volume of the torus obtained.

21. A cylindrical hole is cut symmetrically through a sphere of radius R. If the length of the hole is 2 m, show that the volume remaining is independent of R.

22. Find the positions at time t of particles moving in one dimension and satisfying the following data:

(i) $v = 2$ (for all t); $x = 0$ when $t = 0$.

(ii) $v = 2t + 1$; $x = 3$ when $t = 0$.

(iii) $v = 6t^2 - 6t + 1$; $x = 10$ when $t = 3$.

(iv) $a = 12t - 4$; $x = v = 0$ when $t = 0$.

(v) $a = 12t - 4$; $x = 1$ and $v = 2$ when $t = 3$.

(vi) $a = 8 \sin 2t$; the particle is initially at the origin, and its motion is bounded (i.e. as $t \to \infty$, x remains finite).

(vii) $a = e^{-t}$; the particle is initially at the origin, and its 'terminal velocity' is 10 m/s.

23. Find the distance travelled by these particles in the given time interval:
    (i) $v = 2t$; $t = 1$ to $t = 3$.
    (ii) $v = 1 - t^3$; the first four seconds.
    (iii) $v = \ln 2t$; $t = 2$ to $t = 4$.

24. Show, by integration, that if a particle starts with velocity u and moves with constant acceleration a, then at time t it has travelled a distance $x = ut + \frac{1}{2}at^2$. Find the corresponding expression for x if the particle has a constant rate of change of acceleration b and initial velocity and acceleration u and a.

25. When a particle falls freely under gravity, its acceleration is approximately $9 \cdot 8 \, \text{m/s}^2$ (downwards). A stone is launched from the top of a tower 21 m high. Find its height t seconds later if
    (i) it is dropped from rest;
    (ii) it is thrown upwards with speed $7 \, \text{m/s}$;
    (iii) it is thrown downwards so that it hits the ground 1 second later.

26. A particle P starts from the origin with velocity $3 \, \text{m/s}$ and moves with constant acceleration $4 \, \text{m/s}^2$. A second particle Q starts 2 seconds later from the point $x = 49$ m with velocity $9 \, \text{m/s}$ and moves with constant acceleration $2 \, \text{m/s}^2$. When and where do they subsequently meet?

27. Show that the 'braking distance' of a car moving with constant deceleration is proportional to the square of its speed.

28. A particle has velocity $v = t(t - 1)^2$. Find the distance it travels in
    (i) the first two seconds of its motion;
    (ii) the 3rd second of its motion.

29. A particle starts from rest and moves with an acceleration $a = 6t - 18$. How far does it go before instantaneously coming to rest?

30. A particle moves with acceleration $a = 20 \cos 2t$. It starts from rest at the point $x = -5$. Find its greatest speed and greatest distance from the origin in the subsequent motion. Where does the greatest speed occur?

31. A particle moves with velocity $v = \dfrac{1}{t}$. How far does it go

    (i) in the 2nd second of its motion;
    (ii) between $t = 10$ and $t = 20$;
    (iii) in the second year of its motion.

32. A train travels non-stop between stations A and B (where it does stop), its velocity t seconds after leaving A being $(t - \frac{1}{100}t^2) \, \text{m/s}$. Find
    (i) the distance AB;
    (ii) the distance gone after 25 seconds;
    (iii) the average velocity for the journey.
    Sketch v against t, and shade the region corresponding to your answer to (ii).

33. A particle starts at the origin, and moves with velocity $v = 3t^2$ for 2 seconds.
    (i) Find the average velocity over this interval.
    (ii) Show that $v = 3x^{2/3}$. State the limits for x for this same two-

second interval, and calculate the distance-average of the velocity.

(iii) Find the velocity at intervals of 0·25 s throughout the two-second interval, starting at $t=0$. Find the average of these nine readings to provide a rough estimate of the time-average as found in (i).

(iv) Find the velocity at intervals of 1 m starting at $x=0$. Find the average of these nine readings to give a rough estimate of the distance-average velocity (as found in (ii)).

(v) Explain why the nine values in (iii) are not the same as the nine values in (iv) in view of the fact that each set of values was obtained by dividing the journey into eight equal parts. (This also provides an explanation of the difference between a time-average and a distance-average.)

34. A particle moves according to the law $v=kx^n$ where k and n are constants. The values of x and v at times t_1 and t_2 are x_1, x_2, v_1, v_2 respectively. Show that

$$t_2-t_1 \propto \frac{x_2}{v_2}-\frac{x_1}{v_1}.$$

[*Hint*: use the given equation to express $\frac{dt}{dx}$ in terms of x. Then take a definite integral of each side w.r.t. x.]

35. Obtain the formula for the volume swept out when the area under $y=f(x)$ for $x_1 \leqslant x \leqslant x_2$ is rotated about the x-axis through an angle of $\theta(<2\pi)$.

36. (i) The bowl formed by rotating about the y-axis the part of the curve $y=x^2 2^x$ between $x=0$ and $x=5$ is filled with water at a rate which varies in such a way that the depth increases at a constant rate of 3 units per second. Find the rate at which the water is being poured in at the instant at which the bowl begins to overflow.

(ii) If, instead, the bowl is filled in such a way that the depth satisfies the differential equation

$$\frac{dh}{dt}=\frac{1}{1+\sqrt{h}}$$

find the rate at which the volume is increasing when the radius of the surface is 2 units.

Miscellaneous exercise 19

1. A thin metal container has the shape obtained by rotating the part of the curve $y=x^2$, between $(0,0)$ and $(4,16)$, through 2π radians about the y-axis. Calculate the volume of the container, leaving your answer in terms of π.

If the container, held with its axis of symmetry vertical, is filled by adding liquid at the rate of $2\pi \text{ cm}^3 \text{ s}^{-1}$, how long does it take to fill to a depth of 8 cm? Find, also, the approximate increase in the depth of liquid in the next half-second.

[O & C (O)]

2. A particle, moving in a straight line, passes a fixed point O with a velocity of 21 cm s^{-1} and moves so that t second after passing O

its velocity is $(21 + 18t - 3t^2)\,\text{cm s}^{-1}$. Find its distance from O

(i) when its acceleration becomes zero;

(ii) when it reaches a position of instantaneous rest.

[O & C (O)]

3. Find the points of intersection of the curves $y = x^2 + 2$ and $2y = 16 - x^2$. Find the area cut off between the two curves.

[SUJB (O)]

4. A particle starts from rest at a point O. Its speed, v metres per second, t seconds later is given by the formula
$$v = 10\sqrt{t}.$$
Calculate

 (i) the distance travelled by the particle in 9 seconds;

 (ii) the average velocity in the ninth second;

 (iii) the acceleration after nine seconds;

 (iv) the average acceleration in the ninth second.

[Ox (O)]

5. A particle leaves a fixed point O, at time $t = 0$, with a velocity of 15 m/s and travels in a straight line. The acceleration, in m/s^2, of the particle at time t, where time is measured in seconds, is $(4 - 6t)$ in the direction of the initial motion.

(i) Calculate the value of t when the particle comes to instantaneous rest.

(ii) Calculate the time taken for the particle to return to O.

(iii) Calculate the total distance travelled by the particle before it returns to O.

(iv) Sketch a graph showing the relation between the displacement of the particle from O and time for $0 \leqslant t \leqslant 5$.

(v) Sketch a graph showing the relation between the distance the particle has travelled and time for $0 \leqslant t \leqslant 5$.

[JMB (O)]

6. Using the same axes, sketch the graphs of $y = x^2$ and $y = 3 - 2x^2$. Calculate

(i) the area of the region R contained between the curves,

(ii) the volume of the solid obtained by a rotation of R through π radians about the y-axis,

(iii) the volume of the solid obtained by a rotation of R through 2π radians about the x-axis.

(Leave the answers to (ii) and (iii) as multiples of π.)

[AEB (O) '80]

7. Sketch the graphs of $y^2 = 16x$ and $y = x - 5$. Find

 (i) the coordinates of their points of intersection,

 (ii) the area of the finite region enclosed between the graphs.

[AEB '80]

8. If $y = (x - 0 \cdot 5)e^{2x}$, find $\dfrac{dy}{dx}$ and hence or otherwise calculate.

correct to one decimal place, the mean value of xe^{2x} in the interval $0 \leqslant x \leqslant 2 \cdot 5$.

[AEB '80]

9. Find the coordinates of any turning points and points of inflexion on the curve $y = xe^{-x}$. Sketch the curve.

Find the volume of the solid of revolution formed when the

region enclosed by the curve, the x-axis and the line $x=2$ is rotated completely about the x-axis.

[AEB '78]

10. Sketch the curve $y=\log_e(x-2)$.

   The inner surface of a bowl is of the shape formed by rotating completely about the y-axis that part of the x-axis between $x=0$ and $x=3$ and that part of the curve $y=\log_e(x-2)$ between $y=0$ and $y=2$. The bowl is placed with its axis vertical and water is poured in. Calculate the volume of water in the bowl when the bowl is filled to a depth $h\ (<2)$.

   If water is poured into the bowl at a rate of 50 cubic units per second, find the rate at which the water level is rising when the depth of the water is 1·5 units.

[AEB '79]

11. Sketch the graphs of $y=(2x)^{1/2}$ and $y=\frac{1}{2}x$ in the interval $0\leqslant x\leqslant 8$. The area between the graphs in this interval is rotated about the x-axis. Calculate the volume of the solid obtained in the form $k\pi$ where k is rational.

   Find the value t, $0\leqslant t\leqslant 8$, such that, when the area between the same graphs between $x=0$ and $x=t$ is rotated about the y-axis, the volume of the solid formed is $\dfrac{2\pi t^3}{3}$.

[Ox (O)]

20 Vectors

20.1 Definitions and notation

The definition of a vector that follows is adequate for our present purposes, although a more abstract definition will be given in Chapter 34.

A *vector* is an ordered set of real numbers obeying certain algebraic rules which are discussed in the next section (Section 20.2). (It is possible to consider vectors whose entries are not real numbers, but we shall be concerned only with 'real vectors'.)

The number of entries in a vector is called its *dimension*; so, for example, $\begin{pmatrix} 2 \\ 18 \\ -3 \end{pmatrix}$ is a three-dimensional *column vector*, and

$(\sqrt{2} \quad 0)$ is a two-dimensional *row vector*.

Two column vectors are said to be *equal* if and only if they have the same dimension and the same entries in the same order; i.e.

$$\begin{pmatrix} a_1 \\ a_2 \\ \cdot \\ \cdot \\ \cdot \\ a_n \end{pmatrix} = \begin{pmatrix} b_1 \\ b_2 \\ \cdot \\ \cdot \\ \cdot \\ b_m \end{pmatrix} \Leftrightarrow m=n \text{ and } a_i=b_i \text{ for each } i.$$

So, for example, $\begin{pmatrix} 3 \\ 7 \\ 0 \end{pmatrix} \neq \begin{pmatrix} 3 \\ 7 \end{pmatrix}$. Equality of row vectors is defined similarly.

A vector may be denoted by a single letter, thus: $\mathbf{a} = \begin{pmatrix} 1 \\ 2 \end{pmatrix}$. (On the printed page, bold type is used for letters representing vectors; when written, the letter is underlined. The reader should make a habit of always underlining such letters.) The notation $\mathbf{0}$ is used for any of the (row or column) zero vectors (0), $\begin{pmatrix} 0 \\ 0 \end{pmatrix}$, $\begin{pmatrix} 0 \\ 0 \\ 0 \end{pmatrix}$ etc., the dimension being clear from the context.

426 VECTORS

For the rest of this chapter, all vectors should be assumed to be column vectors.

20.2 The algebra of vectors

Part of the definition of a vector is that it must behave in certain ways; i.e. vectors satisfy certain algebraic rules. These are:

(i) Two vectors of the same dimension may be added to give another vector. If

$$\mathbf{a} = \begin{pmatrix} a_1 \\ a_2 \\ \cdot \\ \cdot \\ \cdot \\ a_n \end{pmatrix} \quad \text{and} \quad \mathbf{b} = \begin{pmatrix} b_1 \\ b_2 \\ \cdot \\ \cdot \\ \cdot \\ b_n \end{pmatrix},$$

then

$$\mathbf{a} + \mathbf{b} = \begin{pmatrix} a_1 + b_1 \\ a_2 + b_2 \\ \cdot \\ \cdot \\ \cdot \\ a_n + b_n \end{pmatrix}.$$

(ii) A vector may be multiplied by a number. With $\mathbf{a}$ as above,

$$\lambda\mathbf{a} = \begin{pmatrix} \lambda a_1 \\ \lambda a_2 \\ \cdot \\ \cdot \\ \cdot \\ \lambda a_n \end{pmatrix}.$$

Note

(1) To distinguish vectors from ordinary numbers, the latter are sometimes called *scalars*. Rule (ii) above is known as multiplication by a scalar, λ being the scalar.

(2) From the two rules above, others can be deduced; for example, $\mathbf{a} + \mathbf{a} = 2\mathbf{a}$ and $\lambda(\mathbf{a} + \mathbf{b}) = \lambda\mathbf{a} + \lambda\mathbf{b}$ (i.e. multiplication by a scalar is distributive over vector addition).

(3) Subtraction of vectors is similar to addition. To be strict, we may define $\mathbf{a} - \mathbf{b}$ to mean $\mathbf{a} + (-1\mathbf{b})$.

Qu.1 If $\mathbf{p} = \begin{pmatrix} 1 \\ 4 \end{pmatrix}$ and $\mathbf{q} = \begin{pmatrix} -2 \\ 1 \end{pmatrix}$, evaluate

(i) $\mathbf{p} + \mathbf{q}$;   (ii) $\mathbf{p} + 2\mathbf{q}$;   (iii) $\mathbf{p} - \mathbf{q}$;   (iv) $3\mathbf{p} - 2\mathbf{q}$.

Qu.2 Solve for m and n

$$2\begin{pmatrix} m \\ -4 \end{pmatrix} + 3\begin{pmatrix} -3 \\ n \end{pmatrix} = \begin{pmatrix} 1 \\ 9 \end{pmatrix}.$$

Qu.3 Show that the vector equation

$$\lambda\begin{pmatrix}3\\2\end{pmatrix}+\mu\begin{pmatrix}-1\\4\end{pmatrix}=\begin{pmatrix}7\\-7\end{pmatrix}$$

may be written as a pair of ordinary simultaneous equations in λ and μ.
Hence find λ and μ.

Qu.4 (For discussion.) Why does the method of Qu.3 break down
if the vector $\begin{pmatrix}-1\\4\end{pmatrix}$ is replaced by $\begin{pmatrix}6\\4\end{pmatrix}$?

Qu.5 For fixed n, determine whether or not the set of all vectors of
dimension n forms an abelian group under vector addition.

20.3 The representation of two-dimensional vectors

This section is no more than a brief summary of what the reader may
already know from elementary work on vectors.

A vector may be represented on a graph by a *directed line-segment*.
In Fig. 20.1, the line-segment PQ is given a 'sense' by the arrow. P is

Figure 20.1

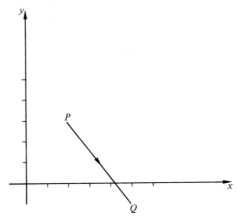

$(2, 3)$ and Q is $(5, -1)$, and so the directed line-segment represents
the vector $\begin{pmatrix}3\\-4\end{pmatrix}$. We write $\mathbf{PQ}=\begin{pmatrix}3\\-4\end{pmatrix}$. The same vector may be
represented by infinitely many different line-segments (Fig. 20.2).

Qu.6 If $\mathbf{RS}=\begin{pmatrix}3\\1\end{pmatrix}$ and R is the point $(4, -2)$, where is S? What is
the vector $\mathbf{SR}$?

Qu.7 If $\mathbf{WZ}=\begin{pmatrix}-4\\3\end{pmatrix}$ and Z is the point $(1, 1)$, where is W?

Qu.8 Let $\mathbf{a}=\begin{pmatrix}-2\\5\end{pmatrix}$. Draw (starting anywhere) directed line-segments
to represent
(i) $\mathbf{a}$; (ii) $2\mathbf{a}$; (iii) $3\mathbf{a}$; (iv) $-\mathbf{a}$; (v) $-\tfrac{1}{2}\mathbf{a}$.

VECTORS

Figure 20.2

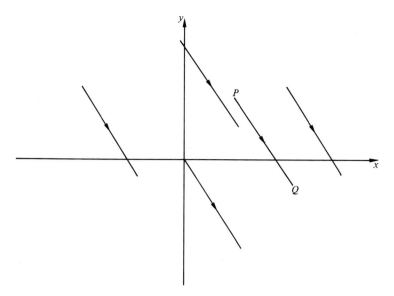

Qu.9 If P is (2, 3), Q is (4, 10) and M is the mid-point of PQ, find the vectors **PQ**, **PM** and **MQ**.

Qu.10 Let $\mathbf{PQ} = \begin{pmatrix} 3 \\ 1 \end{pmatrix}$ and $\mathbf{QR} = \begin{pmatrix} -7 \\ 5 \end{pmatrix}$. Find **PR**, verifying that it does not depend on the actual position of the three points.

The addition of two vectors **a** and **b** may be represented by drawing the corresponding line-segments so that the second starts where the first ends, i.e. so that the arrows 'follow on' (Fig. 20.3).

Figure 20.3

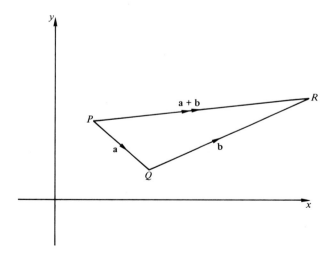

Note

(1) The sum of two or more vectors is called the *resultant* vector and is often illustrated by a double arrow.

(2) In Fig. 20.3, **PQ** + **QR** = **PR**. Note the 'dominoes' pattern.

(3) The addition of three or more vectors may be illustrated similarly (Fig. 20.4).

VECTORS   429

Figure 20.4

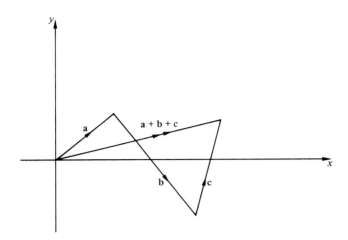

(4) The commutativity of vector addition can be seen geometrically. If we show the additions **a**+**b** and **b**+**a** on the same diagram, and starting from the same point (say the origin), we find that we end up at the same point (Fig. 20.5). The diagram shows why vector addition is sometimes called the parallelogram rule for addition.

Figure 20.5

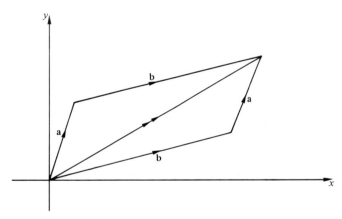

(5) If we wish to illustrate the subtraction **a**−**b**, we must add on to the end of **a** a line-segment representing −**b** (i.e. the line-segment of **b** with the arrow reversed). Care must be taken to ensure that the single arrows do follow on (Figs. 20.6 and 20.7).

Figure 20.6

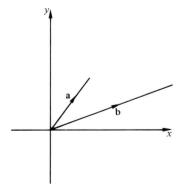

VECTORS

Figure 20.7

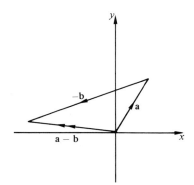

Qu.11 (i) Let $\mathbf{u} = \begin{pmatrix} -4 \\ 3 \end{pmatrix}$ and $\mathbf{v} = \begin{pmatrix} 1 \\ 5 \end{pmatrix}$. Draw on the same diagram line-segments representing $\mathbf{u}$ and $\mathbf{v}$, with each starting at the origin.
(ii) Using a separate diagram for each part, illustrate the following additions, where $\mathbf{u}$ and $\mathbf{v}$ are as above. (Start at the origin in each case.) State the resultant vectors.
(a) $\mathbf{u} + \mathbf{v}$;   (b) $2\mathbf{u} + \mathbf{v}$;   (c) $\mathbf{u} + 2\mathbf{v}$;
(d) $\mathbf{u} - \mathbf{v}$;   (e) $\mathbf{v} - \mathbf{u}$;   (f) $3\mathbf{u} - 2\mathbf{v}$.

20.4 Three dimensions and more

The work of the previous section applies equally to any number of dimensions. We can represent three-dimensional vectors by line-segments in three-dimensional space, taking x-, y- and z-axes in the usual way (see Section 1.6).

In four or more dimensions, we are faced with conceptual problems arising from the fact that we live in mere three-dimensional space. It is still possible, however, to give meanings to geometrical words such as 'distance', 'perpendicular', 'angle' etc. by defining these ideas *algebraically*. We shall meet such definitions in due course.

For the rest of this chapter, the reader should try to interpret results in (at least) three dimensions rather than in just two dimensions.

20.5 The modulus of a vector

Qu.12 A cuboid (rectangular block) has edges of length a, b and c. Show that the length d of a diagonal is given by $d^2 = a^2 + b^2 + c^2$. [*Hint*: find first the diagonal of one face, and then use Pythagoras again.]

Qu.13 Use the result of Qu.12. to find the distance between the points $(3, 0, 4)$ and $(-1, 2, 1)$.

Definition

The modulus of the vector $\mathbf{a} = \begin{pmatrix} a_1 \\ a_2 \\ . \\ . \\ . \\ a_n \end{pmatrix}$ is defined to be

$\sqrt{(a_1^2 + a_2^2 + \ldots + a_n^2)}$. It is denoted by $|\mathbf{a}|$.

Qu.14 Find the modulus of the vectors

(i) $\begin{pmatrix} 3 \\ -1 \\ 1 \\ -2 \end{pmatrix}$; (ii) $\begin{pmatrix} 6 \\ 2 \\ -3 \end{pmatrix}$; (iii) $\begin{pmatrix} 3 \\ 4 \end{pmatrix}$;

(iv) (7); (v) (-7); (vi) $\begin{pmatrix} 2 \\ 0 \\ 0 \\ 0 \end{pmatrix}$; (vii) $\begin{pmatrix} 1 \\ 1 \\ 1 \end{pmatrix}$.

Qu.15 If $|\mathbf{a}| = 2$, find $|3\mathbf{a}|$ and $|-4\mathbf{a}|$.

Note
(1) For two- or three-dimensional vectors, the modulus corresponds to the length of a line-segment.
(2) For four or more dimensions, the definition gives a meaning to the word 'length'.
(3) The modulus of a one-dimensional vector corresponds to the modulus of a number (as defined in Section 2.7).
(4) As illustrated by Qu.15, $|\lambda \mathbf{a}| = |\lambda| \, |\mathbf{a}|$.
(5) If $\mathbf{a} = \mathbf{0}$, then $|\mathbf{a}| = 0$; if $\mathbf{a} \neq \mathbf{0}$, then $|\mathbf{a}| > 0$.

Qu.16 What is the distance between the points $P(2, 0, -6, 1)$ and $Q(4, 1, 2, -2)$?

Definition
A *unit vector* is a vector of modulus 1.

Note
(1) A useful convention is that a circumflex accent (or 'hat') may be used to indicate a unit vector, thus: $\hat{\mathbf{a}}$.
(2) The letters $\mathbf{i}$, $\mathbf{j}$ and $\mathbf{k}$ are used to denote unit vectors in the directions of the x-, y- and z-axes respectively. So $\mathbf{i} = \begin{pmatrix} 1 \\ 0 \\ 0 \end{pmatrix}$,

$\mathbf{j} = \begin{pmatrix} 0 \\ 1 \\ 0 \end{pmatrix}$ and $\mathbf{k} = \begin{pmatrix} 0 \\ 0 \\ 1 \end{pmatrix}$. Any vector $\begin{pmatrix} a_1 \\ a_2 \\ a_3 \end{pmatrix}$ can then be written in the

form $a_1 \mathbf{i} + a_2 \mathbf{j} + a_3 \mathbf{k}$. In two dimensions, we use $\mathbf{i}$ and $\mathbf{j}$ for $\begin{pmatrix} 1 \\ 0 \end{pmatrix}$ and

$\begin{pmatrix} 0 \\ 1 \end{pmatrix}$ respectively.

(3) If $\mathbf{u}$ is any vector, then $\dfrac{\mathbf{u}}{|\mathbf{u}|}$ is a unit vector in the direction of $\mathbf{u}$.

(4) In kinematics, it is sometimes convenient to write a vector as the product of its modulus and a unit vector in the same direction. If the modulus of $\mathbf{r}$ is denoted by r, and the unit vector by $\hat{\mathbf{r}}$, then $\mathbf{r} = r\hat{\mathbf{r}}$.

Qu.17 Find the modulus of $18\mathbf{i} + 12\mathbf{j} - 9\mathbf{k}$. Hence write down a unit vector in the same direction.
Qu.18 Find a unit vector in the same direction as $\mathbf{i} + \mathbf{j} + \mathbf{k}$.

20.6 Linear combinations of vectors

Given n vectors $\mathbf{v}_1, \mathbf{v}_2, \ldots, \mathbf{v}_n$, a vector of the form

$$\lambda_1 \mathbf{v}_1 + \lambda_2 \mathbf{v}_2 + \ldots + \lambda_n \mathbf{v}_n$$

where the λ_i are scalars, is called a *linear combination of* $\mathbf{v}_1, \mathbf{v}_2, \ldots, \mathbf{v}_n$.

Suppose that $\mathbf{u}$ and $\mathbf{v}$ are two non-zero two-dimensional vectors. We pose the following question: which vectors can be expressed as linear combinations of $\mathbf{u}$ and $\mathbf{v}$, i.e. which vectors can be written in the form $\lambda\mathbf{u} + \mu\mathbf{v}$? This question is approached algebraically in Exercise 20a, question 16, but here we give an intuitive geometrical argument. (The reader should think about the problem in terms of line-segments.) There are two cases:

(i) If $\mathbf{u}$ and $\mathbf{v}$ are parallel (with the same or opposite senses), then by adding multiples of $\mathbf{u}$ and $\mathbf{v}$ we can move any distance, but only in one direction, and so we can reach only the points on a line. Any linear combination of $\mathbf{u}$ and $\mathbf{v}$ will be parallel to $\mathbf{u}$ and $\mathbf{v}$. (See Qu.4.)

(ii) If $\mathbf{u}$ and $\mathbf{v}$ are non-parallel, then we may 'slide about' in two different directions, and so reach any point in the plane. Thus *any* two-dimensional vector can be expressed in the form $\lambda\mathbf{u} + \mu\mathbf{v}$.

Corresponding results for higher dimensions will emerge in later chapters.

Example 1
$ABCDEF$ is a regular hexagon. Let $\mathbf{AB} = \mathbf{p}$ and $\mathbf{CD} = \mathbf{q}$. Express
(i) $\mathbf{BC}$, (ii) $\mathbf{AE}$,
in terms of $\mathbf{p}$ and $\mathbf{q}$.

Solution
We make use of the geometrical properties of the hexagon.
(i) In Fig. 20.8,

Figure 20.8

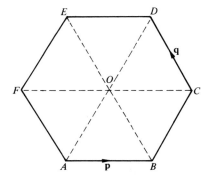

$$AO = AB + BO.$$

But $$BO = q,$$

so $$AO = p + q.$$

So $$BC = p + q \text{ also.}$$

(ii) $$AE = AB + BE$$
$$= p + 2q.$$

Qu.19 Express
(i) **FD**,   (ii) **FB**,
in terms of **p** and **q**.

Exercise 20a

1. Let $\mathbf{u} = 2\mathbf{i} + \mathbf{j}$, $\mathbf{v} = -3\mathbf{i} + 2\mathbf{j}$ and $\mathbf{w} = \mathbf{i} + 4\mathbf{j}$. Simplify
   (i) $\mathbf{u} + \mathbf{v}$;   (ii) $\mathbf{u} - \mathbf{v}$;   (iii) $4\mathbf{u} + \mathbf{v}$;   (iv) $3\mathbf{u} - 5\mathbf{v}$.
   (v) Solve $s\mathbf{u} + t\mathbf{v} = \mathbf{w}$ for s and t.

2. (i) Illustrate the addition $(-6\mathbf{i} + 10\mathbf{j}) + (3\mathbf{i} - 4\mathbf{j})$ and the subtraction $(-6\mathbf{i} + 10\mathbf{j}) - 3(3\mathbf{i} - 4\mathbf{j})$.
   (ii) Use a graphical method to find p where
   $(-6\mathbf{i} + 10\mathbf{j}) + \lambda(3\mathbf{i} - 4\mathbf{j}) = (8\mathbf{i} + p\mathbf{j})$. [*Hint*: illustrate the addition on the L.H.S. and also draw the line $x = 8$.]
   (iii) Find p in part (ii) algebraically.

3. $ABCD$ is a square. If $\mathbf{AB} = \mathbf{p}$ and $\mathbf{AD} = \mathbf{q}$, express in terms of **p** and **q**:
   (i) **BC**;   (ii) **DC**;   (iii) **AC**;   (iv) **DB**.

4. (i) ABC is a triangle, and L, M and N are the mid-points of BC, AC and AB respectively. Let $\mathbf{AB} = \mathbf{p}$ and $\mathbf{AC} = \mathbf{q}$.
   Express in terms of **p** and **q**:
   (a) **BC**;   (b) **CN**;   (c) **AL**;   (d) **NM**.
   (ii) Show how two of your answers in part (i) provide a proof of the mid-point theorem: the line joining the mid-points of two sides of a triangle is parallel to, and half the length of, the third side.

5. A cuboid has a base $ABCD$ with vertices P, Q, R and S above A, B, C and D respectively. If $\mathbf{AB} = 4\mathbf{i}$, $\mathbf{AD} = 2\mathbf{j}$ and $\mathbf{AP} = \mathbf{k}$, find the vectors
   (i) **AS**;   (ii) **AQ**;   (iii) **AR**;
   (iv) **SP**;   (v) **SB**;   (vi) **CP**;
   (vii) **QM** where M is the mid-point of DC.

6. Find the modulus of
   (i) $2\mathbf{i} + 3\mathbf{j}$;   (ii) $-2\mathbf{i}$;   (iii) $\mathbf{i} + 2\mathbf{j} - 3\mathbf{k}$;
   (iv) $\begin{pmatrix} 1 \\ -1 \\ 3 \\ 4 \end{pmatrix}$.

7. Find unit vectors in the same direction as
   (i) $-2\mathbf{i}$;   (ii) $3\mathbf{i} - 4\mathbf{j}$;   (iii) $\mathbf{j} - \mathbf{k}$.

8. (i) Let $\mathbf{u} = -2\mathbf{i} - \mathbf{j}$ and $\mathbf{v} = 6\mathbf{i} + 5\mathbf{j}$. Show that there are two values of λ such that $\mathbf{u} + \lambda\mathbf{v}$ is a unit vector, but no value of μ such that $\mathbf{v} + \mu\mathbf{u}$ is a unit vector. (You are not asked to find the values of λ.)
   (ii) Illustrate (i) diagrammatically.

9. (i) **r** is a two-dimensional vector of modulus 2 making an angle of 30° with the positive x-axis (measured anticlockwise as usual). Find **r** in Cartesian component form, i.e. in the form $a\mathbf{i}+b\mathbf{j}$.

(ii) Repeat (i) for the vector **s** where $|\mathbf{s}|=\sqrt{2}$, and **s** is at an angle of $-45°$.

10. A force **P** is of magnitude 6 Newton due North, and a force **Q** is of magnitude 10 Newton in a direction S 30° W.

(i) Find the magnitude and direction of the resultant force by accurate drawing.

(ii) Find the resultant in component form, by first writing each of **P** and **Q** in component form.

(iii) Check that your answers to (i) and (ii) agree approximately.

11. (The triangle inequality.)

(i) By illustrating the addition $\mathbf{a}+\mathbf{b}$, show geometrically that $|\mathbf{a}+\mathbf{b}|\leqslant|\mathbf{a}|+|\mathbf{b}|$. When will equality occur?

(ii) By taking $\mathbf{a}=\begin{pmatrix} a_1 \\ a_2 \\ \cdot \\ \cdot \\ \cdot \\ a_n \end{pmatrix}$ and $\mathbf{b}=\begin{pmatrix} b_1 \\ b_2 \\ \cdot \\ \cdot \\ \cdot \\ b_n \end{pmatrix}$, give an

algebraic proof of the triangle inequality. [*Hint*: express the inequality in terms of the a_i and b_i, square each side, and show that it reduces to Cauchy's inequality (see Section 16.23).] Another proof is given in Exercise 20c, question 13. The triangle inequality is considered further in Chapter 31.

12. (i) Let **a** and **b** be variable vectors subject to $|\mathbf{a}|=a$ and $|\mathbf{b}|=b$. What are the greatest and least values of $|\mathbf{a}+\mathbf{b}|$?

(ii) For what value of t is the modulus of $(4\mathbf{i}+2\mathbf{j})+t(3\mathbf{i}-\mathbf{j})$ a minimum? Draw a diagram to show why, for this value of t, the vectors $3\mathbf{i}-\mathbf{j}$ and $(4\mathbf{i}+2\mathbf{j})+t(3\mathbf{i}-\mathbf{j})$ are perpendicular.

13. $ABCDEFGH$ is a regular octagon with $\mathbf{AB}=\mathbf{p}$ and $\mathbf{BC}=\mathbf{q}$. Express in terms of **p** and **q**:

  (i) **AD**;   (ii) **CD**;   (iii) **BE**;

  (iv) **DE**;   (v) **DH**.

14. (i) Verify that a rotation of a three-dimensional object may be represented by a line-segment, explaining how the angle of rotation and the axis are given by the line-segment.

(ii) Verify that (finite) rotations performed consecutively do not add vectorially. [*Hint*: it is sufficient to show that two rotations do not commute. Use a book or any cuboidal object.]

15. Let $\mathbf{u}=\mathbf{i}+\mathbf{j}$ and $\mathbf{v}=2\mathbf{i}+\mathbf{j}$. Express as linear combinations of **u** and **v**:

  (i) $3\mathbf{i}+\mathbf{j}$;   (ii) $3\mathbf{j}$;

  (iii) $-4\mathbf{i}-\mathbf{j}$;   (iv) $5\mathbf{i}+5\mathbf{j}$.

16. (i) Show that, given constants a, b, c, d, p, q, the simultaneous

equations

$$\begin{cases} p = \lambda a + \mu c \\ q = \lambda b + \mu d \end{cases}$$

can certainly be solved for λ and μ provided that $ad \neq bc$.

(ii) Show that if the non-zero vectors $\begin{pmatrix} a \\ b \end{pmatrix}$ and $\begin{pmatrix} c \\ d \end{pmatrix}$ are not parallel, then $ad \neq bc$.

(iii) Deduce that, given two non-parallel non-zero vectors $a\mathbf{i} + b\mathbf{j}$ and $c\mathbf{i} + d\mathbf{j}$, any vector $p\mathbf{i} + q\mathbf{j}$ can be expressed as a linear combination of them.

20.7 Position vectors

If O is the origin, and P is a general point, then there is a natural correspondence between the coordinates of P and the elements of the vector $\mathbf{OP}$; if P is the point (α, β, γ), then $\mathbf{OP} = \alpha\mathbf{i} + \beta\mathbf{j} + \gamma\mathbf{k}$.

Definition
The vector $\mathbf{OP}$ (where O is the origin) is called the *position vector* of P.

Example 2
If $\mathbf{p}$ and $\mathbf{q}$ are the position vectors of P and Q respectively, express $\mathbf{PQ}$ in terms of $\mathbf{p}$ and $\mathbf{q}$.
Solution

Figure 20.9

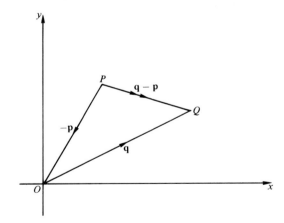

$$\begin{aligned} \mathbf{PQ} &= \mathbf{PO} + \mathbf{OQ} \quad \text{(see Fig. 20.9)} \\ &= -\mathbf{OP} + \mathbf{OQ} \\ &= -\mathbf{p} + \mathbf{q}. \end{aligned}$$

So
$$\mathbf{PQ} = \mathbf{q} - \mathbf{p}.$$

(This general result should be remembered.)

Qu.20 Let P, Q and R be the points $(2, 1, 4)$, $(3, 0, 6)$ and $(6, -3, 12)$ respectively. Find $\mathbf{PQ}$ and $\mathbf{QR}$ and deduce that P, Q and R are collinear. State the ratio of PQ to QR.

Example 3 (*The section formula.*)

(This result is also known as the *ratio theorem.*) Let **a** and **b** be the position vectors of A and B, and let C divide AB in the ratio $\lambda : \mu$. Find an expression for the position vector **c** of C.

Note

If C divides AB *externally*, then either λ or μ is negative, since the ratio $AC:CB$ involves a change in direction. For example, if C divides AB in the ratio $-1:3$, then A, B and C are as shown in Fig. 20.10. (If C cuts AB in the ratio $-1:1$, then C is the 'point at infinity'.)

Figure 20.10

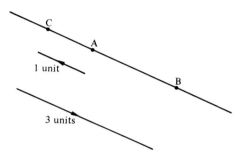

Solution

(Fig. 20.11 shows C between A and B, but the proof is valid for external C also.)

Figure 20.11

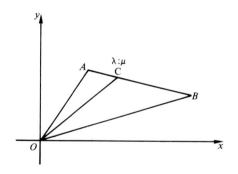

$$\mathbf{c} = \mathbf{OC}$$
$$= \mathbf{OA} + \mathbf{AC}$$
$$= \mathbf{OA} + \frac{\lambda}{\lambda + \mu}\, \mathbf{AB}$$
$$= \mathbf{a} + \frac{\lambda}{\lambda + \mu}\, (\mathbf{b} - \mathbf{a})$$
$$= \frac{\lambda}{\lambda + \mu}\, \mathbf{b} + \left(1 - \frac{\lambda}{\lambda + \mu}\right) \mathbf{a}.$$

So
$$\mathbf{c} = \frac{\lambda \mathbf{b} + \mu \mathbf{a}}{\lambda + \mu}.$$

(This result should be remembered.)

Note

(1) Care must be taken in writing down the section formula. In the numerator, λ is the coefficient of **b**, not of **a**.

(2) If C is the mid-point of AB, so that the ratio is 1:1 (i.e. $\lambda = \mu$) then $\mathbf{c} = \frac{1}{2}(\mathbf{a} + \mathbf{b})$, which is as we should expect.

(3) If we scale λ and μ down so that $\lambda + \mu = 1$, then C cuts AB in the ratio $\lambda : 1 - \lambda$, and $\mathbf{c} = \lambda \mathbf{b} + (1 - \lambda)\mathbf{a}$.

(4) The point C can be regarded as a 'weighted average' of B and A, λ and μ being the weights. (If masses λ and μ are placed at B and A respectively, then C will be the centre of mass.)

(5) In two dimensions, if A is the point (x_1, y_1) and B is (x_2, y_2), then the coordinates of C are

$$\left(\frac{\lambda x_2 + \mu x_1}{\lambda + \mu}, \frac{\lambda y_2 + \mu y_1}{\lambda + \mu} \right),$$

and similarly for higher dimensions.

Example 4 (*The centroid of a triangle.*)

Prove that the medians of a triangle are concurrent (i.e. they all pass through a point). (A *median* of a triangle is a line from a vertex to the mid-point of the opposite side.)

Note

We give two versions of the same proof; the second version may be omitted at a first reading.

First Solution

Let the triangle be ABC, and let P be the mid-point of BC (Fig. 20.12). Then, by the section formula,

$$\mathbf{p} = \frac{1}{2}(\mathbf{b} + \mathbf{c}).$$

Figure 20.12

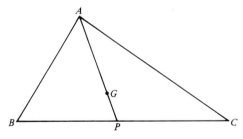

Now consider the point G which cuts AP in the ratio 2:1. Its position vector **g** is given by

$$\mathbf{g} = \frac{2}{3}\mathbf{p} + \frac{1}{3}\mathbf{a} \quad \text{(using the section formula again)}$$

$$= \frac{2}{3} \cdot \frac{1}{2}(\mathbf{b} + \mathbf{c}) + \frac{1}{3}\mathbf{a}$$

$$= \frac{1}{3}(\mathbf{a} + \mathbf{b} + \mathbf{c}).$$

The symmetry of this result indicates that we should obtain the same result if we were to take G to be the point one third of the way up either of the other medians. Thus G, defined by $\mathbf{g} = \frac{1}{3}(\mathbf{a} + \mathbf{b} + \mathbf{c})$, lies on all three medians, and the result is proved. The point G is called the *centroid* of the triangle, and we have shown that it cuts each median in the ratio $2:1$.

Second Solution

This second version shows that the line 'Now consider the point G...' in the first solution is not essential (i.e. this flash of inspiration—or memory—is optional). The proof becomes correspondingly more difficult.

As above, $\mathbf{p} = \frac{1}{2}(\mathbf{b} + \mathbf{c})$. Now consider the point G that cuts AP in the ratio $\lambda:(1 - \lambda)$. Then

$$\begin{aligned} \mathbf{g} &= \lambda\mathbf{p} + (1 - \lambda)\mathbf{a} \\ &= \tfrac{1}{2}\lambda(\mathbf{b} + \mathbf{c}) + (1 - \lambda)\mathbf{a}. \end{aligned} \tag{1}$$

Similarly, if G cuts the median BQ in the ratio $\mu:(1 - \mu)$, then

$$\mathbf{g} = \tfrac{1}{2}\mu(\mathbf{a} + \mathbf{c}) + (1 - \mu)\mathbf{b}. \tag{2}$$

If G is the intersection of these two medians, then (1) and (2) are both true for some λ and μ, and so subtracting,

$$\begin{aligned} &\tfrac{1}{2}(\lambda - \mu)\mathbf{c} + \tfrac{1}{2}(\lambda + 2\mu - 2)\mathbf{b} + \tfrac{1}{2}(2 - 2\lambda - \mu)\mathbf{a} = \mathbf{0} \\ \Rightarrow\ &(\lambda - \mu)\mathbf{c} = (2 - \lambda - 2\mu)\mathbf{b} + (2\lambda + \mu - 2)\mathbf{a}. \end{aligned} \tag{3}$$

We now consider two cases.
(i) $\lambda = \mu$.

Then
$$\begin{aligned} &(2 - 3\lambda)\mathbf{b} + (3\lambda - 2)\mathbf{a} = \mathbf{0} \\ \Rightarrow\ &(2 - 3\lambda)(\mathbf{b} - \mathbf{a}) = \mathbf{0} \\ \Rightarrow\ &\mathbf{a} = \mathbf{b} \text{ or } \lambda = \tfrac{2}{3}. \end{aligned}$$

Since A and B are distinct points, $\mathbf{a} \neq \mathbf{b}$, so $\lambda = \frac{2}{3}$, so that $\mathbf{g} = \frac{1}{3}(\mathbf{a} + \mathbf{b} + \mathbf{c})$.

By the symmetry of this result, G is the intersection of any pair of medians. Hence result.
(ii) $\lambda \neq \mu$.
Then (3) becomes

$$\mathbf{c} = \frac{(2 - \lambda - 2\mu)\mathbf{b} + (2\lambda + \mu - 2)\mathbf{a}}{\lambda - \mu};$$

but this is the point that cuts AB in the ratio $(2 - \lambda - 2\mu):(2\lambda + \mu - 2)$ which implies that C lies on AB. This is not true, so $\lambda \neq \mu$ cannot be true.

Exercise 20b

1. Points P and Q have position vectors $\mathbf{p}$ and $\mathbf{q}$ respectively. Use the section formula to write down the position vectors of
   (i) the mid-point of PQ;
   (ii) the point of trisection of PQ nearer to P;
   (iii) the point X on PQ produced, such that $PX = 2QX$;
   (iv) the point Y on QP produced, such that $YP = \frac{1}{2}QP$.

2. Let P and Q be the points $(2, 7)$ and $(-4, -2)$. Find the coordinates of each of the points described in question 1.

3. ABC is a triangle whose vertices have position vectors $\mathbf{a}$, $\mathbf{b}$ and $\mathbf{c}$ respectively. Find the position vectors of
  (i) M, the mid-point of AB;
  (ii) X, the mid-point of the median CM;
  (iii) the point cutting AK in the ratio $2:3$, where K cuts BC in the ratio $3:1$.

4. Show that the points with position vectors $(\mathbf{u}+\mathbf{v})$, $(2\mathbf{u}-\mathbf{v})$ and $(4\mathbf{u}-5\mathbf{v})$ are collinear.

5. With triangle ABC as in question 3, let P and Q be respectively the points of trisection of CB and CA nearer C. Show that the lines BQ and AP cut one another in the ratio $3:1$. Show further that they intersect on the median through C.

6. Can question 5 be generalized to show that if P and Q lie on CB and CA and $\dfrac{CP}{CB}=\dfrac{CQ}{CA}$ then BQ and AP meet on the median through C?

7. Let P and Q have position vectors $\mathbf{p}$ and $\mathbf{q}$. Show that the direction of $\lambda\mathbf{p}+\mu\mathbf{q}$ is given by the line OX where X cuts PQ in the ratio $\mu:\lambda$.

8. (i) Show that the resultant of two unit vectors bisects the angle between them.
(ii) Deduce that $|\mathbf{q}|\mathbf{p}+|\mathbf{p}|\mathbf{q}$ bisects the angle POQ.
(iii) Hence show, using the result of question 7, that this angle bisector cuts PQ in the ratio $|\mathbf{p}|:|\mathbf{q}|$. (This is the *angle bisector theorem* of elementary geometry.)

9. (i) Let A, B, C be three points with position vectors $\mathbf{a}$, $\mathbf{b}$, $\mathbf{c}$. Let D_1 be the point such that $ABCD_1$ is a parallelogram. Find the position vector of D_1.
(ii) Find the position vectors of the points D_2 and D_3 such that ABD_2C and AD_3BC are parallelograms. Verify that the centroids of the triangles ABC and $D_1D_2D_3$ are the same point.
(iii) Show also that the sides of $D_1D_2D_3$ are parallel to, and twice the length of, the sides of ABC.

10. The vertices of the tetrahedron $ABCD$ have position vectors $\mathbf{a}$, $\mathbf{b}$, $\mathbf{c}$, $\mathbf{d}$. Find the position vector of the point dividing AK in the ratio $3:1$, where K is the centroid of the triangle BCD. Deduce that the four lines joining the vertices to the centroids of the opposite faces are concurrent. (The point where they meet is called the *centroid* of the tetrahedron.)

20.8 The projection of a vector on a line or plane

Let P be a point in two or three dimensions, and l a line. Then the *projection* of the point P on l is the foot of the perpendicular from P to l. If P' and Q' are the projections on l of the points P and Q respectively (Fig. 20.13), then the line-segment $P'Q'$ is the projection of the line-segment PQ. If PQ is directed, then $P'Q'$ becomes correspondingly directed.

The length and direction of the projection $P'Q'$ are unchanged if PQ is moved parallel to itself, and the length of $P'Q'$ is $|PQ\cos\theta|$, where θ is the angle between PQ and l (Fig. 20.14).

Now let $\mathbf{v}$ be a vector in two or three dimensions. If P and Q are

Figure 20.13

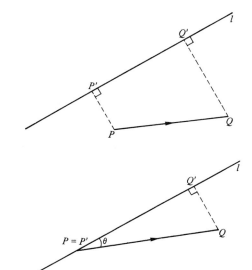

Figure 20.14

two points such that $\mathbf{PQ} = \mathbf{v}$, then the vector $\mathbf{P'Q'}$ is the *projection of* $\mathbf{v}$ *on l*. (Note that we define the projection of a vector on to a line to be another vector. Some books define, instead, the projection of a vector on to another vector, the projection being a scalar.)

In three dimensions, we may define similarly the projections of points, line-segments and vectors on to a plane. The projection of the vector $\mathbf{PQ}$ on a plane π is the vector $\mathbf{P'Q'}$ where P' and Q' are the feet of the perpendiculars from P and Q to π (Fig. 20.15).

Figure 20.15

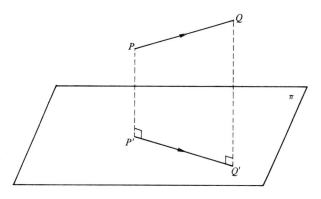

An important result (proved in Theorem 20.2) may now be illustrated diagrammatically for three-dimensional vectors. If l is a line, and $\mathbf{v}$ a vector, then $\mathbf{v}$ may be expressed in the form

$$\mathbf{v} = \mathbf{a} + \mathbf{b}$$

where $\mathbf{a}$ is in the direction of l and $\mathbf{b}$ is perpendicular to l. Fig. 20.16 shows that if π is a plane perpendicular to l, then $\mathbf{a}$ may be interpreted as the projection of $\mathbf{v}$ on l, and $\mathbf{b}$ as the projection of $\mathbf{v}$ on π.

Figure 20.16

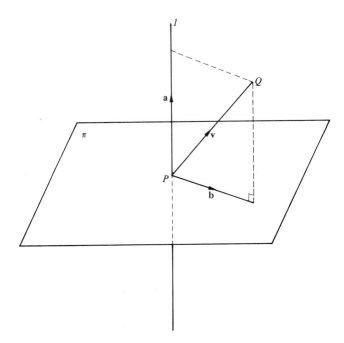

20.9 The scalar product

We now introduce an important extension of the algebra of vectors, by defining a binary operation on vectors of the same dimension which enables us to 'multiply' them. The operation is not closed, since the 'product' is a real number (scalar) and not a vector.

Definition

Let $\mathbf{a} = \begin{pmatrix} a_1 \\ a_2 \\ \cdot \\ \cdot \\ \cdot \\ a_n \end{pmatrix}$ and $\mathbf{b} = \begin{pmatrix} b_1 \\ b_2 \\ \cdot \\ \cdot \\ \cdot \\ b_n \end{pmatrix}$. The *scalar product* (or dot product) of

$\mathbf{a}$ and $\mathbf{b}$ is written $\mathbf{a} . \mathbf{b}$, and is defined by

$$\mathbf{a} . \mathbf{b} = a_1 b_1 + a_2 b_2 + \ldots + a_n b_n.$$

Several properties of scalar products follow immediately from the definition:

(1) The scalar product is commutative, i.e. $\mathbf{a} . \mathbf{b} = \mathbf{b} . \mathbf{a}$.

(2) The scalar product is distributive over addition, i.e. $\mathbf{a} . (\mathbf{b} + \mathbf{c}) = \mathbf{a} . \mathbf{b} + \mathbf{a} . \mathbf{c}$.

(3) $\mathbf{a} . \mathbf{a} = \Sigma(a^r)^2 = |\mathbf{a}|^2$; i.e. any vector 'dotted' with itself gives the square of its modulus.

(4) If λ is a scalar, then $(\lambda \mathbf{a}) . \mathbf{b} = \lambda(\mathbf{a} . \mathbf{b})$ and $\mathbf{a} . (\lambda \mathbf{b}) = \lambda(\mathbf{a} . \mathbf{b})$.

(5) If $\mathbf{a} = \mathbf{0}$ or $\mathbf{b} = \mathbf{0}$, then $\mathbf{a} . \mathbf{b} = 0$. (Note that the converse is false.)

Qu.21 Let $\mathbf{g} = -2\mathbf{i} - \mathbf{j} + 3\mathbf{k}$ and $\mathbf{h} = -\mathbf{i} + 4\mathbf{j} + 2\mathbf{k}$.
(i) Evaluate $\mathbf{g} \cdot \mathbf{h}$. (ii) Evaluate $(2\mathbf{g}) \cdot \mathbf{h}$.
(iii) Write out the vector $(\mathbf{g} \cdot \mathbf{h})\mathbf{h}$.
(iv) Solve for λ the equation $(\mathbf{g} + \lambda\mathbf{h}) \cdot \mathbf{g} = 0$.

Qu.22 Let $\hat{\mathbf{a}}$ be a unit vector, and $\mathbf{b}$ any vector. Let $\mathbf{u} = 2\hat{\mathbf{a}} + \mathbf{b}$ and $\mathbf{v} = 2\hat{\mathbf{a}} - \mathbf{b}$. If $\mathbf{u} \cdot \mathbf{v} = 0$, find $|\mathbf{b}|$.
[*Hint*: multiply out $\mathbf{u} \cdot \mathbf{v}$, checking that each step is valid for scalar products. Then use the third property listed above.]

Theorem 20.1
If $\mathbf{a}$ and $\mathbf{b}$ are vectors in two or three dimensions, then

$$\mathbf{a} \cdot \mathbf{b} = |\mathbf{a}| \, |\mathbf{b}| \cos \theta$$

where θ is the angle between $\mathbf{a}$ and $\mathbf{b}$.

$\Big($ In four or more dimensions, we may *define* the angle between two

vectors to be $\arccos \dfrac{\mathbf{a} \cdot \mathbf{b}}{|\mathbf{a}| \, |\mathbf{b}|}.\Big)$

Proof
(This proof uses the cosine rule; an alternative proof is outlined in question 7 of Exercise 20c.)
In Fig. 20.17, $\mathbf{PQ} = \mathbf{b} - \mathbf{a}$, and so

$$\begin{aligned}
|\mathbf{PQ}|^2 &= (\mathbf{PQ}) \cdot (\mathbf{PQ}) \\
&= (\mathbf{b} - \mathbf{a}) \cdot (\mathbf{b} - \mathbf{a}) \\
&= \mathbf{b} \cdot \mathbf{b} + \mathbf{a} \cdot \mathbf{a} - 2\mathbf{a} \cdot \mathbf{b} \\
&= |\mathbf{OQ}|^2 + |\mathbf{OP}|^2 - 2\mathbf{a} \cdot \mathbf{b}.
\end{aligned}$$

Figure 20.17

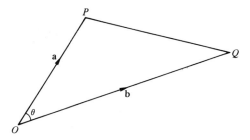

But by the cosine rule,

$$(PQ)^2 = (OP)^2 + (OQ)^2 - 2(OP)(OQ) \cos \theta.$$
So
$$2\mathbf{a} \cdot \mathbf{b} = 2(OP)(OQ) \cos \theta$$
$$\Rightarrow \mathbf{a} \cdot \mathbf{b} = |\mathbf{a}| \, |\mathbf{b}| \cos \theta.$$

Corollary 1
If $\mathbf{a}$ and $\mathbf{b}$ are non-zero vectors, then

$$\mathbf{a} \cdot \mathbf{b} = 0 \Leftrightarrow \mathbf{a} \text{ and } \mathbf{b} \text{ are perpendicular.}$$

Proof

$$\begin{aligned}
\mathbf{a} \cdot \mathbf{b} = 0 &\Leftrightarrow |\mathbf{a}| \, |\mathbf{b}| \cos \theta = 0 \\
&\Leftrightarrow \cos \theta = 0 \text{ (since } |\mathbf{a}| \neq 0 \text{ and } |\mathbf{b}| \neq 0) \\
&\Leftrightarrow \mathbf{a} \text{ and } \mathbf{b} \text{ are perpendicular.}
\end{aligned}$$

Definition

Two non-zero vectors **a** and **b** are said to be *orthogonal* if and only if
a.b$=0$. (Thus in a geometrical context, 'orthogonal' is synonymous
with 'perpendicular'.)

Corollary 2

$$|\mathbf{a}.\mathbf{b}| \leqslant |\mathbf{a}||\mathbf{b}|.$$

Proof

From the Theorem,

$$|\mathbf{a}.\mathbf{b}| = |\mathbf{a}||\mathbf{b}||\cos\theta|.$$

But $|\cos\theta| \leqslant 1$, and so

$$|\mathbf{a}.\mathbf{b}| \leqslant |\mathbf{a}||\mathbf{b}|.$$

Qu.23 Let $\mathbf{a} = \begin{pmatrix} a_1 \\ a_2 \\ \cdot \\ \cdot \\ \cdot \\ a_n \end{pmatrix}$ and $\mathbf{b} = \begin{pmatrix} b_1 \\ b_2 \\ \cdot \\ \cdot \\ \cdot \\ b_n \end{pmatrix}$.

Write out the inequality in Corollary 2 above in terms of these coordinates. Verify that it is Cauchy's inequality.

Example 5

Find the angle between the vectors $\mathbf{a} = \mathbf{i} + \mathbf{j} - 3\mathbf{k}$ and $\mathbf{b} = 3\mathbf{i} - 3\mathbf{j} + 2\mathbf{k}$.

Solution

$$\mathbf{a}.\mathbf{b} = |\mathbf{a}||\mathbf{b}|\cos\theta$$

$$\Rightarrow \cos\theta = \frac{\mathbf{a}.\mathbf{b}}{|\mathbf{a}||\mathbf{b}|}.$$

Now

$$\mathbf{a}.\mathbf{b} = 1 \times 3 + 1 \times (-3) + (-3) \times 2$$
$$= -6,$$

and

$$|\mathbf{a}| = \sqrt{11}$$

and

$$|\mathbf{b}| = \sqrt{22}.$$

So

$$\cos\theta = \frac{-6}{\sqrt{11}\sqrt{22}}$$

$$= \frac{-3\sqrt{2}}{11}.$$

So

$$\theta \approx 112\cdot7°.$$

Qu.24 Show that the angle between $(3\mathbf{i} - 4\mathbf{j})$ and $(7\mathbf{i} - \mathbf{j})$ is $\dfrac{\pi}{4}$.

Example 6
Show that the altitudes of a triangle are concurrent. (An altitude is a line through one vertex perpendicular to the opposite side. The point where they meet is called the *orthocentre* of the triangle.)
Solution

Figure 20.18

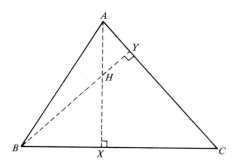

Suppose that the altitudes AX and BY meet at H. It is sufficient to show that CH is perpendicular to AB. Let A, B, C, H have position vectors $\mathbf{a}$, $\mathbf{b}$, $\mathbf{c}$, $\mathbf{h}$. Then

$$AH \text{ is perpendicular to } BC$$
$$\Rightarrow (\mathbf{h}-\mathbf{a}).(\mathbf{c}-\mathbf{b})=0$$
$$\Rightarrow \mathbf{h}.\mathbf{c}-\mathbf{a}.\mathbf{c}-\mathbf{h}.\mathbf{b}+\mathbf{a}.\mathbf{b}=0. \tag{1}$$

Similarly,

$$BH \text{ perpendicular to } AC$$
$$\Rightarrow (\mathbf{h}-\mathbf{b}).(\mathbf{c}-\mathbf{a})=0$$
$$\Rightarrow \mathbf{h}.\mathbf{c}-\mathbf{b}.\mathbf{c}-\mathbf{h}.\mathbf{a}+\mathbf{b}.\mathbf{a}=0. \tag{2}$$

Subtracting (2) from (1),

$$\mathbf{b}.\mathbf{c}+\mathbf{h}.\mathbf{a}-\mathbf{a}.\mathbf{c}-\mathbf{h}.\mathbf{b}=0$$
$$\Rightarrow (\mathbf{c}-\mathbf{h}).(\mathbf{b}-\mathbf{a})=0$$
$$\Rightarrow HC \text{ is perpendicular to } AB.$$

So the altitudes are concurrent.

20.10 Scalar products and projections

We have seen that if $\mathbf{a}$ and $\mathbf{b}$ are two vectors, then the projection of $\mathbf{b}$ on a line in the direction of $\mathbf{a}$ is a vector in the direction of $\mathbf{a}$. Moreover, the number $|\mathbf{b}| \cos \theta$ gives both the length of this projection and also its sense. (A negative sign indicates that θ is obtuse, so that the projection of $\mathbf{b}$ has the opposite sense to $\mathbf{a}$.) Thus the projection of $\mathbf{b}$ on this line is

$$(|\mathbf{b}| \cos \theta)\hat{\mathbf{a}},$$

where $\hat{\mathbf{a}}$ is a unit vector in the direction of $\mathbf{a}$. This expression may also be written $(\mathbf{b}.\hat{\mathbf{a}})\hat{\mathbf{a}}$, and so we have the result†
the projection of $\mathbf{b}$ in the direction of $\mathbf{a}$ is the vector $(\mathbf{b}.\hat{\mathbf{a}})\hat{\mathbf{a}}$.
Similarly the projection of $\mathbf{a}$ in the direction of $\mathbf{b}$ is $(\mathbf{a}.\hat{\mathbf{b}})\hat{\mathbf{b}}$.

---

†The 'projection of $\mathbf{b}$ on $\mathbf{a}$' is sometimes defined to be the scalar $\mathbf{b}.\hat{\mathbf{a}}$.

Figure 20.19

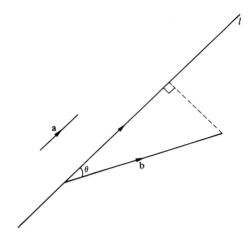

Theorem 20.2

If **a** is a non-zero vector, then any vector **v** may be expressed uniquely in the form $\lambda\mathbf{a}+\mathbf{b}$, where **b** is orthogonal to **a**.

Proof

If **v** can be written in the required form, then

$$\mathbf{b}=\mathbf{v}-\lambda\mathbf{a}$$
$$\Rightarrow \mathbf{b}.\mathbf{a}=\mathbf{v}.\mathbf{a}-\lambda\mathbf{a}.\mathbf{a}.$$

But **b** is orthogonal to **a**, and so

$$0=\mathbf{v}.\mathbf{a}-\lambda\mathbf{a}.\mathbf{a}$$
$$\Rightarrow \lambda=\frac{\mathbf{v}.\mathbf{a}}{\mathbf{a}.\mathbf{a}}.$$

This gives

$$\mathbf{b}=\mathbf{v}-\left(\frac{\mathbf{v}.\mathbf{a}}{\mathbf{a}.\mathbf{a}}\right)\mathbf{a}$$

or

$$\mathbf{b}=\mathbf{v}-(\mathbf{v}.\hat{\mathbf{a}})\hat{\mathbf{a}}.$$

If we choose λ and **b** to be as found above, then $\mathbf{v}=\lambda\mathbf{a}+\mathbf{b}$ and $\mathbf{a}.\mathbf{b}=0$ as required. Also, the method of finding λ shows that this choice of λ (and hence of **b**) is unique.

Qu.25 Express the vector $7\mathbf{i}+2\mathbf{k}$ as the sum of two vectors, one parallel to $\mathbf{i}+\mathbf{j}+\mathbf{k}$, and one perpendicular to it. [*Hint*: take the parallel vector to be $\lambda\mathbf{i}+\lambda\mathbf{j}+\lambda\mathbf{k}$, and find λ by expressing the fact that $(7-\lambda)\mathbf{i}-\lambda\mathbf{j}+(2-\lambda)\mathbf{k}$ is perpendicular to $\mathbf{i}+\mathbf{j}+\mathbf{k}$.]

Exercise 20c

1. Show that the vectors $\mathbf{u}=\mathbf{i}+2\mathbf{j}+4\mathbf{k}$, $\mathbf{v}=-2\mathbf{j}+\mathbf{k}$ and $\mathbf{w}=10\mathbf{i}-\mathbf{j}-2\mathbf{k}$ are mutually perpendicular.
2. (i) For what values of λ are the vectors $6\mathbf{i}+\mathbf{j}$ and $(2\mathbf{i}+3\mathbf{j})+\lambda(-\mathbf{i}+4\mathbf{j})$ perpendicular?

   (ii) Show that if $(\mathbf{a}+\lambda\mathbf{b})$ is perpendicular to **c**, then $\lambda=-\dfrac{\mathbf{a}.\mathbf{c}}{\mathbf{b}.\mathbf{c}}$.

3. Find the angles between the following pairs of vectors:
    (i) $(\mathbf{i}+2\mathbf{j})$ and $(-\mathbf{i}+2\mathbf{j})$;
    (ii) $(3\mathbf{i}-2\mathbf{j}+6\mathbf{k})$ and $(\mathbf{i}+\mathbf{j}+\mathbf{k})$;

    (iii) $\begin{pmatrix} 1 \\ 1 \\ 1 \end{pmatrix}$ and $\begin{pmatrix} 1 \\ -1 \\ -1 \end{pmatrix}$;

    (iv) $\begin{pmatrix} 1 \\ 0 \\ 2 \\ 1 \end{pmatrix}$ and $\begin{pmatrix} 4 \\ -2 \\ 0 \\ -2 \end{pmatrix}$.

4. Write down the 'gradients' of the vectors $r\mathbf{i}+s\mathbf{j}$ and $t\mathbf{i}+u\mathbf{j}$, and express the fact that their product is -1. Show directly that this is equivalent to $(r\mathbf{i}+s\mathbf{j}).(t\mathbf{i}+u\mathbf{j})=0$.

5. If $\mathbf{a}=2\mathbf{i}+3\mathbf{j}+\mathbf{k}$ and $\mathbf{b}=-2\mathbf{j}+4\mathbf{k}$, find a value of λ such that $\mathbf{a}+\lambda\mathbf{b}$ and $\mathbf{a}-\lambda\mathbf{b}$ are perpendicular.

6. Show that

$$|\mathbf{a}|=|\mathbf{b}| \;\Rightarrow\; (\mathbf{a}-\mathbf{b}) \text{ and } (\mathbf{a}+\mathbf{b}) \text{ are perpendicular.}$$

   How can this result be interpreted geometrically if $\mathbf{a}$ and $\mathbf{b}$ are represented by adjacent sides of a parallelogram?

7. (i) Show directly that, for non-zero vectors $\mathbf{a}$ and $\mathbf{b}$,
    $$\mathbf{a}.\mathbf{b}=0 \;\Leftrightarrow\; \mathbf{a} \text{ and } \mathbf{b} \text{ are perpendicular.}$$
    (Use a simplified version of the proof of Theorem 20.1, with Pythagoras's Theorem in place of the cosine rule.)
    (ii) Extend the diagrammatic argument at the end of Section 20.8 to show that if $\mathbf{a}$ is a non-zero vector, then $\mathbf{v}=\lambda\hat{\mathbf{a}}+\mathbf{b}$ where $\lambda=|\mathbf{v}|\cos\theta$, and θ is the angle between $\mathbf{a}$ and $\mathbf{v}$.
    (iii) Deduce from (i) and (ii) that $\mathbf{v}.\mathbf{a}=\lambda\mathbf{a}.\mathbf{a}$, and hence that $\mathbf{v}.\mathbf{a}=|\mathbf{v}||\mathbf{a}|\cos\theta$.

8. Show, by the method below, that the opposite edges of an equilateral tetrahedron are perpendicular.
    (i) Take one vertex as the origin (to simplify the algebra) and the others as A, B, C, with position vectors $\mathbf{a}$, $\mathbf{b}$, $\mathbf{c}$. Express in terms of $\mathbf{a}$, $\mathbf{b}$, $\mathbf{c}$ the fact that $BC^2=AC^2$.
    (ii) Expand and simplify, using the fact that $OA^2=OB^2$.
    (iii) Deduce that OC and AB are perpendicular.

9. AP, BQ, CR and DS are the vertical edges of a cuboid whose base is $ABCD$. $\mathbf{AB}=3\mathbf{i}$, $\mathbf{AD}=\mathbf{j}$ and $\mathbf{AP}=2\mathbf{k}$. M and N are the midpoints of SR and AB respectively.
    (i) By expressing the vectors $\mathbf{AQ}$ and $\mathbf{AR}$ in terms of $\mathbf{i}$, $\mathbf{j}$ and $\mathbf{k}$, find the cosine of the angle QAR, and find the angle to the nearest degree.
    Find similarly the angles
    (ii) SAQ; (iii) RAD; (iv) MPN.

10. A and B are two fixed points, and X is a variable point such that $(\mathbf{a}-\mathbf{x})^2=(\mathbf{b}-\mathbf{x})^2$, where the notation $\mathbf{u}^2$ means $\mathbf{u}.\mathbf{u}$. Explain this statement, and hence state the locus of X. Show algebraically that $(\mathbf{x}-\frac{1}{2}(\mathbf{a}+\mathbf{b})).(\mathbf{a}-\mathbf{b})=0$ and interpret this result.

11. A and B are fixed points, and R is a variable point. Find the vector equation of the circle with diameter AB by writing down a condition which is satisfied if and only if R lies on this circle. [*Hint*: think about which geometrical property of the circle to use.]

12. Let ABC be a triangle and M be the mid-point of BC. Apollonius's Theorem states that

$$AB^2 + AC^2 = 2(AM^2 + BM^2).$$

Prove this vectorially.

13. Use the inequality $|\cos \theta| \leqslant 1$ to show that

$$|\mathbf{a} + \mathbf{b}|^2 \leqslant (|\mathbf{a}| + |\mathbf{b}|)^2.$$

Deduce the triangle inequality.

14. (For discussion.) Verify that $(\mathbf{i} + \mathbf{j} + \mathbf{k})$, $(\mathbf{i} - \mathbf{j} - \mathbf{k})$, $(-\mathbf{i} + \mathbf{j} - \mathbf{k})$ and $(-\mathbf{i} - \mathbf{j} + \mathbf{k})$ are all inclined at the same angle to one another. Is it possible to find a set of more than four vectors (in three dimensions) with this property?

15. Two planes, inclined at an angle θ, meet in a line l whose direction is given by the vector $\hat{\mathbf{u}}$. A pair of lines, one in each plane, intersect on l and make angles ϕ_1 and ϕ_2 with l. We wish to find the angle between this pair of lines. Let $\hat{\mathbf{p}}$ and $\hat{\mathbf{q}}$ be in the directions in each plane which are perpendicular to l.
   (i) Show that the directions of the two lines are given by

$$(\cos \phi_1)\hat{\mathbf{u}} \pm (\sin \phi_1)\hat{\mathbf{p}}$$
   and
$$(\cos \phi_2)\hat{\mathbf{u}} \pm (\sin \phi_2)\hat{\mathbf{q}}.$$

   (ii) Show that these are unit vectors.
   (iii) Show that the angle between these vectors is
$$\arccos (\cos \phi_1 \, \cos \phi_2 \pm \sin \phi_1 \, \sin \phi_2 \, \cos \theta).$$

20.11 The differentiation of vectors

Suppose that a vector $\mathbf{r}$ depends on the value of some real number t. An example would be

$$\mathbf{r} = 3t\mathbf{i} + t^2\mathbf{j} + \mathbf{k}. \tag{1}$$

(This dependence on t may be emphasized when necessary by writing $\mathbf{r}(t)$ instead of merely $\mathbf{r}$.) We have, in effect, a functional relationship between t (the independent variable) and $\mathbf{r}$ (the dependent variable). The domain is the set of real numbers, and the codomain is the set of n-dimensional vectors for some n.

Perhaps the simplest illustration of a vector function of a scalar is to take $\mathbf{r}$ to be the position vector of some point P, and t to be time. As time passes, the point moves.

Qu.26 If the position of P at time t is given by (1) above, sketch the path of P. (Try to give a three-dimensional diagram.) Find the length of the chords joining the points given by
(i) $t = 2$ and $t = 3$; (ii) $t = 2$ and $t = 2 \cdot 1$;
(iii) $t = 2$ and $t = 2 + h$.

In Chapter 5, we defined the derivative of a function from R to R,

VECTORS

and the same definition may be applied to mappings from R to a set of vectors. Thus we define the derivative of $\mathbf{r}$ with respect to t by

$$\frac{d\mathbf{r}}{dt} = \lim_{h \to 0} \frac{1}{h} \{\mathbf{r}(t+h) - \mathbf{r}(t)\}.$$

In our example (1), we then have

$$\frac{d\mathbf{r}}{dt} = \lim_{h \to 0} \frac{1}{h} \{3(t+h)\mathbf{i} + (t+h)^2\mathbf{j} + \mathbf{k} - (3t\mathbf{i} + t^2\mathbf{j} + \mathbf{k})\}$$

$$= \lim_{h \to 0} \frac{1}{h} \{3h\mathbf{i} + (2th + h^2)\mathbf{j}\}$$

$$= 3\mathbf{i} + 2t\mathbf{j}.$$

It is clear, therefore, that when a vector is given in Cartesian form (as in (1)), we may differentiate it by differentiating each coordinate separately; i.e. the derivative of

$$f(t)\mathbf{i} + g(t)\mathbf{j} + h(t)\mathbf{k}$$
is $\quad f'(t)\mathbf{i} + g'(t)\mathbf{j} + h'(t)\mathbf{k}.$

Qu.27 If $\mathbf{r} = \sin 2t\, \mathbf{i} + \cos t\, \mathbf{j} + (t+1)\mathbf{k}$, find

(i) $\dfrac{d\mathbf{r}}{dt}$; (ii) $\dfrac{d^2\mathbf{r}}{dt^2}$ when $t = \dfrac{\pi}{2}$.

Now consider the geometrical significance of $\dfrac{d\mathbf{r}}{dt}$. In Fig. 20.20, P and Q are the positions of a particle at times t and $t + \delta t$, and

$$\mathbf{PQ} = \delta\mathbf{r}$$
$$= \mathbf{r}(t + \delta t) - \mathbf{r}(t).$$

Figure 20.20

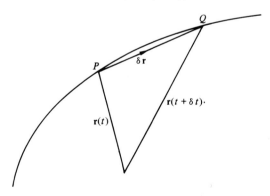

So $\dfrac{\delta\mathbf{r}}{\delta t}$ is a vector in the direction of PQ, whose magnitude is equal to $\dfrac{\text{length of } PQ}{\text{time to travel from } P \text{ to } Q}$, i.e. it gives, in effect, the average speed along PQ. In the limit, as $\delta t \to 0$ and Q approaches P,

$\dfrac{d\mathbf{r}}{dt}$ will give the direction of motion of the particle (i.e. the direction of the tangent to the curve at P) and its speed along the curve. Thus $\dfrac{d\mathbf{r}}{dt}$ is the *velocity* of the particle, and we write $\mathbf{v} = \dfrac{d\mathbf{r}}{dt}$.

Qu.28 (i) The position at time t of a particle moving in two dimensions is given by $\mathbf{r} = t^2\mathbf{i} + t^3\mathbf{j}$. Find $\mathbf{v}$ and the angle between $\mathbf{r}$ and $\mathbf{v}$ when $t = 1$. Illustrate your answer by a sketch of the path of the particle.

(ii) From your sketch, state what happens to the angle between $\mathbf{r}$ and $\mathbf{v}$ as $t \to \infty$. Check your answer by showing that the angle at time t is given by

$$\cos\theta = \frac{2 + 3t^2}{\sqrt{(1 + t^2)}\,\sqrt{(9t^2 + 4)}}$$

and finding the limit of $\cos\theta$ as $t \to \infty$.

Now suppose that $\mathbf{u}$ and $\mathbf{v}$ are each functions of t. Can we apply the ordinary product rule to differentiate the scalar product $\mathbf{u}.\mathbf{v}$? Fortunately, the algebra of vectors sufficiently resembles the algebra of real numbers that the answer is yes, since

$$\frac{d}{dt}(\mathbf{u}.\mathbf{v}) = \lim_{\delta t \to 0} \frac{1}{\delta t}\{(\mathbf{u} + \delta\mathbf{u}).(\mathbf{v} + \delta\mathbf{v}) - \mathbf{u}.\mathbf{v}\}$$

$$= \lim_{\delta t \to 0} \frac{1}{\delta t}(\mathbf{u}.\delta\mathbf{v} + \mathbf{v}.\delta\mathbf{u} + \delta\mathbf{u}.\delta\mathbf{v})$$

$$= \lim_{\delta t \to 0} \left(\mathbf{u}.\frac{\delta\mathbf{v}}{\delta t} + \mathbf{v}.\frac{\delta\mathbf{u}}{\delta t} + \delta\mathbf{u}.\frac{\delta\mathbf{v}}{\delta t} \right)$$

$$= \mathbf{u}.\frac{d\mathbf{v}}{dt} + \mathbf{v}.\frac{d\mathbf{u}}{dt}.$$

Example 7

A particle moves in two dimensions so that its acceleration is always perpendicular to its velocity. Show that it moves with constant speed.

Solution

(We wish to show that $|\mathbf{v}|$ is constant; it is algebraically more convenient to show that $|\mathbf{v}|^2$ is constant.)

$$\frac{d}{dt}|\mathbf{v}|^2 = \frac{d}{dt}(\mathbf{v}.\mathbf{v})$$

$$= 2\mathbf{v}.\mathbf{a}$$

(where $\mathbf{a} = \dfrac{d\mathbf{v}}{dt} = $ acceleration). But $\mathbf{v}.\mathbf{a} = 0$ since $\mathbf{a}$ and $\mathbf{v}$ are perpendicular. So

$$\frac{d}{dt}|\mathbf{v}|^2 = 0$$

$$\Rightarrow |\mathbf{v}|^2 \text{ is constant}$$
$$\Rightarrow |\mathbf{v}| \text{ is constant},$$

i.e. the particle moves with constant speed.

Qu.29 Prove that if λ is a (scalar) function of t, and $\mathbf{u}$ is a vector function of t, then

$$\frac{d}{dt}(\lambda\mathbf{u}) = \frac{d\lambda}{dt}\mathbf{u} + \lambda\frac{d\mathbf{u}}{dt}.$$

Exercise 20d

1. A particle's position at time t is given by

$$\mathbf{r} = \sin 2t\,\mathbf{i} + \cos 2t\,\mathbf{j} + 3t\mathbf{k}.$$

   (i) Sketch the path traced out by the particle. (This is an example of a *helix*—not a spiral, which is a two-dimensional curve.) Does it have a right-hand or a left-hand 'thread'?
   (ii) Find the velocity vector $\mathbf{v}$ and the speed of the particle.
   (iii) By finding the angle between $\mathbf{v}$ and the z-axis, show that the curve has the same 'steepness' at all points.
   (iv) Find the 'radius' of the helix, the time taken to go round one turn of the helix, and the vertical distance between successive turns. (This last distance is called the *pitch* of the helix.)

2. (Hard. See question 1.) A particle starts at the point $(a, 0, 0)$, and moves along a helix of radius a and pitch p, at a constant speed u. If the axis of the helix is the z-axis, and it winds in a right-hand screw sense, and the particle moves in the direction of increasing z, find the position vector of the particle at time t.

3. The path of a particle is given by $\mathbf{r} = t^2\mathbf{i} + t\mathbf{j}$. Find, in terms of t,

   (i) $\mathbf{v}$; (ii) $\mathbf{a}$; (iii) $|\mathbf{r}|$; (iv) $\dfrac{d}{dt}|\mathbf{r}|$; (v) $\left|\dfrac{d\mathbf{r}}{dt}\right|$.

4. (i) Find the position vector at time t of a particle which traces out the same curve as the particle in question 3, but which takes twice as long to reach any point.
   (ii) Show that it is not true that at any instant the speeds of the two particles are in the ratio $2:1$.
   (iii) In what sense *are* the speeds in the ratio $2:1$?

5. Sketch or plot the spiral $\mathbf{r} = (e^t\cos t)\mathbf{i} + (e^t\sin t)\mathbf{j}$. If P is a point on the curve, find the angle between OP and the tangent at P, showing that it does not depend on t. (This is an example of an *equiangular spiral*.)

6. (i) Show (algebraically) that $\dfrac{d}{dt}|\mathbf{r}| = \hat{\mathbf{r}}.\mathbf{v}$ where $\hat{\mathbf{r}}$ is a unit radial vector. [*Hint:* consider $\dfrac{d}{dt}|\mathbf{r}|^2$.]

   (ii) Explain diagrammatically why $\dfrac{d}{dt}|\mathbf{r}| = |\mathbf{v}|\cos\theta$ for suitable θ.

7. (i) A particle moves in two dimensions so that its velocity is always perpendicular to its position vector. Show that it moves on a circle centre O.
   (ii) A particle P moves in three dimensions so that its velocity is always perpendicular to AP where A is a fixed point. What can be said about the path of the particle? (Prove any statement you make.)

8. Let $\mathbf{p}=(1+t)\mathbf{i}+\sin t\,\mathbf{j}+\cos t\,\mathbf{k}$ and
$\mathbf{q}=(1-t)\mathbf{i}-\cos t\,\mathbf{j}+\sin t\,\mathbf{j}+\sin t\,\mathbf{k}$. Find

(i) $\mathbf{p}.\mathbf{q}$ and $\dfrac{d}{dt}(\mathbf{p}.\mathbf{q})$,

(ii) $\dfrac{d\mathbf{p}}{dt}.\mathbf{q}+\mathbf{p}.\dfrac{d\mathbf{q}}{dt}$,

and hence verify the rule for differentiating a scalar product.

9. Explain what is meant by $\left|\dfrac{d\mathbf{r}}{dt}\right|$ and $\left|\dfrac{d|\mathbf{r}|}{dt}\right|$ and also why they are not in general equal. Determine whether
   (i) they can ever be equal;
   (ii) they are connected by any inequality.

10. Two particles P and Q move in two dimensions so that at all times the velocity of each is perpendicular to the position vector of the other.
    (i) If the particles are placed initially at $(1, 1)$ and $(1, -1)$, explain in general terms how they will move.
    (ii) Show that $\mathbf{p}=2\cos t\,\mathbf{i}+\sin t\,\mathbf{j}$ and $\mathbf{q}=\cos t\,\mathbf{i}+2\sin t\,\mathbf{j}$ satisfy the given condition, and sketch these curves on the same axes.

11. A particle's position is given by

$$\mathbf{u}=A\cos\omega t\,\mathbf{i}+A\sin\omega t\,\mathbf{j}.$$

    (i) Sketch the path of the particle.
    (ii) Find its speed, angular velocity and time for one revolution.
    (iii) Show that its acceleration is directed towards the centre.

12. Two ships are manoeuvring so that their distance apart is constant. Show that the relative velocity of one with respect to the other is perpendicular to its relative position.

13. Particles U and V are moving with the same speed along the paths $\mathbf{u}(t)$ and $\mathbf{v}(t)$, and a third particle W moves so that it is always at the mid-point of the line UV. Show that, at any instant, the direction of motion of W bisects the directions of motion of U and V.

Miscellaneous exercise 20

1. Relative to an origin O the points A and B have position vectors $\mathbf{a}$ and $\mathbf{b}$ such that $|\mathbf{a}|=3$, $|\mathbf{b}|=1$ and $A\hat{O}B=60°$.
   A point C has position vector $k\mathbf{b}$ relative to O. By considering the scalar product $\mathbf{AB}.\mathbf{AC}$ determine the value of k such that $B\hat{A}C=90°$. [It may be assumed that $\mathbf{p}.(\mathbf{q}+\mathbf{r})=\mathbf{p}.\mathbf{q}+\mathbf{p}.\mathbf{r}$.]

[C (O)]

2. (i) $\mathbf{F}$ is a force acting at O of magnitude $10\,\text{N}$ on a bearing of $060°$ while $\mathbf{G}$ is a force acting at O which has components $10\,\text{N}$ due north and $5\,\text{N}$ due east.
   Calculate the magnitude and direction of
   (a) $\mathbf{F}+\mathbf{G}$ and (b) $\mathbf{F}-\mathbf{G}$.
   (ii) $\mathbf{P}$ and $\mathbf{Q}$ are two forces acting at O, and the magnitude of $\mathbf{P}$ is $10\,\text{N}$. The forces $\mathbf{P}+\mathbf{Q}$ and $\mathbf{P}-\mathbf{Q}$ are in perpendicular directions. Calculate the magnitude of $\mathbf{Q}$.

[Ox (O)]

3. The position vectors of A and B with respect to O as origin are $2\mathbf{i}+4\mathbf{j}$ and $9\mathbf{i}$ respectively. The point C is such that $OACB$ is a parallelogram with OC and AB as diagonals. M is the mid-point of BC, and AB and OM meet at K.

   If $\overrightarrow{OK}=\lambda\overrightarrow{OM}$ and $\overrightarrow{KB}=\mu\overrightarrow{AB}$, find $\overrightarrow{OC}$, $\overrightarrow{OM}$ and $\overrightarrow{AB}$ in the form $a\mathbf{i}+b\mathbf{j}$ and show that

$$\lambda(10\mathbf{i}+2\mathbf{j})+\mu(7\mathbf{i}-4\mathbf{j})=9\mathbf{i}.$$

   Hence find the ratio in which OM divides AB.

   [AEB (O) '80]

4. In the triangle ABC, $\mathbf{AB}=\mathbf{p}$, $\mathbf{AC}=\mathbf{q}$, M is the mid-point of CB, and N is the mid-point of AM.

   Express in terms of $\mathbf{p}$ and $\mathbf{q}$
   (i) $\mathbf{CM}$,
   (ii) $\mathbf{AN}$,
   (iii) $\mathbf{CN}$.

   If $\mathbf{AR}=k\mathbf{p}$, express $\mathbf{CR}$ in terms of k, $\mathbf{p}$ and $\mathbf{q}$, and hence find the value of k for which R lies on CN produced.

   [MEI (O)]

5. If points A and B have position vectors $\mathbf{a}$ and $\mathbf{b}$ with respect to an origin O, show that $(AB)^2=(\mathbf{b}-\mathbf{a}).(\mathbf{b}-\mathbf{a})$.

   Prove that the sum of the squares of the sides of any quadrilateral (not necessarily plane) is equal to the sum of the squares of its diagonals added to four times the square of the length of the line segment joining the mid-points of the diagonals.

   To what does this result reduce if the quadrilateral is a parallelogram?

   [O & C]

6. The vector $\mathbf{x}$ satisfies

$$\mathbf{x}+\frac{\lambda}{a^2}(\mathbf{a}.\mathbf{x})\mathbf{a}=\mathbf{b},$$

   where $a=|\mathbf{a}|\neq0$. Given that $\lambda\neq-1$, show that

$$\mathbf{a}.\mathbf{x}=\frac{\mathbf{a}.\mathbf{b}}{1+\lambda}.$$

   Hence, or otherwise, find $\mathbf{x}$ in terms of a, $\mathbf{a}$, $\mathbf{b}$ and λ. In the case when $\lambda=-1$, show that either $\mathbf{b}=\mathbf{0}$ or $\mathbf{a}$ and $\mathbf{b}$ are perpendicular.

   [JMB]

7. (i) In the quadrilateral $ABCD$, X and Y are the mid-points of the diagonals AC and BD respectively. Show that
   (a) $\overrightarrow{BA}+\overrightarrow{BC}=2\overrightarrow{BX}$,
   (b) $\overrightarrow{BA}+\overrightarrow{BC}+\overrightarrow{DA}+\overrightarrow{DC}=4\overrightarrow{YX}$.
   (ii) The point P lies on the circle through the vertices of a rectangle $QRST$. The point X on the diagonal QS is such that $\overrightarrow{QX}=2\overrightarrow{XS}$. Express $\overrightarrow{PX}$, $\overrightarrow{QX}$ and $(\overrightarrow{RX}+\overrightarrow{TX})$ in terms of $\overrightarrow{PQ}$ and $\overrightarrow{PS}$.

   [L]

8. The position vectors with respect to a point O of four non-collinear points A, B, C, D are $\mathbf{a}$, $\mathbf{b}$, $\frac{2}{3}\mathbf{a}$, $\frac{1}{3}\mathbf{b}$ respectively. Show that

any point on AD has position vector given parametrically by $(1-t)\mathbf{a}+\frac{1}{3}t\mathbf{b}$ and write down the position vector of any point on BC. Hence find the position vector of the point of intersection E of BC and AD.

If $\triangle\,OAB$ is equilateral and has sides of unit length, prove that $\mathbf{a}.\mathbf{b}=\frac{1}{2}$ and, by considering the scalar product of $\mathbf{ED}$ and $\mathbf{EC}$, calculate $C\hat{E}D$.

[O & C]

9. A particle P moves in a plane so that its co-ordinates at time t ($t\geqslant 0$), referred to rectangular axes Ox, Oy, are given by $x=a(p^2-1)$, $y=2ap$, where a is a positive constant, and p is a parameter. Find the velocity components of P, parallel to Ox and Oy respectively, in terms of a, p and dp/dt. Show that, if the motion is such that $y\dfrac{dx}{dt}-x\dfrac{dy}{dt}$ has the constant value h,

then

$$\frac{dp}{dt}=\frac{h}{2a^2(1+p^2)}.$$

Express the velocity and acceleration components in terms of a, p and h. Show that the acceleration is directed towards the origin and is of magnitude $h^2/(2ar^2)$, where r is the distance from the origin.

[JMB]

10. The position vectors, relative to an origin O, of the vertices A, B, C, D of a tetrahedron are $\mathbf{a}$, $\mathbf{b}$, $\mathbf{c}$, $\mathbf{d}$ respectively. Using vector methods, or otherwise, show that the lines joining the mid-points of the opposite edges are concurrent in a point M.

Given also that AB is perpendicular to CD and that AC is perpendicular to BD, prove that
(i) AD is perpendicular to BC,
(ii) $AB^2+CD^2=AC^2+BD^2=AD^2+BC^2$.
Given also that A, B, C, D lie on a sphere with centre O and that X is the point such that $\overrightarrow{OX}=2\overrightarrow{OM}$, prove that AX is perpendicular to BC and to CD.

Deduce that the altitudes of the tetrahedron $ABCD$ are concurrent. (An *altitude* is a line drawn from a vertex perpendicular to the opposite face.)

[JMB (S)]

11. At time t the position vector of a particle is given by

$$\mathbf{r}=a\cos(t^2)\,\mathbf{i}+a\sin(t^2)\,\mathbf{j},$$

where a is a constant. Show that the rate of change with time of the speed of the particle is constant.

Show also that the scalar product of the velocity vector and the acceleration vector equals $4a^2t$, and express the tangent of the angle between these vectors in terms of t.

[L]

12. $ABCD$ is a plane quadrilateral having no parallel sides. AB and DC intersect at a point E, and AD and BC intersect at a point F.

If P, Q, R are the mid-points of AC, BD, and EF respectively, prove by a vector method that P, Q, R lie on a straight line.

[W (S)]

13. Two points, Q and R, have position vectors $\mathbf{q}$ and $\mathbf{r}$ respectively. Show that if $\mathbf{p}$ is the position vector of a point on the line through Q and R, then

$$\mathbf{p} = \lambda q + \lambda' r$$

where $\lambda + \lambda' = 1$. What can be deduced about three vectors $\mathbf{a}$, $\mathbf{b}$ and $\mathbf{c}$ if

$$\alpha \mathbf{a} + \beta \mathbf{b} + \gamma \mathbf{c} = 0$$

where $\alpha + \beta + \gamma = 0$?

ABC and DEF are two triangles whose vertices have position vectors $\mathbf{a}$, $\mathbf{b}$, $\mathbf{c}$, $\mathbf{d}$, $\mathbf{e}$, $\mathbf{f}$ respectively. Show that if the lines AD, BE, CF intersect in a point with position vector $\mathbf{h}$, then there is a relationship of the form

$$l\mathbf{a} + l'\mathbf{d} = m\mathbf{b} + m'\mathbf{e} = n\mathbf{c} + n'\mathbf{f} = \mathbf{h}$$

where $l + l' = m + m' = n + n' = 1$.

Show further, that

$$\frac{m\mathbf{b} - n\mathbf{c}}{m - n} = \frac{m'\mathbf{e} - n'\mathbf{f}}{m' - n'}.$$

Denoting this vector by $\mathbf{p}$, show that $\mathbf{p}$ is the position vector of the point of intersection of the lines BC and EF. Write down similar expressions for $\mathbf{q}$ and $\mathbf{r}$, the position vectors of the points of intersection of AB and DE, and AC and DF respectively. Deduce that the three points of intersection determined by $\mathbf{p}$, $\mathbf{q}$ and $\mathbf{r}$ are collinear.

[MEI (S)]

21 The algebra of matrices

21.1 Definitions A *matrix* is a rectangular array of numbers called the elements or entries of the matrix. Thus $A = \begin{pmatrix} 2 & 0 & -1 \\ \frac{1}{2} & -12 & 4 \end{pmatrix}$ is a matrix with 2 rows and 3 columns; we say that $\mathbf{A}$ is a 2×3 matrix, and we call this its *order*. A matrix which is 2×2 or 3×3 or, in general, $n \times n$ is called a *square matrix*.

Two matrices are *equal* if and only if they have the same order and have the same elements in corresponding positions.

The *transpose* of a matrix $\mathbf{M}$, denoted by $\mathbf{M}^T$ or $\mathbf{M}'$, is the matrix $\mathbf{M}$ with its rows and columns interchanged; so with $\mathbf{A}$ as above, $\mathbf{A}^T$ would be the 3×2 matrix $\begin{pmatrix} 2 & \frac{1}{2} \\ 0 & -12 \\ -1 & 4 \end{pmatrix}$. Clearly, for any matrix $\mathbf{M}$, $(\mathbf{M}^T)^T = \mathbf{M}$. A matrix $\mathbf{M}$ is said to be *symmetric* if $\mathbf{M}^T = \mathbf{M}$.

Qu.1 What can be said about the order of a symmetric matrix?
Qu.2 Write down a 4×4 symmetric matrix and indicate the 'axis of symmetry'.

In a square matrix, the *leading diagonal* is the diagonal line of numbers from the top left corner to the bottom right corner of the matrix.

21.2 Addition and multiplication by a scalar Two matrices of the same order may be added or subtracted, and any matrix may be multiplied by a scalar (i.e. a number). These operations are defined in precisely the way that we might expect; for example

$$\begin{pmatrix} 1 & 2 & -2 \\ -3 & 0 & 1 \end{pmatrix} + \begin{pmatrix} 0 & 4 & 6 \\ 1 & -3 & -3 \end{pmatrix} = \begin{pmatrix} 1 & 6 & 4 \\ -2 & -3 & -2 \end{pmatrix}$$

and $\quad 3 \begin{pmatrix} 1 & 2 \\ -3 & 6 \end{pmatrix} = \begin{pmatrix} 3 & 6 \\ -9 & 18 \end{pmatrix}$.

Qu.3 Let $\mathbf{A} = \begin{pmatrix} 2 & 1 & 4 \\ 1 & -1 & -3 \end{pmatrix}$, $\mathbf{B} = \begin{pmatrix} 2 & 0 & 0 \\ 1 & 3 & -4 \end{pmatrix}$

and $\mathbf{C} = \begin{pmatrix} -1 & 1 \\ 2 & -1 \\ 3 & 3 \end{pmatrix}$.

(i) Evaluate where possible:
   (a) $\mathbf{A}+\mathbf{B}$; (b) $2\mathbf{A}-3\mathbf{B}$; (c) $\mathbf{A}+\mathbf{C}$;
   (d) $2(\mathbf{A}-\mathbf{C}^T)$; (e) $(\mathbf{A}^T+\mathbf{B}^T)^T$.
(ii) Solve for matrix $\mathbf{X}$:
   (a) $\mathbf{A}+\mathbf{X}=\mathbf{B}$; (b) $3\mathbf{B}^T-2\mathbf{X}=4\mathbf{C}$;
   (c) $\mathbf{A}+\mathbf{X}=\mathbf{A}$.

Qu.4 A matrix $\mathbf{M}$ is said to be *skew-symmetric* if $\mathbf{M}^T=-\mathbf{M}$. Show that any such matrix is square. What can be said about the entries on its leading diagonal?

Qu.5 Is it true that $(\mathbf{A}+\mathbf{B})^T=\mathbf{A}^T+\mathbf{B}^T$?

Qu.6 Show that the set of all 2×3 matrices forms an abelian group under matrix addition. (Note that any matrix whose entries are all zero is called the *zero matrix*.) Show also that multiplication by a scalar is distributive over matrix addition.

Qu.7 (i) Let $\mathbf{M}$ be a square matrix. Use the result of Qu.5. to simplify $(\mathbf{M}+\mathbf{M}^T)^T$. Deduce that $\mathbf{M}+\mathbf{M}^T$ is symmetric.
(ii) Prove or disprove that $\mathbf{M}-\mathbf{M}^T$ is skew-symmetric.

21.3 Matrix multiplication

Unlike matrix addition, matrix multiplication is not defined in an 'obvious' way. If $\mathbf{A}$ and $\mathbf{B}$ are two matrices, we may form the product $\mathbf{AB}$ if and only if the number of columns in $\mathbf{A}$ is the same as the number of rows in $\mathbf{B}$. So if $\mathbf{A}$ has order $m\times n$, and $\mathbf{B}$ has order $n\times p$, then $\mathbf{AB}$ exists and (as we shall see) has order $m\times p$. (Note the 'dominoes' rule here.)

The product $\mathbf{AB}$ is defined as follows: the element in the ith row and jth column of $\mathbf{AB}$ is the scalar product (in the vector sense) of the ith row of $\mathbf{A}$ and the jth column of $\mathbf{B}$.

Example 1

Let $\mathbf{A}=\begin{pmatrix} 2 & 1 & 3 \\ 0 & -1 & 7 \end{pmatrix}$ and $\mathbf{B}=\begin{pmatrix} 4 & -2 & 6 \\ -1 & 0 & 2 \\ 3 & 3 & -5 \end{pmatrix}$. Find $\mathbf{AB}$.

Solution

$\mathbf{A}$ is 2×3 and $\mathbf{B}$ is 3×3, so $\mathbf{AB}$ exists (and is 2×3).

$$\mathbf{AB}=\begin{pmatrix} 2 & 1 & 3 \\ 0 & -1 & 7 \end{pmatrix}\begin{pmatrix} 4 & -2 & 6 \\ -1 & 0 & 2 \\ 3 & 3 & -5 \end{pmatrix}$$

$$=\begin{pmatrix} 16 & 5 & -1 \\ 22 & 21 & -37 \end{pmatrix}.$$

(The 5 in the 1st row and 2nd column of the answer, for example, is

the scalar product of $(2 \ \ 1 \ \ 3)$ and $\begin{pmatrix} -2 \\ 0 \\ 3 \end{pmatrix}$.)

If $\mathbf{C}$ and $\mathbf{D}$ are two matrices, it may be that $\mathbf{CD}$ exists but $\mathbf{DC}$ does not. Even when both products exist, they will not normally be equal.

Qu.8 Let $\mathbf{A} = \begin{pmatrix} 1 & 2 \\ 0 & -1 \end{pmatrix}$, $\mathbf{B} = \begin{pmatrix} 1 & 2 \\ 1 & -3 \\ 0 & 4 \end{pmatrix}$ and

$\mathbf{C} = \begin{pmatrix} 2 & 0 & 3 & 0 \\ -1 & 1 & 4 & 5 \end{pmatrix}$. Evaluate where possible:

   (i) **AB**;   (ii) **AC**;   (iii) **BA**;   (iv) **CA**;
   (v) $\mathbf{AA}^T$;   (vi) $\mathbf{A}^T\mathbf{A}$;   (vii) $\mathbf{A}^2$;   (viii) $\mathbf{A}^3$.

Qu.9 Let $\mathbf{L} = \begin{pmatrix} a & b \\ c & d \end{pmatrix}$, $\mathbf{M} = \begin{pmatrix} p & q \\ r & s \end{pmatrix}$, $\mathbf{N} = \begin{pmatrix} w & x \\ y & z \end{pmatrix}$.

By expanding each side, determine which of the following are true.
  (i) **LM** = **ML** (Commutativity);
  (ii) **(LM)N** = **L(MN)** (Associativity);
  (iii) **L(M + N)** = **LM** + **LN** and **(M + N)L** = **ML** + **NL**
(Distributivity of multiplication over addition);
  (iv) $(\mathbf{LM})^T = \mathbf{L}^T\mathbf{M}^T$;
  (v) $(\mathbf{LM})^T = \mathbf{M}^T\mathbf{L}^T$.

Qu.10 (i) Show that $\begin{pmatrix} 1 & 0 \\ 0 & 1 \end{pmatrix}$ is the identity element for 2×2
matrices under matrix multiplication.
(ii) Find the corresponding identity element for 3×3 matrices.
(The letter **I** is usually used for these, and larger, identity elements.)

The properties of 2×2 matrices established in parts (ii), (iii) and (v) of Qu.9 are true for matrices of any size, although the general proofs are simplest if the suffix notation is used. This is introduced in Chapter 35. We can, however, prove the following general result.

Theorem 21.1
$(\mathbf{AB})^T = \mathbf{B}^T\mathbf{A}^T$.
Proof
We show that for any i and j the elements in the ith row and jth column of each side are equal.

For $(\mathbf{AB})^T$:
    element in ith row and jth column
      = element in jth row and ith column of **AB**
      = scalar product of jth row of **A** with ith column of **B**.   (1)

For $\mathbf{B}^T\mathbf{A}^T$:
    element in ith row and jth column
      = scalar product of ith row of $\mathbf{B}^T$ with jth column of $\mathbf{A}^T$
      = scalar product of ith column of **B** with jth row of **A**.   (2)

The expressions (1) and (2) are equal, so the elements in $(\mathbf{AB})^T$ are the same as in $\mathbf{B}^T\mathbf{A}^T$. Hence $(\mathbf{AB})^T = \mathbf{B}^T\mathbf{A}^T$.

Qu.11   Use the result of the theorem to prove that $(\mathbf{PQR})^T = \mathbf{R}^T\mathbf{Q}^T\mathbf{P}^T$. [*Hint*: write $\mathbf{PQR}$ as $(\mathbf{PQ})\mathbf{R}$ first.]

21.4 Determinants

The *determinant* of a square matrix is a scalar (i.e. a number rather than a matrix) which acts as a 'measure' of the matrix, in much the same way as the modulus measures the 'size' of a vector. The method of calculating the determinant of a 3×3 matrix is moderately complicated and requires practice; but we shall begin with 2×2 matrices.

Let $\mathbf{M} = \begin{pmatrix} a & b \\ c & d \end{pmatrix}$. The determinant of $\mathbf{M}$ is denoted by det $\mathbf{M}$ or $|\mathbf{M}|$ or $\begin{vmatrix} a & b \\ c & d \end{vmatrix}$ or simply Δ (capital delta), and is defined to be $ad - bc$. So, for example,

$$\begin{vmatrix} 2 & 3 \\ -2 & -5 \end{vmatrix} = (2 \times -5) - (3 \times -2)$$

$$= -4.$$

Qu.12   Evaluate

(i) $\begin{vmatrix} 2 & 1 \\ -2 & 3 \end{vmatrix}$;   (ii) $\begin{vmatrix} 2 & 3 \\ 4 & 6 \end{vmatrix}$.

21.5 The determinant of a 3×3 matrix

Consider the matrix $\mathbf{M} = \begin{pmatrix} a & b & c \\ d & e & f \\ g & h & k \end{pmatrix}$. We define the *minor* of an element of $\mathbf{M}$ to be the 2×2 determinant obtained by crossing out the row and column containing that element. For example, the minor of f would be found thus:

$\begin{pmatrix} a & b & c \\ d & e & f \\ g & h & k \end{pmatrix}$ leaving $\begin{vmatrix} a & b \\ g & h \end{vmatrix} = ah - bg$.

The *cofactor* of an element is the minor of the element multiplied by $+1$ or -1, the sign being determined by the

pattern $\begin{pmatrix} + & - & + \\ - & + & - \\ + & - & + \end{pmatrix}$. Cofactors are conveniently denoted by the corresponding capital letter, so that, for example, the cofactors of c and d are

$$C = + \begin{vmatrix} d & e \\ g & h \end{vmatrix} = dh - eg$$

and
$$D = - \begin{vmatrix} b & c \\ h & k \end{vmatrix} = ch - bk.$$

Qu.13 Find
(i) A; (ii) H.

The new matrix formed by replacing each element of a matrix **M** by its cofactor is called the *matrix of cofactors* and may be denoted by cof **M**.

Qu.14 Let $S = \begin{pmatrix} 1 & 0 & 2 \\ -1 & -3 & 2 \\ 0 & 4 & 5 \end{pmatrix}$. Find the matrix cof **S**.

We are now in a position to define the determinant of a 3×3 matrix; the definition is included in the following theorem which is of considerable importance for later work.

Theorem 21.2
(i) Given a 3×3 matrix **M,** the scalar product of any row or column with the corresponding row or column of cof **M** gives a fixed number called the *determinant of* **M** (i.e. the value of this scalar product does not depend on which row or column is chosen.)
(ii) The scalar product of any row of **M** with a 'wrong' row of cof **M** is zero. Similarly, the scalar product of a column of **M** with a wrong column of cof **M** is zero. When the 'wrong' cofactors are used in this way, they are called *alien cofactors.*
Proof

As before, let **M** be the general matrix $\begin{pmatrix} a & b & c \\ d & e & f \\ g & h & k \end{pmatrix}$ so that

cof $M = \begin{pmatrix} A & B & C \\ D & E & F \\ G & H & K \end{pmatrix}$. Each part of the theorem may be proved by direct evaluation.
(i) Using the top row, the determinant is given by

$$\begin{aligned} aA &+ bB + cC \\ &= a(ek - fh) + b(fg - dk) + c(dh - eg) \\ &= aek - afh + bfg - bdk + cdh - ceg. \end{aligned}$$

The reader is left to verify that the same result is obtained using any other row or column. (Note that each term contains three factors, no two of which are in the same row or column.)
(ii) Using the top row of **M** and the second row of cof **M**,

$$\begin{aligned} aD &+ bE + cF \\ &= a(ch - bk) + b(ak - cg) + c(bg - ah) \\ &= ach - abk + bak - bcg + cbg - cah \\ &= 0. \end{aligned}$$

Again the reader is left to complete the proof.

Example 2

Evaluate
$$\begin{vmatrix} 1 & -3 & 1 \\ 1 & 2 & -2 \\ -5 & 4 & 0 \end{vmatrix}.$$

Solution
We shall expand by the third column. (The zero in this column simplifies the calculation.) The cofactors of the 1 and the -2 in this column are 14 and 11 respectively, so

$$\Delta = (1 \times 14) + (-2 \times 11)$$
$$= -8.$$

Qu.15 Obtain the same answer by using
(i) the top row; (ii) the third row.

Qu.16 Let $\mathbf{M} = \begin{pmatrix} a & b \\ c & d \end{pmatrix}$. Defining the cofactor of an element to be the number remaining when the row and column containing the element is deleted, with the appropriate sign attached according to the pattern $\begin{pmatrix} + & - \\ - & + \end{pmatrix}$, show that if the method of finding the determinant of a 3×3 matrix is used, the correct result (i.e. $ad - bc$) is obtained.

21.6 Determinants of larger square matrices

All the work of the previous section for 3×3 matrices applies also to larger square matrices. A cofactor in a 4×4 matrix is the 3×3 determinant obtained by deleting the row and column containing the element in question, together with a sign according to the pattern

$$\begin{pmatrix} + & - & + & - \\ - & + & - & + \\ + & - & + & - \\ - & + & - & + \end{pmatrix}.$$

So evaluating a 4×4 determinant involves calculating four 3×3 determinants. Larger square determinants may be calculated similarly.

21.7 Inverse matrices

We have seen that $\begin{pmatrix} 1 & 0 \\ 0 & 1 \end{pmatrix}$ and $\begin{pmatrix} 1 & 0 & 0 \\ 0 & 1 & 0 \\ 0 & 0 & 1 \end{pmatrix}$ are the multiplicative identity matrices for 2×2 and 3×3 matrices respectively. But we have not yet considered the question of whether inverse matrices exist; i.e.

given a square matrix $\mathbf{M}$, is there a matrix $\mathbf{M}^{-1}$ such that

$$\mathbf{M}\mathbf{M}^{-1} = \mathbf{M}^{-1}\mathbf{M} = \mathbf{I}?$$

Definition

A square matrix whose determinant is zero is said to be *singular*. All other square matrices are said to be *non-singular*.

We shall now see that any non-singular matrix has an inverse.

21.8 The inverse of a 2×2 matrix

Let $\mathbf{M} = \begin{pmatrix} a & b \\ c & d \end{pmatrix}$ be a non-singular matrix (i.e. $ad - bc \neq 0$). Then the inverse of $\mathbf{M}$ is given by

$$\mathbf{M}^{-1} = \frac{1}{\det \mathbf{M}} \begin{pmatrix} d & -b \\ -c & a \end{pmatrix}$$

$$= \begin{pmatrix} \dfrac{d}{ad-bc} & \dfrac{-b}{ad-bc} \\[2mm] \dfrac{-c}{ad-bc} & \dfrac{a}{ad-bc} \end{pmatrix}.$$

In practice, the method for finding a 2×2 inverse may be remembered in three stages:

(i) Starting with the original matrix, interchange 'a' and 'd'.

(ii) Change the signs of 'b' and 'c'.

(iii) Divide by the determinant.

So, for example, if $\mathbf{S} = \begin{pmatrix} 2 & 1 \\ -3 & 4 \end{pmatrix}$ then

$$\mathbf{S}^{-1} = \frac{1}{11} \begin{pmatrix} 4 & -1 \\ 3 & 2 \end{pmatrix}$$

$$= \begin{pmatrix} 4/11 & -1/11 \\ 3/11 & 2/11 \end{pmatrix}.$$

Qu.17 Check this result by evaluating $\mathbf{S}^{-1}\mathbf{S}$ and $\mathbf{S}\mathbf{S}^{-1}$.

Qu.18 Find where possible the inverses of the following. Check your answers.

(i) $\begin{pmatrix} -1 & 4 \\ 0 & 2 \end{pmatrix}$; (ii) $\begin{pmatrix} 2 & 3 \\ -3 & -4 \end{pmatrix}$; (iii) $\begin{pmatrix} 1 & 2 \\ -2 & -4 \end{pmatrix}$.

Qu.19 Let $\mathbf{A} = \begin{pmatrix} 4 & 3 \\ 6 & 4 \end{pmatrix}$. Find $\mathbf{A}^{-1}$ and $(\mathbf{A}^{-1})^{-1}$. Comment on your second answer.

Qu.20 Let $\mathbf{T} = \begin{pmatrix} p & q \\ r & s \end{pmatrix}$. Write down $\mathbf{T}^{-1}$ and check by direct multiplication that $\mathbf{T}\mathbf{T}^{-1} = \mathbf{T}^{-1}\mathbf{T} = \mathbf{I}$.

THE ALGEBRA OF MATRICES

Qu.18(iii) shows how the method of finding an inverse fails if $\Delta=0$. The fact that singular matrices have no inverse is analogous to the fact that the number zero has no inverse under ordinary multiplication.

21.9 The inverse of a 3×3 matrix

Definition

Let **M** be a 3×3 matrix. The *adjoint* or *adjugate* of **M**, denoted by adj **M**, is defined by

$$\text{adj } \mathbf{M} = (\text{cof } \mathbf{M})^T.$$

So if $\mathbf{M} = \begin{pmatrix} a & b & c \\ d & e & f \\ g & h & k \end{pmatrix}$ then adj $\mathbf{M} = \begin{pmatrix} A & D & G \\ B & E & H \\ C & F & K \end{pmatrix}$.

Theorem 21.3

$$\mathbf{M} \,(\text{adj } \mathbf{M}) = (\text{adj } \mathbf{M})\, \mathbf{M} = \begin{pmatrix} \Delta & 0 & 0 \\ 0 & \Delta & 0 \\ 0 & 0 & \Delta \end{pmatrix}$$

where $\Delta = \det \mathbf{M}$.

Proof

Let **M** be as above. The result follows immediately from the two parts of Theorem 21.2, since in evaluating the product

$$\mathbf{M} \,(\text{adj } \mathbf{M}) = \begin{pmatrix} a & b & c \\ d & e & f \\ g & h & k \end{pmatrix} \begin{pmatrix} A & D & G \\ B & E & H \\ C & F & K \end{pmatrix}$$

each row of **M** is multiplied either by its own cofactors (giving Δ) or by alien cofactors (giving zero). Similarly for (adj **M**) **M**.

Now since **M** (adj **M**) = Δ**I** (by the theorem), it follows that $\mathbf{M}\left(\dfrac{1}{\Delta}\text{adj } \mathbf{M}\right) = \mathbf{I}$, and similarly $\left(\dfrac{1}{\Delta}\text{adj } \mathbf{M}\right)\mathbf{M} = \mathbf{I}$. So $\dfrac{1}{\Delta}\text{adj } \mathbf{M}$ is the inverse of **M**, i.e.

$$\mathbf{M}^{-1} = \frac{1}{\Delta}\text{adj } \mathbf{M}.$$

The method for finding the inverse of a 3×3 matrix may therefore be summarized thus:

(i) Find the matrix of cofactors, cof **M**; also find Δ.
(ii) Transpose to get adj **M**.
(iii) Divide through by Δ to get $\mathbf{M}^{-1}$.

This method may be extended to find inverses of 4×4 and higher order matrices.

Example 3

Find the inverse of $\mathbf{M} = \begin{pmatrix} 1 & -2 & 0 \\ 2 & -5 & 1 \\ 4 & -4 & -2 \end{pmatrix}$.

Solution

Calculating cofactors of each element gives

$\text{cof } \mathbf{M} = \begin{pmatrix} 14 & 8 & 12 \\ -4 & -2 & -4 \\ -2 & -1 & -1 \end{pmatrix}$, and expanding by the top row

gives $\Delta = -2$. (A useful way of checking the cofactors at this stage is to find Δ using each of the other two rows.)

Transposing cof $\mathbf{M}$ gives $\text{adj } \mathbf{M} = \begin{pmatrix} 14 & -4 & -2 \\ 8 & -2 & -1 \\ 12 & -4 & -1 \end{pmatrix}$.

Then

$$\mathbf{M}^{-1} = \frac{1}{\Delta} \text{adj } \mathbf{M}$$

$$= \frac{1}{-2} \begin{pmatrix} 14 & -4 & -2 \\ 8 & -2 & -1 \\ 12 & -4 & -1 \end{pmatrix}$$

$$= \begin{pmatrix} -7 & 2 & 1 \\ -4 & 1 & \frac{1}{2} \\ -6 & 2 & \frac{1}{2} \end{pmatrix}.$$

Qu.21 Check this answer by evaluating $\mathbf{M}\mathbf{M}^{-1}$.

Qu.22 Find, if possible, the inverses of

(i) $\begin{pmatrix} 7 & -2 & -3 \\ -4 & 1 & 1 \\ 3 & -1 & 1 \end{pmatrix}$; (ii) $\begin{pmatrix} 0 & 2 & 3 \\ -3 & 5 & -1 \\ 3 & -3 & 4 \end{pmatrix}$.

Qu.23 (See also Qu.16.) Let $\mathbf{M} = \begin{pmatrix} a & b \\ c & d \end{pmatrix}$. Find adj $\mathbf{M}$ and verify that $\mathbf{M}^{-1} = \frac{1}{\Delta} \text{adj } \mathbf{M}$.

Qu.24 Show that
(i) $(\mathbf{A}^{-1})^{-1} = \mathbf{A}$; (ii) $(\mathbf{A}\mathbf{B})^{-1} = \mathbf{B}^{-1}\mathbf{A}^{-1}$.
(There is no need for heavy algebra: the proofs are outlined in questions 8 and 15 of Exercise 1d.)

21.10 Solving matrix equations

In solving equations involving square matrices, it must be remembered that matrices do not commute under multiplication. If one side of an equation is pre-multiplied by **M**, the other side must also be pre-multiplied by **M** and not post-multiplied.

Example 4
Solve for **X** the equation **AX**=**B**, where each of the matrices is square and **A** is non-singular.
Solution
Pre-multiplying each side of the equation by A^{-1},

$$A^{-1}AX = A^{-1}B$$
$$\Rightarrow IX = A^{-1}B$$
$$\Rightarrow X = A^{-1}B.$$

Qu.25 Factorize where possible:
(i) **XA**+2**XB**; (ii) **XA**+**X**; (iii) **AX**+**XB**.
Qu.26 Solve for **X** the following. You may assume that any matrix whose inverse you need is non-singular.
(i) **AX**−**B**=**C**; (ii) **AXA**$^{-1}$=**B**;
(iii) **AX**+**B**=**CX**.

Theorem 21.4
If **A** is a non-singular square matrix, then

$$(A^T)^{-1} = (A^{-1})^T.$$

Proof
Using Theorem 21.1,

$$(AA^{-1})^T = (A^{-1})^T A^T.$$

But

$$(AA^{-1})^T = I^T = I,$$

so

$$(A^{-1})^T A^T = I.$$

Post-multiplying each side by $(A^T)^{-1}$ gives

$$(A^{-1})^T = (A^T)^{-1}$$

as required.

Exercise 21a   **1.** Let $P = \begin{pmatrix} 1 & 3 \\ 2 & -2 \end{pmatrix}$, $Q = \begin{pmatrix} -2 & 0 & -3 \\ 1 & -2 & 2 \end{pmatrix}$ and

$$R = \begin{pmatrix} 3 & -2 \\ 0 & 4 \\ 8 & 0 \end{pmatrix}.$$ Find, where possible,

(i) **P**+**Q**; (ii) **PQ**; (iii) **QR**;
(iv) **Q**+**R**^{T}; (v) **R**^{T}**P**; (vi) **P**^{3}; (vii) **Q**^{2};
(viii) **RR**^{T}; (ix) 2**P**+**QR**; (x) **P**^{T}+**RQ**;
(xi) **PQR**; (xii) **RQR**; (xiii) **QRQ**.

2. Let $\mathbf{A}=(p \quad q)$ and $\mathbf{B}=\begin{pmatrix} r \\ s \end{pmatrix}$. Evaluate

(i) $\mathbf{AB}$;   (ii) $\mathbf{BA}$.

3. Suppose that $\mathbf{A}$ and $\mathbf{B}$ are two matrices such that both products $\mathbf{AB}$ and $\mathbf{BA}$ can be formed.

   (i) Show that the sum $\mathbf{A}+\mathbf{B}^T$ can be formed.

   (ii) Is the converse true?

   (iii) What can be said about the matrices $\mathbf{AB}$ and $\mathbf{BA}$?

4. Prove that, for any matrix $\mathbf{A}$, $\mathbf{AA}^T$ exists and is symmetric. [*Hint*: use the method of Qu.7 in the text.]

5. By considering the possible order of $\mathbf{I}$, show that if $\mathbf{A}$ is a non-square matrix, then no matrix $\mathbf{I}$ exists such that $\mathbf{AI}=\mathbf{IA}=\mathbf{A}$.

6. Simplify (i) $(\mathbf{A}^T\mathbf{B}^T\mathbf{C}^T)^T$;   (ii) $\{(\mathbf{AB}^T)^T\mathbf{C}^T\}^T$.

7. Let $\mathbf{S}$ be a 2×3 matrix, and let $\lambda \in R$. Find a matrix $\mathbf{X}$ such that $\mathbf{XS}=\lambda\mathbf{S}$, and a matrix $\mathbf{Y}$ such that $\mathbf{SY}=\lambda\mathbf{S}$.

8. (i) Prove that if $\mathbf{S}$ is both symmetric and skew-symmetric then it is the zero matrix.

   (ii) Let $\mathbf{M}$ be square. Show that $\mathbf{M}+\mathbf{M}^T$ is symmetric, and that $\mathbf{M}-\mathbf{M}^T$ is skew symmetric.

   (iii) Show that any square matrix $\mathbf{M}$ is the sum of a symmetric matrix and a skew-symmetric matrix. [*Hint*: use part (ii).]

   (iv) Prove that the representation obtained in (iii) is unique.

9. Prove that two symmetric matrices commute if and only if their product is symmetric.

10. Let $\mathbf{A}=\begin{pmatrix} 7 & -4 \\ -6 & 3 \end{pmatrix}$ and $\mathbf{B}=\begin{pmatrix} -5 & -2 \\ 3 & 1 \end{pmatrix}$. Find

(i) $\mathbf{A}^{-1}$;   (ii) $\mathbf{B}^{-1}$;   (iii) $\mathbf{B}^{-1}\mathbf{A}^{-1}$;   (iv) $\mathbf{AB}$;   (v) $(\mathbf{AB})^{-1}$.
Hence verify directly that

$$(\mathbf{AB})^{-1}=\mathbf{B}^{-1}\mathbf{A}^{-1}.$$

Show by direct calculation that

$$(\mathbf{A}^T)^{-1}=(\mathbf{A}^{-1})^T.$$

11. Let $\mathbf{A}=\begin{pmatrix} 8 & 3 & 0 \\ -1 & -1 & 2 \\ 3 & 1 & 0 \end{pmatrix}$ and $\mathbf{B}=\begin{pmatrix} 2 & 0 & 1 \\ 3 & 1 & 4 \\ -4 & 1 & 0 \end{pmatrix}$.

Evaluate

(i) $(\mathbf{A}+\mathbf{B})^2$;   (ii) $\mathbf{A}^2+2\mathbf{AB}+\mathbf{B}^2$.

Explain why these are not equal, giving the correct expansion of (i).

12. Find the determinants of the following matrices:

(i) $\begin{pmatrix} 3 & 1 & -5 \\ -7 & -4 & 8 \\ -1 & -2 & -1 \end{pmatrix}$;   (ii) $\begin{pmatrix} -2 & -9 & 3 \\ 4 & -2 & 1 \\ 0 & 3 & -1 \end{pmatrix}$;

(iii) $\begin{pmatrix} 4 & 0 & 1 \\ -3 & 2 & 0 \\ 5 & -1 & -7 \end{pmatrix}$;   (iv) $\begin{pmatrix} -6 & 3 & 2 \\ 0 & 4 & -3 \\ 1 & 2 & 1 \end{pmatrix}$.

13. Find the inverses of each of the matrices in question 12.

14. With **A** and **B** as in question 11, find $\mathbf{A}^{-1}$ and $\mathbf{B}^{-1}$.

15. Find the inverse of $\mathbf{A} = \begin{pmatrix} 2 & 1 & -1 \\ 1 & 2 & 3 \\ 6 & 4 & 0 \end{pmatrix}$, and hence find

a matrix **M** such that $\mathbf{A}^{-1}\mathbf{M}\mathbf{A} = \begin{pmatrix} 1 & 0 & 1 \\ 0 & 1 & 0 \\ 1 & 0 & 1 \end{pmatrix}$.

16. Factorize where possible:
(i) $\mathbf{PQ} + \mathbf{PR}$; (ii) $\mathbf{PQP} + \mathbf{RP}$; (iii) $\mathbf{A}^2 - \mathbf{B}^2$;
(iv) $\mathbf{A}^2 + \mathbf{BA} - \mathbf{AC} - \mathbf{BC}$; (v) $\mathbf{A}^2 - \mathbf{I}$; (vi) $\mathbf{A}^2\mathbf{B}^2 - 4\mathbf{I}$.

17. Solve for **X**:
(i) $\mathbf{A} + \mathbf{X} = \mathbf{B}$;   (ii) $\mathbf{AX} = \mathbf{B}$;   (iii) $\mathbf{XA} = \mathbf{B}$;
(iv) $\mathbf{AXA} = \mathbf{B}$;   (v) $\mathbf{AX} + \mathbf{X} = \mathbf{B}$;   (vi) $\mathbf{AXB} = \mathbf{I}$;
(vii) $\mathbf{ABX} = \mathbf{I}$;   (viii) $\mathbf{XAB} = \mathbf{I}$.

18. (Hard.) Show that there are infinitely many matrices **A** satisfying $\mathbf{A}^2 = \mathbf{I}$.

Find all solutions of $\mathbf{A}^2 = \begin{pmatrix} 1 & 1 \\ 2 & 3 \end{pmatrix}$.

19. Find the inverse of $\begin{pmatrix} 1 & 0 & -2 & 0 \\ 1 & 1 & 0 & 2 \\ 0 & -3 & 4 & 0 \\ 0 & 0 & -1 & 0 \end{pmatrix}$.

(A more suitable method is given in Section 21.11 (Qu.31), or see Exercise 27c, question 15.)

21.11 Properties of determinants

Most of the rules below are fairly simple, and their proofs are left to the reader. It will help to consider the determinant $\begin{vmatrix} a & b & c \\ d & e & f \\ g & h & k \end{vmatrix}$,

although the rules are valid for determinants of any size.

Rule 1. If two rows (or columns) of a determinant are equal, then the determinant is zero. [*Hint*: expand by the remaining row or column; all cofactors are zero.]

Rule 2. If one row is a multiple of another row, then the determinant is zero. Similarly for columns.

Rule 3. If two rows (or columns) are interchanged, the determinant changes sign. [Expand by the remaining row (column); the cofactors change sign.]

Rule 4. If one row or column is multiplied by a scalar λ, then the determinant is multiplied by λ.

Rule 5. If a multiple of one row is added to another, the determinant is unchanged. (Similarly for columns.) To show this, suppose that λ times the 2nd column is added to the 1st column, giving

$$\begin{vmatrix} a+\lambda b & b & c \\ d+\lambda e & e & f \\ g+\lambda h & h & k \end{vmatrix}.$$ Expanding by the first column, it may be shown

that this is equal to $\begin{vmatrix} a & b & c \\ d & e & f \\ g & h & k \end{vmatrix} + \begin{vmatrix} \lambda b & b & c \\ \lambda e & e & f \\ \lambda h & h & k \end{vmatrix}$, and this second

determinant is zero by rule 2.

Rule 6. If a matrix is transposed, its determinant is unchanged.

Qu.27   If det $\mathbf{M} = \Delta$, express det $(\lambda \mathbf{M})$ in terms of λ and Δ when $\mathbf{M}$ is
(i) 2×2;   (ii) 3×3;   (iii) $n \times n$.

Qu.28   If the rows of a 3×3 matrix are interchanged cyclicly, what is the effect on the determinant? [*Hint*: use rule 3 above as many times as is necessary.]

The manipulations described in rules 3, 4 and 5 above are known as *elementary row and column operations*. They can sometimes be used to simplify a determinant before evaluating it. The aim is to reduce (some of) the entries to a more manageable size; in particular, it is useful to obtain a zero entry, or, ideally, two zeros in one row or column. But note that it is not worth spending an hour on these manipulations to save two minutes.

The following example illustrates the method and also a useful check.

Example 5

Evaluate $\begin{vmatrix} 64 & 51 & 34 \\ 58 & -41 & 29 \\ -40 & 12 & -19 \end{vmatrix}.$

Solution
(We use r for row and c for column.)

$$\Delta = 2 \begin{vmatrix} 32 & 51 & 34 \\ 29 & -41 & 29 \\ -20 & 12 & -19 \end{vmatrix} \quad \text{(common factor from } c_1\text{)}$$

$$= 2 \begin{vmatrix} 32 & 51 & 2 \\ 29 & -41 & 0 \\ -20 & 12 & 1 \end{vmatrix} \quad \text{(taking } c_1 \text{ from } c_3\text{)}$$

$$= 2 \begin{vmatrix} 72 & 27 & 0 \\ 29 & -41 & 0 \\ -20 & 12 & 1 \end{vmatrix} \quad \text{(taking } 2 \times r_3 \text{ from } r_1\text{)}$$

$$= 18 \begin{vmatrix} 8 & 3 & 0 \\ 29 & -41 & 0 \\ -20 & 12 & 1 \end{vmatrix} \quad \text{(common factor from } r_1\text{)}$$

$$= 18 \times 1 \times \{8 \times (-41) - 3 \times 29\}$$
$$= -7470.$$

A useful check is to reduce the original problem modulo m for any convenient m and recalculate. Working modulo 5, we have

$$\begin{vmatrix} 4 & 1 & 4 \\ 3 & 4 & 4 \\ 0 & 2 & 1 \end{vmatrix} \equiv 4(4-8) - 3(1-8) \bmod 5$$

$$\equiv 0 \bmod 5.$$

Since -7470 is also zero mod 5, we have (limited) confirmation of our answer.

Qu.29 Using elementary row and column operations where appropriate, evaluate the following. Check your answer to (i) by reducing mod 3 and also mod 7.

$$\text{(i)} \quad \begin{vmatrix} 123 & 83 & -20 \\ 56 & 57 & -49 \\ 12 & 66 & 36 \end{vmatrix}; \quad \text{(ii)} \quad \begin{vmatrix} -92 & -15 & 74 \\ 83 & 16 & -28 \\ 65 & 18 & 64 \end{vmatrix}.$$

Qu.30 (i) Let $\mathbf{M}$ be any 3×3 matrix. Show that pre-multiplying $\mathbf{M}$

by $\begin{pmatrix} 1 & 0 & 0 \\ 0 & 0 & 1 \\ 0 & 1 & 0 \end{pmatrix}$ has the effect of interchanging the second and third rows

of $\mathbf{M}$. What is the effect of post-multiplying by the same matrix?
(ii) Write down the matrix which will interchange the first two rows of $\mathbf{M}$ when $\mathbf{M}$ is pre-multiplied by it.
(iii) Investigate the effect of pre-multiplying $\mathbf{M}$ and post-multiplying $\mathbf{M}$ by the following:

$$\text{(a)} \begin{pmatrix} 2 & 0 & 0 \\ 0 & 1 & 0 \\ 0 & 0 & 1 \end{pmatrix}; \quad \text{(b)} \begin{pmatrix} 1 & 0 & 0 \\ 0 & 1 & -3 \\ 0 & 0 & 1 \end{pmatrix}.$$

(These matrices, which induce elementary row and column operations on $\mathbf{M}$, are examples of *elementary matrices*.)

Row operations may be used to find the inverse of a matrix $\mathbf{A}$. The method is to apply a sequence of row operations to each of $\mathbf{A}$ and $\mathbf{I}$. If the matrix that starts as $\mathbf{A}$ can be reduced to $\mathbf{I}$, then the matrix that starts as $\mathbf{I}$ will become $\mathbf{A}^{-1}$. This can be shown by observing that if the elementary matrices corresponding to the operations are $\mathbf{E}_1$, $\mathbf{E}_2$,

$$\ldots \mathbf{E}_n, \text{ then}$$

$$\mathbf{E}_n \mathbf{E}_{n-1} \ldots \mathbf{E}_2 \mathbf{E}_1 (\mathbf{A}) = \mathbf{I}$$
$$\Rightarrow \mathbf{E}_n \mathbf{E}_{n-1} \ldots \mathbf{E}_2 \mathbf{E}_1 = \mathbf{A}^{-1}$$
$$\Rightarrow \mathbf{E}_n \mathbf{E}_{n-1} \ldots \mathbf{E}_2 \mathbf{E}_1 (\mathbf{I}) = \mathbf{A}^{-1}.$$

The method is known as *row reduction*.

Qu.31 Let $\mathbf{M}$ be as in Example 3 of this chapter. By applying a sequence of row operations to $\mathbf{A}$ and $\mathbf{I}$ to reduce $\mathbf{A}$ to $\mathbf{I}$, find $\mathbf{A}^{-1}$. [*Hint*: start by subtracting twice the top row from the second row, and four times the top row from the bottom row. This puts the first column into the required form. Now deal with the second column by adding or subtracting multiples of the second row.]

We now state an important theorem on determinants. A proof for 3×3 matrices is outlined in question 3 of Exercise 21b, and see question 2 for 2×2 matrices. Another proof for 3×3 matrices is given in Chapter 34 (corollary to Theorem 34.5), but the general proof is beyond the scope of this book.

Theorem 21.5
If $\mathbf{A}$ and $\mathbf{B}$ are square matrices of the same size, then

$$\det (\mathbf{AB}) = (\det \mathbf{A})(\det \mathbf{B}).$$

Corollary 1
The product of two non-singular matrices is non-singular.

Corollary 2
If $\mathbf{A}$ is non-singular, then

$$\det (\mathbf{A}^{-1}) = \frac{1}{\det \mathbf{A}}.$$

This follows immediately from

$$(\det \mathbf{A}^{-1})(\det \mathbf{A}) = \det (\mathbf{A}^{-1}\mathbf{A})$$
$$= \det \mathbf{I}$$
$$= 1.$$

21.12 Factorizing algebraic determinants

The properties of determinants discussed above, together with basic algebraic ideas from Chapter 10, can sometimes allow an algebraic determinant to be factorized without direct evaluation. Example 6 illustrates some of the most common techniques.

Example 6
Factorize

$$\text{(i)} \quad \begin{vmatrix} 1 & x & x^3 \\ 1 & y & y^3 \\ 1 & z & z^3 \end{vmatrix} ; \quad \text{(ii)} \quad \begin{vmatrix} a & b & c & 0 \\ b & a & 0 & c \\ c & 0 & a & b \\ 0 & c & b & a \end{vmatrix}.$$

Solution

(i) We notice the following features of the determinant Δ:

(a) Each term will be of degree 4, since it is a product of one entry from each column.

(b) We may regard Δ as a polynomial in x. Putting $x = y$ gives two equal rows so that $\Delta = 0$; so y is one root and $(x - y)$ is a factor. Similarly, $(y - z)$ and $(z - x)$ are factors.

(c) Δ is unchanged by a cyclic interchange of x, y and z.

From these observations, it follows that

$$\Delta \equiv (x - y)(y - z)(z - x)Q$$

where Q is a first degree expression in x, y and z which is unchanged by cyclic interchange. Q must therefore be of the form $k(x + y + z)$ for some $k \in R$. Thus

$$\Delta \equiv k(x - y)(y - z)(z - x)(x + y + z).$$

The value of k may be found by (for example) equating coefficients of $x^3 y$:

$$-k = -1 \Rightarrow k = 1.$$

So $\qquad\qquad \Delta \equiv (x - y)(y - z)(z - x)(x + y + z).$

(ii) We observe that

(a) Each term is of degree 4.

(b) Adding row 1, row 2 and row 3 to row 4 gives a common factor of $(a + b + c)$ in the new row 4.

(c) Adding row 1 and subtracting rows 2 and 3 from row 4 gives a common factor of $(b + c - a)$.

(d) Similarly, by taking suitable linear combinations of rows, $(c + a - b)$ and $(a + b - c)$ are factors.

So Δ is of the form

$$k(a + b + c)(b + c - a)(c + a - b)(a + b - c).$$

Putting $a = 1$ and $b = c = 0$ gives

$$1 = k(1)(-1)(1)(1) \Rightarrow k = -1.$$

So $\qquad\qquad \Delta \equiv -(a + b + c)(b + c - a)(c + a - b)(a + b - c)$

i.e. $\qquad\qquad \Delta \equiv (a + b + c)(a - b - c)(b - c - a)(c - a - b).$

Exercise 21b

1. (i) Let $\det \mathbf{M} = \Delta \ (\neq 0)$. Use the fact that $\mathrm{adj}\,\mathbf{M} = \Delta \mathbf{M}^{-1}$ to express $\det(\mathrm{adj}\,\mathbf{M})$ in terms of Δ when $\mathbf{M}$ is

(a) 3×3;   (b) $n \times n$.

[*Hint*: see Qu.27 in the text.]

(ii) Simplify $\mathrm{adj}\,(\mathrm{adj}\,\mathbf{M})$ when $\mathbf{M}$ is non-singular and is

(a) 3×3;   (b) $n \times n$.

2. Let $\mathbf{A} = \begin{pmatrix} p & q \\ r & s \end{pmatrix}$ and $\mathbf{B} = \begin{pmatrix} w & x \\ y & z \end{pmatrix}$. Prove by direct computation that $\det(\mathbf{AB}) = (\det \mathbf{A})(\det \mathbf{B})$.

3. This question outlines a proof that det $\mathbf{AB}=\det \mathbf{A}\det \mathbf{B}$ for 3×3 matrices. Let $\mathbf{A}=\begin{pmatrix} a_1 & b_1 & c_1 \\ a_2 & b_2 & c_2 \\ a_3 & b_3 & c_3 \end{pmatrix}$ and $\mathbf{B}=\begin{pmatrix} p_1 & p_2 & p_3 \\ q_1 & q_2 & q_3 \\ r_1 & r_2 & r_3 \end{pmatrix}$.

(i) Write out the product $\mathbf{AB}$ in full.

(ii) Show, by splitting the first column of the determinant $\mathbf{AB}$ (as in the proof of rule 5 in the text), that it may be expressed as the sum of three determinants, and that, by similarly splitting the other columns, det $\mathbf{AB}$ is the sum of 27 determinants.

(iii) Show that 21 of these 27 determinants are zero, and that each of the remaining six factorizes by removing a common factor from each column.

(iv) Show that the sum of these six determinants is det $\mathbf{A}$ det $\mathbf{B}$.

4. Use the method of question 3 to show that if $\mathbf{A}$ is 3×2 and $\mathbf{B}$ is 2×3, then $\mathbf{AB}$ is singular. [*Hint*: det $\mathbf{AB}$ expands as eight determinants.]

5. Prove that if $\mathbf{A}$ is an $n\times n$ skew-symmetric matrix with n odd, then det $\mathbf{A}=0$.

6. If $\begin{vmatrix} a_1 & a_2 & a_3 \\ b_1 & b_2 & b_3 \\ c_1 & c_2 & c_3 \end{vmatrix}=\Delta$, express $\begin{vmatrix} b_3 & c_3 & a_3 \\ b_1 & c_1 & a_1 \\ b_2 & c_2 & a_2 \end{vmatrix}$ in terms of Δ. (It is not necessary to expand either determinant.)

7. Show that the equation $\mathbf{A}^2=\begin{pmatrix} 0 & 0 & 1 \\ 0 & 1 & 0 \\ 1 & 0 & 0 \end{pmatrix}$ has no (real) solution for $\mathbf{A}$.

8. Write down the roots of $\begin{vmatrix} 1 & x & x^2 \\ 1 & 3 & 9 \\ 1 & 4 & 16 \end{vmatrix}=0$, and hence factorize this determinant.

9. Write down the sum of the roots of $\begin{vmatrix} 1 & x & x^3 \\ 1 & 2 & 8 \\ 1 & 3 & 27 \end{vmatrix}=0$, and factorize the determinant.

10. Factorize $\begin{vmatrix} 1 & x & x^2 & x^4 \\ 1 & 2 & 4 & 16 \\ 1 & -3 & 9 & 81 \\ 1 & 4 & 16 & 256 \end{vmatrix}$.

11. Factorize (i) $\begin{vmatrix} 1 & x & x^2 \\ 1 & y & y^2 \\ 1 & z & z^2 \end{vmatrix}$; (ii) $\begin{vmatrix} x & x^2 & x^3 \\ y & y^2 & y^3 \\ z & z^2 & z^3 \end{vmatrix}$.

12. Factorize $\begin{vmatrix} 1 & x^2 & x^3 \\ 1 & y^2 & y^3 \\ 1 & z^2 & z^3 \end{vmatrix}$. (Try your own method, and then try

also the following. Remove a factor of $x^3 y^3 z^3$, and use the result of Example 6 (i) with x replaced by $1/x$ etc.)

13. Factorize $\begin{vmatrix} x & 1+x^2 & x^3 \\ y & 1+y^2 & y^3 \\ z & 1+z^2 & z^3 \end{vmatrix}$ by first writing it as the sum of two

determinants.

14. Factorize (i) $\begin{vmatrix} x & y & z \\ yz & zx & xy \\ y+z & z+x & x+y \end{vmatrix}$; (ii) $\begin{vmatrix} x & y & z \\ x^2 & y^2 & z^2 \\ y+z & z+x & x+y \end{vmatrix}$.

15. Factorize $\begin{vmatrix} a & b & c & x \\ a & b & x & c \\ a & x & b & c \\ x & a & b & c \end{vmatrix}$.

1. (a) Given that $\mathbf{A} = \begin{pmatrix} 1 & -2 & -6 \\ 1 & -1 & 2 \\ 2 & -3 & -1 \end{pmatrix}$ and $\mathbf{B} = \begin{pmatrix} 7 & 16 & -10 \\ 5 & 11 & -8 \\ -1 & -1 & 1 \end{pmatrix}$

determine the matrix product $\mathbf{AB}$ and hence write down the matrix $\mathbf{A}^{-1}$. Use the matrix $\mathbf{A}^{-1}$ to obtain the values of x, y

and z given that $\mathbf{A} \begin{pmatrix} x \\ y \\ z \end{pmatrix} = \begin{pmatrix} 1 \\ 2 \\ 3 \end{pmatrix}$.

(b) Given that $\mathbf{C} = \begin{pmatrix} 3 & 1 \\ -1 & 2 \end{pmatrix}$, find the values of m and n for

which

$$\mathbf{C}^2 + m\mathbf{C} + n\mathbf{l} = \mathbf{0}.$$

[C (O)]

2. Square matrices $\mathbf{A}$ and $\mathbf{B}$ are such that $\mathbf{AB} = \mathbf{B}^{-1}$. Express $\mathbf{A}^{-1}$ in terms of $\mathbf{B}$.

Given that $\mathbf{B} = \begin{pmatrix} 2 & -1 \\ 2 & 0 \end{pmatrix}$, determine

THE ALGEBRA OF MATRICES

473

(i) the matrix $\mathbf{A}$,

(ii) the value of k for which $k\mathbf{A} - 2\mathbf{B}^{-1} + \mathbf{I} = \mathbf{0}$.

[C (O)]

3. If
$$\mathbf{A} = \begin{pmatrix} 1 & 2 & 3 \\ 0 & 1 & 2 \\ -1 & 1 & -1 \end{pmatrix}$$

show that
$$\mathbf{A}^3 = a\mathbf{A}^2 + b\mathbf{I}$$

where a, b are numbers to be determined. Deduce that $\mathbf{A}$ is non-singular.

Hence, or otherwise, find the matrix $\mathbf{X}$ such that
$$\mathbf{A}\mathbf{X} = \begin{pmatrix} 1 & 0 \\ 0 & 1 \\ 1 & 1 \end{pmatrix}.$$

[O & C]

4. Let
$$\mathbf{A} = \begin{pmatrix} 1 & 0 & 1 \\ 2 & 1 & 2 \\ 0 & -1 & 1 \end{pmatrix}.$$

Find elementary matrices $\mathbf{E}_1$, $\mathbf{E}_2$, $\mathbf{E}_3$ such that

(i) $\mathbf{E}_1\mathbf{A}$ has only one non-zero entry in the first column,

(ii) $\mathbf{E}_2\mathbf{E}_1\mathbf{A}$ has only one non-zero entry in each of the first two columns,

(iii) $\mathbf{E}_3\mathbf{E}_2\mathbf{E}_1\mathbf{A}$ is the identity matrix.

Use these results to find $\mathbf{A}^{-1}$.

[SMP]

5. $\mathbf{A}$ and $\mathbf{B}$ are 3×3 matrices such that $\mathbf{A} + \mathbf{B} = \mathbf{A}\mathbf{B}$, and $\mathbf{A}$ is non-singular. Prove that $\mathbf{B}$ is non-singular, and show that
$$\mathbf{A}^{-1} + \mathbf{B}^{-1} = \mathbf{I},$$

where $\mathbf{I}$ is the unit 3×3 matrix.

Given
$$\mathbf{A} = \begin{pmatrix} 2 & 2 & 0 \\ 0 & 2 & 0 \\ 0 & 0 & 2 \end{pmatrix},$$

find $\mathbf{B}$ so that $\mathbf{A}$ and $\mathbf{B}$ satisfy the above conditions.

[W]

6. Factorise the determinant
$$\begin{vmatrix} a^3 & a^2 & 1 \\ b^3 & b^2 & 1 \\ c^3 & c^2 & 1 \end{vmatrix}.$$

Given that a, b, c are unequal and the $bc + ca + ab \neq 0$, solve for x

the equation

$$\begin{vmatrix} a^3 & a^2 & (x+a) \\ b^3 & b^2 & (x+b) \\ c^3 & c^2 & (x+c) \end{vmatrix} = 0.$$

[JMB]

7. (i) Find all 3×3 matrices which commute with the matrix

$$\begin{pmatrix} 1 & 1 & 1 \\ 0 & 1 & 1 \\ 0 & 0 & 1 \end{pmatrix}$$ and show that they all commute with each other.

(ii) Prove that the only 3×3 matrices which commute with all 3×3 matrices are $\lambda \mathbf{I}$ for some value of λ.

(iii) If the matrices $\mathbf{C}$, $\mathbf{D}$ are defined by $\mathbf{C} = \mathbf{AB}$, $\mathbf{D} = \mathbf{BA}$ for 3×3 matrices $\mathbf{A}$, $\mathbf{B}$ prove that $\sum_{i=1}^{3} c_{ii} = \sum_{i=1}^{3} d_{ii}$. Deduce that it is impossible to find matrices $\mathbf{A}$, $\mathbf{B}$ such that

$$\mathbf{AB} - \mathbf{BA} = \begin{pmatrix} 1 & 1 & 1 \\ 0 & 1 & 1 \\ 0 & 0 & 1 \end{pmatrix}.$$

(iv) Prove that, if the 3×3 matrices $\mathbf{A}$, $\mathbf{B}$ both commute with $\mathbf{AB} - \mathbf{BA}$, then $\mathbf{AB}$ and $\mathbf{BA}$ commute.

[O & C (S)]

8.† (i) Show that

$$\begin{bmatrix} \lambda & 1 & 0 \\ 0 & \lambda & 1 \\ 0 & 0 & \lambda \end{bmatrix}^n = \begin{bmatrix} \lambda^n & n\lambda^{n-1} & \frac{1}{2}n(n-1)\lambda^{n-2} \\ 0 & \lambda^n & n\lambda^{n-1} \\ 0 & 0 & \lambda^n \end{bmatrix}$$

for all positive integers n.

(ii) Show that if $\mathbf{A}$ is any square matrix, than $\mathbf{A}^2$ commutes with $\mathbf{A}$ under matrix multiplication. Show that if

$$\mathbf{X}^2 = \begin{pmatrix} 4 & 1 & 0 \\ 0 & 4 & 1 \\ 0 & 0 & 4 \end{pmatrix},$$

then $\mathbf{X}$ has the form

$$\mathbf{X} = \begin{pmatrix} a & b & c \\ 0 & a & b \\ 0 & 0 & a \end{pmatrix}.$$

Hence find all possible such $\mathbf{X}$.

[MEI (S)]

---

†Square brackets, as in this MEI question, are a standard alternative to round brackets for a matrix. Do not confuse with the notation for a determinant.

22 Probability I

22.1 Terminology

In the study of probability, the word *experiment* means any process which leads to the noting or recording of a result. For example, an experiment might consist of selecting a card from a pack and noting whether it is or is not a spade; or it might consist of throwing a die and recording the score.

An experiment has a number of possible *outcomes*, and the set of all possible outcomes is called the *outcome space* or *sample space* for the experiment. A single experiment may give rise to a number of different outcome spaces, the choice depending on what feature of the experiment is of particular interest. Thus if two coins are tossed, and we ensure that the coins are distinguishable (perhaps by marking one of them), then possible outcome spaces would be:

(a) $\{(H\ H), (H\ T), (T\ H), (T\ T)\}$;
(b) {no heads, one head, two heads};
(c) {same, different}.

Of these three, (a) is perhaps the most useful, because it specifies the result in more detail than (b) or (c), and also because, unlike (b), the outcomes are all equally likely.

The essential features of an outcome space are that
(i) every possible result of the experiment must be included;
(ii) no two of the outcomes must 'overlap'.
Thus

{first coin a head, second a head, neither a head}

is not an outcome space for the experiment above, since the first two of the three outcomes overlap in the case of two heads.

An *event* is a subset of an outcome space; so if a die is thrown, and the natural outcome space $\{1, 2, 3, 4, 5, 6\}$ is selected, we may wish to refer to the events

E_1: an even number is scored,

i.e. $E_1 = \{2, 4, 6\}$;

and E_2: the score is greater than 3,

i.e. $E_2 = \{4, 5, 6\}$.

Set theory notation may be used for events, so that E_1' would mean 'an odd number is scored', and $E_1 \cap E_2$ would be 'an even number greater than 3 is scored'. In general, $\cap$ can be read as 'and' or 'both', and $\cup$ will mean 'or'.

22.2 Bernoulli trials

In performing an experiment, we are often interested in whether something specific does or does not happen. For example, in many children's games it is necessary to throw a six with a die in order to start. At the beginning of the game, we are not interested in the actual score, but merely in whether or not a six is thrown. We mentally classify the result according to the outcome space {six, not a six}. Any experiment for which there are just two possible outcomes is called a *Bernoulli trial*, and the terms 'success' and 'failure' are often used for the two outcomes.

22.3 Probability   We all have an intuitive idea of probability: we might say that the chances of something happening are 'fifty-fifty' or 'one in three' or 'seven to two against'. In mathematics, a probability is expressed as a real number p satisfying $0 \leqslant p \leqslant 1$, so that the non-mathematical terms above would convert thus:

$$\text{fifty-fifty} \to p = \tfrac{1}{2};$$
$$\text{one in three} \to p = 1/3;$$
$$\text{seven to two against} \to p = 2/9.$$

The value $p = 0$ means that the outcome or event *cannot* happen, and $p = 1$ means that it *must* happen.

Qu.1   Write down the probability of
 (i) scoring a 6 with a die;
 (ii) drawing a heart from an ordinary pack of cards;
(iii) drawing a card which is not a heart.

In many cases, what we intuitively mean by saying that the probability of some outcome is p is that if the experiment were repeated n times where n is large, and the number of successes (i.e. when the outcome *is* obtained) were s, then we should expect that

$$\frac{s}{n} \approx p.$$

More precisely, we might claim that we mean

$$\lim_{n \to \infty} \frac{s}{n} = p,$$

so that, for example, we could estimate the probability that a drawing-pin will land point-up when dropped, by dropping many such pins (or the same pin many times).

Some experiments, however, are not repeatable in any sense. For example, if we talk of the probability of there being life somewhere else in the universe, there is no question of our being able to check any estimate by testing for life in several hundred other universes.

Too detailed an examination of what we mean by probability can lead to difficult philosophical questions, so we shall be content here with our intuitive ideas. These lead immediately to a number of rules which we list below. They refer to an outcome space Ω with members (i.e. outcomes) ω_i. To each ω_i we attach a probability p_i.

(1) For each i, $0 \leqslant p_i \leqslant 1$.

(2) $\sum_{\omega_i \in \Omega} p_i = 1$ (i.e. the total probability is 1).

(3) For an event E, the probability of E, denoted by $P(E)$, is given by

$$P(E) = \sum_{\omega_i \in E} p_i$$

(i.e. it is the sum of the probabilities of the individual outcomes within E).

(4) $P(E') = 1 - P(E)$. This follows directly from (2) and (3).

22.4 Probability and our 'state of knowledge'

Qu.2 Discuss this argument: The probability of scoring a six with a (fair) die is said to be $1/6$; but the die either will or will not show a six, so the probability must be 0 or 1.

It is important to appreciate that the probability that we assign to an event depends on our 'state of knowledge'. Suppose, for example, that three men are asked to give the probability that a coin which has been tossed (and which is now covered over) has come up heads. The first man, suspecting nothing, gives the answer $\frac{1}{2}$. The second man, who provided the coin, knows that it is biased in favour of heads and says $\frac{2}{3}$. The third man, who arrived first in the room and looked at the coin, knows that it has come up tails; he says zero. Although three different answers are given, none of the men can be said to be wrong; they all have different states of knowledge. In a sense, they have all performed slightly different experiments.

The idea that a probability changes as more information becomes available is the basis of conditional probability (Section 22.8).

Qu.3 What is the probability that a six is scored with a die if
 (i) no information is available;
 (ii) the score is known to be even;
 (iii) the score is even and greater than 3?

22.5 Equiprobable outcomes

In many cases it is possible to select an outcome space in which all the outcomes are equally likely or *equiprobable*. If there are n possible outcomes, the probability of each will be $1/n$, and the probability of an event which contains s of these outcomes is s/n. This is expressed in the formula

$$P(E) = \frac{\text{number of successful outcomes}}{\text{total number of outcomes}}. \tag{1}$$

Example 1
Find the probability that
(i) a total score of 8 is obtained when two dice are thrown;
(ii) a card chosen at random from a pack is a queen or a heart.
Solution
(i) We may take the outcome space to be the 36 ordered pairs

$$(1\ 1)\quad(1\ 2)\ldots(1\ 6)$$
$$(2\ 1)\quad(2\ 2)\ldots(2\ 6)$$
$$\cdot\qquad\cdot\qquad\cdot$$
$$\cdot\qquad\cdot\qquad\cdot$$
$$\cdot\qquad\cdot\qquad\cdot$$
$$(6\ 1)\quad(6\ 2)\ldots(6\ 6).$$

These are all equally likely (since there is no reason why any one of them should be more likely than any other). Of these 36 possibilities, 5 give a total score of 8; so the required probability is $\frac{5}{36}$.

(ii) Of the 52 cards, 16 satisfy the given condition.

So $$P(\text{queen or heart})=\tfrac{16}{52}=\tfrac{4}{13}.$$

Example 2

A bag contains five white balls, three red and two black. The ten balls are drawn consecutively and not replaced. Find

 (i) P(first ball is red);

(ii) P(second ball is red);

(iii) P(last ball is red).

Solution

(i) All ten balls are equally likely to be chosen on the first draw, so

$$P(\text{first ball is red})=\frac{3}{10}.$$

(ii) Knowing nothing about the first draw, we can argue that each of the ten balls is equally likely to be the one that gets chosen on the second draw. So

$$P(\text{second is red})=\frac{3}{10}.$$

(iii) Similarly, each of the ten balls has a probability of $\frac{1}{10}$ of being the last; so

$$P(\text{last is red})=\frac{3}{10}.$$

Qu.4 (See Example 2.) Find the probability that the second ball is red given that

(i) the first was red;   (ii) the first was white.

   The formula (1) above means that many probability questions are essentially problems of counting; a knowledge of permutations and combinations is therefore needed (Chapter 9).

Example 3

Five men, A, B, C, D and E, sit in a random order in a row. Find the probability that A is next to B.

Solution

The number of ways of seating five men in a row is 5!, and the number of successful ways (i.e. with A next to B) is $8 \times 3!$. So

$$P(A \text{ is next to } B) = \frac{8 \times 3!}{5!} = \frac{2}{5}.$$

Qu.5 Write down $P(A \text{ is not next to } B)$.

22.6 Adding and multiplying probabilities

We saw in Chapter 9 that for subsets A and B of some universal set

$$n(A \cup B) = n(A) + n(B) - n(A \cap B).$$

By a similar argument (using a Venn diagram), it is clear that, for events E_1 and E_2,

$$P(E_1 \cup E_2) = P(E_1) + P(E_2) - P(E_1 \cap E_2). \qquad (2)$$

Thus to find the probability that one event or another occurs, we may add the individual probabilities, provided that we subtract the probability of both events' occuring, to avoid counting it twice. As in Chapter 9, the result (2) may be extended to the case of three events (or more):

$$P(E_1 \cup E_2 \cup E_3) = P(E_1) + P(E_2) + P(E_3) - P(E_1 \cap E_2) - \\ - P(E_1 \cap E_3) - P(E_2 \cap E_3) + P(E_1 \cap E_2 \cap E_3).$$

Definition
If two events E_1 and E_2 do not overlap, i.e. if $P(E_1 \cap E_2) = 0$, they are said to be *mutually exclusive*. The meaning of this phrase is easy to remember, since it means that each event excludes the possibility of the other, i.e. they cannot both happen.

Qu.6 Two dice are thrown. Which pair of the following are mutually exclusive?
  A = the total score is 7;
  B = at least one die shows an odd score;
  C = the product of the scores is odd.

It follows immediately from (2) and the definition above that *for mutually exclusive events.*

$$P(E_1 \cup E_2) = P(E_1) + P(E_2).$$

Qu.7 Find the probability that in Qu.6 either A or C occurs.

We now consider the probability of *both* of two events' occuring, i.e. $P(E_1 \cap E_2)$. As an example, suppose that a coin is tossed and a die thrown. By considering the appropriate equiprobable outcome spaces, we know that
  for the coin, $P(\text{head}) = \frac{1}{2}$;
  for the die, $P(\text{six}) = \frac{1}{6}$;
  for the combined experiment, $P(\text{head and six}) = \frac{1}{12}$.

So, in this case,

$$P(\text{head and six}) = P(\text{head}) \times P(\text{six}).$$

This leads us to wonder whether, in general,

$$P(E_1 \cap E_2) = P(E_1) \times P(E_2).$$

This question is answered formally in Sections 22.8 and 22.9. For the present, we shall assume that we can multiply probabilities to find $P(E_1 \cap E_2)$ provided that either

(1) the two events do not affect one another (as in the case of the coin and the die above),

or (2) in determining $P(E_2)$, we allow for the fact that E_1 must occur (as in Example 5 below).

As an example of why we cannot multiply probabilities indiscriminately, suppose we choose a card from a pack, then

$$P(\text{heart}) = \tfrac{1}{4}$$

and $P(\text{black card}) = \tfrac{1}{2}$,

but $P(\text{heart and black}) \neq \tfrac{1}{4} \times \tfrac{1}{2}$.

Example 4

In a certain school, 12% of boys wear glasses, and 35% are over 6 feet tall. Find the probability that a boy chosen at random is either under 6 feet or wears glasses (or both).

Solution

(We assume that there is no connection between height and eyesight. Without this reasonable assumption, the problem is insoluble as it stands. We also assume that the probability of a boy's being exactly 6 feet tall is negligible; see Chapter 32.)

$$P(\text{boy is under 6 feet or wears glasses})$$

$$= 1 - P(\text{boy is over 6 feet and does not wear glasses}).$$

(A Venn diagram clarifies this.)

Now $P(\text{boy is over 6 feet}) = 0.35$

and $P(\text{he does not wear glasses}) = 0.88$.

So $P(\text{over 6 feet and no glasses}) = 0.35 \times 0.88$
$$= 0.308.$$

Then

$$P(\text{under 6 feet or wears glasses}) = 1 - 0.308$$
$$= 0.692.$$

Qu.8 Draw a Venn diagram to illustrate the two sets in Example 4, showing the probability of each of the four regions.

Example 5

A bag contains three white and five black balls. Three are drawn at random without replacement. Find the probability that all three are the same colour.

Solution

$$P(\text{all same}) = P(\text{all white}) + P(\text{all black}),$$

since 'all white' and 'all black' are mutually exclusive.
Now

$$P(\text{first is white}) = \tfrac{3}{8}$$

and $P(\text{second is white, assuming the first was}) = \tfrac{2}{7}$,

and $P(\text{third is white if the first two are}) = \tfrac{1}{6}$.

So $P(\text{all white}) = \tfrac{3}{8} \times \tfrac{2}{7} \times \tfrac{1}{6} = \tfrac{1}{56}$.

Similarly,

$$P(\text{all black}) = \tfrac{5}{8} \times \tfrac{4}{7} \times \tfrac{3}{6} = \tfrac{10}{56}.$$

So $$P(\text{all same}) = \tfrac{1}{56} + \tfrac{10}{56} = \tfrac{11}{56}.$$

Qu.9 Solve Example 5 if each ball is replaced after it is drawn

Qu.10 Three balls are drawn from the bag in Example 2. Find the probability that they are all the same colour if the drawing is (i) without replacement;   (ii) with replacement.

22.7 Probability trees Problems with several stages (such as Example 5 above) can often be solved with the aid of probability trees.

Example 6

In a certain town, each day is classified as wet or dry according to whether or not more than a certain amount of rain falls. Experience has shown that if one day is wet, then the probability that the next day is wet is $\tfrac{1}{2}$; and if one day is dry, then the probability that the next is wet is $\tfrac{1}{3}$. Given that Monday is dry, find the probability that Thursday is wet.

Solution

We draw a tree showing all possible sequences of weather and the probabilities at each stage (Fig. 22.1).

Figure 22.1

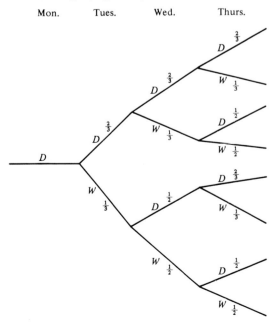

There are eight possible 'routes', four of which lead to a wet Thursday. Multiplying along these four routes,

$$\tfrac{1}{3} \times \tfrac{1}{2} \times \tfrac{1}{2} = \tfrac{1}{12}$$

$$\tfrac{1}{3} \times \tfrac{1}{2} \times \tfrac{1}{3} = \tfrac{1}{18}$$

$$\tfrac{2}{3} \times \tfrac{1}{3} \times \tfrac{1}{2} = \tfrac{1}{9}$$

$$\tfrac{2}{3} \times \tfrac{2}{3} \times \tfrac{1}{3} = \tfrac{4}{27}.$$

Now these four routes are (pair-wise) mutually exclusive, so we may add the four probabilities to give the required answer:

$$\tfrac{1}{12} + \tfrac{1}{18} + \tfrac{1}{9} + \tfrac{4}{27} = \tfrac{43}{108}.$$

Qu.11 A bag contains 3 white, 4 red and 5 black balls. Two are chosen at random without replacement. Draw a tree to illustrate this experiment, and hence find the probability that the balls are the same colour. (Note that at each 'node' there are three branches corresponding to the three colours.)

Qu.12 (See Example 6.) Draw a tree to find the probability that Thursday is wet given that Monday is wet.

Exercise 22a

1. A card is drawn from a pack at random. Find the probability that it is
   (i) a spade;   (ii) not a spade;   (iii) a king;
   (iv) either a king or a spade;
   (v) neither a king nor a spade;
   (vi) not the king of spades.

2. Two dice are thrown. Find the probability that
   (i) the total score is 10;
   (ii) the total score is less than 6;
   (iii) the two scores are equal;
   (iv) the product of the scores is at least 16;
   (v) the larger score (or the common score if they are equal) is 4 or more.

3. Five coins are tossed. By considering all possible sets of ordered 5-tuplets such as (H, H, T, H, T), find, for $i = 0$ to 5, the values of p_i where

   $$p_i = P(\text{exactly } i \text{ heads are obtained}).$$

   Check that $\Sigma p_i = 1$.

4. Balls numbered 1 to 10 are placed in a bag and two are drawn consecutively without replacement.
   (i) How many outcomes are there for this experiment?
   (ii) Find the probability that the numbers on the two balls differ by exactly 4.
   (iii) *Write down* the probability that the second ball drawn has a higher score than the first.
   (iv) What difference would be made to the answer to (ii) if the balls were drawn simultaneously?

5. The thirteen spades from a pack of cards are placed in a line at

random. Find the probability that the queen is adjacent to both the king and the jack.

6. Three married couples sit at random in a railway compartment in which three seats face three others. Find the probability that
   (i) each man sits opposite his wife;
   (ii) no man sits on the same side as his wife.

7. In a game of bridge, before play begins, South has decided that his two opponents East and West (who hold thirteen cards each) have six spades between them. He wishes to know the probability that they each hold 3 spades. Leave the answers to the following in ^nC_r form.
   (i) In how many ways can 26 cards be shared between East and West?
   (ii) In how many ways can the 26 cards be shared so that East has 3 spades and 10 non-spades?
   (iii) Deduce that the probability that each hold 3 spades is $\dfrac{^6C_3 \times {}^{20}C_{10}}{{}^{26}C_{13}} \approx 0.36$.

8. (See question 7.) Find the probability that the six spades are split 4–2. (Remember to count separately the cases when East holds 4 and when West holds 4.)

9. A bag contains n balls, k of which are white and the rest black. Find the probability that if m ($\leqslant n$) are drawn without replacement exactly r are white. (Assume that $r \leqslant k$ and $r \leqslant m$.)

10. It is known that three minor allergies affect 20%, 50% and 60% of the population, and it is also known that the three allergies are completely unconnected. A person is chosen at random. Find the probability that he has at least one of these allergies. [*Hint*: find first the probability that he has none of them.]

11. A man has four children. Estimate the probability that at least one of them is a girl.

12. Two players A and B take turns to throw a die, the winner being the first to throw a six. A throws first. Find the probability that
    (i) A wins on his second throw;
    (ii) A wins on one of his first three throws.
    Write down an infinite geometric series giving the probability that A wins, and find this probability.

13. (See question 12.) Another method of finding A's probability of winning is as follows. Let this probability be p, so that B's probability of winning is $1-p$.
    (i) Explain *why* the sum of the probabilities should be 1.
    (ii) By considering the fact that immediately after his first throw A has either won or is in the same situation as B was in initially, show that
    $$p = \frac{1}{6} + \frac{5}{6}(1-p),$$
    and hence find p.

14. If three players play the game in question 12, find their probabilities of winning.

15. A certain country has three political parties A, B and C, whose supporters represent 10%, 30% and 60% of the population respectively. Four men are selected at random and asked which party they support. Let E_1 be the event 'none of the four supports A', and let E_2 and E_3 be defined similarly for B and C.
(i) Describe in words as simply as possible the event $(E_1 \cup E_2 \cup E_3)'$.
(ii) Expand, and hence calculate, $P(E_1 \cup E_2 \cup E_3)$.
(iii) Hence find the probability that each party has at least one supporter among the four men.

16. A rat in a maze is faced with a series of T-junctions, and it behaves as follows: at the first junction, it takes either path with a probability $\frac{1}{2}$; thereafter, there is a probability of $\frac{3}{4}$ that it will turn the same way (right or left) as at the previous junction.
(i) Draw a probability tree to find the probability that after 3 junctions it has turned right exactly once.
(ii) If its first choice was left, find the probability that its fourth choice is left.

17. Two boxes contain balls as follows:
  Box 1: 5 white, 5 black;
  Box 2: 3 white, 6 black.
A ball is selected at random from box 1 and put into box 2. A ball is then selected from box 2 and put into box 1.
(i) Using a tree or otherwise, find the probability that the colour distribution is as it was initially.
(ii) Repeat (i) if the transfers had been in the reverse order (i.e. 2 to 1 then 1 to 2).

18. A bag initially contains one white and one black ball. A ball is drawn and replaced, and one extra black ball is added. This process is repeated n times in all. Find the probability that the white ball is not chosen in any of the n drawings.

19. (See question 10.) Let p_i be the probability that a person has precisely i of the allergies. Find p_0, p_1, p_2, p_3. Check that $\Sigma p_i = 1$.

20. It is known that exactly 10% of a large batch of bullets are faulty. Two are selected and fired. What is the probability that they are both faulty? Explain why this problem could not be solved if the batch were not 'large'.

21. Three men each have three children. Find
 (i) P(at least one man has at least one son);
 (ii) P(at least one man has at most one son).

22. (i) Four distinct points lie on the circumference of a circle. Two of the points are selected and joined by a chord; the other pair are also joined by a chord. Find the probability that the chords intersect (internally).
(ii) If six points are chosen and three chords drawn, find the probabilities of 0, 1, 2 and 3 intersections. Check that the sum of the probabilities is 1. (Assume that the points are such that three chords cannot be concurrent.)

23. (i) If n pegs numbered 1 to n are placed at random in n holes also numbered 1 to n, find the probability that no peg is in its 'correct' hole when n is

(a) 2; (b) 3; (c) 4.

(ii) (For discussion; see question 24.) What would you expect to happen to this probability as $n \to \infty$?

24. (i) Prove by induction or otherwise that for events E_1, E_2, ... E_n,

$$P(E_1 \cup E_2 \cup \ldots \cup E_n) = \sum P(E_i) - \sum P(E_i \cap E_j) +$$
$$+ \sum P(E_i \cap E_j \cap E_k) - \ldots + (-1)^{n-1} P(E_1 \cap E_2 \cap \ldots E_n),$$

where, under each summation, no two of i, j, k, ... are equal.

(ii) (See question 23.) Let E_r be the event 'the rth peg is in the correct hole'. Use (i) to show that the probability that all pegs are in the wrong holes is

$$1 - 1 + \frac{1}{2!} - \frac{1}{3!} + \frac{1}{4!} - \ldots + \frac{(-1)^n}{n!}.$$

We shall see in Chapter 28 that this series tends to e^{-1} (≈ 0.368) as $n \to \infty$.

25. A certain system has four states: A, B, Win, Lose (see Fig. 22.2). The system starts in state A and moves repeatedly along the lines until Win or Lose is reached. If the choice of route out of A and B is random, find the probability of an eventual win.

Figure 22.2

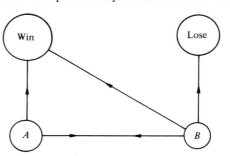

22.8 Conditional probability

We have already seen that we usually need to change the probability attached to an event when further information becomes available. We now examine this idea in more detail.

Notation
We write $P(A|B)$ to mean 'the probability that the event A occurs given that the event B occurs'; it is read as 'the probability of A given B'. Then, for example, $P(A|B')$ means the probability of A given that B does not occur. In this context, $P(A)$ means the probability of A given no information about the event B.

Qu.13 State the value of $P(A|B)$ if
(i) A and B are mutually exclusive;   (ii) $B \subseteq A$.

Now consider a Venn diagram showing two events A and B (Figure 22.3). The total probability of the whole outcome space Ω is 1. (It may now help to think of area as representing probability.) Suppose that we are given that the event B occurs. Then the probability

Figure 22.3

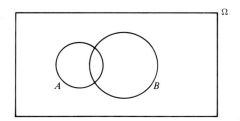

that A occurs is given by the probability of that part of A inside B as a fraction of the probability of B; i.e.

$$P(A|B) = \frac{P(A \cap B)}{P(B)}.$$ (3)

Another way of obtaining this result is to argue that once it is known that B occurs, the probabilities of all the outcomes inside B are 'scaled up' so that their total is 1; i.e. each probability is increased by the factor $\dfrac{1}{P(B)}$. (The probability of any outcome outside B is, of course, reduced to zero.) Then

$$P(A|B) = P(A \cap B) \times \frac{1}{P(B)}.$$

The important result (3) may also be written

$$P(A \cap B) = P(B) . P(A|B),$$

which says that the probability that both A and B occur is the probability of B multiplied by the probability of A modified to allow for the fact that B occurs. This is the rule that we used in the previous section (Example 5).

Qu.14 In a certain community of 500 people, 200 are male, of whom a quarter are over 40 years old; the rest are female of whom one in three are over 40. Draw a Venn diagram to illustrate this information, and hence find
 (i) P(a person chosen at random is over 40);
 (ii) P(a person over 40 is male);
 (iii) P(a person not over 40 is female).

22.9 Independent events

Qu.15 A card is chosen at random from a pack. Find
 (i) P(it is an Ace);
 (ii) P(it is an Ace|it is red).

Definition
Two events A and B are said to be *independent* if knowing that one occurs does not affect the probability that the other occurs, i.e. if (and only if)

$$P(A|B) = P(A) \text{ and } P(B|A) = P(B).$$

Theorem 22.1

Two events A and B are independent if and only if

$$P(A \cap B) = P(A).P(B).$$

Proof

(i) If A and B are independent,

$$\begin{aligned}
P(A \cap B) &= P(A|B).P(B) \\
&= P(A).P(B) \quad \text{(by the independence)}.
\end{aligned}$$

(ii) Now suppose that

$$P(A \cap B) = P(A).P(B).$$

Then

$$\begin{aligned}
P(A) &= \frac{P(A \cap B)}{P(B)} \\
&= P(A|B) \quad \text{by (3)}.
\end{aligned}$$

Similarly, $P(B) = P(B|A)$, and so the events are independent.

(Note that the condition stated in the theorem is often taken as the *definition* of independence.)

Theorem 22.1 justifies the assumption made in the last section, and used in Example 4, that we may multiply probabilities to obtain the probability of both of two events' occuring, provided that the two events do not affect one another.

Exercise 22b

1. An integer from 1 to 12 (inclusive) is selected at random. Draw a Venn diagram to represent the outcome space, showing the two events

   A: a prime number is chosen (note that 2 is the smallest prime);
   B: the score is less than 7.

   Mark each of the twelve numbers in the appropriate region, and hence write down
   (i) $P(A)$; (ii) $P(A|B)$; (iii) $P(A|B')$; (iv) $P(B)$;
   (v) $P(B|A)$; (vi) $P(B'|A)$; (vii) $P(A'|B')$; (viii) $P(B'|A')$.

2. A man has two children. Estimate
   (i) P(both are girls);
   (ii) P(both are girls|at least one is a girl);
   (iii) P(both are girls|the older is a girl).

3. Two dice are thrown. Find
   (i) P(total score is 9|total score is at least 9);
   (ii) P(total score is at least 9|at least one score is 5).

4. A man has three children. Estimate

   $$P(\text{he has a son}|\text{he has a daughter}).$$

5. Two dice are thrown. Let A be the event 'the first score is 5' and B be 'the total score is 7'. Evaluate $P(A)$, $P(B)$ and $P(A \cap B)$, and deduce that A and B are independent. If C is 'the total score is 8', show that A and C are not independent.

PROBABILITY I

6. Two events A and B are independent and mutually exclusive. What can be deduced?

7. Prove or disprove with a counter-example the suggestion that if A, B and C are three events then

A and B independent and A and C independent
$\Rightarrow$ B and C independent.

8. Let A and B be two events. Prove or disprove that
(i) $P(A|B)=P(B|A) \Rightarrow P(A)=P(B)$.
(ii) A and B independent $\Rightarrow$ A and B' independent.
[*Hint*: draw a Venn diagram and let the probabilities of the four regions be x, y, z and $1-x-y-z$.]
(iii) $P(A|B)>P(A) \Rightarrow P(A|B')<P(A)$.
Explain this last result in words.

9. The outcome space for an experiment is the set of nine ordered pairs

$$\{(i,j):i,j\in\{1,2,3\}\}.$$

(i) If the associated probabilities are given by

$$P\{(i,j)\}=\frac{ij}{36},$$

verify that the total probability is 1, and show that the two coordinates are independent, i.e. that for each choice of i and j,

$P(\text{1st coordinate is } i)\times P(\text{2nd coordinate is } j)=P\{(i,j)\}.$

(ii) If, instead,

$$P\{(i,j)\}=\frac{i+j}{36},$$

verify that the total probability is 1, and show by a specific counter-example that the coordinates are not independent.

10. During a battle, a general sends a runner to deliver a message, and tells him to return to confirm that the message was delivered. Assuming that the probability that the runner does not complete the outward journey is p, and that the return journey is equally dangerous, find in terms of p
(i) the probability that the runner returns to the general;
(ii) the probability that the message was delivered, given that the runner fails to return.
What happens to the answer to (ii) as $p\to1$ and as $p\to0$? What is the best value of p in (ii) from the general's point of view?

11. (See Exercise 22a question 12.) Find
(i) $P(A \text{ won}|\text{not more than 4 throws took place})$;
(ii) $P(\text{not more than 4 throws took place}|A \text{ won})$;
(iii) $P(\text{exactly 4 throws took place}|\text{not more than 4 throws took place})$.

12. Two coins are tossed. Find three events which are pair-wise independent, but which do not form an independent set (i.e. $P(A\cap B\cap C)\neq P(A).P(B).P(C)$).

22.10 Bayes's Theorem and Bayesian problems

Bayes's Theorem is one of the most important theorems in conditional probability. The idea behind the theorem is rather simpler than its formal statement, and so Bayesian problems are often solved more easily by working 'from first principles' than by explicit use of the theorem. We shall therefore delay the statement and proof of the theorem until we have seen some applications.

Example 7

Three bags A, B and C contain white and black balls as follows:

Bag A: 3 white, 9 black;
Bag B: 1 white, 9 black;
Bag C: 8 white, 2 black.

A bag is chosen at random and a ball is drawn from it and found to be white. What is the probability that bag C was chosen?

Solution

Before drawing the ball, we should have said that the probability that bag C was chosen was $\frac{1}{3}$. Knowing that the ball chosen was white, however, seems to make it more likely that C was chosen, since C has a larger proportion of whites than A or B. We can illustrate precise probabilities on a Venn diagram in which we take area to represent probability.

Starting with a rectangle measuring 1×1, we divide it into three strips of width $\frac{1}{3}$ to represent the initial or *a priori* probabilities of the three bags (Fig. 22.4). The part of each strip corresponding to the extra information (a white was drawn) is now shaded; i.e. we shade

Figure 22.4

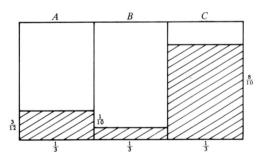

3/12 of the A-strip, 1/10 of the B-strip, and 8/10 of the C-strip. We then seek the probability that the outcome is in the shaded part of the C-strip given that it is in one of the shaded regions. This is given by

$$\frac{8/10 \times 1/3}{(3/12 \times 1/3) + (1/10 \times 1/3) + (8/10 \times 1/3)} = \frac{16}{23}.$$

Qu.16 (See Example 7.) In the experiment above, find
(i) P(bag B was chosen|a white ball was drawn);
(ii) P(bag C was chosen|a black ball was drawn).

In Example 7 the *a priori* probabilities were each $\frac{1}{3}$ since the bag was chosen 'at random'. If the bag had been chosen by some other method, this might alter the initial probabilities and hence the widths of the three strips in Fig. 22.3.

Qu.17 (See Example 7.) Suppose that the initial choice of bag was made as follows: a card is drawn from a pack; if it is red, bag A is chosen; if it is a spade, bag B is chosen; and if it is a club, bag C is chosen. Draw a suitable diagram to find

$$P(C \text{ was chosen}|\text{a white ball was drawn}).$$

Example 8

In a certain hereditary disease, women who carry the disease show no symptoms themselves, but pass the disease on to any offspring with probability $\frac{2}{3}$. Female offspring having the disease will merely become carriers; male offspring having the disease will die in infancy. Figure 22.5 shows a simplified family tree in which the M_i are males and the F_i females. It is known that

Figure 22.5

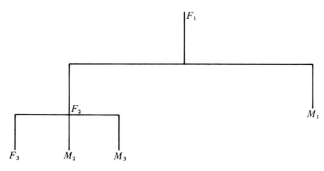

    (i) M_1 died of the disease;

    (ii) M_2 and M_3 do not have the disease.

Find the probability that F_3 is a carrier.

Solution

Since M_1 had the disease, F_1 was certainly a carrier. Now without knowing that M_2 and M_3 do not have the disease, the probability that the disease was passed down to F_3 would be $\frac{2}{3} \times \frac{2}{3} = \frac{4}{9}$. The extra information suggests that the disease did not even reach F_2, thereby reducing this figure of $\frac{4}{9}$. Figure 22.6 illustrates the problem of finding whether F_2 was a carrier or not (N.B. F_2, not F_3).

    The *a priori* probability of F_2's being a carrier is $\frac{2}{3}$. The shaded region indicates the event that both M_2 and M_3 do not have the disease. So

Figure 22.6

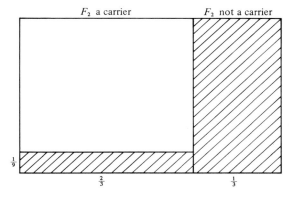

F_2 a carrier   F_2 not a carrier

$$P(F_2 \text{ is a carrier} \mid M_2 \text{ and } M_3 \text{ are disease-free})$$

$$=\frac{1/9 \times 2/3}{(1/9 \times 2/3)+(1 \times 1/3)}$$

$$=\frac{2}{11}.$$

So $\quad P(F_3 \text{ is a carrier} \mid M_2 \text{ and } M_3 \text{ are disease-free})$

$$=\frac{2}{11} \times \frac{2}{3}$$

$$=\frac{4}{33}.$$

Qu.18 (See Example 8.) Find the probability that F_3 is a carrier if she has a third disease-free brother M_4.

Theorem 22.2 (Bayes).

Let H_1, H_2, ... H_n be n events which partition an outcome space, i.e. they cover the whole space without overlapping. Let E be another event in the same outcome space (Fig. 22.7). Then if H_k is one of the H_i,

$$P(H_k \mid E)=\frac{P(E \mid H_k).P(H_k)}{\sum_i P(E \mid H_i).P(H_i)}$$

(In remembering this theorem, it may help to note that the denominator is the sum of n terms of which the numerator is just one.)

Proof

$$P(H_k \mid E)=\frac{P(H_k \cap E)}{P(E)}$$

$$=\frac{P(H_k \cap E)}{\sum P(H_i \cap E)}$$

(since the $H_i \cap E$ are mutually exclusive). But for each i, $P(H_i \cap E)=P(E \mid H_i).P(H_i)$. So

$$P(H_k \mid E)=\frac{P(E \mid H_k).P(H_k)}{\sum_i P(E \mid H_i).P(H_i)}.$$

Figure 22.7

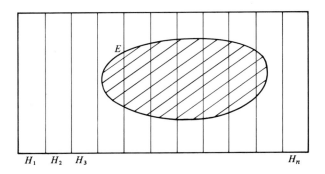

$H_1 \quad H_2 \quad H_3 \qquad\qquad\qquad\qquad\qquad\qquad H_n$

To see how this theorem is related to (and used in) the examples above, it is helpful to compare Figures 22.4 and 22.7. In Example 7, the H_i would be the three events 'bag A is chosen', 'bag B is chosen' and 'bag C is chosen'; and the event E would be 'a white ball is drawn'.

Qu.19 Solve Example 7 by substituting into Bayes's Theorem.

Exercise 22c
1. A certain component is produced by two machines A and B. Machine A produces 60% of the output, and, of these, 5% are faulty. Machine B produces 40% of the output, of which 10% are faulty.
   (i) Find the probability that a component chosen at random is faulty.
   (ii) A component is tested and found to be faulty. Find the probability that it came from machine A.
   (iii) Another component is tested and found not to be faulty. Find the probability that it came from A.
2. Of three coins, two are unbiased, and the third comes up heads twice as often as tails. One of the coins is selected and tossed four times. Find
   (i) P(it is the biased coin | 4 heads were obtained);
   (ii) P(it is the biased coin | 3 tails and 1 head were obtained).
3. Two boxes contain black and white balls thus:
       Box I :   2 white, 4 black;
       Box II: 1 white, 3 black.
   A ball is moved from box I to box II, and then a ball is chosen at random from box II and moved to box I. Find
   (i) the probability that the final distribution of colours is the same as the original distribution;
   (ii) the probability that the first ball moved was white given that the final distribution of colours is the same as the original distribution.
4. An anthropologist knows that the ratio of the sexes on a certain island is 3:1, but he cannot remember which way round, i.e. whether there are more men or more women. Find
   (i) P(women predominate | the first person he meets is a woman):
   (ii) P(women predominate | the first three people he meets are all women).
5. If P(a certain day is wet | previous day was dry)$=\frac{1}{4}$, and P(a certain day is wet | previous day was wet)$=\frac{1}{2}$, find P(Monday was wet | the Sunday before and the Wednesday after were both dry).
6. In Figure 22.8, it is known that M_1 had the disease described in Example 8, and that M_2 does not have the disease.
   (i) Find the probability that F_2 carries the disease. Hence find the probability that F_3 carries the disease.
   (ii) If F_3 now has a son who does not have the disease, find a new estimate for the probability that she carries the disease.

Figure 22.8

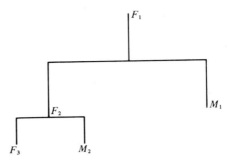

Miscellaneous exercise 22

1. (a) Three balls are drawn at random without replacement from a box containing 4 red and 6 black balls. Calculate the probabilities that
    (i) all 3 balls will be red,
    (ii) all 3 balls will have the same colour,
    (iii) there will be exactly 2 red balls.
(b) When fired at a ship a certain type of torpedo has probability $\frac{1}{4}$ of sinking the ship, probability $\frac{1}{4}$ of damaging the ship, and probability $\frac{1}{2}$ of missing the ship. Two damaging shots are sufficient to sink the ship.
When 3 torpedoes are fired independently at a ship, calculate the probabilities that
    (i) all 3 will miss the ship,
    (ii) the ship will not be sunk.

[JMB (O)]

2. (a) The pages of a book are numbered from 1 to 200. If a page is chosen at random, what is the probability that its number will contain just two digits?
(b) The probability that a January night will be icy is $\frac{1}{4}$. On an icy night the probability that there will be a car accident at a certain dangerous corner is $\frac{1}{25}$. If it is not icy, the probability of an accident is $\frac{1}{100}$. What is the probability that
    (i) 13 January will be icy and there will be an accident,
    (ii) there will be an accident on 13 January?

[C (O)]

3. A girl has two ordinary six-sided dice and one special one, which has two sides showing two, three sides showing three and one side showing four. She throws the three dice and notes the three scores.
    Determine the probability that
    (i) the sum of the three scores is 5,
    (ii) the three scores are the same.
    Calculate the probability that all three scores are odd and hence, or otherwise, find the probability that the product of the three scores is even.

[SUJB (O)]

4. Two pianists, X and Y, both give a recital on the same night. The independent probabilities that two newspapers A and B publish reviews of their recitals are set out in the table.

	Probability of review	
	A	B
Recital by X	$\frac{1}{2}$	$\frac{2}{5}$
Recital by Y	$\frac{1}{4}$	$\frac{3}{5}$

(a) If X buys both newspapers, what is the probability that both the papers will contain a review of his recital?

(b) If Y buys both newspapers, what is the probability that only one paper will contain a review of his recital?

(c) If X buys one paper at random, what is the probability that it will contain reviews of both recitals?

[MEI (O)]

5. The events A and B are such that

$$P(A) = \tfrac{1}{2},$$
$$P(A \text{ or } B \text{ but not both } A \text{ and } B) = \tfrac{1}{3},$$
$$P(B) = \tfrac{1}{4}.$$

Calculate $P(A \cap B)$, $P(A' \cap B)$, $P(A \mid B)$ and $P(B \mid A')$, where A' is the event 'A does not occur'.

State, with reasons, whether A and B are (i) independent, (ii) mutually exclusive.

[C]

6. In Camelot it never rains on Friday, Saturday, Sunday or Monday. The probability that it rains on a given Tuesday is $\tfrac{1}{5}$. For each of the remaining two days, Wednesday and Thursday, the conditional probability that it rains, given that it rained the previous day, is α, and the conditional probability that it rains, given that it did not rain the previous day, is β.

(i) Show that the (unconditional) probability of rain on a given Wednesday is $\tfrac{1}{5}(\alpha + 4\beta)$, and find the probability of rain on a given Thursday.

(ii) If X is the event that, in a randomly chosen week, it rains on Thursday, Y is the event that it rains on Tuesday, and $\bar{Y}$ is the event that it does not rain on Tuesday, show that

$$P(X \mid Y) - P(X \mid \bar{Y}) = (\alpha - \beta)^2.$$

(iii) Explain the implications of the case $\alpha = \beta$.

[C]

7. In a building programme the event that all the materials will be delivered at the correct time is M, and the event that the building programme is completed on time is F. Given that $P(M) = 0.8$ and $P(M \cap F) = 0.65$, explain in words the meaning of $P(F \mid M)$ and calculate its value. If $P(F) = 0.7$, find the probability that the building programme will be completed on time if all the materials are not delivered at the correct time.

[L]

8. Twelve people, of whom A and B are two, stand in a line. Find the probability that
(a) A and B are next to one another,
(b) there are exactly 4 people between A and B,
(c) there are not more than 3 people between A and B.

When the twelve people are seated round a table, find the probability that

(d) A and B sit next to one another,

(e) A and B have exactly one person between them.

[L (S)]

9. (a) An unbiased die, with faces numbered one to six, is thrown repeatedly. Find the probability that a six is obtained in fewer than eight throws.

(b) A deputation of two women and one man is to be chosen at random from a group composed of 18 women and 10 men. If there are three Annes and two Johns in the group find the probabilities that the deputation contains

 (i) only one Anne,

 (ii) no John,

 (iii) only one Anne and one John.

  Calculate the probability that the deputation contains at least one of the five named either Anne or John.

[AEB '78]

10. Two cards are drawn without replacement from a pack of playing cards. Using a tree diagram, or otherwise, calculate the probability

(a) that both cards are aces,

(b) that one (and only one) card is an ace,

(c) that the two cards are of different suits.

  Given that at least one ace is drawn, find the probability that the two cards are of different suits.

[L]

11. A class contains $N+1$ pupils. One pupil tells a rumour to a second pupil who passes it on to a third and so on. At each stage the hearer is chosen at random from the N possible hearers. Find the probability that the rumour will be told r times without (a) it being repeated to the originator, (b) it being repeated to any pupil.

  Repeat the problem when at each stage a pupil tells the rumour to a group of n pupils chosen at random from the N possible hearers, and show that in this case the probabilities are

(a) $(1-n/N)^{r-1}$, (b) $[(N-n)!]^r/[(N!)^{r-1}(N-rn)!]$ for $rn < N$.

[Ox (S)]

12. (a) A committee of school prefects consists of 4 senior prefects, of whom one is a girl, and 10 junior prefects, of whom 3 are girls. A subcommittee of 5 of these prefects is selected at random. Calculate the probability that the subcommittee will include exactly one senior prefect and at least one girl.

(b) A and B are two independent events each having a probability of occurring which is greater than $\frac{1}{2}$. The probability that A will occur with B not occurring is $\frac{3}{25}$ and the probability that B will occur with A not occurring is $\frac{8}{25}$. Calculate the individual probabilities of A and B occurring.

[W (S)]

Appendix

A.1 Introduction   This appendix outlines an alternative approach to the exponential and logarithmic functions exp x and ln x. Some of the work below assumes a knowledge of Chapter 17 and Sections 18.5 and 18.8 of Chapter 18.

A.2 The function ln x   **Definition**
For $x>0$,

$$\ln x = \int_1^x \frac{1}{t}\,dt.$$

Graphically, this means that ln x is the area under the curve $y=1/t$ from 1 to x (Fig. A.1). Note that ln x is not defined for $x \leqslant 0$.

Figure A.1

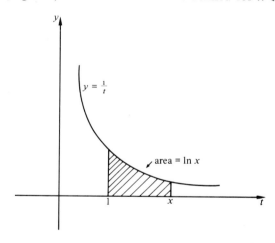

With the aid of Fig. A.1, the following properties of ln x are seen to follow from the definition:
(i) ln $1=0$, and ln x is positive or negative according as $x>1$ or $x<1$;
(ii) ln x is an increasing function;
(iii) ln x is continuous;
(iv) $\dfrac{d}{dx}\ln x = \dfrac{1}{x}$.

Qu.1 (For discussion.) Explain clearly why we cannot deduce from properties (i), (ii) and (iii) above that $\exists\, x$ such that ln $x=1$.

The central problem with which we now have to deal is that of showing that ln x is in fact a logarithmic function (in the sense of Chapter 4), i.e. that it is the inverse of some power function. Before showing this, however, we can derive some further properties of ln x.

Theorem A.1
(i) Ln $(pq) = \ln p + \ln q$.
(ii) Ln $(x^n) = n \ln x$.

Proof

(i)
$$\ln pq = \int_1^{pq} \frac{1}{t} \, dt$$

$$= \int_1^p \frac{1}{t} \, dt + \int_p^{pq} \frac{1}{t} \, dt.$$

The first integral on the R.H.S. is ln p, and applying the substitution $t = pu$ to the second gives

$$\int_1^q \frac{1}{pu} p \, du = \int_1^q \frac{1}{u} \, du$$

$$= \ln q.$$

Hence the required result.

(ii) $\ln x^n = \int_1^{x^n} \frac{1}{t} \, dt.$

Applying the substitution $u^n = t$ gives

$$\int_1^x \frac{1}{u^n} n u^{n-1} \, du = n \int_1^x \frac{1}{u} \, du$$

$$= n \ln x.$$

We can now prove our central result.

Theorem A.2
The function ln x is a logarithmic function.
Proof
There certainly exists a value of x such that ln $x = 1$, since (for example)

$$\ln (3^{1/\ln 3}) = \frac{1}{\ln 3} \ln 3 \quad \text{(by Theorem A.1 (ii))}$$

$$= 1.$$

Let e be this value of x, so that ln $e = 1$.
Now any $x > 0$ may be written as $x = e^p$, where $p = \log_e x$. Then

$$\ln x = \ln (e^p)$$
$$= p \ln e \quad \text{(by Theorem A.1 (ii))}$$
$$= p$$
$$= \log_e x.$$

Thus the function ln x is identical to the function $\log_e x$, where e is the number satisfying $\int_1^e \frac{1}{t} \, dt = 1$. The value of e is approximately $2 \cdot 71828$; this is evaluated in Chapter 28.

A.3 The exponential function

Definition

The function $\exp x$, known as *the exponential function*, is defined by

$$\exp x = e^x,$$

where e is as defined in the previous section.

The fundamental property of $\exp x$ can be deduced from a property of its inverse function $\ln x$.

Let
$$y = \exp x,$$
then
$$\ln y = x.$$

Differentiating with respect to x,

$$\frac{1}{y}\frac{dy}{dx} = 1$$

$$\Rightarrow \frac{dy}{dx} = y$$

$$= \exp x.$$

Thus
$$\frac{d}{dx}(\exp x) = \exp x.$$

Qu.2 (i) Sketch the area corresponding to the integral $\int_1^\infty \frac{1}{t}dt$, and write down the infinite series giving the area under the corresponding lower step function evaluated at integer values of t. Deduce that $\ln x \to \infty$ as $x \to \infty$. (Use Exercise 2e question 6 (ii).)

(ii) Show also that $\ln x \to -\infty$ as $x \to 0$.

Answers

Chapter 1

Qu.1 (i) T; (ii) F; (iii) F; (iv) T; (v) F; (vi) F; (vii) F;
(viii) T; (ix) T; (x) F; (xi) F; (xii) F; (xiii) T;
(xiv) F; (xv) T; (xvi) F; (xvii) T; (xviii) T;
(xix) T.

Qu.2 (i) T; (ii) F; (iii) T; (iv) F.

Qu.3 (i) (a) $P \cup Q$; (b) P';
(ii) (a) $P \subseteq Q$; (b) $Q \subseteq P$; (c) $P = Q$.

Qu.6 If Peter is not a bird, then he is not a sparrow.

Exercise 1a

1.

p	q	p or q	q'	p and q'	p' or q
1	1	1	0	0	1
1	0	1	1	1	0
0	1	1	0	0	1
0	0	0	1	0	1

2. (i) A quadrilateral is a square $\Rightarrow$ its diagonals are perpendicular. (ii) If the diagonals of a quadrilateral are perpendicular, then it is a square. (iii) Any kite.

4. (iii) (a) $(p$ and $q)' = p'$ or q';
(b) $(p$ or $q)' = p'$ and q'.

6. (i) F; (ii) F; (iii) T; (iv) F; (v) T.

9. (i) Yes; (ii) No; $(n+1) = 2 \Rightarrow (n+1)^2 = 4$ is not reversible; (iii) Yes.

10. (i) P and Q are equal if and only if every member of P is in Q and every member of Q is in P;
(ii) Yes; (iii) $x \in P \Rightarrow x \in Q$.

11. (i) $\varnothing$, $\{a\}$, $\{b\}$, $\{c\}$, $\{a, b\}$, $\{a, c\}$, $\{b, c\}$, $\{a, b, c\}$;
(ii) (a) 4; (b) 16; (c) 2^k.

12. (i), (iv), (v), (vi).

Qu.8 $\dfrac{123}{999}$.

Qu.9 (i) $\dfrac{61}{99}$; (ii) $\dfrac{3435}{9999}$; (iii) $\dfrac{100}{999}$; (iv) $\dfrac{34798}{99000}$.

Exercise 1b

1. (i) $\{1, 2, 3, 4, 5\}$; (ii) $\{4, 5, 6, 7, 8, 9, 10\}$.

5. (i) Rational; (ii) rational; (iii) irrational;
(iv) either (u might be zero); (v) either;
(vi) either.

6. (i) No; (ii) yes; (iii) yes; (i) and (iii).

7. (i) There isn't one: $P(n)$ is true for all natural numbers n; (ii) T includes all odd natural numbers.

8. (i) True; (ii) true.

9. (i) is false and (ii) is trivially true.

10. (i) There isn't one; (iii) 2; (v) Yes, yes.

12. (ii) The word 'alternately' suggests that we can speak of the 'next' number: we can't. (What, for example, is the next number after 2?)

13. (ii) The H.C.F.

Qu.11 (i) $\{(x, a), (y, a), (x, b), (y, b), (x, c), (y, c)\}$;
(ii) $\{(x, x), (x, y), (y, x), (y, y)\}$.

Qu.14 $\{(2, 1), (6, 1), (20, 1), (6, 2), (20, 2), (20, 4)\}$.

Qu.15 (i) not r, not s; (ii) yes; (iii) not r, not s;
(iv) yes; (v) yes; (vi) not r nor s nor t;
(vii) not r, not t; (viii) not r, not t; (ix) yes.

Qu.18 (ii) $\{2, 7, 12, 17\}$; (iii) $\{1, 6, 11, 16\}$, $\{3, 8, 13, 18\}$, $\{4, 9, 14, 19\}$, $\{5, 10, 15, 20\}$.

Qu.19 (iii) Footballers and cricketers may overlap;
(iv) some will be too young to be in either subset.

Qu.21 (i) 2; (ii) 3; (iii) 0.

Qu.22 (i) 2; (ii) 0; (iii) 3; (iv) 2.

Qu.23 (i) 4; (ii) 51.

Exercise 1c

2. (ii) requires the extra condition 'and neither P nor Q is empty'.

4. (i) 5; (ii) 15.

5. (i) is less than; (ii) is a multiple of;
(iii) is the wife of; (iv) is a square root of.

6. (i) (a) Yes; (b) no; (c) yes;
(ii) (a) yes; (b) yes; (c) yes;
(iii) (a) no; (b) yes; (c) no;
(iv) (a) yes; (b) yes; (c) no;
(v) (a) yes; (b) no; (c) yes;
(vi) (a) no; (b) yes; (c) yes.

7. 'Choose some y': there may not be one.

9. 12.

10. (i) 1; (ii) 4; (iii) 6.

Qu.26 p; $q^{-1} = r$, $r^{-1} = q$, rest are self-inverse.

Qu.28 $x = b * a^{-1}$.

Qu.30 $(x+2)^2 = (2x+1)^2 \not\Rightarrow (x+2) = (2x+1)$.

Qu.31 No: $x = 4$ or 0.

Exercise 1d

1. Proves associativity.

2.   (i) (a) Y; (b) Y; (c) N; (d) –; (e) Y;
(ii) (a) Y; (b) Y; (c) Y; (d) Y; (e) Y;
(iii) (a) Y; (b) Y; (c) Y; (d) N; (e) Y;
(iv) (a) Y; (b) Y; (c) Y; (d) N; (e) Y;
(v) (a) Y; (b) Y; (c) Y; (d) N; (e) Y;
(vi) (a) N; (b) –; (c) Y; (d) N; (e) Y;

(vii) (a) Y; (b) Y; (c) Y; (d) Y; (e) Y;
(viii) (a) Y; (b) N; (c) N; (d) –; (e) Y;
(ix) (a) Y; (b) N; (c) N; (d) –; (e) N;
(x) (a) Y; (b) Y; (c) Y; (d) Y; (e) Y;
(xi) (a) Y; (b) Y; (c) Y; (d) Y; (e) Y;
(xii) (a) Y; (b) Y; (c) Y; (d) N; (e) N;
(xiii) (a) Y; (b) Y; (c) Y; (d) Y; (e) Y;
(xiv) (a) Y; (b) Y; (c) Y; (d) Y; (e) Y;
(xv) (a) Y; (b) Y; (c) Y; (d) N; (e) Y;
(xvi) (a) Y; (b) Y; (c) Y; (d) N; (e) Y;
(xvii) (a) Y; (b) Y; (c) Y; (d) N; (e) Y;
(xviii) (a) Y; (b) Y; (c) Y; (d) Y; (e) Y.

3. (iii) No.
5. No ((0, 1) is not in the set).
6. $x = a^{-1} * c * b^{-1}$.
9. (ii) (a) No; (b) Yes.
10. 5 ways.
11. (ii) Yes.
12. $A^2 + AB + BA + B^2$.
13.

	e	c	a	u	v	w
e	e	c	a	u	v	w
c	c	a	e	w	u	v
a	a	e	c	v	w	u
u	u	v	w	e	c	a
v	v	w	u	a	e	c
w	w	u	v	c	a	e

14. (i) Not associative; (ii) non-abelian group.
15. (i) e; $b^{-1} * a^{-1}$; (iii) $c^{-1} * b^{-1} * a^{-1}$.
17. (i) Not in the set; (ii) 2; 3.

Miscellaneous exercise 1

1. (0, 1), $(-x, 1/y)$.

Chapter 2

Qu.1 (i) 17; (ii) 26; (iii) 23.
Qu.2 (i) Yes; (ii) No: zero has no image;
  (iii) Yes; (iv) No: not defined for $x < 0$;
  (v) Yes; (vi) No: not clear what happens to $1\frac{1}{2}$, $2\frac{1}{2}$,
  etc.; (vii) Yes; (viii) Yes.
Qu.3 $-\frac{1}{3}$.
Qu.5 (i) Neither; (ii) 1 – 1 and onto; (iii) onto but not
  1 – 1.
Qu.6 (i) x^4; (ii) $4x + 3$.
Qu.7 Yes.

Exercise 2a

1. (i) 11; (ii) 1; (iii) 1; (iv) $x^2 - x - 1$; (v) $4x^2 + 6x + 1$.
2. (i) No; (ii) no; yes.
3. (i) Both; (ii) 1 – 1; (iii) neither; (iv) both.
4. (i) $n(B) \geqslant n(A)$; (ii) $n(A) \geqslant n(B)$; (iii) $n(A) = n(B)$.
6. (i) $10x + 7$; (ii) $10x + 17$.
8. (i) $h^{-1}(x) = \dfrac{x+3}{2}$; (ii) $u^{-1}(x) = \dfrac{3x+1}{1-2x}$ for $x \neq \frac{1}{2}$.
9. $\forall x \in A, f(x) = g(x)$.
10. (ii) μ^λ.
11. If U is a finite subset of T, we could take $f(U) = $ the
  sum of the members of U.
12. $k(k-1)(k-2) \ldots 3.2.1$.

13. (i) Stretch, by a factor 2, parallel to the x-axis;
  (ii) Reflection in the y-axis;
  (iii) Rotation about O, 90° anticlockwise.
14. Points drop perpendicularly on to the x-axis.

Exercise 2b

5. 3; none if k is odd; $3 \times 5 \times 7 \times \ldots \times (k-1)$ if k is even.
8. (ii) Yes; (iii) yes; (iv) $\theta^{-1}(p, q) = \left(\dfrac{p-q}{2}, \dfrac{p+q}{2}\right)$.
Qu.9 $x = 0$.
Qu.12 (ii) (a) Any horizontal line meets the graph at most
  once; (b) any horizontal line meets the graph at least
  once.
Qu.13 (i) $[-1, 2)$; (ii) $y = \begin{cases} x & \text{for } -1 \leqslant x < 0 \\ \sqrt[3]{\dfrac{x}{2}} & \text{for } 0 \leqslant x < 2. \end{cases}$

Exercise 2c

1. It is not discontinuous anywhere.
2. $x = 0$.
3. (ii) No; (iii) yes.
5. (i) Periodic with period na for any $n \in Z^+$.
  (ii) Periodic with period 1.
  (iii) Periodic with period 3.
6. (i) $\frac{1}{2}$ and 1; (ii) infinitely many.
7. (ii) $y = 0$ for all x.
8. Function is odd (even) if n is odd (even); yes.
9. Yes.
10. (i) $a_1 = a_3 = 0$; (ii) $a_0 = a_2 = 0$.
11. (i) Odd; (ii) even; (iii) odd; (iv) neither; (v) odd;
  (vi) odd; (vii) even.
13. (i) $-\frac{1}{2} < x < \frac{1}{2}$; (iii) $-3 < x < -1$.
15. $-1 < x \leqslant 1$.
16. (ii) $y = \begin{cases} \frac{1}{2}x & \text{for } x < 0 \\ \sqrt{x} & \text{for } 0 \leqslant x \leqslant 4 \\ x - 2 & \text{for } x > 4. \end{cases}$
17. $f^{-1}(y) = 1 \pm \sqrt{(2-y)}$; (i) $y > 2$; (ii) $y < 2$.
18. Everywhere.
Qu.17 $(x-1)^3$; triple root at $x = 1$.
Qu.18 $a < 0$ and $b^2 - 4ac < 0$.

Exercise 2d

1. 4 and 5; 8 and 9.
2. (i) 3 and 4; (ii) 0 and 1; 9 and 10.
3. (ii) -9; 7; (iii) less.
4. $1008 < c < 1340$.
5. (i) None; (ii) 2; (iii) 1; (iv) 2.
6. $c < \frac{25}{8}$.
7. (ii) $y = \dfrac{20x}{21}$.
8. $b = 0$, $a = 0$ or 4.
Qu.19 (i) 0; (ii) 3; (iii) 0; (iv) no limit; (v) no finite limit;
  we say that the sequence tends to infinity; (vi) no
  limit.
Qu.20 -1; 1; not defined.
Qu.21 -0.035.
Qu.22 (i) 1; (ii) ∞.

Exercise 2e

1. (i) No limit (the sequence tends to ∞):
   (ii) $0\dot{3} = \frac{1}{3}$; (iii) $\frac{2}{3}$; (iv) no limit; (v) 0.
2. (i) -1; (ii) undefined; 0.43 (2 s.f.); same.
3. (i) $\frac{2}{3}$; (ii) 1; (iii) 0.
4. (i) 0.01745; (ii) -0.000152.
5. (i) (a) $\frac{2}{3}$; (b) $\frac{5}{8}$; (c) $\frac{11}{20}$; (ii) $\frac{1}{2}$.
6. (i) 2.
Qu.24 (ii) Countable.

Exercise 2f

4. Countable.

Miscellaneous exercise 2

1. $f^{-1}: x \to \dfrac{1-x}{x}$; $\quad ff: x \to \dfrac{x+1}{x+2}$.
2. $k \geqslant 10$ or $k \leqslant -10$; (ii) any member of R.
3. (i) $f(x) = 1 + x$; (ii) $f(x) = 3 - x$.
5. (a) $[0, 3]$.
6. $1 - \dfrac{1}{x}$; $\dfrac{2(1 + x - x^2)}{1 - x}$.
7. Range of f is $\{y : y \geqslant 0,\ y \neq 2r - 1,\ r \in Z_+ \}$; f is not one-one; $g^{-1}: y \to y - \frac{1}{2}[y]$; domain of g^{-1} is $\{y : 2r - 2 < y < 2r - 1,\ r \in Z_+ \}$.

Chapter 3

Qu.1 (i) $\sqrt{5}$; (ii) $\sqrt{61}$; (iii) $\sqrt{325}$.
Qu.2 $\sqrt{10}$; $\sqrt{40}$; $\sqrt{50}$.
Qu.3 (i) $(5, 7\frac{1}{2})$; (ii) $(-\frac{1}{2}, -\frac{1}{2})$; (iii) $(-1\frac{1}{2}, -7\frac{1}{2})$.

Exercise 3a

1. (i) $\sqrt{65}$; $(1, -1\frac{1}{2})$; (ii) 17; $(-2\frac{1}{2}, -1)$;
   (iii) 10; (p, q); (iv) $\sqrt{(a^2 + b^2)}$; $\left(\dfrac{a}{2}, \dfrac{b}{2}\right)$.
3. $\sqrt{26}$; $\sqrt{6\frac{1}{2}}$.
5. $D(-3, -1)$; (i) $\sqrt{40}$; (ii) $\sqrt{5}$; (iii) $(\frac{1}{2}, -1)$.
6. $\sqrt{50}$; $(\frac{1}{2}, 11\frac{1}{2})$.
7. $\sqrt{132\frac{13}{16}}$; $\sqrt{7\frac{13}{16}}$; $\sqrt{125}$.
8. $S(14, 5)$; $T(-10, -7)$.
9. (iii) P is $(2, 3)$ or $(2, -11)$.
Qu.5 (i) 4; (ii) -1; (iii) infinite; (iv) $\frac{5}{4}$; (v) $b + a$.
Qu.6 $63.4°$; (ii) $29.7°$; (iii) 1 or -1.
Qu.7 14.
Qu.8 $-3\frac{1}{2}$.
Qu.9 $y_1 = y_0 + m(x_1 - x_0)$.
Qu.11 (ii) (a) $y = -4$; (b) $x = 1$.
Qu.12 (i) $y + 2x = 2$; (ii) $3y + 4x = 31$; $(-12\frac{1}{2}, 27)$.
Qu.13 (iv) (a) $-\frac{3}{2}$; (b) $-\frac{5}{2}$; (c) $-\frac{16}{13}$; (v) $x + 4y = 12$.

Exercise 3b

1. Both $\frac{2}{5}$; collinear; $\frac{2}{5}$.
2. (i) -1; (ii) $-\frac{4}{3}$; (iii) $p + q$; (iv) $6 + h$.
3. (i) $p = 0$; (ii) $q = \frac{4}{3}$; (iii) $c = mh + b$.
4. (i) $y = 2x - 5$; (ii) $y + 2x + 1 = 0$;
   (iii) $5y + 2x + 26 = 0$; (iv) $y = 3$; (v) $x = 3$;
   (vi) $2x + 5y = 11$; (vii) $11y = 3x + 2$; (viii) $3y = x$.
5. (i) $(-2, 3)$.

6. (i) $(-\frac{11}{4}, -\frac{7}{4})$; (ii) $(-4, 2)$;
   (iii) no solution; lines are parallel.
8. (i) $71.6°$; (ii) $45°$.
10. $(8, 2)$.
Qu.14 (i) $3y = 4x + 11$; (ii) $(-\frac{4}{5}, \frac{13}{5})$; (iii) 2.
Qu.15 $\frac{67}{13}$.
Qu.17 (i) Perpendicular bisector of AB; (ii) circle with diameter AB; (iii) line through A perpendicular to AB; (iv) circle centre A through B; (v) the half-plane on the side of (i) containing B; (vi) pair of parallel lines.

Exercise 3c

1. (i) $y = 2x$; (ii) $2y = 3x$; (iii) $5y + 4x = 0$.
2. (i) $2y - 5x + 11 = 0$; (ii) $x + y + 2 = 0$; (iii) $y = x - 4$;
   (iv) $y = -3$.
3. (i) $2y + x = 0$; (ii) $3y + 2x = 0$; (iii) $4y = 5x$.
4. (i) $5y + 2x + 13 = 0$; (ii) $y = x - 4$; (iii) $y + x + 2 = 0$;
   (iv) $x = 1$.
5. (i) $(-\frac{1}{2}, 3)$; (ii) $\frac{2}{3}$; $4y + 6x = 9$.
6. Rhombus.
7. $(\frac{35}{3}, -\frac{37}{3})$.
8. Perpendicular bisector of RS.
9. $2y + x = 6$.
10. $2x(a - c) + 2y(b - d) = a^2 + b^2 - c^2 - d^2$.
12. (i) $(\frac{25}{7}, \frac{16}{7})$; $\frac{5}{7}\sqrt{26}$; (ii) $(\frac{8}{3}, \frac{7}{3})$; (iii) $(\frac{6}{7}, \frac{17}{7})$.
13. (i) $\dfrac{8}{\sqrt{10}}$; (ii) $\dfrac{5}{\sqrt{13}}$; (iii) 0.
15. (i) $-\frac{13}{8}$; (ii) $\dfrac{2k - 17}{5k - 2}$.
16. $(-\frac{17}{5}, \frac{11}{5})$.
20. $\frac{30}{17}$.
21. $2x^2 + 2y^2 - 18y + x + 35 = 0$.
22. $3x^2 - y^2 + 4x - 8y - 20 = 0$.
23. $x^2 - y^2 = 2c^2$.
Qu.18 (i) $x^2 + y^2 - 2x - 4y = 11$; (ii) $x^2 + y^2 = a^2$;
   (iii) $x^2 + y^2 + 2x + 2y = 0$.

Exercise 3d

1. $x^2 + y^2 - 30x + 16y = 0$; circle passes through origin.
2. $x^2 + y^2 - 2x - 6y = 15$; (i) $(-3, 0)$ and $(5, 0)$;
   (ii) $(1 - \sqrt{21}, 5)$ and $(1 + \sqrt{21}, 5)$; (iii) $(4, 7)$ and $(-\frac{2}{5}, -\frac{9}{5})$.
3. (i) $x^2 + y^2 + 2x + 10y + 13 = 0$; (ii) $2y = 3x - 7$;
   (iii) $3y + 2x + 4 = 0$.
4. (i) $(-\frac{1}{2}, 3)$; (ii) 13; $x^2 + y^2 + x - 6y = 33$.
5. (i) $\dfrac{y - 6}{x - 2}$; $\dfrac{y - 2}{x + 1}$; (iii) $x^2 + y^2 - x - 8y + 10 = 0$; circle with AB as diameter.
6. (i) $x^2 + y^2 + 5x + 3y = 14$;
   (iii) $(y - y_0)(y - y_1) + (x - x_0)(x - x_1) = 0$.
9. (i) $p = q = 0$; (ii) the single point $(3, -1)$.
10. $x^2 + y^2 \pm 2xr = 0$.
11. $x^2 + y^2 - 14x - 2y = 0$.
12. $x^2 + y^2 - 10x - 10y + 30 = 0$.
13. $3x^2 + 3y^2 - 14x - 6y = 64$.
14. $(3, -1)$ and $(-\frac{4}{3}, \frac{23}{3})$.
15. (i) $x^2 + y^2 - 10x - 4y + 9 = 0$;
   (ii) $x^2 + y^2 + 20x + 6y - 61 = 0$; $(1, 4)$ and $(3, -2)$.

Exercise 3e

2. (i) $x + 2y = 5$; (ii) $x = y^2 + 2y$;
(iii) $x^2 + y^2 = y$; (iv) $x^5 = y(x^3 + y^3)$.
3. (i) $pqy + x = c(p + q)$.
4. (i) $t = 4$ or -1; (ii) $t = -\frac{1}{2}$, i.e. $(\frac{4}{5}, -\frac{2}{5})$.
5. $(0, 1)$.

Miscellaneous exercise 3

1. 20; $P(3, 6)$, $Q(2, -1)$.
2. $x^2 + y^2 + 2x - 16y + 45 = 0$; $2y + x = 5$;
$R(3, 6)$; $S(1, 2)$.
3. (i) $y = x + 3$; (ii) $(2\frac{1}{2}, 5\frac{1}{2})$.
4. $3x = 2y$.
5. $x^2 + y^2 - 8x - 6y = 0$; $3y = 4x - 32$.
6. $x^2 + y^2 - 12x - 6y + 20 = 0$;
$x^2 + y^2 - 28x - 18y + 252 = 0$;
$3y + 4x = 58$, $4y - 3x - 19 = 0$, $4y - 3x + 31 = 0$.
7. $x^2 + y^2 - 10x - 8y + 16 = 0$; $(10, -6)$; 10; $53°$.
8. (i) $(1, 1)$, $(4, 5)$; 2, 3; (ii) -32; (iii) $(2\frac{1}{5}, 2\frac{3}{5})$.
9. $x^2 + y^2 - 5x - 6y + 9 = 0$; $y = 2x + 3$.
10. $AD = \sqrt{45}$; $35m^2 + 20m - 64 = 0$; $106° 41'$.

Chapter 4

Qu.4 (i) $3\sqrt{5}$; (ii) $6\sqrt{6}$; (iii) 40; (iv) $4\sqrt{2}$;
(v) 12; (vi) $5 + 2\sqrt{6}$; (vii) $2\sqrt[3]{3}$.
Qu.5 (i) $\dfrac{4\sqrt{5}}{5}$; (ii) $3\sqrt{2}$; (iii) $\dfrac{\sqrt{6}}{4}$;
(iv) $\sqrt{3} - \sqrt{2}$; (v) $\dfrac{(3 + \sqrt{3})}{2}$.
Qu.6 $\dfrac{\sqrt[3]{4}}{2}$.
Qu.7 (ii) (a) $\frac{1}{3}(4 + 2\sqrt[3]{5} + \sqrt[3]{25})$;
(b) $\frac{1}{13}(4 - 2\sqrt[3]{5} + \sqrt[3]{25})$.
Qu.8 (ii) $\sqrt{(2x + 5)} - \sqrt{(x + 6)} = 3$.

Exercise 4a

1. (i) $10\sqrt{2}$; (ii) $3\sqrt{10}$; (iii) $4\sqrt{6}$; (iv) $4\sqrt{15}$;
(v) $50\sqrt{2}$; (vi) $100\sqrt{5}$.
2. (i) $\sqrt{12}$; (ii) $\sqrt{18}$; (iii) $\sqrt{175}$; (iv) $\sqrt{24}$.
3. (i) $10\sqrt{2}$; (ii) $3\sqrt{5}$; (iii) 30; (iv) $\sqrt{2}$; (v) 0.
4. (i) 12; (ii) $8\sqrt{3}$; (iii) 18; (iv) $54\sqrt{2}$;
(v) $128\sqrt{2}$; (vi) $\sqrt{3}$; (vii) $\dfrac{2\sqrt{3}}{3}$.
5. (i) $3 + 2\sqrt{2}$; (ii) $22 - 12\sqrt{2}$; (iii) $5 + 2\sqrt{6}$;
(iv) $5 + 3\sqrt{2}$; (v) $3\sqrt{2} - \sqrt{6} - 4 + 4\sqrt{3}$;
(vi) $26 - 15\sqrt{3}$.
6. (i) $\dfrac{\sqrt{2}}{2}$; (ii) $\dfrac{5\sqrt{3}}{3}$; (iii) $2\sqrt{2}$; (iv) $\dfrac{3\sqrt{6}}{2}$; (v) $\sqrt{6}$;
(vi) $\dfrac{(\sqrt{10} + 2\sqrt{5})}{10}$; (vii) $\dfrac{\sqrt{2}}{2}$; (viii) $\dfrac{\sqrt{15}}{5}$.
7. (i) $\dfrac{(\sqrt{7} + \sqrt{5})}{2}$; (ii) $2\sqrt{5} + 2\sqrt{2}$; (iii) $\dfrac{(2\sqrt{3} - \sqrt{7})}{5}$;
(iv) $\dfrac{(12 + 3\sqrt{3})}{13}$, (v) $\sqrt{2} - 1$;
(vi) $\frac{1}{7}(\sqrt{10} + 1)(3 - \sqrt{2})$; (vii) $6 - \sqrt{35}$;
(viii) $\frac{1}{6}(3\sqrt[3]{3} + 3 + \sqrt[3]{9})$.
8. (i) $x = 8$; (ii) no solution.

9. (i) $x = 5$; (ii) $x = -6$.
10. (i) $x = 1$ or -3; (ii) $x = 2$ or -2; (iii) $x = -1$.
12. (i) $3 - 2\sqrt{2}$; (ii) $5 + 2\sqrt{6}$; (iii) $\sqrt{3} + 5\sqrt{2}$.
13. $\frac{97}{56}$.
14. $\frac{3}{2}$, $\frac{17}{12}$, $\frac{577}{408}$.
15. $\dfrac{1}{2\sqrt{x}}$.
Qu.9 (i) 2^{2n}; (ii) $2^{3(n+1)}$; (iii) 2^5.

Exercise 4b

1. (i) $\frac{1}{4}$; (ii) 2; (iii) $\frac{1}{2}$; (iv) 1; (v) 9; (vi) $\frac{16}{625}$; (vii) $\frac{8}{27}$;
(viii) $\frac{1024}{243}$; (ix) 1.
2. (i) $16\sqrt{3}$; (ii) $3\sqrt{2}$; (iii) $55\sqrt{2}$.
3. (i) $x^{1/2}$; (ii) $x^{3/2}$; (iii) $x^{3/2}$; (iv) $x^{-5/2}$; (v) $x^{2/3}$;
(vi) $x^{1/2}$; (vii) $x^{-5/2}$.
4. (i) $(x + 2)(x + 1)^{1/2}$; (ii) $(a + 1)a^{-1/2}$;
(iii) $(4p^2 + 10p + 7)(2p + 3)^{-1/2}$.
5. (i) 2^n; (ii) 2^n; (iii) 3^{n+3}; (iv) 10^n.
6. (i) $(x^2 + 1)^{-1/2}$; (ii) $(x - 1)^{1/2}$; (iii) $4(x^2 + 2)^{-3/2}$;
(iv) $(x - 4)(x - 2)^{-1/3}$.
Qu.10 Reflection in the y-axis.
Qu.11 (ii) 1; 4; 256; (iii) $\to \infty$; $\to 0$; (iv) $\to \infty$.
Qu.12 (i) $\log_2 64 = 6$; (ii) $\log_{10} 1000 = 3$; (iii) $\log_8 512 = 3$;
(iv) $\log_9 3 = \frac{1}{2}$; (v) $\log_3(\frac{1}{9}) = -2$; (vi) $\log_{1/4}(1/8) = 3/2$.
Qu.13 (i) $2^7 = 128$; (ii) $10^4 = 10000$; (iii) $8^2 = 64$;
(iv) $8^{1/3} = 2$; (v) $2^{-4} = \frac{1}{16}$; (vi) $(\frac{4}{9})^{3/2} = \frac{8}{27}$.
Qu.14 (i) 4; (ii) $\frac{1}{4}$; (iii) $\frac{3}{4}$; (iv) 0; (v) $\frac{1}{2}$; (vi) $-\frac{3}{2}$; (vii) -1.
Qu.16 (i) 0·77815; (ii) 0·90309; (iii) 0·17609; (iv) 1·85733;
(v) 0·69897; (vi) $-0·60206$; (vii) 2·57403.

Exercise 4c

1. (i) 4; (ii) -4; (iii) $-\frac{2}{3}$; (iv) $\frac{3}{2}$; (v) $-\frac{1}{2}$; (vi) 0.
2. (i) 2; (ii) $\frac{3}{2}$; (iii) 4; (iv) $-\frac{1}{2}$.
3. (i) $p + q$; (ii) $p - q$; (iii) $2p$; (iv) $3p - 2q$; (v) $1 + p$;
(vi) $p - q - 1$; (vii) $\frac{1}{3}(p - 2q)$.
4. (i) $\log 2$; (ii) $\log 15$; (iii) $\log 9$; (iv) $\log 2$;
(v) $\log 20$; (vi) $\log 4\sqrt{3}$.
5. (i) 1; (ii) 2; (iii) 0; (iv) $\frac{1}{2}$; (v) $1\frac{1}{2}$; (vi) 2; (vii) $\frac{2}{3}$.
6. (i) 2·3219; (ii) 1·6902; (iii) $-2·0959$.
7. (i) $x = 2$; (ii) $x = -1$ or 3; (iii) $x = 2·585$; (iv) $x = 0$
or -1.
8. (i) $x = \frac{5}{2}$; (ii) $x = 18$; (iii) 0·6300 or 8; (iv) 0·1640 or
54·89; (v) no solution.
10. $A = 10^c$.
11. 2·322.
12. Move the y-axis 3 units to the right.
13. (i) $\log(x + 1)$; (ii) $\log 2$.
14. (i) $\log\{(x + 1)^2(2x - 3)^3\}$; (ii) $\log \dfrac{x}{2\sqrt{(x + 1)}}$.
15. (i) $\log_3 \sqrt{x}$; (ii) $\log_3(2x + 1)^2$; (iii) $\log_3(1/x^2)$.
16. (i) $\frac{1}{8}$; (ii) 8 or 17; (iii) no solution.
17. £$100 \times (1·1)^n$; just over 41 years.
18. (i) True; (ii) true; (iii) false.
Qu.19 $a \approx 0·270$, $b \approx -4·18$.
Qu.20 $a = -7·3$, $b = 3·8$.
Qu.21 (i) $\log y$ against x; (ii) gradient is $\log a$; intercept is
$\log k$.

ANSWERS

Exercise 4d

1. (i) $p=0.4$, $q=13$; (ii) the fourth; (a) 4·71; (b) 20·2.
2. (i) $\dfrac{1}{L-1}$ against $\dfrac{1}{M}$; (ii) $a=2.8$, $b=2.5$; (iii) and (iv) 3·39; 1·48.
3. $a=-1.2$, $b=3$.
4. $A=0.85$, $n=0.5$.
5. $a=1.34$, $k=1700$.
6. $A=6$, $n=-2.3$; $h=54$.

Miscellaneous exercise 4

1. (i) $\frac{2}{5}$; (ii) 1·3; (iii) $-\frac{2}{3}$.
2. (i) $z \propto x^{-3/2}$, $\frac{1}{72}$; (ii) $a=10^4$; $b=10^3$.
3. (a) (i) 3; (ii) $\frac{1}{4}$; (iii) $\frac{1}{16}$; (iv) -1; (b) 0 or 1; (c) 1.
4. 1·58.
5. (i) 2; (ii) 1; (iii) 0·683.
6. (i) 80°C, 57·8°C, (ii) (a) 15 min. (b) 23·8 min.
7. (a) (i) 1·204 120 0; (ii) $-0.954 242 6$; (iii) 0·359 727 1; (b) (i) 3; (ii) 2; (c) log $x=2.6$, log $y=0.6$; (d) -0.30.
8. $a=19.05$ to 22·4, $n=2.5$.
9. $x=7$ and $y=2$ or $x=-\frac{1}{2}$, $y=\frac{1}{2}$.
10. $x=y=27$ or $x=9$, $y=81$.

Chapter 5

Qu.1 (i) 8; (ii) (a) 7; (b) $6\frac{1}{2}$; (c) 6·1; (d) 6·01; (e) 6·001; (iii) 6; (iv) (a) 5; (b) $5\frac{1}{2}$; (c) 5·9; (d) 5·99; (e) 5·999; yes.

Qu.2 3.

Exercise 5a

1. (i) 20; (ii) 17; (iii) 14·3; (iv) 14·03; (v) 14·0003; 14.
2. (i) 13·97; (ii) 13·9997.
3. (i) 0·015038; (ii) 0·015107; the second.
4. (i) 0·69; (ii) 0·6937.
6. 0·29.

Qu.4 $3x^2$.

Qu.5 $\dfrac{1}{2\sqrt{x}}$.

Exercise 5b

1. Gradient is -3.
2. (i) $6x$; (ii) $12x^2$; (iii) 0; (iv) $2x+1$; (v) 3; (vi) $4ax^3$.
3. (i) 18; (ii) 48; (iii) 48.
4. $-\frac{1}{2}x^{-3/2}$.
5. $\frac{1}{3}x^{-2/3}$.

Qu.6 (i) $7x^6$; (ii) $-3x^{-4}$; (iii) $\frac{5}{2}x^{3/2}$; (iv) $\frac{1}{3}x^{-2/3}$; (v) $-4x^{-5}$; (vi) $-\frac{7}{3}x^{-10/3}$; (vii) 0; (viii) 1.

Qu.7 (i) $8x$; (ii) $-\frac{1}{2}x^{-4}$; (iii) $2x^{-2}$; (iv) $\dfrac{1}{\sqrt{x}}$; (v) $-\dfrac{14}{3x^3}$; (vi) 5; (vii) 0; (viii) $\dfrac{x}{2}$.

Qu.8 0.
Qu.9 m.
Qu.10 (i) $18x^2+6x$; (ii) $28x^3+28x^{-5}$; (iii) 5; (iv) m; (v) $2ax+b$; (vi) $3x^2-6x+2$.

Qu.11 (i) $\frac{1}{2}+\dfrac{4}{x^2}$; (ii) $x-\dfrac{3}{4x^2}+\dfrac{1}{x^3}$.

Qu.13 (i) 12; (ii) 4; (ii) $\frac{2}{3}$; (iv) 0.

Qu.14 (i) 13; (ii) -0.0997; $\dfrac{\delta y}{\delta x}=9.97$; $\dfrac{dy}{dx}=10$.

Qu.15 1·79 litre per second.

Exercise 5c

1. (i) $4x^3-2x+7$; (ii) $-2x^{-3}$; (iii) $\frac{2}{3}x^{-1/3}$; (iv) $-14x^{-3}+24x^{-4}$; (v) $6\sqrt{x}$; (vi) $-3x^{-2}+12x^{-4}$; (vii) $-\frac{9}{10}x^{-8/5}$; (viii) $\dfrac{5\sqrt{2}}{2}x^{3/2}$; (ix) $6x^2-8x+6$; (x) $-2x^{-2}-6x^{-3}$; (xi) $x^{-1/2}+6x^{1/2}-\frac{15}{2}x^{3/2}$; (xii) $-\frac{1}{2}x^{-2}+\frac{1}{2}$; (xiii) $\frac{5}{6}x^{3/2}-x^{1/2}$.
2. -5.
3. $(3, 7)$, gradient 4; $(-1, 7)$, gradient -4.
4. $(\frac{2}{3}, \frac{17}{3})$.
7. x^3+2x^2+2x+ any constant.
9. (i) x; (ii) $\pm\frac{1}{4}x^{-3/2}$.
10. $(2, 1)$.
11. $(3, -17)$ and $(-\frac{1}{3}, \frac{41}{27})$.
12. $(3, 19)$ and $(-1, -1)$.
13. No.
15. The gradients are equal.
16. (i) $-\frac{3}{10}$ degree per metre; (ii) $-\frac{11}{20}$ degree per metre.

Chapter 6

Qu.2 $x=1$ only.

Exercise 6a

1. $x=0$ or 1 or -2.
2. $x=-1$ or $\frac{1}{2}$ or -2.
3. $x=6$ or -1.
4. $x=-2$ only.
5. $x=-1$ or $-2\frac{1}{2}$ or $\frac{1}{3}$.
6. $x=\frac{3}{2}$ only.
7. $x=-3$ or 2 or 5.
8. (i) $(x-1)(x^2+x+1)$.
9. (i) $x=2$ or -2; (ii) no solution.

Qu.3 (i) No.
Qu.4 $(\frac{3}{2}, \frac{7}{4})$.
Qu.5 Max at $(-2, 44)$, min at $(6, -212)$.
Qu.6 $36x^2+42x$.
Qu.7 Max at $(0, 6)$, minima at $(-2, -26)$ and $(5, -369)$.
Qu.8 Minimum; zero.

Exercise 6b

1. $(-\frac{5}{4}, -\frac{17}{8})$.
3. (i) None; (ii) inflexion at $(1, 5)$; (iii) minima at $(-2, -17)$ and $(\frac{7}{3}, -100\frac{3}{16})$, max at $(0, 7)$; (iv) max at $(-\frac{4}{5}, \frac{256}{3125})$, min at $(0, 0)$; (v) min at $(-\frac{3}{4}, -\frac{27}{256})$, inflexion at $(0, 0)$.
4. $\frac{2}{3}$.
5. $\dfrac{4ac-b^2}{4a}$ provided $a>0$.
7. Max 7 (at $x=-2$), min -17 (at $x=-4$).
8. $-\frac{375}{16}$.
9. Either $y=6x^{-2}$ or $y=0$ (for all x).
10. A and D negative; B and C positive.

Qu.10 $\frac{8}{3}$.

1. Min at $x=\frac{3}{2}$.
2. Min at $x=-\frac{2}{3}$.
3. Max at $x=\frac{1}{2}$.
4. Min at $x=\frac{7}{4}$.
5. Max at $x=\frac{2}{3}$, min at $x=2$.
6. Max at $x=-4$, min at $x=\frac{2}{3}$.
7. Inflexion at $x=2$.
8. Min at $x=2$.
9. Minima at $x=0$ and $1\cdot7$, max at $x=1$.
10. (i) $k<0$ or $k>\frac{4}{27}$; (ii) $k=0$ or $\frac{4}{27}$; (iii) $0<k<\frac{4}{27}$.
11. None if $k>1536$; one if $k=1536$; two if $\frac{11}{16}<k<1536$ or $k<0$; three if $k=0$ or $\frac{11}{16}$; four if $0<k<\frac{11}{16}$.
Qu.12 $y+3x+2=0$; $3y-x=4$.
Qu.13 $(2,-3)$.

Exercise 6d

1. $y+26x+46=0$; $(-\frac{23}{13},0)$.
2. $3y=x-34$; $(-\frac{1}{3},-11\frac{4}{9})$.
3. $(1\frac{1}{2},-6\frac{1}{4})$; $4y+8x+13=0$.
4. $2y=2x+3$.
5. $y=5x-20$; $y+5x+5=0$.
6. $y=4x+3$; $128y+32x=129$.
7. $\frac{7}{6}$; $y=3x-5$ and $18y+6x=5$; $(\frac{19}{12},-\frac{1}{4})$.
8. $4y+x=4$; $2y=8x-15$.
9. $3py=2x+p^3$; $2y+3px=2p^2+3p^4$.
10. $y+2x=13$; $(7,-1)$.
11. $x=3$ or -3.
12. $y+x=5$; $(\frac{1}{2},4\frac{1}{2})$.
13. $y=x+1$; $(2,3)$ and $(5,6)$.
14. (i) $y=x-2$; $(-3,-5)$, $(-\frac{3}{2},-\frac{7}{2})$; (ii) $11y+x+57=0$; no other points.
15. $y=(2a+b)x-a+c$.
17. $y=2px-p^2$; $y=14x-49$, $y+2x+1=0$.
18. At $x=2a+at^2$ (for $t\neq0$).
19. $t^3+\frac{1}{3}t^{-1}$.
20. $(-\frac{1}{2},-1)$ and $(1,2)$; acute angle is about $72°$.

Exercise 6e

3. (i) $x^2+\dfrac{480}{x}$; $x=\sqrt[3]{240}$; (ii) $x=\sqrt[3]{120}$.
4. $(a-2x)x\tan60°$; $\dfrac{a^2}{8}\tan60°$.
5. $\dfrac{1000\pi}{3\sqrt3}$.
6. $2\cdot11$.
7. $\frac{9}{4}\pi a^3$.
Qu.17 (i) 5; -3; (ii) 12; (iii) $t=2$ and $t=6$; (iv) 12 m; (v) 6 ms^{-1}; (vi) $t=9$ and $t=-1$; (vii) 0.
Qu.18 -2 ms^{-1}; moving backwards.

Exercise 6f

1. (i) $t=1,3,4$; (ii) $v=6,-2,3$; $a=-10,2,8$; (iii) $1\cdot78$ and $3\cdot55$; (vi) $-\frac{7}{3}$ ms^{-1} when $t=\frac{8}{3}$.
2. (i) 1 ms^{-1}; (ii) 0 ms^{-1}; (iii) 1 ms^{-2}; (iv) $-\frac{1}{8}$ ms^{-1}.
3. (iii) $\frac{108}{3125}$ m.

5. $1\cdot6$ ms^{-1}.
6. (i) 5 ms^{-1}; (ii) $t=3$; (iii) 60 m.
8. $t<\frac{8}{3}$.
9. $\frac{2}{3}$ ms^{-1}; $t=\sqrt3$.
10. (i) $10\frac{1}{2}$ ms^{-1}; (ii) $13\frac{1}{2}$ ms^{-1}.

Miscellaneous exercise 6

1. $p^2y+x=6p$; 18; (i) $(5\frac{5}{8},0)$; (ii) $\frac{153}{32}$.
2. $y=3x-3$, $3\frac{2}{3}$.
3. 24 s.
4. $(\frac{1}{2},2)$, $4x+y=4$, $(-1,8)$, $8x-y=2$, $(-\frac{1}{4},-4)$.
5. (i) $-\frac{2}{3}$; (ii) $7\frac{2}{3}$.
6. $\dfrac{75-x^2}{4x}$ m; $\frac{1}{4}(75x-x^3)$ m^3; 5 m $\times5$ m $\times2\frac{1}{2}$ m.
7. $y=x+\dfrac{1}{x}+1$; $6y+8x=37$; $(-2,-\frac{3}{2})$; $4y=3x$.
8. $4y+3x=24$; $3y+4x=-24$.
9. (i) 18 cm s^{-1}, 30 cm s^{-2}; (ii) 3 s, 81 cm; (iii) $4\cdot7$ s.
10. $x+ty=2t+t^3$; $16t^3-13t+3=0$; $(\frac{1}{2},\frac{1}{16})$, $(\frac{3}{2},\frac{9}{16})$.
11. 614 cm^3.

Chapter 7

Qu.1 (i) $9x^2+4x+3$; (ii) $10x^4-4x^3+18x^2-4x$; (iii) $2x$; (iv) $5x^4$.
Qu.2 (i) $\dfrac{2x}{(x^2+1)^2}$; (ii) $\dfrac{1}{(x+2)^2}$; (iii) $\dfrac{-4x}{(2x^2+3)^2}$; (iv) 1; (v) $(m-n)x^{m-n-1}$.

Exercise 7a

1. (i) $\frac{3}{2}\sqrt{x}-\dfrac{1}{2\sqrt{x}}-2x$; (ii) $2x+3x^2+\dfrac{1}{x^2}$; (iii) $4x+3-\dfrac{3}{x^2}-\dfrac{4}{x^3}$.
2. (i) $\dfrac{1-x^2}{(1+x^2)^2}$; (ii) $\dfrac{-4x^2+14x+2}{(x^2-2x+4)^2}$; (iii) $\dfrac{4x^2}{(x^3+2)^2}$; (iv) $\dfrac{6x^4+45x^2}{(2x^2+5)^2}$; (v) $\dfrac{\sqrt{2}-1}{\sqrt{(2x)}(1+\sqrt{(2x)})^2}$.
3. $\frac{1}{7}$; $y\to0$.
4. 1.
5. (ii) $vwz\dfrac{du}{dx}+uwz\dfrac{dv}{dx}+uvz\dfrac{dw}{dx}+uvw\dfrac{dz}{dx}$.
Qu.4 (i) $20(2x+3)^9$; (ii) $-24x^3(3x^4+5)^{-3}$; (iii) $\dfrac{x}{\sqrt{(x^2+1)}}$.
Qu.5 (i) $\dfrac{-1}{(x+1)^2}$; (ii) $\dfrac{-6}{(2x+1)^2}$; (iii) $21(3x+4)^6$; (iv) $-20x(2x^2-3)^{-6}$; (v) $\dfrac{6x^2}{\sqrt{(4x^3-3)}}$.

Exercise 7b

1. $\dfrac{x}{\sqrt{(1+x^2)}}$.
2. $5(2x^2-3x+1)^4(4x-3)$.
3. $\frac{12}{5}(3x-7)^{-1/5}$.

4. $10(2x+3)(x^2+3x-2)^9$.

5. 1.

6. $-x(x^2-1)^{-4/3}$.

7. $8x^2(x^3+2)^{-1/3}$.

8. $\dfrac{3x^2+3x-5}{2(1+x)^{3/2}}$.

9. $(x+1)^{-1/2}(3x+2)$.

10. $\dfrac{-3x^2-4x+1}{2\sqrt{(x+1)(x^2+1)^2}}$.

11. $2(x^2-3)^9(19x^2+10x+3)(2x+1)^{-2}$.

12. $\frac{4}{5}(x+1)^{-9/5}x^{-1/5}$.

13. $-\frac{1}{11}$.

14. $y=\pm\sqrt{(4-x^2)}$; $\dfrac{dy}{dx}=\mp\dfrac{x}{\sqrt{(4-x^2)}}$; $py=x\sqrt{(4-p^2)}$.

15. $\frac{12}{5}$; $\frac{192}{125}$.

16. (i) $3t$; $|100-2t|$; (ii) $\sqrt{(9t^2+(100-2t)^2)}$; (iii) $t=\frac{200}{13}$.

17. $(1+p^2)^2 y+2px=1+3p^2$; $y=\dfrac{1+3p^2}{(1+p^2)^2}$; $p=\pm\sqrt{\frac{1}{3}}$.

19. $-\dfrac{11}{(x-4)^2}$.

Chapter 8

Qu.1 (i) (a) 0; (b) 4; (c) no; (ii) (a) 2, 3, $3\frac{1}{2}$, $3\frac{3}{4}$; yes, 4; (b) 3, $6\frac{1}{4}$, $10\frac{1}{4}$, $14\frac{1}{8}$; no; (c) 1, 0, 1, 0; no.

Qu.2 (i) Sequence converges to zero; (ii) no.

Qu.3 (i) u_5; (ii) u_n; (iii) $u_{10}-u_9$; (iv) s_{n+1}; (v) $2s_n u_{n+1}$.

Qu.4 (i) 1, 4, 7, 10, 13; (ii) 2, $\frac{3}{4}$, $\frac{4}{9}$, $\frac{5}{16}$, $\frac{6}{25}$; (iii) 4, 7, 10, 13, 16; (iv) 1, 2, 6, 24, 120.

Qu.5 1, 1, 2, 3, 5, 8, 13, 21, 34, 55.

Qu.7 5050.

Exercise 8a

1. (i) 3, 5, 7, 9, 11, 13; (ii) 1, 3, 5, 7, 9, 11;
(iii) 0, $\frac{1}{2}$, $\frac{2}{3}$, $\frac{3}{4}$, $\frac{4}{5}$, $\frac{5}{6}$; (iv) 0, 1, 1, 1, 2, 2;
(v) $-1, 1, -1, 1, -1, 1$;
(vi) $-1, 4, -9, 16, -25, 36$;
(vii) $2, -4, 6, -8, 10, -12$;
(viii) $4\frac{1}{4}$, $5\frac{1}{4}$, $4\frac{7}{8}$, $5\frac{1}{16}$, $4\frac{31}{32}$, $5\frac{1}{64}$.

2. (i) 1, 2, 4, 8, 16; (ii) 1, $1\frac{1}{3}$, $\frac{17}{12}$, 1·4142, 1·4142;
(iii) $\sqrt{2}, \sqrt{2}, \sqrt{2}, \sqrt{2}, \sqrt{2}$; (iv) 1, $-\frac{1}{2}$, $\frac{1}{4}$, $-\frac{1}{8}$, $\frac{1}{16}$;
(v) 1, $-\frac{1}{2}$, $-\frac{1}{4}$, $\frac{1}{8}$, $\frac{1}{16}$.

3. (i) 1; (ii) 2, $3\frac{1}{2}$, $4\frac{3}{4}$, $5\frac{7}{8}$, $6\frac{15}{16}$; (iii) no.

4. (1) (i) No, no; (ii) no, no; (iii) 1, no; (iv) no, no;
(v) no, no; (vi) no, no; (vii) no, no; (viii) 5,
no.
(2) (i) No, no; (ii) $\sqrt{2}$, no; (iii) $\sqrt{2}$, no; (iv) 0, yes;
(v) 0, yes.

5. (i) $2n$; (ii) $2n-1$; (iii) n^2; (iv) $(n+1)^2$; (v) $(2n-1)^3$.

6. (i) $(n+1)^2$; (ii) $2n-3$; (iii) $3n^2+6n+1$; (iv) $\dfrac{n+1}{n+2}$;
(v) $(2n+3)^2$; (vi) $\frac{1}{6}(n+1)(n+2)(2n+3)$.

7. (i) $2n$; (ii) $2n+2$; (iii) 2^{n-1}; (iv) $(\frac{1}{3})^{n-1}$;
(v) $(-1)^n(2n+1)$; (vi) $3(-2)^{n-1}$; (vii) $(2n-1)^2$;
(viii) $n(n+1)$; (ix) $2n(2n-1)$.

8. (i) $2n+2$; (ii) $2n+4$; (iii) 2^n; (iv) $(\frac{1}{3})^n$;
(v) $(-1)^{n+1}(2n+3)$; (vi) $3(-2)^n$; (vii) $(2n+1)^2$;

(viii) $(n+1)(n+2)$; (ix) $(2n+2)(2n+1)$.

9. (i) 7, 3, 5; (ii) $(n-1)^2+6$; $u_n=2n-1$; because $s_0\neq0$.

10. (i) 3, 3, 3; 3; (ii) 1, 7, 19; $(3n^2-3n+1)$; (iii) $\frac{1}{2}$, $\frac{1}{6}$, $\frac{1}{12}$; $1/\{n(n+1)\}$; (iv) 1, 2, 4; 2^{n-1}.

Qu.8 (i) 1, 3, 5, 7; 199; (ii) 13, 17, 21, 25; 409;
(iii) $7\frac{1}{2}$, $8\frac{1}{2}$, $9\frac{1}{2}$, $10\frac{1}{2}$; $106\frac{1}{2}$;
(iv) 12, $8\frac{3}{4}$, $5\frac{1}{2}$, $2\frac{1}{4}$; $-309\frac{3}{4}$.

Qu.9 (i) 1, 3; 58; (ii) 6, $2\frac{1}{2}$; $53\frac{1}{2}$; (iii) 10, -6; -104;
(iv) -8, $-\frac{1}{2}$; $-17\frac{1}{2}$.

Qu.10 (i) 46; (ii) 90; (iii) 1005.

Qu.11 (i) $2a+n-(n-1)d$; (ii) $\frac{1}{2}(n+1)(4a+n-(n-2)d)$.

Exercise 8b

1. (i) 99; 34947; (ii) 175; $72\,537\frac{1}{2}$; (iii) 54; -10611;
(iv) 69; $69p-4209q$.

2. (i) 5860; (ii) 4430; (iii) -5340; (iv) $40p-1280q$.

3. (i) 2500; (ii) 2550; (iii) 50.

4. (i) $-23\frac{1}{2}$, -19; (ii) $4\frac{1}{2}n-28$; (iii) $9n-28$.

5. 6.

6. 425.

7. (i) The 1246th; (ii) 46.

8. 1522·5.

9. Nothing.

10. (i) 1775; (ii) 21050; (iii) 19275.

11. 32.

12. -131.

13. $\frac{1}{2}(p+q)$.

14. $s_n=\frac{1}{2}n(21-(n-1)d)$.

Qu.12 (i) 2, 6, 18, 54; (ii) 7, -14, 28, -56; (iii) 11, $\frac{11}{3}$, $\frac{11}{9}$, $\frac{11}{27}$;
(iv) 36, -27, $\frac{81}{4}$, $-\frac{243}{16}$; (v) 7, 7, 7, 7.

Qu.13 (i) $\frac{1}{2}$; $\frac{5}{256}$; (ii) -2; -512; (iii) -1; -6; (iv) $\frac{2}{5}$; $\frac{512}{3125}$.

Qu.14 (i) 16; (ii) 7; (iii) $k+2$; (iv) $\frac{3}{2}p-2$.

Qu.15 (i) $3^{20}-1$; (ii) $\frac{243}{5}(1-(\frac{2}{3})^{10})$; (iii) $\frac{3}{2}(1-(\frac{1}{3})^7)$;
(iv) $\dfrac{x^3(1+2^{49}x^{98})}{1+2x^2}$.

Qu.16 There isn't one.

Qu.17 (i) $(a+1)r^6$; (ii) $(a+1)r^{2n-2}$; (iii) $(a+1)r^{4n-2}$;
(iv) $\dfrac{(a+1)(1-r^{2n+2})}{1-r^2}$.

Exercise 8c

1. (i) $(\frac{2}{3})^9$; $3(1-(\frac{2}{3})^{20})$; (ii) $-\frac{1}{32}$; $\frac{32}{3}(1-(\frac{1}{2})^{20})$;
(iii) $48(\frac{3}{4})^9$; $-\frac{192}{7}(1-(\frac{3}{4})^{20})$; (iv) $\dfrac{q^{10}}{p^6}$;
$\dfrac{p^4 q}{p-q}(1-(q/p)^{20})$; (v) x^{-15}; $\dfrac{x^{40}-1}{x^{35}(x^2-1)}$.

2. $k=14$.

3. $-\frac{20}{27}$.

4. -10, 2, $-0\cdot4$.

5. $-\frac{4}{3}$, $\frac{2}{3}$, $-\frac{1}{3}$, $\frac{1}{6}$.

6. 2, -4, 8, or 2, 2, 2.

7. 4, $2\sqrt{2}$, 2; 4, $-2\sqrt{2}$, 2; $\frac{31}{4}\sqrt{2}$.

8. Both $63\frac{31}{32}$; they contain the same terms.

9. £2000$\times(1\cdot05^8-1)\approx$£955.

10. (ii)$\pm\sqrt{(pq)}$.

11. (i) $1-x^{n+1}$; (ii) $1-(-x)^{n+1}$.

12. (i) $\frac{2}{3}, \frac{2}{9}, \frac{2}{27}$; (ii) $\frac{2}{3} + \frac{1}{3} \cdot \frac{2}{3} + (\frac{1}{3})^2 \frac{2}{3} + (\frac{1}{3})^3 \frac{2}{3} + \ldots + (\frac{1}{3})^{n-1} \frac{2}{3}$; (iii) $\frac{1}{3}, \frac{1}{9}, \frac{1}{27}$; (iv) $(\frac{1}{3})^n$.

13. 14.

14. £1295.

15. £3094.

16. $\dfrac{(r+1)(r^{33}-1)}{r^3-1}$.

17. (i) $b=90$ and either $a=30$ and $c=270$
or $a=-30$ and $c=-270$;
(ii) $x_1 = \sqrt[3]{(a^2 b)}$, $x_2 = \sqrt[3]{(ab^2)}$;
(iii) it is an A.P..

Qu.18 (i) $1\frac{1}{2}$; (ii) $\frac{3}{4}$.

Qu.19 (i) $|x| < \frac{1}{2}$; (ii) $|x| < 2$; (iii) $|x| < 1$.

Qu.20 $\frac{3}{11}$.

Exercise 8d

1. (i) 2; (ii) $\frac{2}{3}$; (iii) 18; (iv) $8\frac{1}{3}$; (v) no; (vi) no; (vii) no.

2. (i) $\dfrac{1}{1+3x}$; $|x| < \frac{1}{3}$; (ii) $\dfrac{1}{1-4x^2}$; $|x| < \frac{1}{2}$; (iii) 1; $0 < x < 2$; (iv) $\dfrac{x^4}{x-1}$; $|x| > 1$.

3. (i) $1 + x + x^2 + x^3 + \ldots$; (ii) $1 - x + x^2 - x^3 + \ldots$.

4. $\frac{8}{5}$.

5. (i) 6; 3.

8. (i) $u_n = \dfrac{3}{2^{n-1}}$; (ii) 6; (iii) $4\frac{1}{2}, 1\frac{1}{8}, \frac{9}{32}, \frac{9}{128}$; (iv) 6.

9. 1936.

10. 14 or 18; $r=1$ or $\frac{4}{3}$.

11. The series is constant.

12. $1, \dfrac{1+\sqrt{5}}{2}, \dfrac{1-\sqrt{5}}{2}$.

13. $na + (r^n d - rd)/(r-1)$.

14. $(1-x)^{-2}$; $|x| < 1$.

Qu.21 (i) $0 + 1 + 8 + 27 + \ldots + 1000$;
(ii) $1 + \frac{1}{2} + \frac{1}{3} + \ldots + \frac{1}{100}$;
(iii) $\sqrt{20} + \sqrt{21} + \ldots + \sqrt{30}$;
(iv) $a + (a+d) + (a+2d) + \ldots + (a+nd)$;
(v) $a + ar + ar^2 + ar^3 + \ldots$.

Qu.22 (i) 5050; (ii) 468; (iii) 3; (iv) 10.

Qu.23 (i) $\displaystyle\sum_{r=51}^{100} u_r$; (iii) yes.

Qu.25 $\displaystyle\sum_{r=1}^{99} \dfrac{1}{r(r+1)}$.

Exercise 8e

1. (i) 14; (ii) 17; (iii) 682; (iv) 36; (v) 10 000; (vi) 168; (vii) 168; (viii) $\frac{3}{5}$.

2. (i) $1 + x + x^2 + \ldots + x^{100}$; (ii) $x + \dfrac{x^2}{4} + \dfrac{x^3}{9} + \dfrac{x^4}{16} + \ldots$;

(iii) $\dfrac{x}{2.3} + \dfrac{x^2}{3.4} + \dfrac{x^3}{4.5} + \ldots + \dfrac{x^{19}}{20.21}$.

3. Note: the answers given here are not the only correct forms.

(i) $\displaystyle\sum_{0}^{50}(2r+1)$; (ii) $\displaystyle\sum_{0}^{33}(3r+1)(3r+2)$; (iii) $\displaystyle\sum_{1}^{49}\dfrac{r}{r+1}$;

(iv) $\displaystyle\sum_{1}^{25}\dfrac{2r-1}{2r}$; (v) $\displaystyle\sum_{0}^{\infty}(\frac{1}{2})^r$; (vi) $\displaystyle\sum_{0}^{\infty}(-\frac{1}{2})^r$;

(vii) $\displaystyle\sum_{r=1}^{\infty}\dfrac{(-x)^r}{r}$; (viii) $\displaystyle\sum_{r=1}^{\infty}rx^{r-1}$; (ix) $\displaystyle\sum_{r=0}^{\infty}(r+1)^2(-x)^r$;

(x) $\displaystyle\sum_{r=2}^{\infty}\dfrac{rx^{r-1}}{(r+1)(r+2)}$; (xi) $\displaystyle\sum_{r=1}^{\infty}(-1)^{r+1}\dfrac{2r-1}{4r(r+1)}$.

Qu.27 (ii)$P(1)$ is false here.

Exercise 8f

9. (i) $\begin{pmatrix} 1 & n \\ 0 & 1 \end{pmatrix}$; (ii) $\begin{pmatrix} 1 & 1-(\frac{1}{2})^n \\ 0 & (\frac{1}{2})^n \end{pmatrix}$.

Miscellaneous exercise 8

1. (a) 4.

2. (i) $\frac{64}{3}$; (ii) 4.

3. (a) $\frac{1}{2}$; 20, 40; (b) $3\frac{1}{4}$; 425.

4. (i) 3, $1\frac{1}{2}$; (ii) the 20th.

6. (a) 1 200 000; (b) 72.

7. $-2 < x < 2$; $n=6$.

8. (i) $\frac{4}{27}$.

9. $r = -\frac{1}{2}$, first term = 8. Sum to infinity $\frac{16}{3}$.

10. $\frac{1}{20}n(81-n)$, $41\frac{2}{5}$m.

11. (i) $\dfrac{a^n-1}{a-1}$; (ii) $a^{\frac{1}{2}n(n-1)}$;

(iii) $\dfrac{1}{(a-1)^2}\{a^{n+1}-(n+1)a+n\}$.

12. (i) 2 160 600; $\frac{1199}{60}$; (ii) $250\{1-(\frac{4}{5})^n\}$.

Chapter 9

Qu.4 $n(A) + n(B) + n(C) + n(D) - n(A \cap B) - n(A \cap C) - n(A \cap D) - n(B \cap C) - n(B \cap D) - n(C \cap D) + n(A \cap B \cap C) + n(A \cap B \cap D) + n(A \cap C \cap D) + n(B \cap C \cap D) - n(A \cap B \cap C \cap D)$.

Qu.7 36.

Qu.8 21.

Qu.10 16.

Qu.11 7.

Qu.13 (i) 120; (ii) 720.

Qu.15 (i) $n^{10} < 10^n < n!$; (ii) $0! = 1$.

Qu.16 (i) 11880; (ii) $\dfrac{10!}{7!}$.

Qu.17 240.

Exercise 9a

1. (i) 720; (ii) 72; (iii) 120; (iv) $\frac{2}{3}$; (v) 198.

2. (i) 8!; (ii) 5!.

3. (i) $9 \times 9!$; (ii) $239 \times 14!$; (iii) $(n+2)(n-1)(n-1)!$; (iv) $(n^2+1)(n-1)!$.

4. (i) $(n+1)!$; (ii) $\dfrac{n+2}{(n+1)!}$; (iii) $\dfrac{n!(n-r+2)}{(n-r+1)!}$.

5. $\dfrac{(n+1)!}{4!(n-3)!}$.

6. (i) 60; (ii) 720; (iii) 9900.

7. 36.

8. (i) 336; (ii) 512.

9. (i) 243; (ii) 240; (iii) 48.

10. 576.

11. 31.

12. 96.

13. (i) 720; (ii) 120; (iii) 600.

14. (i) 120; (ii) 60; (iii) 12; (iv) 168.

15. 8.

16. 120; (i) 24; (ii) 24; (iii) 24.

17. 1320.

18. 504.

19. (i) 24×6^4; (ii) 6×24^3.

20. (i) 2880; (ii) 5760.

21. (i) 6720; (ii) 1440; (iii) 3840.

22. $90 \times 11!$.

23. $49! \times 48 \times 47 \times 46$.

24. 2880.

Qu.19 360

Qu.20 (ii) 8.

Qu.22 3.

Qu.24 12.

Qu.25 2.

Exercise 9b

1. (i) 360; (ii) 20; (iii) $9 \times 7!$; ·(iv) 3780.

2. 120.

3. 3360; 1680.

4. 1260.

5. $\dfrac{15!}{6^5}$.

6. (i) 1260; (ii) 360; (iii) 120; (iv) 600; (v) 660.

7. $4 \times 7!$.

8. $8!/12$.

9. $7!$.

10. (i) 2520; (ii) 42; (iii) 1.

11. (i) 720; (ii) 1440.

Qu.26 10.

Qu.28 (i) 2^n; (ii) ^nC_r.

Exercise 9c

1. (i) 10; (ii) 10; (iii) 10; (iv) 70; (v) 4950.

2. (i) $\frac{1}{2}n(n-1)$; (ii) n; (iii) $\frac{1}{2}n(n+1)$;
   (iv) $\frac{1}{6}(n+1)(n+2)(n+3)$; (v) 1.

3. 56.

4. 100.

5. 9450.

6. 21.

7. 2024; 1771.

8. 15.

9. (i) 45; (ii) 120.

10. $\dfrac{24!}{4!8!12!}$.

11. $^{13}C_6 \times {}^{39}C_7$.

Qu.31 n^r.

Qu.32 21.

Exercise 9d

1. $4.^{13}C_6({}^{39}C_7 - 39.^{13}C_6)$.

2. 606.

3. (i) 192; (ii) 13056.

4. (i) 10; (ii) 57.

5. (i) 50; (ii) 16; (iii) 3.

7. 265.

8. (i) 18; (ii) 15.

9. (i) None; (ii) 45; (iii) 240; (iv) 1890.

10. 461.

12. 5.

14. (i) $(2^n)!/2^{2^n-1}$.

Qu.34

1		6		15		20		15		6		1	
1		7		21		35		35		21		7	1
1		8		28		56		70		56		28	8	1.

Qu.35 $x^5 + 5x^4 y + 10x^3 y^2 + 10x^2 y^3 + 5xy^4 + y^5$.

Qu.37 $1 + 5x + 5x^2 - 10x^3$.

Exercise 9e

1. (i) $x^6 + 6x^5 y + 15x^4 y^2 + 20x^3 y^3 + 15x^2 y^4 + 6xy^5 + y^6$;
   (ii) $x^4 - 4x^3 y + 6x^2 y^2 - 4xy^3 + y^4$;
   (iii) $x^3 + 6x^2 + 12x + 8$;
   (iv) $32x^5 + 80x^4 + 80x^3 + 40x^2 + 10x + 1$;
   (v) $16x^4 - 16x^3 y + 6x^2 y^2 - xy^3 + \frac{1}{16}y^4$;
   (vi) $32 - 240x^2 + 720x^4 - 1080x^6 + 810x^8 - 243x^{10}$;
   (vii) $x^8 + 8x^6 + 28x^4 + 56x^2 + 70 + 56x^{-2} + 28x^{-4} + 8x^{-6} + x^{-8}$;
   (viii) $x^6 - 18x^4 + 135x^2 - 540 + 1215x^{-2} - 1458x^{-4} + 729x^{-6}$.

2. (i) $1 + 10x + 45x^2 + 120x^3 + 210x^4$;
   (ii) $256 - 256x + 112x^2 - 28x^3 + \frac{35}{8}x^4$;
   (iii) $1 + 120x^2 + 7020x^4$.

3. (i) 1·082 432 16; (ii) 1·020 181 0; (iii) 0·980 179 0.

4. (i) 31·207 960 099 9; (ii) 243·081 010 800 7;
   (iii) 0·480 002 16.

5. (i) -360; (ii) 7920; (iii) 84; (iv) 0; (v) -30.

6. (i) $\frac{160}{27}$; (ii) 3360.

7. (i) $1 - 10x^2$; (ii) $1 - \frac{5}{4}x^2$;
   (iii) $1 - 11x + 48x^2 - 100x^3$; (iv) $-162 - 135x + 30x^3$.

8. $256a^8 - 5120a^7 b + 44800a^6 b^2$; $2·509 248 \times 10^{10}$.

9. $a = \pm\frac{1}{2}$.

10. (i) $1 + 7x + 28x^2 + 77x^3$; (ii) $1 - 5x + 20x^2 - 50x^3$;
   (iii) $16 - 96x + 184x^2 - 72x^3$.

11. (i) $17 + 12\sqrt{2}$; (ii) $29\sqrt{2} - 45$; (iii) $56 - 24\sqrt{5}$;
   (iv) $\sqrt{2}/8$; (v) $45 + 36\sqrt[3]{2} + 27\sqrt[3]{4}$.

12. $p = -1\frac{1}{2}$, $q = -2$; $-21x^3$.

13. $a_r = {}^{10}C_r 5^{10-r} 2^r$; a_3; 75 000 000.

14. Coefficient of $x^7 = {}^{12}C_5 3^7 2^5$.

18. (i) £$100(1·04)^{10}$; (ii) £148.

19. 4480.

Miscellaneous exercise 9

1. $1 + 14x + 84x^2 + 280x^3$; $a = 1$, $b = -3$, $c = 2$.

2. (a) 168; 240; (b) 181 440.

3. (a) 15 120; (b) 0·0005.

4. 2304; 18.

5. (i) 30 240; (ii) 720; (iii) 5040 (iv) 3003 (v) 2142.
6. (a) $4xy(3x^2+y^2)(x^2+3y^2)$; (b) (i) $(n-2)[(n-1)!]$;
  (ii) $(n-3)[(n-2)!]$.
7. (i) (a) 113 400; (b) 945.

Chapter 10

Qu.1 (i) $4x^2+6x-8$; (ii) 2; (iii) -3; (iv) 3; (v) -6.
Qu.2 (i) $m+n$; (iii) m.
Qu.3 $a=1$, $b=8$, $c=-24$.
Qu.4 $p=q=r=0$.
Qu.6 $a=\frac{7}{5}$, $b=\frac{4}{15}$, $c=-\frac{2}{3}$.
Qu.9 $Q(x)\equiv 2x^4+3x^3-x^2-10x-15$; $R(x)\equiv 22x+78$.
Qu.10 Quotient continues $+63x^4-189x^5$; rem. $567x^6$.
Qu.11 (i) rem. x^5; (ii) x^{n+1}; (iii) $|x|<1$.
Qu.12 (i) rem. $-64x^5$; (ii) $2(-2x)^{n+1}$; (iii) $|x|<\frac{1}{2}$.

Exercise 10a

1. (i) 23·564; (ii) $2x^3-3x^2+7x-1$;
  (iii) $X=1\cdot2$, $N=4$, $A=1$, -2, 1, -7, -9;
  (iv) $(((x-2)x+1)x-7)x-9$; (v) $-17\cdot3424$.
2. (i) $a=-2$, $b=5$; (ii) $x=5\frac{3}{4}$.
3. $a=3$, $b=3$, $c=-8$.
4. $a=1$, $b=1$, $c=2$.
5. $p=q=0$; r may take any value.
6. $a=2$, $b=1$, $c=3$.
7. (i) $x+\frac{1}{2}$; $-4\frac{1}{2}x+3\frac{1}{2}$; (ii) $2x^2-3x-4$; no rem.;
  (iii) x^2-3; -2; (iv) $2x^3+x^2+\frac{1}{2}x-\frac{1}{4}$; $-\frac{5}{4}$.
8. (i) $x^2-3x+\frac{3}{2}$; $-7\frac{1}{2}$; (ii) $-1-\frac{4}{3}x-\frac{1}{9}x^2$; $\frac{20}{9}x^3$.
9. (i) x^3-2x^2-3x+4; (ii) the same.
12. (i) $\frac{1}{2}Q(x)$; (ii) $R(x)$.
Qu.13 (i) 43; (ii) 3.
Qu.15 (i) $5\frac{1}{2}$; (ii) $4\frac{13}{27}$.
Qu.17 $\{x-(2-\sqrt5)\}\{x-(2+\sqrt5)\}$.

Exercise 10b

1. (i) -46; (ii) 4; (iii) 0.
2. (i) $-2\frac{7}{8}$; (ii) $\frac{5}{3}$; (iii) $-0\cdot3439$.

3. (i) (a) $\dfrac{1}{\lambda}Q(x)$; (b) R(x);

  (ii) (a) $\lambda Q(x)$; (b) $\lambda R(x)$.
4. (i) $c=8$; (ii) $a=-4$.
5. $a=3$, $b=-2$.
6. (i) $\{x-\frac{1}{2}(5+\sqrt{13})\}\{x-\frac{1}{2}(5-\sqrt{13})\}$; (ii) x^2-3x+5;
  (iii) $(x-1)(x-1-\sqrt5)(x-1+\sqrt5)$;
  (iv) $(x-1)(x^2+x+1)$; (v) $x(x-1)^2$.
7. (i) $1023x-1022$; (ii) $122x-123$.
8. (i) $-11x-9$; (ii) $\frac{1}{2}x-\frac{3}{8}$.

10. (i) $\lambda=\dfrac{L(b)}{b-a}$, $\mu=\dfrac{L(a)}{a-b}$.

11. $\dfrac{aP(b)-bP(a)}{a-b}$.

12. (ii) $\frac{1}{2}P''(a)(x-a)^2+P'(a)(x-a)+P(a)$.
13. k odd; $x^{k-1}-x^{k-2}a+x^{k-3}a^2-\ldots+a^{k-1}$.

15. $y=\dfrac{y_1(x-x_2)(x-x_3)(x-x_4)}{(x_1-x_2)(x_1-x_3)(x_1-x_4)}+$ etc. (3 more terms).

16. $y=3x^2-5x-1$.

17. (i) $x=\frac{1}{2}$ or $-\frac{9}{4}$; (ii) $x=-6$ or -2 or 3.
20. (i) Yes; (ii) no.
Qu.22 $x=-1$.
Qu.23 Yes.

Exercise 10c

12. (i) $x=\frac{1}{2}$; (ii) $x=-\frac{5}{3}$ and $x=-1$; (iii) none.
Qu.24 (i) $\frac{2}{3}$; (ii) $-\frac{4}{3}$.
Qu.25 $3x^2-4x-6=0$.
Qu.26 $a^3x^2+(b^3-3abc)x+c^3=0$.
Qu.28 $+\alpha\gamma+\beta\delta$.

Qu.30 (i) $\dfrac{1}{\alpha}+\dfrac{1}{\beta}+\dfrac{1}{\gamma}$; (ii) $\dfrac{\alpha}{\beta}+\dfrac{\beta}{\alpha}+\dfrac{\alpha}{\gamma}+\dfrac{\gamma}{\alpha}+\dfrac{\beta}{\gamma}+\dfrac{\gamma}{\beta}$;

  (iii) $\alpha^2\beta^2+\alpha^2\gamma^2+\beta^2\gamma^2$;
  (iv) $\alpha^2\beta^3+\beta^2\alpha^3+\alpha^2\gamma^3+\gamma^2\alpha^3+\beta^2\gamma^3+\gamma^2\beta^3$.
Qu.31 (i) 12; (ii) 12; (iii) 24; (iv) 12.
Qu.32 (i) $2\sum\alpha\beta$; (ii) $\sum\alpha^3$; (iii) $\sum\alpha^3-3\alpha\beta\gamma$; (iv) 0.

Qu.34 $x^3-14x^2+52x-30=0$.
Qu.35 $y^4-12y^3-32y^2+784y-2208=0$.
Qu.36 3, -2, 10; -11, 42.
Qu.37 $s_6=1298$.

Exercise 10d

1. (i) p^2-2q; (ii) pq; (iii) p^3-3pq; (iv) p/q;
  (v) $q^{-2}(p^2-2q)$; (vi) $q^3+4q-4q^2-2p^4+8p^2q$.
2. (i) $4x^2-69x+25=0$; (ii) $x^2-7x-10=0$;
  (iii) $2x^2-3x-10=0$.
4. (i) $x^2-(p^2-2q)x+q^2=0$; (ii) $x^2-2px+4q=0$;
  (iii) $x^2-(p-2)x+(q-p+1)=0$.
6. (i) $x^2-9x+23=0$; (ii) $x^2-2x-3=0$;
  (iii) $x^2+17x+207=0$.
7. (i) $+$ $+$; (ii) $+$ $-$; (iii) none; (iv) $+$ $-$;
  (v) $-$ $-$; (vi) $+$ $-$; (vii) $+$ $+$.
8. Converse is true.
10. (i) $x^3-6x^2-6x+81=0$;
  (ii) $3x^3+30x^2+94x+97=0$;
  (iii) $9x^3-2x^2-6x+3=0$;
  (iv) $3x^3+2x^2-18x-27=0$;
  (v) $3x^3-12x^2+10x-5=0$;
  (vi) $27x^3+18x^2-2x-3=0$.
11. (i) $ax^3+3bx^2+9cx+27d=0$;
  (ii) $ax^3+(12a+b)x^2+(48a+8b+c)x$
    $+(64a+16b+4c+d)=0$;
  (iii) $dx^3+cx^2+bx+a=0$;
  (iv) $a^2x^3-acx^2+bdx-d^2=0$;
  (v) $a^2x^3+2abx^2+(b^2+ac)x+(bc-ad)=0$;
  (vi) $d^2x^3-bdx^2+acx-a^2=0$.
15. $q^3=rp^3$.
16. $a^2x^3+a(b-c)x^2+c(a-b)x-c^2=0$.
17. (i) $8x^4-28x^3-4x^2-x+4=0$;
  (ii) $x^4-x^3-16x^2-448x+512=0$;
  (iii) $x^4-53x^3+6x^2-33x+64=0$.
18. $(ax^2+bx+c)(px^2+qx+r)=0$.
19. $\lambda=666$; 12, 15, 18.
21. (i) $s_3=-13$, $s_{-3}=-48$; (ii) $x^3+13x^2+48x+1=0$.
22. 23 168.
23. -3, 5, -9.
24. -3, $\frac{5}{2}$, 5.

25. (i) 3, $\frac{1}{3}$, $-\frac{2}{3}$, $-\frac{3}{2}$; (iii) $\frac{2}{5}$, $\frac{5}{2}$, $-\frac{1}{5}$, -5.

26. 2, $-\frac{1}{2}$, $\frac{2}{3}$, $-\frac{3}{2}$.

Qu.39 $A=1$, $B=15$, $C=19$.

Qu.42 $\dfrac{8}{5(x-2)^2} - \dfrac{1}{25(x-2)} + \dfrac{2}{25(2x+1)}$.

Exercise 10e

1. $\dfrac{1}{x-1} + \dfrac{2}{x-3}$.

2. $\dfrac{-3}{x+4} + \dfrac{5}{x-5}$.

3. $\dfrac{3}{2(x+1)} - \dfrac{1}{2(x-1)}$.

4. $\dfrac{3}{2x-3} + \dfrac{4}{x-3}$.

5. $\dfrac{42}{5(3x-5)} + \dfrac{11}{5x}$.

6. $\dfrac{1}{2x-1} + \dfrac{2}{3(2x+1)}$.

7. $\dfrac{1}{2(x-1)} + \dfrac{2}{x+2} - \dfrac{3}{2(x+3)}$.

8. $\dfrac{-4}{2x+7} + \dfrac{2}{x-3} + \dfrac{3}{5(x+1)}$.

9. $\dfrac{1}{(x+2)^2} - \dfrac{2}{x+2} + \dfrac{1}{x-1}$.

10. $\dfrac{3}{(x+1)^2} + \dfrac{1}{x+1}$.

11. $\dfrac{4}{(2x+5)^2} + \dfrac{1}{x+3}$.

12. $\dfrac{1}{(x-1)^3} + \dfrac{2}{(x-1)^2} - \dfrac{1}{x-1}$.

13. $\dfrac{4}{(x+3)^3} - \dfrac{1}{x+3} + \dfrac{2}{(x+1)^2} + \dfrac{2}{x+1}$.

14. $\dfrac{3x+1}{x^2+2} - \dfrac{1}{x+1}$.

15. $\dfrac{2x-3}{x^2+2x+3} + \dfrac{2}{x-4}$.

16. $\dfrac{x}{x^2-4x+10} - \dfrac{1}{x}$.

17. $\dfrac{-2}{x^2+1} + \dfrac{1}{(x+1)^2} + \dfrac{2}{x+1}$.

18. $2 + \dfrac{3}{x-1} - \dfrac{6}{x+1}$.

19. $3x-1 - \dfrac{2}{2x+3} + \dfrac{1}{x-4}$.

20. $2x^2 + \dfrac{1}{x-1}$.

21. $\dfrac{dy}{dx} = \dfrac{-2}{(x-3)^2} - \dfrac{1}{(x+1)^2}$.

Qu.44 9.

Qu.45 (i) $(x+5)^2 - 30$; (ii) $(x-\frac{7}{2})^2 - \frac{1}{4}$.

Qu.46 (i) $(x-4)^2 + 3$; (ii) $(x-\frac{5}{2})^2 - 9\frac{1}{4}$.

Qu.47 (i) $2(x-\frac{1}{4})^2 + \frac{7}{8}$; (ii) $(2x-1)^2$; (iii) $\frac{5}{4} - (x+\frac{1}{2})^2$.

Qu.49 $\frac{11}{20}$.

Qu.50 Max is $-\frac{11}{4}$ at $x=\frac{1}{4}$.

Qu.53 (ii) 3; (iv) result remains valid (one root).

Qu.54 (i) 3; (ii) $x = y - \dfrac{a}{3}$.

Exercise 10f

1. (i) -4 at $x=-1$; (ii) 4 at $x=1$; (iii) -6 at $x=4$; (iv) $13\frac{3}{4}$ at $x=-2\frac{1}{2}$; (v) -17 at $x=2$; (vi) $-8\frac{1}{3}$ at $x=-\frac{2}{3}$; (vii) $-\frac{1}{24}$ at $x=\frac{1}{12}$.

2. (i) $2\frac{1}{4}$; (ii) $2\frac{1}{4}$; (iii) $7\frac{4}{5}$; (iv) $2\frac{1}{4}$.

3. (i) $(x^2+2)^2+6$; 10; (ii) $(x^2-2)^2+6$; 6.

4. (i) $3\pm\sqrt{6}$; (ii) -9 or -4; (iii) $\frac{1}{4}(1\pm\sqrt{41})$.

5. (i) $(2,\,-1)$; 3; (ii) $(-2\frac{1}{2},\,3\frac{1}{2})$; $\sqrt{3}$.

6. (i) The point $(3,4)$ only; (ii) no point at all.

7. (i) $(-5,1)$; $\sqrt{2}$; (ii) $2\sqrt{2}$; (iii) 4.

8. $(\frac{8}{3},\,-\frac{2}{3})$; $4\sqrt{2}/3$.

9. $\left(\dfrac{2\lambda^2}{\lambda^2-1},\,\dfrac{-2}{\lambda^2-1}\right)$; $\left|\dfrac{2\sqrt{2}\lambda}{\lambda^2-1}\right|$.

10. Three.

Miscellaneous exercise 10

1. $k^2=8$.

2. (a) $4x^2+33x+4=0$; (b) $a=-19$, $b=30$; $x\in\{-5,\,-2,\,1,\,3\}$.

3. (i) $a=0$, $b=-7$; $(x-1)(x-2)(x+3)$; (ii) $a=4$, $b=-11$; $(x-1)^2(x+6)$.

4. (a) 2, 1, 3; 103; (b) -1, -1, 1.

5. (a) 4, 5; (b) 3.

6. (a) (i) $\frac{5}{8}$; (ii) 9; (b) $k \geqslant -1$.

7. $a=-9$, $b=7$; $(x-2)(x-3)(2x+1)$.

8. $p=10$, $q=17$; 100.

9. (a) -5, $\frac{1}{2}$, 6; (b) (i) $5\frac{1}{4}$; (ii) $-8\frac{3}{8}$; (iii) $2x^3-5x^2-x+7=0$.

10. $x^2-47x+1=0$.

11. (b) $3(x-y)(y-z)(z-x)$.

12. (a) $\frac{2}{3}$, $\frac{2}{3}$, $\frac{2}{3}$, $-\frac{1}{2}$; (b) $ac(n-1)^2 + 4b^2(m-1)(m-n)=0$.

Chapter 11

Qu.1 (i) $30°$; (ii) $240°$; (iii) $540°$; (iv) $114\cdot6°$.

Qu.2 (i) $\dfrac{2\pi}{3}$; (ii) $\dfrac{3\pi}{4}$; (iii) $0\cdot01745$; (iv) 4π.

Qu.4 10 cm.

Qu.5 (i) $-0\cdot4$; (ii) $0\cdot9$.

Qu.6 (i) 0; (ii) -1; (iii) -1; (iv) 1; (v) 1; (vi) -1; (vii) -1; (viii) 0; (ix) -1; (x) 0.

Qu.7 (i) Reflection in x-axis; (ii) $(\lambda,\,-\mu)$; (iii) $\sin(-\theta) = -\sin\theta$.

Qu.8 All true for all θ.

Qu.9 (i) $\dfrac{\pi}{2} + 2\pi n$, for $n\in Z$ (i.e. $\dfrac{\pi}{2}$ with any multiple of 2π added);

(ii) $\pm\dfrac{\pi}{2} + 2\pi n \left(=\dfrac{\pi}{2} + \pi n\right)$.

Qu.11 (i) 0; (ii) undefined; (iii) -1; (iv) 0;
(v) 1; (vi) undefined; (vii) 0; (viii) 0.

Qu.12 (i) $\tan\theta$; (ii) 1; (iii) $\operatorname{cosec}\theta$; (iv) 1.

Qu.13 (i) 1st; (ii) 2nd; (iii) 2nd; (iv) 4th; (v) 2nd;
(vi) 4th; (vii) 1st; (viii) 2nd.

Qu.14 $155°$, $205°$, $335°$.

Qu.15 (i) -0.3420; (ii) 0.7660; (iii) -0.1763; (iv) 5.6713;
(v) -5.7588; (vi) -1.5557; (vii) 0.7660.

Qu.16 (i) $\dfrac{2}{\sqrt{3}}$; (ii) 2; (iii) 2; (iv) $\dfrac{2}{\sqrt{3}}$.

Qu.17 (i) $\sqrt{3}$; (ii) $-\sqrt{2}$; (iii) -2.

Exercise 11a

1. (i) $135°$; (ii) $300°$; (iii) $630°$; (iv) $1800°$; (v) $112\frac{1}{2}°$;
(vi) $260°$.

2. (i) $\dfrac{5\pi}{4}$; (ii) $\dfrac{4\pi}{3}$; (iii) $\dfrac{7\pi}{6}$; (iv) $\dfrac{3\pi}{2}$; (v) $\dfrac{\pi}{8}$; (vi) $\dfrac{10\pi}{3}$.

3. (i) 2nd; (ii) 1st; (iii) 2nd; (iv) 4th; (v) 3rd;
(vi) 2nd; (vii) 3rd.

4. (i) $60°$; (ii) $\dfrac{\pi}{4}$; (iii) $70°$; (iv) $\dfrac{2\pi}{5}$; (v) $\dfrac{\pi}{3}$; (vi) $10°$;
(vii) $80°$.

5. (i) 0.6428; (ii) -0.1736; (iii) -0.9272;
(iv) -0.9563; (v) 0.8894; (vi) -0.6051.

6. (i) $\dfrac{1}{\sqrt{2}}$; (ii) $\dfrac{\sqrt{3}}{2}$; (iii) $\dfrac{\sqrt{3}}{2}$; (iv) $\dfrac{1}{\sqrt{2}}$; (v) $\dfrac{-\sqrt{3}}{2}$;
(vi) $\dfrac{-\sqrt{3}}{2}$.

7. (i) -1; (ii) $\dfrac{1}{\sqrt{3}}$; (iii) 1.0154; (iv) -1.3054;
(v) -0.4663; (vi) -0.4142; (vii) $-\sqrt{2}$; (viii) $\dfrac{-2}{\sqrt{3}}$.

8. $(-\mu, \lambda)$; (i) $\cos\theta$; (ii) $-\sin\theta$; (iii) $-\cot\theta$.

10. $\sqrt{2}-1$.

11. (i) 1.1π; (ii) (a) 7.3×10^{-5}; (b) 2×10^{-7}.

12. $\sqrt{5}$; $\dfrac{1}{\sqrt{5}}$; $\dfrac{2}{\sqrt{5}}$.

13. (i) $\dfrac{1}{\sqrt{8}}$; $\dfrac{3}{\sqrt{8}}$; (ii) $\dfrac{-\sqrt{5}}{2}$; $\dfrac{-\sqrt{5}}{3}$.

14. $75°$.

15. θ can be at $50°$, $170°$ or $290°$.

16. (ii) $\frac{1}{2}r^2(\theta-\sin\theta)$; (iii) $\pi r^2 - \frac{1}{2}r^2\theta + \frac{1}{2}r^2\sin\theta$; yes.

17. (i) 0.167%; (ii) 0.00667%; $\theta < 30.8°$.

Qu.19 (i) $\left.\begin{matrix}60°\\120°\end{matrix}\right\}+180°$; (ii) $\left.\begin{matrix}240°\\480°\end{matrix}\right\}+720n°$.

Qu.20 (i) $10°$, $50°$, $130°$, $170°$, $250°$, $290°$;
(ii) $5°$, $23°$, $41°$, $59°$, $77°$.

Qu.21 (i) $153.4°$ or $333.4°$; (ii) $197.6°$ or $342.4°$;
(iii) 0, $30°$, $150°$, $180°$; (iv) 0, $90°$, $180°$, $270°$;
(v) $30°$, $60°$, $120°$, $150°$, $210°$, $240°$, $300°$, $330°$;
(vi) $90°$; $270°$.

Qu.24 (i) $y = -\sin x$; (ii) $y = -\cos x$; (iii) $y = \sin x$.

Qu.25 $\left(\dfrac{\pi}{4}, 1\right)$ and $\left(\dfrac{3\pi}{4}, -1\right)$.

Qu.27 (i) $\dfrac{\pi}{3}$; (ii) $\dfrac{-\pi}{6}$; (iii) $\dfrac{\pi}{4}$; (iv) $\dfrac{3\pi}{4}$.

Qu.28 (i) $\dfrac{-\pi}{2} < y < \dfrac{\pi}{2}$; (ii) $0 < y < \pi$; (iii) $0 \leqslant y \leqslant \pi$;
(iv) $\dfrac{-\pi}{2} \leqslant y \leqslant \dfrac{\pi}{2}$.

Exercise 11b

1. (i) $\left.\begin{matrix}30°\\150°\end{matrix}\right\}+360n°$; (ii) $\left.\begin{matrix}120°\\240°\end{matrix}\right\}+360n°$;
(iii) $\left.\begin{matrix}30°\\150°\end{matrix}\right\}+180n°$; (iv) same as (iii);

2. (i) $\left.\begin{matrix}\dfrac{\pi}{12}\\[4pt]\dfrac{5\pi}{12}\end{matrix}\right\}+\pi n$; (ii) $\left.\begin{matrix}\dfrac{5\pi}{18}\\[4pt]\dfrac{7\pi}{18}\end{matrix}\right\}+\dfrac{2\pi n}{3}$;
(iii) $\left.\begin{matrix}\dfrac{\pi}{30}\\[4pt]\dfrac{\pi}{6}\end{matrix}\right\}+\dfrac{2\pi n}{5}$; (iv) $(4n+2)\pi$;
(v) $\dfrac{\pi}{8}+\dfrac{n\pi}{2}$; (vi) $\dfrac{2\pi}{3}+4\pi n$.

3. (i) $13.6°$ or $146.4°$; (ii) $120.5°$ or $339.5°$;
(iii) $13.3°$, $103.3°$, $193.3°$, $283.3°$;
(iv) $40°$, $160°$, $280°$.

4. (i) $\pm\dfrac{\pi}{6}$, $\pm\dfrac{\pi}{3}$, $\pm\dfrac{2\pi}{3}$, $\pm\dfrac{5\pi}{6}$;
(ii) $-\dfrac{7\pi}{9}$, $-\dfrac{4\pi}{9}$, $-\dfrac{\pi}{9}$, $\dfrac{2\pi}{9}$, $\dfrac{5\pi}{9}$, $\dfrac{8\pi}{9}$;
(iii) $-\dfrac{7\pi}{9}$, $-\dfrac{\pi}{3}$, $-\dfrac{\pi}{9}$, $\dfrac{\pi}{3}$, $\dfrac{5\pi}{9}$, π;
(iv) no solution.

5. (i) $\dfrac{\pi}{3}$, $\dfrac{4\pi}{3}$; (ii) $\dfrac{\pi}{6}$, $\dfrac{5\pi}{6}$, $\dfrac{3\pi}{2}$;
(iii) 0, $\dfrac{\pi}{3}$, $\dfrac{2\pi}{3}$, π, $\dfrac{4\pi}{3}$, $\dfrac{5\pi}{3}$;
(iv) 0, $\dfrac{\pi}{3}$, π, $\dfrac{5\pi}{3}$; (v) no solution.

6. Two.

8. (i) $\dfrac{2\pi}{3}$; (ii) 4π; (iii) 2π; (iv) not periodic.

10. Infinitely many.

11. (i) $0 \leqslant \theta \leqslant \dfrac{\pi}{6}$ or $\dfrac{5\pi}{6} \leqslant \theta < 2\pi$;
(ii) $0 \leqslant \theta \leqslant \dfrac{\pi}{6}$ or $\dfrac{5\pi}{6} \leqslant \theta \leqslant \dfrac{7\pi}{6}$ or $\dfrac{11\pi}{6} \leqslant \theta < 2\pi$.

12. (i) $\dfrac{\pi}{6}$; (ii) $\dfrac{\pi}{2}$; (iii) $\dfrac{\pi}{3}$; (iv) $\dfrac{2\pi}{3}$; (v) $\dfrac{-\pi}{4}$.

13. (i) $\dfrac{7\pi}{18}$; (ii) $\dfrac{\pi}{18}$; (iii) $\dfrac{\pi}{2} - \theta$.

15. (i) $0, \dfrac{\pi}{5}, \dfrac{3\pi}{5}, \pi, \dfrac{7\pi}{5}, \dfrac{9\pi}{5}, 2\pi$;

(ii) $85°$ or $265°$; (iii) $\dfrac{\pi}{10}, \dfrac{3\pi}{5}, \dfrac{11\pi}{10}, \dfrac{8\pi}{5}$;

(iv) $35°, 155°, 255°, 275°$; (v) $\theta = \dfrac{\pi}{3}$ or $\dfrac{4\pi}{3}$.

Qu.29 $\theta = 0\cdot464$ or $\dfrac{\pi}{4}$.

Qu.31 (i) $\pm\dfrac{\sqrt{(\lambda^2-1)}}{\lambda}$; (ii) $\pm\dfrac{1}{\sqrt{(\lambda^2-1)}}$.

Qu.33 $216\cdot9°$ is false; squaring introduces solutions of $-2\cos\theta = 1 - \sin\theta$.

Exercise 11c

1. (i) $\sin A$; (ii) $\pm\sec A$; (iii) $\cot A$; (iv) $2\sec A$;
(v) $\tan A \tan B$; (vi) $2\tan A$.
3. (i) $30°, 150°, 221\cdot8°, 318\cdot2°$; (ii) $22\cdot5°, 157\cdot5°$;
(iii) $48\cdot6°, 131\cdot4°, 270°$.

4. (i) $\pm\dfrac{\sqrt{99}}{10}$; $\pm\sqrt{99}$; (ii) $\pm\dfrac{\sqrt{13}}{2}$; $\pm\dfrac{2}{\sqrt{13}}$;

(iii) $\pm\dfrac{\sqrt{17}}{4}$; $\pm\dfrac{1}{\sqrt{17}}$; (iv) $\pm\dfrac{3}{\sqrt5}$; (v) $\pm\dfrac{\sqrt8}{3}$; $\pm\dfrac{1}{\sqrt8}$.

5. (i) $|3\cos 2\theta|$; (ii) $|\tfrac12\cos\tfrac14\theta|$; (iii) $\pm2\sec 3\theta$.

6. $\pm\dfrac{1}{\sqrt{(t^2+1)}}$; $\pm\sqrt{(t^2+1)}$; $\pm\dfrac{\sqrt{(t^2+1)}}{t}$.

7. $\pm\dfrac{\sqrt{(z^2-4)}}{z}$; $\pm\dfrac{2}{\sqrt{(z^2-4)}}$; $\pm\dfrac{z}{\sqrt{(z^2-4)}}$.

8. (i) $\theta = \arcsin\tfrac12 y$; (ii) $\theta = \tfrac12\arcsin y$;
(iii) $\theta = \pm\tfrac13\arcsin\tfrac12\sqrt y$.
9. (i) $(x-1)^2 + y^2 = 1$; (ii) $4(y-1)^2 = (x+3)^2 + 4$;

(iii) $\left(\dfrac{y}{a}\right)^2 + \left(\dfrac{x}{b}\right)^2 = 1$; (iv) $\left(\dfrac{x}{a}\right)^2 = \left(\dfrac{y-b}{c}\right)^2 + 1$;

(v) $x^2 = y^2(1-x^2)$; (vi) $\left(\dfrac{q}{y-p}\right)^2 + \left(\dfrac{p}{x-q}\right)^2 = 1$.

10. (i) 1st or 4th; (ii) positive; (iii) $\sqrt{(1-p^2)}$;
(iv) $\sqrt{(1-p^2)}$.

Chapter 12

Qu.3 (i) $\dfrac{\sqrt3+1}{2\sqrt2}$; (ii) $\dfrac{\sqrt3-1}{2\sqrt2}$.

Qu.5 $\dfrac{\sqrt3-1}{\sqrt3+1}$.

Qu.6 $\tfrac12\sqrt{(2-\sqrt2)}$; $\tfrac12\sqrt{(2+\sqrt2)}$.
Qu.7 $\sqrt2-1$; $\tan 112\tfrac12°$.
Qu.8 1st or 3rd.
Qu.9 (i) $-\cos\theta$; (ii) $\operatorname{cosec}\theta$; (iii) $-\operatorname{cosec}\theta$; (iv) $\cot\theta$;
(v) $-\tan\theta$.

Exercise 12a

1. (i) $\sqrt{2}(\sqrt3-1)$; (ii) $\sqrt{2}(\sqrt3-1)$; (iii) $-(2+\sqrt3)$.
2. (i) $3\sin A - 4\sin^3 A$; (ii) $4\cos^3 A - 3\cos A$.
3. (i) $\sin A\cos B\cos C + \cos A\sin B\cos C + \cos A$
$\cos B\sin C - \sin A\sin B\sin C$;
(ii) $\cos A\cos B\cos C - \cos A\sin B\sin C$
$-\sin A\cos B\sin C - \sin A\sin B\cos C$;
(iii) $\dfrac{\tan A + \tan B + \tan C - \tan A\tan B\tan C}{1 - \tan A\tan B - \tan A\tan C - \tan B\tan C}$.

4. (i) $\sin\left(x+\dfrac{\pi}{3}\right)$ or $\cos\left(\dfrac{\pi}{6}-x\right)$; (ii) $\sin 3x$;

(iii) $\tan A$.

6. $\dfrac{1}{\sqrt2}(\cos\theta - \sin\theta)$.

7. (i) $\tfrac{16}{65}$; (ii) $\tfrac{33}{65}$; (iii) $-\tfrac{16}{63}$; (iv) $-\tfrac{25}{7}$.

8. (i) $\tfrac12\sqrt{(2-\sqrt3)} = \dfrac{\sqrt3-1}{2\sqrt2}$; (ii) $\tfrac12\sqrt{(2-\sqrt2)}$;

(iii) $\tfrac12\sqrt{(2+\sqrt3)}$.

9. $\dfrac{\sec A\sec B}{1 - \tan A\tan B} \equiv \dfrac{\operatorname{cosec} A\operatorname{cosec} B}{\cot A\cot B - 1}$.

11. (i) $\tfrac{2}{11}$; (iii) $\left|\dfrac{m_1 - m_2}{1 + m_1 m_2}\right|$.

12. (i) $\pm2s\sqrt{(1-s^2)}$; (ii) $\pm\dfrac{2s\sqrt{(1-s^2)}}{1-2s^2}$.

13. (i) $\pm4c(2c^2-1)\sqrt{(1-c^2)}$; (ii) $1 - 8s^2 + 8s^4$.
14. (i) $\sin 2\theta$; (ii) $\cos 2\theta$; (iii) $\tan 2\theta$.
15. (i) $2\sin\tfrac12\theta\cos\tfrac12\theta$; (ii) $|\sqrt2\cos\tfrac12\theta|$;
(iii) $\tfrac12(\cot\tfrac12\theta - \tan\tfrac12\theta)$; (iv) $\cot\tfrac12\theta$;
(v) $|\cos\tfrac12\theta + \sin\tfrac12\theta|$; (vi) $|\cos\tfrac12\theta - \sin\tfrac12\theta|$.

16. (i) $-\tfrac19$; (ii) $\pm\dfrac{1}{\sqrt6}$.

26. (i) $4(a^2 - ab + b^2) = 3$; (ii) $m(1+n^2) = 2n$;
(iii) $2q = 6 - 3p^2$; (iv) $x(b^2 - y^2) = 2yba$.
27. (ii) $z = x\sqrt{(1-y^2)} + y\sqrt{(1-x^2)}$.
28. (i) $-\sec\theta$; (ii) $\operatorname{cosec}\theta$; (iii) $-\tan\theta$; (iv) $-\sin\theta$;
(v) $-\sin\theta$; (vi) $-\tan\theta$.
29. (ii) Zero.

30. (i) $18°$; $36°$; (iii) $\dfrac{\sqrt5-1}{4}$.

Qu.11 $\tfrac12(\cos 30° + \cos 110°)$.
Qu.12 (i) $\tfrac12(\sin 110° + \sin 30°)$; (ii) $\tfrac12(\sin 110° - \sin 30°)$.
Qu.13 $2\sin 26°\sin 7°$.
Qu.14 (i) $2\cos 50°\cos 9°$; (ii) $2\sin 50°\sin 9°$;
(iii) $2\sin 50°\cos 9°$.
Qu.15 (i) $2\sin 25°\cos 5°$; (ii) $2\sin x\cos 20°$;

(iii) $2\sin 4A\sin 2A$; (iv) $2\sin\left(\theta - \dfrac{\pi}{4}\right)\cos\dfrac{\pi}{4}$.

Qu.17 (i) $2\sin 15°\cos 53°$; (ii) $-2\sin 19°\cos 36°$.
Qu.18 (i) $\sqrt{34}\sin(\theta - 59\cdot0°)$; (ii) $\sqrt{29}\cos(\theta - 21\cdot8°)$.
Qu.23 $76\cdot7°$ or $209\cdot6°$.
Qu.24 $\tan 3\theta$.

Exercise 12b

1. (i) $\tfrac12(\sin 70° - \sin 10°)$; (ii) $\tfrac12(\cos 30° + \cos 70°)$;
(iii) $\tfrac12(\sin 150° - \sin 10°)$; (iv) $\tfrac12(\cos 10° - \cos 40°)$;

(v) $\frac{1}{2}(\sin 4A - \sin 2A)$;   (vi) $\frac{1}{2}(\cos 2Q - \cos 2P)$;
(vii) $\frac{1}{2}\sin 2A$;   (viii) $\frac{1}{2}(\sin(2\theta - 10^\circ) + \sin 50^\circ)$.

2. (i) $\frac{1}{4}$;   (ii) $\dfrac{\sqrt{3}-1}{4}$;   (iii) $\frac{1}{4}(2 - \sqrt{2})$.

3. (i) $2\sin 35^\circ \cos 15^\circ$;   (ii) $2\sin 19^\circ \cos 61^\circ$;
(iii) $2\sin 48^\circ \sin 16^\circ$;   (iv) $2\cos 19^\circ \cos 1^\circ$;

(v) $-2\sin(A + 15^\circ)\sin 15^\circ$;   (vi) $2\sin\dfrac{\pi}{3}\cos x$;

(vii) $2\sin 10^\circ \cos 30^\circ$;
(viii) $2\cos(\frac{1}{4}\pi + \frac{1}{2}A - \frac{1}{2}B)\cos(\frac{1}{2}A + \frac{1}{2}B - \frac{1}{4}\pi)$;

(ix) 0;   (x) $2\cos\dfrac{\pi}{4}\sin\left(A - \dfrac{\pi}{4}\right)$;   (xi) $2\sin^2 A$.

5. (i) $\sqrt{(\frac{3}{2})}$;   (ii) $\dfrac{1}{\sqrt{2}}$.

8. (ii) $\tan(A + B)(1 - \tan A\,\tan B)$;
(iv) $2\,\mathrm{cosec}\,A\,\mathrm{cosec}\,B\sin\frac{1}{2}(B - A)\cos\frac{1}{2}(B + A)$.

9. (i) $\pm 150^\circ$, $\pm 90^\circ$, $\pm 30^\circ$, 0, 180°;
(ii) 0, 180°, $\pm 80\cdot 4^\circ$;   (iii) 0, 180°;
(iv) 30°, 150°;   (v) 0, 180°.

11. (i) $\sqrt{2}\sin\left(\theta + \dfrac{\pi}{4}\right)$;   (ii) $2\sin\left(\theta - \dfrac{\pi}{6}\right)$;

(iii) $\sqrt{34}\sin(\theta - 211^\circ)$.

12. (i) $5\cos(\theta - 36\cdot 9^\circ)$;   (ii) $\sqrt{2}\cos(\theta + 45^\circ)$;
(iii) $-\sqrt{5}\cos(\theta + 63\cdot 4^\circ)$.

13. $\sqrt{5}$.

14. (i) $65\cdot 7^\circ$ or $204\cdot 3^\circ$;   (ii) $82\cdot 6^\circ$ or $322\cdot 6^\circ$;
(iii) $58\cdot 3^\circ$ or $189\cdot 1^\circ$;   (iv) 90° or $126\cdot 9^\circ$.

16. (i) $c = a + b\cos\varepsilon$, $d = b\sin\varepsilon$;

(iii) (a) $\begin{pmatrix} a\cos\theta \\ a\sin\theta \end{pmatrix}$; $\begin{pmatrix} b\cos(\theta + \varepsilon) \\ b\sin(\theta + \varepsilon) \end{pmatrix}$;

(b) $\begin{pmatrix} a\cos\theta + b\cos(\theta + \varepsilon) \\ a\sin\theta + b\sin(\theta + \varepsilon) \end{pmatrix}$;

(c) $b\sin\varepsilon$; $a + b\cos\varepsilon$;   (d) $\sqrt{(a^2 + b^2 + 2ab\cos\varepsilon)}$;

(e) $\dfrac{b\sin\varepsilon}{a + b\cos\varepsilon}$.

18. (i) $\theta = \dfrac{(2n+1)\pi}{10}$;   (ii) $\theta = \dfrac{n\pi}{5}$ or $\dfrac{(3n\pm 1)\pi}{6}$;

(iii) $\theta = \dfrac{(2n+1)\pi}{2}$ or $\dfrac{(6n\pm 2)\pi}{15}$.

19. (i) $32\cdot 2^\circ + 360n^\circ$;   (ii) $\left.\begin{matrix} 79\cdot 1^\circ \\ -38\cdot 9^\circ \end{matrix}\right\} + 360n^\circ$.

20. $\left.\begin{matrix} 9\cdot 2^\circ \\ 60\cdot 5^\circ \end{matrix}\right\} + 90n^\circ$.

22. (i) $4\sin\dfrac{A+B}{2}\sin\dfrac{A-C}{2}\sin\dfrac{B-C}{2}$;

(ii) $\cos(A + B)\cos(A - B)$.

23. (i) $\frac{3}{4}$; $x = 30^\circ + 180n^\circ$;
(ii) $\frac{1}{2}(1 + \cos 50^\circ)$; $x = 95^\circ + 180n^\circ$;
(iii) $2\cos 20^\circ$; $x = -20^\circ + 360n^\circ$;
(iv) $\sqrt{3}$; $x = 143^\circ + 360n^\circ$.

Qu.26 (i) They add up to 180°;   (ii) $47\cdot 9^\circ$ or $132\cdot 1^\circ$.
Qu.30 $\mathrm{Arctan}\,(\cos\phi\,\tan\theta)$.

2. (i) $10\cdot 2$ cm;   (ii) $57\cdot 1^\circ$;   (iii) $34\cdot 5$ m;   (iv) $7\cdot 65$ cm;
(v) $121\cdot 4^\circ$.
4. (ii) $90\cdot 1^\circ$, $42\cdot 2^\circ$, $47\cdot 7^\circ$.
5. (i) $C = 41\cdot 3^\circ$, $B = 109\cdot 7^\circ$, $b = 11\cdot 3$ cm, or $C = 138\cdot 7^\circ$,
$B = 12\cdot 3^\circ$, $b = 2\cdot 55$ cm; $R = 5\cdot 98$ cm;   (ii) $C = 20\cdot 85^\circ$,
$B = 130\cdot 15^\circ$, $b = 12\cdot 5$ cm; $R = 8\cdot 15$ cm.
6. (ii) About 6° and 9°.

Miscellaneous exercise 12

1. -3, $\frac{3}{2}$, $\frac{3}{2}$.
2. (b) 30°, 150°, 270°.
3. (a) (i) $\frac{3}{5}$, $-\frac{4}{5}$, $-\frac{24}{25}$, $\frac{7}{25}$;   (b) -60°, 0°, 60°.
4. $1 - 2\sin^2\theta$; $\dfrac{2\tan\theta}{1 - \tan^2\theta}$;   (i) $\dfrac{1}{\sqrt{6}}$;   (ii) $\frac{1}{2}$;   (iii) $\frac{1}{3}$.

5. (b) 0°, 90°, 180°, 210°, 270°, 330°, 360°;
(c) $\sqrt{61}\sin(\theta - 50\cdot 2)^\circ$;   (i) $81\cdot 0^\circ$ or $199\cdot 4^\circ$;
(ii) $50\cdot 2^\circ$ or $230\cdot 2^\circ$.
6. (ii) $45 < y < 60$,   (ii) 15, 75, 195, 255.

7. $h = 24\sqrt{5}$; $\arctan\left(\dfrac{\sqrt{5}}{2}\right) \approx 48\cdot 2^\circ$.

8. (a) $R = \sqrt{5}$, $\alpha = 63^\circ 26'$; $x = 16^\circ 16'$, $110^\circ 36'$.
(b) (i) $5\sqrt{3}$;   (ii) $4^\circ 20'$.
9. (b) $42\cdot 29^\circ$, $162\cdot 29^\circ$, $282\cdot 29^\circ$, $53\cdot 13^\circ$.

10. (i) $\theta = (2n+1)\dfrac{\pi}{2}$ or $(2n + \frac{1}{6})\pi$ or $(2n + \frac{5}{6})\pi$;

(ii) $61\cdot 0^\circ$, $73\cdot 8^\circ$, $225\cdot 2^\circ$, $10\cdot 8$ m.

12. (b) $\dfrac{\pi}{4}$.

13. (a) $\theta = n.360^\circ + 38^\circ 58'$, $n.360^\circ - 100^\circ 54'$;
(b) $B = 100^\circ 54'$, $C = 19^\circ 6'$, $a = 6\cdot 6$.
14. $s^3 - 18s^2 + 48s - 32 = 0$.

Chapter 13

Qu.2 $\theta < 0\cdot 551911$.
Qu.6 (i) $\sin x$, $\tan x$, $\sec x$;   (ii) nowhere.
Qu.7 (i) 1;   (ii) -1.
Qu.8 Area of sector OPQ in proof of Theorem 13.1.

Exercise 13

1. $2\cos 2x$.
2. $3\sec(3x + 1)\tan(3x + 1)$.
3. $-2x\,\mathrm{cosec}^2\,x^2$.
4. $\dfrac{3}{x^3}\,\mathrm{cosec}\,\dfrac{3}{2x^2}\cot\dfrac{3}{2x^2}$.
5. $\tan x + x\sec^2 x$.
6. $6x\cos\left(x + \dfrac{\pi}{4}\right) - 3x^2\sin\left(x + \dfrac{\pi}{4}\right)$.

7. $\dfrac{x\cos x - \sin x}{x^2}$.

8. $\sin x\,(1 + \sec^2 x)$.
9. $\sec 3x\,(-2\,\mathrm{cosec}^2\,2x + 3\tan 3x\cot 2x)$.
10. $\mathrm{cosec}\,\dfrac{3x}{2}\left(\cos\dfrac{x}{2} - \frac{1}{2}x\sin\dfrac{x}{2} - \dfrac{3x}{2}\cos\dfrac{x}{2}\cot\dfrac{3x}{2}\right)$.

11. $2(1+x^2)^{-2} \sec x \, (1-x^2+x\tan x+x^3 \tan x)$.

12. $\cot(3x-2) - 3x \operatorname{cosec}^2 (3x-2)$.

13. $6x^2 \operatorname{cosec} 3x^2 \, (1-2x^2 \cot 3x^2)$.

14. $\dfrac{(4x+2)\{(x^2+x)\sec^2(x^2+x) - \tan(x^2+x)\}}{(x^2+x)^2}$.

15. $\dfrac{(1-x)\cos x - (1+x)\sin x - x^2}{(1+x\sin x)^2}$.

16. $\dfrac{x}{\sqrt{(x^2+1)}} \sec\sqrt{(x^2+1)} \tan\sqrt{(x^2+1)}$.

17. $\frac{3}{2}x^{1/2} \cot 5x - 5x^{3/2} \operatorname{cosec}^2 5x$.

18. $2\sin x \cos x$.

19. $3\tan^2 x \sec^2 x$.

20. $-\frac{1}{2}\cot(x+3)\sqrt{\operatorname{cosec}(x+3)}$.

21. $\sec^2 x\,(1+2x\tan x)$.

22. $-3x^{-4}\cos^2 x\,(x\sin x + \cos x)$.

23. $-\frac{2}{3}(\cot x)^{-1/3}\operatorname{cosec}^2 x$.

24. $6\tan\left(3x-\dfrac{3\pi}{4}\right)\sec^2\left(3x-\dfrac{3\pi}{4}\right)$.

25. $24x\sin^2 4x^2 \cos 4x^2$.

26. $16x^{-3}\cos^3\left(\dfrac{2}{x^2}\right)\sin\left(\dfrac{2}{x^2}\right)$.

27. $\dfrac{3\cos 2x}{\sqrt{\sin 2x}}$.

28. $-60x\sqrt{(\operatorname{cosec}^3 4x^2)}\cot 4x^2$.

29. $y = A(\operatorname{cosec}\theta + \cot\theta)$.

30. $\sqrt{2};\ -\sqrt{2}$.

31. Both 1.

33. $5y = 4x + 3 - 4\arcsin\frac{3}{5}$.

34. $\dfrac{\pi}{180}\cos x^\circ$.

35. $2\sqrt{3}$.

36. Both $2\sec^2 x \tan x$.

38. (i) $-\sin\theta$; (ii) $\cos\theta$.

39. $-\dfrac{2\pm\sqrt{3}}{10}$.

40. $-An^2 \sin nt$.

41. $n\pi$ and $n\pi \pm \arctan\sqrt{2}\,(n\in Z)$.

scellaneous exercise 13

1. 3, 1·43, larger.

2. $40\sin\theta$ cm, $20(\sin\theta+\cos\theta)$ cm,
$400\sin\theta(\cos\theta+\sin\theta)$ cm², $200(1+\sqrt{2})\approx 483$ cm².

3. (i) $2\cos(2x+5)$; (ii) $3\cos 3x - 9x\sin 3x$;

(iii) $\dfrac{1}{x^3}(x\sec^2 x - 2\tan x)$.

4. (a) (i) $2\sec^2 2x$; (ii) $9\sin^2 3x\cos 3x$;
(iii) $5\cos x\,(1+\sin x)^4$.

5. $V = \dfrac{1000\pi}{3}\sin^2\theta\cos\theta$, 403 cm³.

6. (a) Minimum at $x = 2^{-1/3}$.

8. $x = \dfrac{\pi}{6},\ \dfrac{5\pi}{6},\ \dfrac{3\pi}{2};\ y = \dfrac{3\sqrt{3}}{2}$(max.),

$y = \dfrac{-3\sqrt{3}}{2}$(min.),

$y = 0$ (neither); (i) $k > \dfrac{1}{\pi}$; (ii) $0 < k \leqslant \dfrac{1}{\pi}$.

Chapter 14

Qu.2 $q \to 2$ and $\dfrac{dq}{dt} \to 0$; hole is not at the bottom.

Qu.3 (i) Yes; (ii) no; someone pulled the plug out perhaps.

Qu.4 $\dfrac{dr}{dt} \to 0$.

Exercise 14a

1. $\dfrac{b}{9}$; $a=1$, $b=2$.

2. (i) $\dfrac{5\pi}{4}$ degrees per second; (ii) $\dfrac{5\pi}{2}$ degrees per second.

3. 4.

4. (i) 128π cm³ s⁻¹; (ii) 96π cm² s⁻¹.

5. $\frac{5}{22}$ cm s⁻¹.

7. $\dfrac{G}{2\pi r_0}$.

9. (i) 10 m; (ii) $\dfrac{90}{\sqrt{13}}$ m s⁻¹; (iii) 30 m s⁻¹ in the limit.

11. (i) $\frac{1}{8}$m s⁻¹; (ii) 3 seconds; (iii) infinitely long.

Qu.5 (i) 0·0175; (ii) 0·0163; (iii) 0·0111;
(iv) 0·005 83; (v) 0·000 152.

Qu.8 ε and $f''(x)$ have the same sign.

Exercise 14b

1. 0·0033; 3·0033.

2. 2·000 75.

3. $\frac{1}{16}$; $2\frac{1}{16}$.

4. $-\frac{1}{36}$; 2·236.

5. $\frac{1}{147}$; 7·0068; 3·5034.

6. $\frac{1}{25}$; 1·26.

7. 0·090 869.

8. 0·4849.

9. 0·874 75.

10. 1·0698.

11. 0·7194.

12. $-1·4389$.

13. 0·111 85.

14. (i) 0·498 75; (ii) 2·996 67; (iii) 2·0025.

15. 8·21.

16. $0·016\pi$.

17. $-\dfrac{100GMm}{6400^3}$; $-\frac{1}{64}$.

18. 6%.

19. $-\frac{3}{4}\%$.

20. $nk\%$

21. $\dfrac{3p}{2}\%$; circular disc.

22. (iii) (a) -1; (b) $-\frac{8}{5}$; (c) $-\frac{3}{5}$; (d) -3.

23. $c = pq$.

Qu.9 $\dfrac{3x}{4y^3}(4y^2-3x^3)$ or equivalent.

Qu.10 (i) $\sec^2 t\,\tan t$; $\dfrac{4\sqrt{3}}{9}$; (ii) $\dfrac{xy}{1-x^2}$.

Qu.12 (i) $\dfrac{-1}{\sqrt{(1-x^2)}}$; (ii) $\dfrac{1}{1+x^2}$; (iii) $\dfrac{-1}{1+x^2}$.

Qu.14 $\dfrac{-1}{|x|\sqrt{(x^2-1)}}$.

Exercise 14c

1. $\dfrac{1}{2y}$.

2. $4-\dfrac{y}{x}$.

3. $\dfrac{2\cos 2x}{3y^2}$.

4. $-(y\cot x+1)$.

5. $\dfrac{y\cos x+\sec^2 x-2xy}{x^2-\sin x}$.

6. $-\dfrac{(1+y^2)^2}{2y}$.

7. $\dfrac{3y-4y^3\sec^2 x\tan x}{6x}$.

8. $-\dfrac{y^2+2xy}{x^2+2xy}$.

9. $\dfrac{\sin y}{4y-x\cos y}$.

10. $\dfrac{1-y\cos xy}{x\cos xy}$.

11. $\dfrac{y+xy^2}{x-y^3}$.

12. $\{3y^2(1+xy)^2\sec^2 y^3+x^2\}^{-1}$.

13. $\dfrac{2+2y}{2\sin y\cos y-3-6y-2x}$.

14. t; $\frac{1}{2}$.

15. $-2t$; -2.

16. $\dfrac{3t}{2}$; $\dfrac{3}{4t}$.

17. $-\dfrac{b}{a}\cot t$; $-\dfrac{b}{a^2}\operatorname{cosec}^3 t$.

18. $\operatorname{cosec} t$; $-\cot^3 t$.

19. $\dfrac{t^2-1}{2t}$; $\dfrac{-(1+t^2)^3}{4t^3}$.

20. $\dfrac{2t^3-1}{3t^2}$; $\dfrac{-2(1+t^3)^3}{9t^5}$.

21. $-\dfrac{1}{t^2}$; $\dfrac{2}{ct^3}$.

22. $-\dfrac{3}{8t^3}$.

23. $\dfrac{3-t}{4}$.

25. $\dfrac{y^2-2xy}{(y-x)^3}=\dfrac{3}{(y-x)^3}$.

26. $\operatorname{Cot}\dfrac{\theta}{2}$; vertically.

28. (i) $\dfrac{2}{\sqrt{(1-4x^2)}}$; (ii) $\dfrac{-2x}{\sqrt{(1-x^4)}}$; (iii) $\dfrac{-1}{1+x^2}$;

(iv) $\dfrac{3x^2}{|1-x^3|\sqrt{(x^6-2x^3)}}$.

29. $\arcsin\left(\dfrac{1}{x}\right)-(x^2-1)^{-1/2}$.

30. $(1-\cos t)^2$.
31. 25, $-\frac{25}{2}$, $24\cdot 9625$.
32. (i) $1\cdot 25\,\text{cm s}^{-1}$; (ii) $20\,\text{cm s}^{-1}$.

33. $-\dfrac{c}{mh^2}$.

34. $11y+8x=46$.

36. (i) $\dfrac{3x^2}{2y}$; (ii) $\frac{3}{2}x^{1/2}$; (iii) $\frac{3}{2}y^{1/3}$.

37. (i) -1.
38. $(1,-4)$.
39. (i) $4y=11x-27$; $11y+4x=234$;
(ii) $y+2x\sin t_0=\cos 2t_0+4\sin^2 t_0$;
$2y\sin t_0-x=2\sin t_0(\cos 2t_0-1)$.

40. $\theta=\arctan\dfrac{3t}{4}$; $\dot\theta=\dfrac{12}{16+9t^2}$; $\frac{3}{4}$ radian per second.

41. (i) $\dfrac{2\cos\phi}{\cos\theta}$; (ii) $\cot\theta\,\sec\phi\,\operatorname{cosec}\phi$.

Miscellaneous exercise 14

1. (i) $\dfrac{10}{\sqrt{\pi}}$, (ii) $\dfrac{1}{(4\sqrt{\pi})}$ cm/s.
2. $3t^2$; $(-2\frac{1}{2},-2)$.
3. (i) $v=\dfrac{150}{t}$, $\delta v\approx 150\dfrac{\delta t}{t^2}$, $2\cdot 88$ km/h, (ii) 10 cm/s.
4. $\dfrac{1}{2\pi}$ m per minute; 63π minutes.
5. (i) $\dfrac{8\pi}{3}$ cm^3 s^{-1}; (ii) $\dfrac{54\pi}{5\sqrt{29}}$ cm^2 s^{-1}.
6. $-2\sin^3\theta\cos\theta$.
8. $\frac{3}{7}x$ m, $\frac{9}{7}$ m/s, $\frac{3}{7}(3-\frac{1}{5}t)$ m/s, $13\cdot 5$ m.
9. $t=2$ and $t=-2$.

Chapter 15

Qu.1 (i) $L(2)\approx 0\cdot 693$; (ii) $L(3)\approx 1\cdot 099$; $L(4)\approx 1\cdot 386$;
(iii) 0.

Qu.3 (i) $4e^{4x}$;　(ii) $-e^{-x}$;
(iii) $\sec x \tan x \exp(\sec x)$;　(iv) $\exp x \exp(\exp x)$;
(v) $2e^x \sec^2(2e^x)$;　(vi) $e^x(1+x)$;　(vii) $\dfrac{e^x(x-1)}{x^2}$.

Qu.4 $5e^{5x}$.
Qu.5 $3\exp 3x$.
Qu.6 $4e^{2x}$;　$8e^{2x}$;　$2^n e^{2x}$.
Qu.7 $x\exp(\tfrac{1}{2}x^2)$.

Exercise 15a

1. (i) e^{2x};　(ii) 1;　(iii) e^2;　(iv) e;　(v) $\exp 2x^2$.
2. (i) $-4e^{-4x}$;　(ii) $4e^x$;　(iii) $-2xe^{-x^2}$;　(iv) 0;
(v) $e^x(\sin x + \cos x)$;　(vi) $2e^x \cos x$;　(vii) $-\tfrac{1}{6}x^3 e^x$;
(viii) $e^x \cos e^x$;　(ix) $e^x(\sec x + \tan x)(1 + x + x\sec x)$.
3. (i) $18x \exp 9x^2$;　(ii) $\tfrac{1}{2}\cos x \exp(\tfrac{1}{2}\sin x)$.
4. (ii) Either $q = \pm 1$ (and p may take any value) or $p = 0$ (and q may take any value).
6. $e^x(5\sin x + 2\cos x)$.
9. (i) $y = e^a(x+1-a)$;　(ii) $(a-1, 0)$.
10. (ii) (a) $e^{2x}(2+D)^n y$;　(b) $e^{-x}(D-1)^n y$;
(iii) $e^x(x^2 + 20x + 90)$.
14. (i) $x = \pm 1$;　(ii) $x = 2$.
18. (ii) 1.
19. (ii) $0\cdot 06$.
Qu.8 (i) ∞;　(ii) $-\infty$.
Qu.9 (i) $\ln x + 1$;　(ii) $\dfrac{x^2 + 1 - 2x^2 \ln x}{x(x^2+1)^2}$;　(iii) $\dfrac{1}{x(1-3\ln x)^2}$.
Qu.10 (i) $\dfrac{6x^2 - 2x + 7}{2x^3 - x^2 + 7x + 3}$;　(ii) $\tan x$;
(iii) $\dfrac{1}{x}\sec(\ln x)\tan(\ln x)$;　(iv) $\dfrac{1}{x}$.
Qu.11 $\dfrac{3}{x}$.
Qu.12 $\dfrac{4}{2x+3} - \dfrac{15}{5x+1}$.
Qu.14 $9\ln 3$.
Qu.15 $x^x(\ln x + 1)$.
Qu.16 $2x.5^{x^2}\ln 5$.

Exercise 15b

1. (i) 1;　(ii) 2;　(iii) 2;　(iv) -1;　(v) $\ln 2$;　(vi) $\tfrac{1}{2}$;
(vii) x^2;　(viii) $\ln(x+2)$.
2. (i) $\dfrac{1}{x}$;　(ii) $\dfrac{4}{x}$;　(iii) $\dfrac{4}{x}$;　(iv) $-\dfrac{1}{x}$;　(v) $\dfrac{1}{1+x}$;
(vi) $\dfrac{2x}{1+x^2}$;　(vii) $\cot x$;　(viii) $\sec x$;　(ix) $\dfrac{3}{x}$;
(x) $\ln x + 1$;　(xi) 1;　(xii) $\dfrac{-3}{4x(\ln x)^2}$;　(xiii) $2x\ln x + x$;
(xiv) $\dfrac{2}{x} - \dfrac{3x^2}{x^3 - 3}$;　(xv) $\dfrac{2\ln x}{x}$;　(xvi) $e^x\left(\ln x + \dfrac{1}{x}\right)$;
(xvii) $(\ln x + 1)\exp(x\ln x) = x^x(\ln x + 1)$;
(xviii) $\dfrac{6}{3x-7} - \dfrac{5}{x+5}$;　(xix) $\dfrac{-2}{x(x^2+1)}$;
(xx) $-\cot x$;　(xxi) $-2\cot x$.

3. (i) $x > 1$;　(ii) $x > e$;　(iii) none.
5. $\dfrac{f'(x)}{f(x)}$;　(i) $\ln(x^2+1)$;　(ii) $\ln(x^3 + 4x + 1)$;
(iii) $\ln(\tan x)$;　(iv) $-\ln\cos x = \ln\sec x$;
(v) $\ln(\arctan x)$.
6. No s.p. if $n = 0$;　max at $\left(e^{-1/n}, -\dfrac{1}{ne}\right)$ if $n < 0$,
min if $n > 0$.
7. $y = \dfrac{x}{e}$;　$y = ex$.
8. (i) and (ii), but not (iii).
9. (i) $4^x \ln 4$;　(ii) $2^x(1 + x\ln 2)$;　(iii) $4^x \ln 4$;
(iv) $\ln 2$;　(v) $x2^{x^2+1}\ln 2$;
(vi) $x^{\sin x}\left(\cos x \ln x + \dfrac{\sin x}{x}\right)$;
(vii) $2x^{\ln x - 1}\ln x$;
(viii) $\exp 3x(\tan 2x)^{\exp 3x}(3\ln\tan 2x + 2\sec x \csc x)$.
10. $\dfrac{\ln 2}{1 - \ln 2}$.
11. $e^{1/2e}$.
12. (i) $\dfrac{(x-1)^2}{(x+2)^3(2x-3)^3}\left(\dfrac{2}{x-1} - \dfrac{3}{x+2} - \dfrac{6}{2x-3}\right)$;
(ii) $\tfrac{1}{2}\sqrt{\left\{\dfrac{(x-2)}{(x+1)(x+2)^3}\right\}}\left(\dfrac{1}{x-2} - \dfrac{1}{x+1} - \dfrac{3}{x+2}\right)$.
13. (i) e;　(ii) b.
14. $(e^{3/2}, \tfrac{3}{2}e^{-3/2})$.
15. (i) $1\cdot 10$;　(ii) $1\cdot 43$;　(iii) 0 or $0\cdot 347$.

Miscellaneous exercise 15

1. $\dfrac{1}{e}$;　$0 < k < \dfrac{1}{e}$;　$1\cdot 63$.
3. (i) (a) $6\sin 3x \cos 3x$;　(b) $\dfrac{1 - 2\ln x}{x^3}$;
(ii) min. at $(-1, -e^{-1/2})$, max. at $(1, e^{-1/2})$.
5. $|a| < 1$ and $a \neq 0$.

Chapter 16

Qu.1 (iv) The max. is below the min.
Qu.2 (i) Max (m, n);　(ii) max $(m, n+1)$.
Qu.3 Sign changes only at an odd-degree asymptote.
Qu.4 (ii) $(-\tfrac{3}{2}, \tfrac{4}{7})$.
Qu.6 $y = x - 4$;　$x = -\tfrac{4}{5}$.

Exercise 16a

3. (i) $x = 1$; $y = 2$;　(ii) $x = -3$; $y = x$;　(iii) $x = \tfrac{1}{2}$;
(iv) $x = -\tfrac{1}{2}$; $y = \tfrac{3}{2}$;　(v) $x = 2$; $x = -2$; $y = 1$;
(vi) $x = 1$;　(vii) $x = 2$; $x = 3$; $y = x + 10$;
(viii) $x = -2$; $y = 1$.
4. (i) $y = \tfrac{1}{2}x^2 + \tfrac{1}{4}x - \tfrac{3}{8}$;　(ii) $y = x^3 + x^2 + x + 1$.
5. (i) $(-1, 1)$;　(iv) $y = \tfrac{1}{2}x$.
8. (ii) $bd = ae$;　(iii) $(-2, 1)$.
Qu.8 (i) S;　(ii) y-axis;　(iii) $y = x$;　(iv) $y = -x$;
(v) x-axis and y-axis (and so S also); (vi) x-axis;
(vii) S.

Qu.9 $\left(-\frac{1}{2}, \pm\frac{1}{\sqrt{3}}\right)$.

Qu.10 Zero.

Qu.11 $\pm 1/\sqrt{2}$.

Qu.12 $1 \leqslant x \leqslant 2$ or $x \geqslant 4$; infinite.

Qu.14 (i) $a_0 = 0$; (ii) $a_0 = a_1 = 0$;
(iii) $a_0 = a_1 = \ldots = a_{k-1} = 0$.

Qu.16 (ii) $\frac{9}{4}$.

Qu.17 (i) $y = x$, $y = -x$; $y = x + \dfrac{x^2}{2}$, $y = -x + \dfrac{x^2}{2}$.

(ii) $x \to \pm \infty$; (iii) no.

Qu.20 The line $y = -x$.

Qu.21 $y \approx -\frac{1}{3}x - \frac{1}{3} - \dfrac{1}{4x+1}$.

Qu.22 $y = 2x - \dfrac{2x+1}{x^2+1}$.

Qu.23 (i) No; (ii) yes; (iii) no.

Exercise 16b

20. (ii) $xy^3 + yx^2 + 1 = 0$;
(iii) (a) $yx^2 - xy^3 = 1$; (b) $yx^3 - xy^2 = 1$;
(c) $yx^3 + xy^2 = 1$; (d) $xy^3 + yx^2 = 1$.
30. $(2x - 3y - 1)(x - y + 3)$.

Exercise 16c

18. $(0, 0)$ and $(2, 0)$.

Qu.25 (i) $(\frac{2}{9}, \frac{4}{9})$ and $(\frac{4}{9}, \frac{2}{9})$;
(ii) $(\frac{3}{26}, -\frac{9}{26})$ and $(-\frac{9}{26}, \frac{3}{26})$.

Qu.26 $t = 1$, $(\frac{1}{2}, \frac{1}{2})$.

Qu.27 In the y-axis.

Exercise 16d

2. (i) $y^2 = 4ax$; (ii) $x^2 + y^2 = a^2$; (iii) $x^2 - y^2 = a^2$;
(iv) $y^2 = x^3$; (v) $xy = a^2$; (vi) $y = x^2 - 2x + 2$;

(vii) $y = \dfrac{x(x-2)}{x-1}$; (viii) $y = \dfrac{x^2}{1-x}$; (ix) $x^2 + y^2 = x$.

Qu.29 (i) $(-1, \sqrt{3})$ and $(-1, 1)$;
(ii) $(\sqrt{2}, 315°)$ and $(2, 150°)$.

Qu.30 (i) (a) $(x^2 + y^2)^3 = y^2$; (b) $x = 2$;
(ii) (a) $r = \sec\theta\tan\theta$; (b) $r^2 = 7$.

Qu.33 Zero.

Qu.34 $1 + \dfrac{2}{\pi}$, $1 + \dfrac{2}{5\pi}$, $1 + \dfrac{2}{9\pi}$.

Qu.36 (ii) Once; twice.

Qu.38 (ii) $(\frac{1}{4}, 138\cdot6°)$, $(\frac{1}{4}, 221\cdot4°)$; (iii) $r = 2$, $\theta = 0$.

Qu.39 $\left(\dfrac{1}{\sqrt{2}}, 45°\right)$ and the pole.

Exercise 16e

1. (i) $r = \dfrac{1}{\sin\theta - 2\cos\theta}$; (ii) $r = \cot^2\theta \, \mathrm{cosec}\,\theta$;

(iii) $r = 2$;
(iv) $r^2 = 4\sec 2\theta$; (v) $r^2 = 2\,\mathrm{cosec}\,2\theta$; (vi) $r^2 = \cos\theta - \sin\theta$.

2. (i) $x^2 + y^2 = 9$; (ii) $x^2 + y^2 = y$; (iii) $x = 2$;
(iv) $x^2 + y^2 = 2x$; (v) $(x^2 + y^2)^2 = 2xy$; (vi) $y = x$;
(vii) $y + \sqrt{3}x = 8$; (viii) $(x^2 - y^2)\sqrt{(x^2 + y^2)} = 1$.

19. (i) (a) $|x| \leqslant 1$; (b) $x = \pm 1$.

20. (i) y-axis; (ii) 1st and 3rd.

32. $\sqrt{\dfrac{2}{3\sqrt{3}}}$.

33. The pole, $(1, \pi/2)$ and $(1, -\pi/2)$.

34. $(5, 0°)$, $(5, 180°)$, $(1, 90°)$, $(1, -90°)$.

35. $r = \theta = 4\pi k$ for $k \in Z^+$ or $k = 0$.

Qu.40 True are: (i), (iii), (v), (vi) and (ix).

Qu.41 (i) No solution; (ii) $x \leqslant 5$.

Qu.42 (i) $2 < x < 3$; (ii) $x \leqslant -1$ or $x \geqslant 1\frac{1}{2}$.

Qu.43 (i) See Section 2.12; (ii) no solution; (iii) 9.

Qu.44 $x \leqslant -2$ or $x = -1$ or $0 \leqslant x \leqslant 2$ or $x \geqslant 2\frac{1}{2}$.

Qu.45 (ii) $x \geqslant \frac{1}{3}$.

Qu.46 $-1 < x < 0$ or $0 < x < 1$.

Qu.47 (ii) It is the empty set (i.e. it has no points).

Qu.48 $-1 < x < 3$.

Qu.49 $x < -3$ or $-1 < x < 0$ or $x > 1$.

Qu.50 $-\sqrt{2} \leqslant x \leqslant 1$ or $\frac{4}{3} < x \leqslant \sqrt{2}$.

Qu.51 $y = -1 \Rightarrow x = -1$; $y = \frac{1}{3} \Rightarrow x = 1$.

Qu.52 (i) $y \neq 1$; (ii) $y = 1$ is an asymptote.

Qu.53 (i) $x > 1$ or $x < -\frac{5}{3}$; (ii) no solution.

Qu.56 (i) 30; (ii) 18.

Exercise 16f

1. (i) $x < 7$; (ii) $x \geqslant 2$; (iii) $x > -1$; (iv) $x \geqslant -7$;
(v) $x \in R$ (i.e. true for all x); (vi) $x < \frac{27}{4}$.

3. (i) $-1 \leqslant x \leqslant 6$; (ii) $x < \frac{1}{2}$ or $x > 3$;
(iii) $x < -2$ or $x > 5$; (iv) $x < -2$ or $x > 2$;
(v) $x \leqslant 0$ or $1 \leqslant x \leqslant 2$; (vi) $x = 1$ or $x \geqslant 2$;
(vii) $-\frac{1}{2} < x < 1$ or $x > 3$; (viii) $-2 \leqslant x \leqslant 0$ or $x \geqslant 2$;
(ix) $x \leqslant -1\frac{1}{2}$ or $1 \leqslant x \leqslant 2$ or $x \geqslant 5$;
(x) $x \in R$; (xi) $x \geqslant -\frac{1}{2}$ or $x \leqslant -4$;
(xii) $x < -2$ or $x > 2$; (xiii) $x < 0$ or $x > 2$;
(xiv) $-3 < x < 1$; (xv) $x \leqslant -1$ or $x \geqslant 2\frac{1}{2}$.

4. (i) $p > 1$; (ii) $-12 < p < 12$; (iii) no value of p;
(iv) $-\frac{3}{2} < p < 2$.

5. No solution.

6. (i) $x < 2$ or $1 \leqslant x < 5$;
(ii) $-4 < x \leqslant -3$ or $2 < x \leqslant \frac{5}{2}$; (iii) $x > 3$;
(iv) $x < \frac{1}{2}$ or $x > 1$; (v) $x \leqslant -2$ or $4 < x \leqslant 9$;
(vi) $\frac{2}{3} < x < 1$ or $2 < x < 5$;
(vii) $x < \frac{17}{16}$ or $2 < x < 3$;

(viii) $0 < x \leqslant \dfrac{1 + \sqrt{5}}{2}$ or $-1 < x \leqslant \dfrac{1 - \sqrt{5}}{2}$;

(ix) $1 < x < 2$ or $3 < x \leqslant 5$;
(x) $x < -1$ or $0 < x < 1$.

8. (i) $y \leqslant -6$ or $y \geqslant -2$;
(ii) $y \leqslant \frac{34}{25}$ or $y \geqslant 2$; (iii) $-\frac{1}{4} \leqslant y \leqslant \frac{1}{2}$.

9. (i) $-4 < x < 1$; (ii) $0 \leqslant x \leqslant 2$;
(iii) $x < -14$ or $x > \frac{6}{5}$; (iv) $x > -\frac{3}{2}$; (v) $x \leqslant -\frac{2}{3}$;
(vi) $x \in R$; (vii) $\frac{1}{3} < x < 7$; (viii) $x < 3$ or $x > 7$;
(ix) $-1 < x < 4$; (x) $-8 < x < \frac{2}{3}$;
(xi) $x \leqslant -14$ or $x \geqslant 6$.

10. (i) $1 < x < 2$ or $4 < x < 5$;
(ii) $c - b < x < c - a$ or $c + a < x < c + b$.

11. (i) $-1 \leqslant x \leqslant \frac{1}{2}$; (ii) $0 < x < 1$ or $3 < x < 4$;
(iii) $-2 \leqslant x \leqslant -1$ or $3 \leqslant x \leqslant 6$.
12. (i) $x < -\sqrt{6\cdot5}$ or $x > \sqrt{6\cdot5}$;
(ii) $x < -\sqrt{12}$ or $x > \sqrt{12}$.
13. (ii) $x = y = z$.
15. $a = b = 1$.
17. (iii) $\frac{2}{3}, \frac{1}{2}, \frac{2}{5}, \frac{1}{3}$; an $A.P.$.
20. $a_1/b_1 = a_2/b_2 = \ldots = a_n/b_n$.

Miscellaneous exercise 16

1. $ax = y^2 - 2ay + 2a^2$; $\dfrac{dy}{dx} = \dfrac{1}{2t}$.

2. $-\frac{1}{3} \leqslant y \leqslant 1$; -1; $-\frac{1}{3} \leqslant f(x) \leqslant 1$; $(-2, -\frac{1}{3})$, $(0, 1)$.

3. $(0, 0)$; $\left(\pm\dfrac{1}{\sqrt{3}}, \dfrac{1}{4}\right)$; $|k| < \frac{1}{2}$ with $k \neq 0$.

4. $(\frac{1}{3}, \frac{9}{2})$, $(3, \frac{1}{2})$; $x = -1$, $x = 1$, $y = 0$.
5. $x = 0$, $x = 4$, $y = 2$; max. at $(1, -1)$, min. at $(-2, \frac{5}{4})$.
6. $0\cdot618$.
7. (i) $y = 1$, $x = 2$; (ii) $y = -\frac{1}{2}x$.
8. $a = 3$, $b = -6$, $c = 8$, $p = -6$, $q = 8$.
9. $\frac{2}{3} < x < \frac{3}{4}$.
10. $(-2, -\frac{1}{2})$, $(\frac{2}{3}, -4\frac{1}{2})$; (i) $\{x \in R : x \neq 0, x \neq 2\}$,
$\{x \in R : x \leqslant -4\frac{1}{2}$ or $x \geqslant -\frac{1}{2}\}$; (ii) $[\frac{2}{5}, 1]$.
11. (ii) $k < -\frac{1}{3}$.
12. $(-2, -\frac{1}{2})$, $(1, 1)$; $0\cdot42$.

Chapter 17
Qu.1 (i) 15; (ii) $4\frac{1}{2}$; (iii) 18; (iv) -6; (v) $10\frac{1}{2}$;
(vi) -12.
Qu.2 (i) 60; (ii) 4; (iii) $9\frac{1}{2}$; (iv) $-7\frac{1}{2}$; (v) -6;
(vi) 12.
Qu.3 (i) 34; 58; (ii) $L_8 = 39\frac{1}{2}$; $U_8 = 51\frac{1}{2}$.
Qu.5 $\frac{1}{4}a^4$.
Qu.6 $63\frac{3}{4}$.
Qu.8 (i) $\tan x + c$; (ii) $-\csc x + c$; (iii) $\frac{1}{8}x^8 + c$;
(iv) $\frac{3}{4}x^4 + c$; (v) $\frac{4}{3}x^3 - \frac{1}{2}x^2 + x + c$; (vi) $-\frac{1}{2}\cos 2x + c$.
Qu.10 (i) $9\frac{1}{3}$; (ii) $-9\frac{1}{3}$.
Qu.11 $3\frac{2}{15}$.

Exercise 17

1. (i) $L_3 = 19$, $U_3 = 30\frac{1}{8}$; (ii) $L_6 = 22\frac{1}{2}$, $U_6 = 28\frac{1}{16}$;

(iii) $\displaystyle\int_0^3 (1 + 9x - 2x^2)\, dx$; $25\frac{1}{2}$.

2. (i) $\frac{3}{2}a^2$; (ii) $10\frac{1}{2}$; (iii) $q - p$.
3. (i) $\frac{1}{6}x^6 + c$; (ii) $-5x^{-1} + c$; (iii) $\frac{1}{2}x^4 + x^{-2} + c$;
(iv) $\frac{1}{2}x^2 + x + c$; (v) $\frac{1}{2}e^{2x} + c$; (vi) $-\frac{1}{3}\cos(3x+1) + c$;
(vii) $2\ln x + c$; (viii) $-\frac{1}{2}\cot(2x-1) + c$.
4. (i) 15; (ii) $\frac{1}{5}$; (iii) 22; (iv) $\frac{2}{3}$; (v) $\ln 2$;
(vi) $32 + 2\ln 3$; (vii) 1; (viii) $1 - \sqrt{3}$.
6. $\frac{1}{2}$.
7. Zero.
9. f is non-negative everywhere (i.e. $\forall t \in R, f(t) \geqslant 0$).
10. $-f(x)$.
12. (i) $\displaystyle\lim_{n \to \infty} \left(\frac{1}{n+1} + \frac{1}{n+2} + \ldots + \frac{1}{2n}\right) = \ln 2$;
(iii) too large.

13. (i) $F(x) = \begin{cases} 5x & \text{for } x \leqslant 2 \\ 10x - 10 & \text{for } x > 2 \end{cases}$;
(iii) $F(x) = 5x$.
15. (i) 12; (ii) 24; (iii) $15\frac{3}{4}$.

Chapter 18
Note. To save space, arbitrary constants have been omitted from answers in this chapter; the reader should always include them.

Exercise 18a

1. $-\dfrac{3}{x} + \dfrac{2}{9x^3}$.

2. $-\dfrac{1}{2x^2} - \dfrac{2}{x}$.

3. $\frac{2}{3}x^{3/2} + \frac{3}{4}x^{4/3}$.

4. $\frac{4}{3}x^{3/2} + \frac{2}{5}x^{5/2} + \frac{6}{7}x^{7/2}$.

5. $-\dfrac{1}{x+3}$.

6. $-\dfrac{1}{6(2x+3)^3}$.

7. $-\frac{1}{3}\cos 3x$.

8. $\frac{1}{4}\sin(4x+1)$.

9. $\frac{1}{3}\tan 3x$.

10. $-2\csc x/2$.

11. $\dfrac{2^x}{\ln 2}$.

12. $\sin^5 x$.

13. $\cos^4 x$.

14. $\frac{1}{7}\sin^7 x$.

15. $-\frac{1}{6}\cos^6 x$.

16. $-\frac{1}{10}\sin^5 2x$.

17. $-\frac{1}{9}\cos^3 3x$.

18. $\frac{1}{5}\tan^5 x$.

19. $-\frac{1}{4}\cot^4 x$.

20. $\frac{1}{24}\tan^6 4x$.

21. $-\frac{1}{6}\cot^3(2x-1)$.

22. $\sec^5 x$.

23. $\csc^3 x$.

24. $\frac{1}{20}\sec^4 5x$.

25. $-\frac{1}{21}\csc^7(3x+4)$.
26. $\frac{1}{2}\sin^2 x$ (or $-\frac{1}{2}\cos^2 x$); these functions
differ by $\frac{1}{2}$, i.e. they have different arbitrary constants.
27. $\frac{1}{2}\tan^2 x$ or $\frac{1}{2}\sec^2 x$.

28. $\frac{1}{6}(x^2+1)^6$.

29. $\frac{1}{42}(2x^3-3)^7$.

30. $\frac{1}{16}(4x+1)^4$.

31. $\frac{1}{3}(x^2+1)^{3/2}$.

32. $\frac{2}{9}(3x-2)^{3/2}$.

33. $-\frac{1}{6}(2x^3+1)^{-3}$.

34. $-\frac{3}{2}(x^2+4)^{-1}$.

35. $\frac{3}{8}(x^2+2x)^{4/3}$.

36. $\frac{1}{2}(2x^2-3)^{1/2}$.

37. e^{x^2}.

38. $\frac{1}{2}e^{2x-1}$.

39. $-\frac{1}{2}\csc x^2$.

40. $\frac{1}{3}\tan x^3$.

41. $\frac{2}{3}(\sin x)^{3/2}$.

42. $-\frac{1}{3}(\sin x - 1)^{-3}$.

43. $e^{\tan x}$.

44. $\frac{1}{2}\ln x$.

45. $\ln(x^2+1)$.

46. $\frac{1}{2}\ln(4x+3)$.

47. $\frac{1}{2}\ln(\sin 2x + 1)$.

48. $2\ln(x^2-3x+2)$.

49. $-\ln(1-e^x)$.

50. $\ln(\sin x)$.

51. $-\ln(\cos x) = \ln(\sec x)$.
52. $\frac{1}{2}\ln\{\tan(2x+1)\}$.

53. $3(x^4+3)^{1/2}$.

54. $\dfrac{-2}{1+\tan x/2}$.

55. $\dfrac{1}{3\cos^3 x}$.

56. $\frac{1}{2}\ln(x^2-4x+5)$.

57. $xe^{\frac{1}{2}x^2}$.

Qu.1 (i) $\dfrac{1}{2\sqrt{2}}\ln\left(\dfrac{x-\sqrt{2}}{x+\sqrt{2}}\right)$; (ii) $2\ln(x-1) - \dfrac{3}{x-1}$.

Qu.2 $\ln(x^2+5x+6)$.
Qu.3 (i) $-\frac{1}{5}\cos^5\theta + \frac{1}{7}\cos^7\theta$;
(ii) $-\cos\theta + \frac{2}{3}\cos^3\theta - \frac{1}{5}\cos^5\theta$.
Qu.4 (i) $\frac{1}{4}\sin^4\theta - \frac{1}{6}\sin^6\theta$ or $-\frac{1}{4}\cos^4\theta + \frac{1}{6}\cos^6\theta$.
Qu.5 $\frac{1}{5}\tan^5\theta + \frac{1}{3}\tan^3\theta$.
Qu.6 (i) $\frac{1}{6}\sec^6\theta - \frac{1}{4}\sec^4\theta$ or $\frac{1}{4}\tan^4\theta + \frac{1}{6}\tan^6\theta$;
(ii) m odd and n even.
Qu.7 $\frac{1}{8}\sin 4x + \frac{1}{20}\sin 10x$.

Qu.8 (i) $\frac{1}{2}\theta + \frac{1}{12}\sin 6\theta$; (ii) $\ln(\sin\theta)$;
(iii) $-2\ln(\mathrm{cosec}\,\frac{1}{2}\theta + \cot\frac{1}{2}\theta)$.

Qu.9 $\ln(\sec\theta)$.

Exercise 18b

1. $\frac{1}{3}\ln(x-1) + \frac{5}{3}\ln(x+2)$.
2. $4\ln(x+1) - \frac{7}{2}\ln(2x+1)$.
3. $\frac{1}{2}\ln(2x-1) + \ln(x-3)$.
4. $\frac{2}{5}\ln(2x-1) + \frac{3}{5}\ln(x+2)$.
5. $-\frac{3}{2}\ln(3x+1) + \frac{5}{3}\ln(x-1)$.
6. $\frac{1}{3}\ln(x-1) + \ln(x+1) - \frac{1}{3}\ln(x+2)$.
7. $\frac{1}{3}\ln x - \ln(x-1) + \frac{1}{2}\ln(x-2)$.
8. $\ln(x-2) - 5(x-2)^{-1}$.
9. $-\ln(x-1) - (x-1)^{-1} + \ln(x-4)$.
10. $\frac{3}{4}\ln x - \frac{3}{4}\ln(x-2) - \frac{3}{2}(x-2)^{-1}$.
11. $\ln(x+1) - 4(x-1)^{-1}$.
12. $\frac{1}{x} + \frac{2}{3(2x-3)} + \frac{8}{9}\ln\left(\frac{2x-3}{x}\right)$.
13. $x - 2\ln(x+2)$.
14. $\frac{3}{2}x + \frac{7}{2}\ln(x-2)$.
15. $\frac{1}{2}x^2 + x + 3\ln(x+1)$.
16. $x - \ln(x^2+5x+4)$.
17. $\frac{1}{2}\ln\left\{\frac{(x-1)^3}{x+1}\right\} + \frac{1}{2}x^2$.
18. $\frac{1}{2}\ln(x^2+4x+5)$.
19. $-\frac{1}{6}\cos^3 2x + \frac{1}{10}\cos^5 2x$.
20. $-\frac{1}{3}\cot^3 x - \frac{1}{5}\cot^5 x$.
21. $-\frac{1}{9}\mathrm{cosec}^3(3x-1)$.
22. $\frac{2}{3}\tan^3\frac{1}{2}x - 2\tan\frac{1}{2}x + x$.
23. $\frac{1}{2}x - \frac{1}{8}\sin 4x$.
24. $\frac{1}{8}x - \frac{1}{32}\sin 4x$.
25. $\frac{3}{8}x - \frac{1}{4}\sin 2x + \frac{1}{32}\sin 4x$.
26. $\frac{1}{4}\sin 2x - \frac{1}{16}\sin 8x$.
27. $-\frac{1}{10}\cos 5x - \frac{1}{2}\cos x$.
28. $-\frac{1}{2}\ln(\cos 2x)$.
29. $-\ln\{\sec(1-x) + \tan(1-x)\}$
$= \ln\{\sec(1-x) - \tan(1-x)\}$.
30. $\frac{1}{2}\ln(\sin x^2)$.

Qu.12 (i) $\frac{2}{\sqrt{21}}\arctan\sqrt{\frac{3}{7}}x$; (ii) $\frac{5}{6}\arctan\frac{2x}{3}$.

Qu.13 (i) $\frac{3}{4}\arctan\frac{1}{2}x$; (ii) $\frac{5\sqrt{3}}{3}\arctan\sqrt{3}x$.

Qu.14 (ii) and (iii) $\frac{1}{2}\arctan\frac{2x+1}{3}$.

Qu.15 (i) $\frac{1}{\sqrt{2}}\arcsin\frac{x}{3}$; (ii) $\frac{3}{\sqrt{2}}\arcsin\sqrt{2}(x+1)$.

Qu.17 $x^2 + x - 3\ln(x^2-4x+6) - \sqrt{2}\arctan\frac{x-2}{\sqrt{2}}$.

Qu.18 (ii) $\frac{\pi}{4}$.

Exercise 18c

1. (i) $\frac{2}{5}(x-1)^{5/2} + \frac{2}{3}(x-1)^{3/2} \equiv \frac{2}{15}(x-1)^{3/2}(3x+2)$.
   (ii) $\frac{1}{7}(x-4)^7 + \frac{5}{6}(x-4)^6 \equiv \frac{1}{42}(x-4)^6(6x+11)$;
   (iii) $\frac{2}{27}(3x+2)^{3/2} - \frac{10}{9}(3x+2)^{1/2} \equiv \frac{2}{27}(3x+2)^{1/2}(3x-13)$;

(iv) $\ln\left(\frac{\sqrt{x-1}}{\sqrt{x+1}}\right)$;

(v) $-\frac{1}{2}(x-3)^{-2} - \frac{4}{3}(x-3)^{-3} - \frac{3}{4}(x-3)^{-4}$
$\equiv -\frac{1}{12}(x-3)^{-4}(6x^2 - 20x + 15)$.

2. (i) $\frac{2}{15}(x+4)^{3/2}(6x-11)$; (ii) $\frac{1}{35}(2x+3)^{3/2}(5x^2-6x+6)$;
   (iii) $\frac{1}{72}(x+1)^8(8x-1)$; (iv) $\frac{1}{20}(3x-1)^{5/3}(5x+1)$;
   (v) $\frac{2}{35}(x-2)^{5/2}(5x+18)$; (vi) $\frac{8-3x}{6(x-2)^3}$;
   (vii) $\frac{2}{7}\sqrt{(x-5)}(x^3+6x^2+40x+400)$;
   (viii) $2\sqrt{x} + 2\ln\left(\frac{\sqrt{x-2}}{\sqrt{x+2}}\right)$.

3. (i) $\frac{2}{3}(x^2+1)^{3/2}$; (ii) $(x^2-1)^{1/2}$; (iii) $\frac{3}{2}\ln(x^2-4)$;
   (iv) $-\frac{1}{2(x^2+3)}$; (v) $\frac{1}{5}\sin^5 x$; (vi) $\frac{1}{8}\tan^4 2x$;
   (vii) $-\frac{1}{3}\cos x^3$; (viii) $\ln(e^x + \sin x)$.

4. (i) $\ln\left(\frac{\sqrt{(x-1)}-1}{\sqrt{(x-1)}+1}\right)$;
   (ii) $2\sqrt{(x+3)} + 3\ln\left(\frac{\sqrt{(x+3)}-3}{\sqrt{(x+3)}+3}\right)$;
   (iii) $-\left(\frac{1-x}{x+1}\right)^{1/2}$.

5. $\ln\left(\frac{x}{\sqrt{(x^2+1)}}\right)$.

6. (i) $2\sqrt{(\tan x)}$; (ii) $\frac{-1}{6(x+1)^3(x-3)^3}$; (iii) $2\sqrt{(\sec x)}$;
   (iv) $\frac{1}{10}\sin^5 2x$; (v) $2(1+\sin x)^{1/2}$;
   (vi) $2\ln(2+\sqrt{x}) + \frac{4(2\sqrt{x}+3)}{(2+\sqrt{x})^2}$;
   (vii) $\frac{1}{4}\ln(2x^2+3)$; (viii) $-2\sqrt{(1-x^2)}$.

7. (i) $\arcsin\frac{x}{2}$; (ii) $\frac{1}{2}\arctan\frac{x}{2}$; (iii) $3\arcsin\frac{x}{\sqrt{3}}$;
   (iv) $4\sqrt{2}\arctan\frac{x}{\sqrt{2}}$; (v) $\frac{1}{2}\arcsin\frac{2x}{3}$;
   (vi) $\frac{1}{12}\arctan\frac{3x}{4}$; (vii) $\frac{7}{\sqrt{75}}\arcsin 5x$;
   (viii) $\frac{5}{4}\arctan 2x$; (ix) $\frac{1}{\sqrt{3}}\arcsin\frac{x-1}{2}$;
   (x) $\frac{1}{2}\arctan\frac{x+2}{3}$; (xi) $\frac{1}{2}\arcsin(2x+3)$;
   (xii) $\frac{1}{3\sqrt{5}}\arctan\frac{3x-1}{\sqrt{5}}$; (xiii) $2\arcsin\frac{x-3}{4}$;
   (xiv) $\arctan\frac{x+4}{5}$; (xv) $2\sqrt{3}\arcsin\frac{3x+1}{2}$;
   (xvi) $2\arctan(5x+2)$.

8. (i) $\frac{1}{2}\arctan\frac{x}{2}$; $\frac{1}{4}\ln\left(\frac{x-2}{x+2}\right)$; $\frac{1}{2}\ln(x^2+4)$;
   (ii) $\arcsin\frac{x}{3}$; $-\sqrt{(9-x^2)}$;
   (iii) $\frac{2}{\sqrt{7}}\arctan\frac{2x+1}{\sqrt{7}}$; $\frac{1}{3}\ln\left(\frac{x-1}{x+2}\right)$.

9. (i) $\frac{1}{2}\text{arcsec}\,\frac{x}{2}$; (ii) $\frac{1}{2}\arctan\frac{\sqrt{(x^2-4)}}{2}$;

(iii) $-\frac{1}{2}\arcsin\frac{2}{x}$.

10. (i) $\sqrt{(x^2+1)}+\frac{1}{2}\ln\left\{\frac{\sqrt{(x^2+1)}-1}{\sqrt{(x^2+1)}+1}\right\}$;

(ii) $\frac{x}{4\sqrt{(x^2+4)}}$; (iii) $-\frac{x}{4\sqrt{(x^2-4)}}$;

(iv) $\arcsin(x-1)-\sqrt{(2x-x^2)}$; (v) $\frac{x}{4\sqrt{(4-x^2)}}$;

(vi) $\frac{1}{2}\arcsin x+\frac{1}{2}x\sqrt{(1-x^2)}$;

(vii) $-\arcsin\frac{1}{2(x-3)}$; (viii) $2\arctan\sqrt{(x-2)}$;

(ix) $17\arcsin\frac{x+3}{4}+\frac{1}{2}\sqrt{(7-6x-x^2)}(9-x)$.

11. (i) $\frac{3}{2}\ln(x^2+4x+13)+\frac{2}{3}\arctan\frac{x+2}{3}$;

(ii) $-\frac{2}{3}(2-x)^{1/2}(x+7)$;
(iii) $\frac{1}{2}x-\ln(2x^2+2x+1)+\frac{13}{2}\arctan(2x+1)$;
(iv) $\frac{1}{2}x^2-2x+\ln(x^2+2x+2)+2\arctan(x+1)$.

12. (i) $23\frac{31}{35}$; (ii) $\frac{3}{14}$; (iii) $\sqrt{2}$; (iv) $2+2\ln\frac{5}{3}$; (v) $\frac{\pi}{2}$;

(vi) $\frac{\pi}{2}$.

Qu.20 (i) $-x\cos x+\sin x$; (ii) xe^x-e^x.
Qu.22 $-x^2\cos x+2x\sin x+2\cos x$.

Qu.23 (i) $x\ln x-x$; (ii) $\frac{1}{\ln 10}(x\ln x-x)$.

Exercise 18d

1. $x\sin(x+1)+\cos(x+1)$.
2. $\frac{1}{2}xe^{2x}-\frac{1}{4}e^{2x}$.
3. $-\frac{1}{3}x\cos 3x+\frac{1}{9}\sin 3x$.
4. $\frac{1}{4}x^2(2\ln x-1)$.
5. $x\tan x+\ln(\cos x)$.
6. $\frac{1}{3}(2x+3)\sec(3x-1)-\frac{2}{9}\{\sec(3x-1)+\tan(3x-1)\}$.
7. $\frac{2}{3}(x-2)\sqrt{(1+x)}$.
8. $\frac{1}{15}(1+x^2)^{3/2}(3x^2-2)$.
9. $x^2\sin x+2x\cos x-2\sin x$.
10. $-e^{-x}(x^2+8x+17)$.
11. $\frac{1}{5}e^x(\sin 2x-2\cos 2x)$.
12. $\frac{1}{2}x\tan(2x+5)+\frac{1}{4}\ln(\cos(2x+5))-\frac{1}{2}x^2$.
13. $\frac{1}{2}\sec x\tan x-\frac{1}{2}\ln(\sec x+\tan x)$.
14. $\frac{1}{9}x^3(3\ln x-1)$.

15. $\frac{x2^x}{\ln 2}-\frac{2^x}{(\ln 2)^2}$.

16. $x\arctan x-\frac{1}{2}\ln(1+x^2)$.
17. $x\,\text{arccot}\,2x+\frac{1}{4}\ln(1+4x^2)$.
18. $(x+\frac{1}{2})\ln(2x+1)-x$.
19. $\frac{1}{2}x\sqrt{(x^2+1)}-\frac{1}{2}\ln\{x+\sqrt{(x^2+1)}\}$.
20. $I_4=\frac{1}{3}\tan\theta\,(2+\sec^2\theta)$;
$I_6=\frac{1}{15}\tan\theta\,(8+4\sec^2\theta+3\sec^4\theta)$.

Qu.24 $\tan\frac{\theta}{2}$.

Qu.25 $\frac{1}{3}\ln\left(\frac{\tan\frac{1}{2}\theta-1}{\tan\frac{1}{2}\theta+2}\right)$.

Qu.26 $\frac{1}{4}\ln\left(\frac{1+2\tan\theta}{1-2\tan\theta}\right)$.

Qu.28 $\frac{1}{2}\arcsin(2x-3)-\sqrt{\{(x-1)(2-x)\}}$.

Exercise 18e

1. $-(-3+4x-x^2)^{1/2}+\arcsin(x-2)$.
2. $\frac{5}{2}\arcsin(x-2)-\frac{1}{2}(x+4)\sqrt{\{(x-1)(3-x)\}}$.
3. $\frac{1}{2}\theta-\frac{1}{2}\ln(\sin\theta+\cos\theta)$.
4. $\frac{3}{5}\theta-\frac{4}{5}\ln(2\cos\theta+\sin\theta)$.

5. $\dfrac{1}{\sqrt{2}}\,\ln\left(\dfrac{\sqrt{2}-1+\tan\frac{1}{2}\theta}{\sqrt{2}+1-\tan\frac{1}{2}\theta}\right)$ or

$\dfrac{1}{\sqrt{2}}\ln\left\{\sec\left(\theta-\dfrac{\pi}{4}\right)+\tan\left(\theta-\dfrac{\pi}{4}\right)\right\}$ or

$\dfrac{1}{\sqrt{2}}\ln\left\{\tan\left(\dfrac{1}{2}\theta+\dfrac{\pi}{8}\right)\right\}$.

6. $\dfrac{-2}{1+\tan\frac{1}{2}\theta}$.

7. $-\frac{1}{2}\ln(3+2\cos\theta)$.

8. $-2\theta+\ln(2-\sin\theta+\cos\theta)+5\sqrt{2}\arctan\left(\dfrac{\tan\frac{1}{2}\theta-1}{\sqrt{2}}\right)$.

9. $\dfrac{1}{\sqrt{10}}\arctan\dfrac{\sqrt{2}\tan\theta}{\sqrt{5}}$.

10. $\frac{1}{5}\theta+\dfrac{\sqrt{3}}{10\sqrt{2}}\ln\left(\dfrac{\sqrt{2}\tan\theta-\sqrt{3}}{\sqrt{2}\tan\theta+\sqrt{3}}\right)$.

Exercise 18f

1. $\frac{1}{12}\sin 6x+\frac{1}{32}\sin 16x$; $\dfrac{1}{12\sqrt{2}}$.

2. $\dfrac{1}{\ln 10}(x\ln x-x)$; $10-\dfrac{9}{\ln 10}$.

3. $e^{\frac{1}{2}x^2}$; e^2-1.
4. $2\ln(x+1)-\frac{1}{3}\ln(3x-4)$; $2\ln 5-2\ln 3-\frac{1}{3}\ln 4$.
5. $\frac{1}{2}x\sqrt{(4-x^2)}+2\arcsin\frac{1}{2}x$; 2π.

6. $\dfrac{1}{1-\tan x}$; $\dfrac{1}{\sqrt{3}-1}$.

7. $\frac{2}{5}x^{5/2}+\frac{2}{3}x^{3/2}$; $\frac{16}{15}$.
8. $2x-3\ln(x+2)$; $4-3\ln 2$.
9. $\text{arcsec}\,\theta$; $\dfrac{\pi}{3}$.

10. $\frac{1}{9}\ln(9x^2-12x+8)-\frac{15}{18}\arctan(\frac{3}{2}x-1)$; $-\frac{1}{9}\ln 2-\dfrac{5\pi}{72}$.

11. $\frac{1}{2}\ln(\sin x^2)$; $\frac{1}{4}\ln 2$.
12. $\frac{1}{105}(2x-1)^{3/2}(15x^2+6x+2)$; $\dfrac{10966}{105}$.
13. $-\frac{1}{12}\text{cosec}^4\,3x$; $\frac{5}{4}$.
14. $\frac{1}{3}\sqrt{(3x^2-1)}$; $\sqrt{2/3}$.
15. $-\frac{1}{4}e^{-2x}(2x^2+2x+1)$; $\frac{1}{4}-\frac{5}{4}e^{-2}$.

16. $\dfrac{1}{\sqrt{2}}\arcsin\sqrt{2}x$; $\dfrac{\pi}{2\sqrt{2}}$.

17. $-\frac{1}{4}\ln(\text{cosec}\,4x+\cot 4x)$; $\frac{1}{4}\ln(1+\sqrt{2})$.

ANSWERS 521

18. $\frac{1}{6}\ln(x^2-x+1)-\frac{1}{3}\ln(x+1)+\frac{1}{\sqrt{3}}\arctan\frac{2x-1}{\sqrt{3}}$;

$-\frac{1}{3}\ln 2+\frac{\pi}{3\sqrt{3}}$.

19. $x\operatorname{arcsec} x-\ln\{x+\sqrt{(x^2-1)}\};\ \frac{2\pi}{3}-\ln(2+\sqrt{3})$.

20. $x^2+3x+\frac{125}{14}\ln(2x-5)+\frac{4}{7}\ln(x+1)$;
$52+\frac{125}{7}\ln 3+\frac{4}{7}\ln 2$.

Miscellaneous exercise 18

1. $\frac{1}{4}\pi-\frac{1}{2}$.

2. (a) (i) $-\frac{1}{2(2x+1)}+C$,

(ii) $-\frac{1}{4(2x+1)}+\frac{1}{8(2x+1)^2}+C$.

3. (a) (i) $\frac{4}{3}$, (ii) $\frac{1}{2}x^2\ln x-\frac{1}{4}x^2+c$,

(b) $\frac{1}{2}\ln\left(\frac{e^x-1}{e^x+1}\right)+c$.

4. (a) $\frac{\pi}{4}$; (b) $e^{2x}(x-\frac{1}{2})+c$;

(c) $\frac{1}{x-1}+\ln(x-1)-\ln(2x+1)+c$.

5. (a) $\frac{\pi}{2}-1$; (b) (i) $-\frac{1}{2a}e^{-ax^2}+c$;

(ii) $\frac{1}{2a^2}-\frac{e^{-aR^2}}{2a^2}(aR^2+1)$.

6. (ii) $x\tan x+\ln\cos x+c$.
7. (a) $\sqrt{3}-1$; (b) $\frac{1}{3}\tan^{-1}3-\frac{1}{18}+\frac{1}{162}\ln 10$.
8. (a) (i) $\ln\sec x+c$; (ii) $\frac{1}{3}\sin^3 x-\frac{1}{5}\sin^5 x+c$;

(iii) $e^{2x}(\frac{1}{2}x-\frac{1}{4})+c$; (iv) $\frac{1}{2}\tan x+c$; (b) $\frac{\pi}{4}$;

(c) $-\frac{\sqrt{(9-x^2)}}{9x}+c$.

9. (i) $2x^{1/2}-3x^{1/3}+6x^{1/6}-6\ln(1+x^{1/6})+c$;

(ii) $-\frac{1}{2}e^{-x}(\sin x+\cos x);\ \frac{(1+e^{-\pi})(1-e^{-(n+1)\pi})}{2(1-e^{-\pi})}$.

10. (i) $\frac{1}{2}\arctan x^2+c$; (ii) $\frac{1}{2}(x^2+1)\arctan x-\frac{1}{2}x+c$;
(iii) $-\ln(1+\cos^2 x)+c$; (iv) $\ln(1+e^x)+c$;
(v) $-\cot x-\operatorname{cosec} x+c$.

Chapter 19

Qu.1 2.

Qu.2 $\left(\frac{\pi}{3},\frac{\sqrt{3}}{2}\right)$; $2\frac{1}{2}$.

Qu.3 7.
Qu.4 Zero.

Qu.5 $\frac{1}{\sqrt{2}}$.

Qu.6 $\frac{\pi}{2}$.

Qu.7 $\frac{\pi}{2}(e^2-1)$.

Qu.8 $\pi\left(1-2\ln(1+\sqrt{2})+\frac{\pi}{4}\right)$.

Qu.9 $44\frac{1}{10}\pi$.

Qu.10 (i) $\frac{\pi}{6}$; (ii) $\frac{\pi^3}{36}$; (iii) $\frac{108}{\pi^3}$.

Qu.11 $6\frac{1}{3}$m.

Exercise 19

1. 21.
2. $1\frac{1}{8}$.
3. (i) $\frac{1}{3}$; (ii) $\frac{2n-1}{2n+1}$; 1.
4. $2\sqrt{2}$.
5. $a^2\left(\frac{2\pi}{3}-\frac{\sqrt{3}}{2}\right)$.
6. (i) $\frac{1}{3}$; (ii) 68; (iii) $\frac{11}{10}\ln 11-1$; (iv) $\frac{2}{\pi}$.
7. (i) $\sqrt{\frac{28}{3}}$; (ii) $\sqrt{2}$; (iii) $\frac{1}{\sqrt{2}}$; (iv) $\frac{6\sqrt[4]{3}}{\sqrt{\pi}}$.
8. (i) $\frac{3}{4}$; (ii) $\sqrt{\frac{2}{3}}$.
9. $27\frac{1}{15}\pi$.
10. π.
11. $\frac{14}{3}\pi$.
12. $4\pi\ln 4$.
13. 71π.
14. $\pi(73-24\ln 2)$.
15. $\frac{2\pi}{15}$.
16. $\frac{1}{3}\pi m^2 h^3$.
17. $k=\sqrt{\frac{21}{5}}$.
20. $2\pi^2 r^2 c$.
22. (i) $x=2t$; (ii) $x=t^2+t+3$;
(iii) $x=2t^3-3t^2+t-20$; (iv) $x=2t^3-2t^2$;
(v) $x=2t^3-2t^2-40t+85$; (vi) $x=-2\sin 2t$;
(vii) $x=e^{-t}+10t-1$.
23. (i) 8 m; (ii) -60m; (iii) $(8\ln 2-2)$m.
24. $x=ut+\frac{1}{2}at^2+\frac{1}{6}bt^3$.
25. (i) $21-4\cdot9t^2$; (ii) $21+7t-4\cdot9t^2$;
(iii) $21-16\cdot1t-4\cdot9t^2$.
26. At $x=119$; 7 s after P departs.
28. (i) $\frac{2}{3}$; (ii) $6\frac{1}{12}$.
29. -108 m.
30. $10\ \mathrm{ms}^{-1}$; 5 m; at the origin.
31. All $\ln 2$.
32. (i) $1\cdot67$ km; (ii) $260\frac{5}{12}$ m; (iii) $16\cdot7\ \mathrm{ms}^{-1}$.
33. (i) $4\ \mathrm{ms}^{-1}$; (ii) 0 to 8; $7\frac{1}{5}\ \mathrm{ms}^{-1}$; (iii) $4\cdot25\ \mathrm{ms}^{-1}$;
(iv) $7\cdot02\ \mathrm{ms}^{-1}$.

35. $\frac{1}{2}\theta\displaystyle\int_{x_1}^{x_2} y^2\, dx.$

36. (i) 75π;   (ii) $\dfrac{4\pi}{5}.$

Miscellaneous exercise 19

1. 128π; 16s, $\frac{1}{8}$cm.
2. (i) 117 cm;   (ii) 245 cm.
3. (2, 6), $(-2, 6)$; 16.
4. (i) 180 m,   (ii) 29·1 m/s,   (iii) $\frac{5}{3}$ m/s^2,   (iv) 1·7 m/s^2.
5. (i) $t=3$;   (ii) 5 seconds;   (iii) 72 m.
6. (i) 4;   (ii) $\dfrac{3\pi}{2}$;   (iii) $\dfrac{56\pi}{5}.$
7. (i) (1, -4), (25, 20);   (ii) 144.
8. $2xe^{2x}$; 59·5.
9. $\left(1, \dfrac{1}{e}\right)$, $\left(2, \dfrac{2}{e^2}\right)$; $\frac{1}{4}\pi\left(1-\dfrac{13}{e^4}\right).$
10. $\pi(\frac{1}{2}e^{2h}+4e^h+4h-4\cdot5)$; 0·379 units per second.
11. $\dfrac{64\pi}{3}$; $\frac{32}{25}.$

Chapter 20

Qu.1 (i) $\begin{pmatrix} -1 \\ 5 \end{pmatrix}$;   (ii) $\begin{pmatrix} -3 \\ 6 \end{pmatrix}$;   (iii) $\begin{pmatrix} 3 \\ 3 \end{pmatrix}$;   (iv) $\begin{pmatrix} 7 \\ 10 \end{pmatrix}$.
Qu.2 $m=5$, $n=\frac{17}{3}$.
Qu.3 $\lambda=1\frac{1}{2}$, $\mu=-2\frac{1}{2}$.
Qu.5 Yes.
Qu.6 (7, -1); $\begin{pmatrix} -3 \\ -1 \end{pmatrix}$.
Qu.7 (5, -2).
Qu.9 $\begin{pmatrix} 2 \\ 7 \end{pmatrix}$; $\begin{pmatrix} 1 \\ 3\frac{1}{2} \end{pmatrix}$; $\begin{pmatrix} 1 \\ 3\frac{1}{2} \end{pmatrix}$.
Qu.10 $\begin{pmatrix} -4 \\ 6 \end{pmatrix}$.
Qu.11 (ii) (a) $\begin{pmatrix} -3 \\ 8 \end{pmatrix}$;   (b) $\begin{pmatrix} -7 \\ 11 \end{pmatrix}$;   (c) $\begin{pmatrix} -2 \\ 13 \end{pmatrix}$;
   (d) $\begin{pmatrix} -5 \\ -2 \end{pmatrix}$;   (e) $\begin{pmatrix} 5 \\ 2 \end{pmatrix}$;   (f) $\begin{pmatrix} -14 \\ -1 \end{pmatrix}$.
Qu.13 $\sqrt{29}$.
Qu.14 (i) $\sqrt{15}$;   (ii) 7;   (iii) 5;   (iv) 7;   (v) 7;   (vi) 2;
   (vii) $\sqrt{3}$.
Qu.15 6; 8.
Qu.16 $\sqrt{78}$.
Qu.17 $3\sqrt{61}$; $\dfrac{6}{\sqrt{61}}\mathbf{i}+\dfrac{4}{\sqrt{61}}\mathbf{j}-\dfrac{3}{\sqrt{61}}\mathbf{k}.$
Qu.18 $\dfrac{1}{\sqrt{3}}\mathbf{i}+\dfrac{1}{\sqrt{3}}\mathbf{j}+\dfrac{1}{\sqrt{3}}\mathbf{k}$
Qu.19 (i) $2\mathbf{p}+\mathbf{q}$;   (ii) $\mathbf{p}-\mathbf{q}$.

Exercise 20a

1. (i) $-\mathbf{i}+3\mathbf{j}$;   (ii) $5\mathbf{i}-\mathbf{j}$;   (iii) $5\mathbf{i}+6\mathbf{j}$;   (iv) $21\mathbf{i}-7\mathbf{j}$;
   (v) $s=2$, $t=1$.
2. (iii) $p=-8\frac{2}{3}$.
3. (i) $\mathbf{q}$;   (ii) $\mathbf{p}$;   (iii) $\mathbf{p}+\mathbf{q}$;   (iv) $\mathbf{p}-\mathbf{q}$.
4. (i) (a) $\mathbf{q}-\mathbf{p}$;   (b) $\frac{1}{2}\mathbf{p}-\mathbf{q}$;   (c) $\frac{1}{2}(\mathbf{q}+\mathbf{p})$;   (d) $\frac{1}{2}(\mathbf{q}-\mathbf{p})$.
5. (i) $2\mathbf{j}+\mathbf{k}$;   (ii) $4\mathbf{i}+\mathbf{k}$;   (iii) $4\mathbf{i}+2\mathbf{j}+\mathbf{k}$;   (iv) $-2\mathbf{j}$;
   (v) $4\mathbf{i}-2\mathbf{j}-\mathbf{k}$;   (vi) $-4\mathbf{i}-2\mathbf{j}+\mathbf{k}$;   (vii) $-2\mathbf{i}+2\mathbf{j}-\mathbf{k}$.
6. (i) $\sqrt{13}$;   (ii) 2;   (iii) $\sqrt{14}$;   (iv) $\sqrt{27}$.
7. (i) $-\mathbf{i}$;   (ii) $\frac{3}{5}\mathbf{i}-\frac{4}{5}\mathbf{j}$;   (iii) $\dfrac{1}{\sqrt{2}}(\mathbf{j}-\mathbf{k})$.
9. (i) $\sqrt{3}\mathbf{i}+\mathbf{j}$;   (ii) $\mathbf{i}-\mathbf{j}$.
10. (i) 5·7 Newton;   S62°W;   (ii) $-5\mathbf{i}-2\cdot66\mathbf{j}$.
11. (i) When $\mathbf{a}$ and $\mathbf{b}$ are in the same direction.
12. (i) $a+b$; $|a-b|$;   (ii) $t=-1$.
13. (i) $(1+\sqrt{2})\mathbf{q}$;   (ii) $\sqrt{2}\mathbf{q}-\mathbf{p}$;
   (iii) $(2+\sqrt{2})\mathbf{q}-(1+\sqrt{2})\mathbf{p}$;   (iv) $\mathbf{q}-\sqrt{2}\mathbf{p}$;
   (v) $-\sqrt{2}(\mathbf{p}+\mathbf{q})$.
15. (i) $2\mathbf{v}-\mathbf{u}$;   (ii) $6\mathbf{u}-3\mathbf{v}$;   (iii) $2\mathbf{u}-3\mathbf{v}$;   (iv) $5\mathbf{u}$.
Qu.20 $\mathbf{i}-\mathbf{j}+2\mathbf{k}$; $3\mathbf{i}-3\mathbf{j}+6\mathbf{k}$;   1:3.

Exercise 20b

1. (i) $\frac{1}{2}\mathbf{p}+\frac{1}{2}\mathbf{q}$;   (ii) $\frac{2}{3}\mathbf{p}+\frac{1}{3}\mathbf{q}$;   (iii) $2\mathbf{q}-\mathbf{p}$;   (iv) $\frac{3}{2}\mathbf{p}-\frac{1}{2}\mathbf{q}$.
2. (i) $(-1, 2\frac{1}{2})$;   (ii) (0, 4);   (iii) $(-10, -11)$;   (iv) $(5, 11\frac{1}{2})$.
3. (i) $\frac{1}{2}(\mathbf{a}+\mathbf{b})$;   (ii) $\frac{1}{4}\mathbf{a}+\frac{1}{4}\mathbf{b}+\frac{1}{2}\mathbf{c}$;   (iii) $\frac{3}{5}\mathbf{a}+\frac{1}{10}\mathbf{b}+\frac{3}{10}\mathbf{c}$.
6. Yes.
9. (i) $\mathbf{a}+\mathbf{c}-\mathbf{b}$;   (ii) $\mathbf{b}+\mathbf{c}-\mathbf{a}$;   (iii) $\mathbf{a}+\mathbf{b}-\mathbf{c}$.
10. $\frac{1}{4}(\mathbf{a}+\mathbf{b}+\mathbf{c}+\mathbf{d})$.
Qu.21 (i) 4;   (ii) 8;   (iii) $-4\mathbf{i}+16\mathbf{j}+8\mathbf{k}$;   (iv) $-\frac{7}{2}$.
Qu.22 2.
Qu.25 $3\mathbf{i}+3\mathbf{j}+3\mathbf{k}$ and $4\mathbf{i}-3\mathbf{j}-\mathbf{k}$.

Exercise 20c

2. (i) $\lambda=7\frac{1}{2}$.
3. (i) 53·1°;   (ii) 54·7°;   (iii) 109·5°;   (iv) 80·4°.
4. $\dfrac{s}{r}$ and $\dfrac{u}{t}$.
5. $\sqrt{\frac{7}{10}}$.
6. Diagonals of a rhombus are perpendicular.
8. (i) $(\mathbf{c}-\mathbf{b})^2=(\mathbf{c}-\mathbf{a})^2$;   (ii) $-2\mathbf{c}.\mathbf{b}=-2\mathbf{c}.\mathbf{a}$.
9. (i) $\mathbf{AQ}=3\mathbf{i}+2\mathbf{k}$, $\mathbf{AR}=3\mathbf{i}+\mathbf{j}+2\mathbf{k}$; $\angle QAR=16°$;
   (ii) 60°;   (iii) 74°;   (iv) 60°.
10. Perpendicular bisector of AB; XM perpendicular to AB, where M is the mid-point of AB.
11. $(\mathbf{a}-\mathbf{r}).(\mathbf{b}-\mathbf{r})=0$.
Qu.26 (i) $\sqrt{34}$;   (ii) 0·508;   (iii) $h\sqrt{(25+8h+h^2)}$.
Qu.27 (i) $(2\cos 2t)\mathbf{i}-(\sin t)\mathbf{j}+\mathbf{k}$;   (ii) 0.
Qu.28 (i) $\mathbf{v}=2t\mathbf{i}+3t^2\mathbf{j}$;   11·3°;   (ii) $\cos\theta\rightarrow 1$, i.e. $\theta\rightarrow 0$.

Exercise 20d

1. (i) Left;   (ii) $(2\cos 2t)\mathbf{i}-(2\sin 2t)\mathbf{j}+3\mathbf{k}$; $\sqrt{13}$;
   (iii) $\arccos\dfrac{3}{\sqrt{13}}$;   (iv) 1; π; 3π.
2. $\mathbf{r}=\left(a\cos\dfrac{2\pi ut}{\sqrt{(4\pi^2 a^2+p^2)}}\right)\mathbf{i}$

$$+\left(a\sin\frac{2\pi ut}{\sqrt{(4\pi^2a^2+p^2)}}\right)\mathbf{j}+\left(\frac{put}{\sqrt{(4\pi^2a^2+p^2)}}\right)\mathbf{k}.$$

3. (i) $2t\mathbf{i}+\mathbf{j}$; (ii) $2\mathbf{i}$; (iii) $t\sqrt{(1+t^2)}$; (iv) $\dfrac{1+2t^2}{\sqrt{(1+t^2)}}$;

(v) $\sqrt{(1+4t^2)}$.

4. (i) $\frac{1}{4}t^2\mathbf{i}+\frac{1}{2}t\mathbf{j}$; (iii) as the particles pass the same point on the curve.

5. $\dfrac{\pi}{4}$.

7. (ii) It moves on the surface of a sphere centre A.

8. (i) $1-t^2$; $-2t$.

9. (i) When $\mathbf{r}$ and $\dfrac{d\mathbf{r}}{dt}$ are parallel (not necessarily in the same sense); (ii) $\left|\dfrac{d\mathbf{r}}{dt}\right|\geqslant\left|\dfrac{d|\mathbf{r}|}{dt}\right|$. (Think of the former as 'speed' and the latter as 'rate of moving towards or away from the origin'.)

10. (i) Along the lines $y=\pm x$.

11. (ii) $A\omega$; ω; $\dfrac{2\pi}{\omega}$.

Miscellaneous exercise 20

1. 15.

2. (i) (a) 20·3 N at 42·3°, (b) 6·2 N at 143·8°, (ii) 10 N.

3. $11\mathbf{i}+4\mathbf{j}$; $10\mathbf{i}+2\mathbf{j}$; $7\mathbf{i}-4\mathbf{j}$; $2:1$.

4. (i) $\frac{1}{2}(\mathbf{p}-\mathbf{q})$; (ii) $\frac{1}{4}(\mathbf{p}+\mathbf{q})$; (iii) $\frac{1}{4}(\mathbf{p}-3\mathbf{q})$; $k=\frac{1}{3}$.

6. $\mathbf{b}-\dfrac{\lambda}{a^2(1+\lambda)}(\mathbf{a}.\mathbf{b})\mathbf{a}$.

7. (ii) $\frac{1}{3}(\overrightarrow{PQ}+2\overrightarrow{PS})$, $\frac{1}{3}(2\overrightarrow{PS}-2\overrightarrow{PQ})$, $\frac{1}{3}(\overrightarrow{PS}-\overrightarrow{PQ})$.

8. $(1-s)\mathbf{b}+\frac{2}{3}s\mathbf{a}$; $\frac{1}{7}\mathbf{b}+\frac{4}{7}\mathbf{a}$; $120°$.

9. $2ap\dfrac{dp}{dt}$, $2a\dfrac{dp}{dt}$; $\dfrac{ph}{a(1+p^2)}$, $\dfrac{h}{a(1+p^2)}$;

$\dfrac{h^2(1-p^2)}{2a^3(1+p^2)^3}$, $\dfrac{-ph^2}{a^3(1+p^2)^3}$.

11. $2t^2$.

Chapter 21

Qu.1 It is square.

Qu.3 (i) (a) $\begin{pmatrix}4&1&4\\2&2&-7\end{pmatrix}$; (b) $\begin{pmatrix}-2&2&8\\-1&-11&6\end{pmatrix}$;

(c) —; (d) $\begin{pmatrix}6&-2&2\\0&0&-12\end{pmatrix}$; (e) see (a);

(ii) (a) $\begin{pmatrix}0&-1&-4\\0&4&-1\end{pmatrix}$; (b) $\begin{pmatrix}5&-\frac{1}{2}\\-4&6\frac{1}{2}\\-6&-12\end{pmatrix}$;

(c) $\begin{pmatrix}0&0&0\\0&0&0\end{pmatrix}$.

Qu.4 All zero.

Qu.5 Yes

Qu.7 (ii) True.

Qu.8 (i) —; (ii) $\begin{pmatrix}0&2&11&10\\1&-1&-4&-5\end{pmatrix}$;

(iii) $\begin{pmatrix}1&0\\1&5\\0&-4\end{pmatrix}$; (iv) —; (v) $\begin{pmatrix}5&-2\\-2&1\end{pmatrix}$;

(vi) $\begin{pmatrix}1&2\\2&5\end{pmatrix}$; (vii) $\begin{pmatrix}1&0\\0&1\end{pmatrix}$; (viii) $\begin{pmatrix}1&2\\0&-1\end{pmatrix}$.

Qu.9 (ii), (iii) and (v).

Qu.10 (ii) $\begin{pmatrix}1&0&0\\0&1&0\\0&0&1\end{pmatrix}$.

Qu.12 (i) 8; (ii) 0.

Qu.13 (i) $ek-fh$; (ii) $cd-af$.

Qu.14 $\begin{pmatrix}-23&5&-4\\8&5&-4\\6&-4&-3\end{pmatrix}$.

Qu.18 (i) $\begin{pmatrix}-1&2\\0&\frac{1}{2}\end{pmatrix}$; (ii) $\begin{pmatrix}-4&-3\\3&2\end{pmatrix}$;

(iii) no inverse.

Qu.19 $\begin{pmatrix}-2&1\frac{1}{2}\\3&-2\end{pmatrix}$; same as A.

Qu.22 (i) $-\frac{1}{3}\begin{pmatrix}2&5&1\\7&16&5\\1&1&-1\end{pmatrix}$; (ii) no inverse.

Qu.25 (i) $X(A+2B)$; (ii) $X(A+I)$; (iii) —.

Qu.26 (i) $X=A^{-1}(B+C)$; (ii) $X=A^{-1}BA$; (iii) $X=(C-A)^{-1}B$.

Exercise 21a

1. (i) —; (ii) $\begin{pmatrix}1&-6&3\\-6&4&-10\end{pmatrix}$;

(iii) $\begin{pmatrix}-30&4\\19&-10\end{pmatrix}$ (iv) $\begin{pmatrix}1&0&5\\-1&2&2\end{pmatrix}$;

(v) —; (vi) $\begin{pmatrix}1&27\\18&-26\end{pmatrix}$; (vii) —;

(viii) $\begin{pmatrix}13&-8&24\\-8&16&0\\24&0&64\end{pmatrix}$; (ix) $\begin{pmatrix}-28&10\\23&-14\end{pmatrix}$;

(x) —; (xi) $\begin{pmatrix}27&-26\\-98&28\end{pmatrix}$;

(xii) $\begin{pmatrix} -128 & 32 \\ 76 & -40 \\ -240 & 32 \end{pmatrix}$; (xiii) $\begin{pmatrix} 64 & -8 & 98 \\ -48 & 20 & -77 \end{pmatrix}$.

2. (i) $(pr+qs)$; (ii) $\begin{pmatrix} rp & rq \\ sp & sq \end{pmatrix}$.

3. (ii) Yes; (iii) both square.

6. (i) $\mathbf{CBA}$; (ii) $\mathbf{CAB}^{\mathrm{T}}$.

7. $\mathbf{X} = \begin{pmatrix} \lambda & 0 \\ 0 & \lambda \end{pmatrix}$; $\mathbf{Y} = \begin{pmatrix} \lambda & 0 & 0 \\ 0 & \lambda & 0 \\ 0 & 0 & \lambda \end{pmatrix}$.

10. (i) $-\frac{1}{3}\begin{pmatrix} 3 & 4 \\ 6 & 7 \end{pmatrix}$; (ii) $\begin{pmatrix} 1 & 2 \\ -3 & -5 \end{pmatrix}$;

(iii) $\begin{pmatrix} -5 & -6 \\ 13 & 15\frac{2}{3} \end{pmatrix}$; (iv) $\begin{pmatrix} -47 & -18 \\ 39 & 15 \end{pmatrix}$; (v) as (iii).

11. (i) $\begin{pmatrix} 105 & 32 & 28 \\ 14 & 18 & 2 \\ -6 & -3 & 11 \end{pmatrix}$; (ii) $\begin{pmatrix} 111 & 28 & 48 \\ -34 & 7 & -5 \\ 36 & 11 & 16 \end{pmatrix}$;

$\mathbf{A}^2 + \mathbf{AB} + \mathbf{BA} + \mathbf{B}^2$.

12. (i) -5; (ii) 2; (iii) -63; (iv) -77.

13. (i) $-\frac{1}{5}\begin{pmatrix} 20 & 11 & -12 \\ -15 & -8 & 11 \\ 10 & 5 & -5 \end{pmatrix}$; (ii) $\begin{pmatrix} -\frac{1}{2} & 0 & -\frac{3}{2} \\ 2 & 1 & 7 \\ 6 & 3 & 20 \end{pmatrix}$;

(iii) $-\frac{1}{63}\begin{pmatrix} -14 & -1 & -2 \\ -21 & -33 & -3 \\ -7 & 4 & 8 \end{pmatrix}$;

(iv) $-\frac{1}{77}\begin{pmatrix} 10 & 1 & -17 \\ -3 & -8 & -18 \\ -4 & 15 & -24 \end{pmatrix}$.

14. $\begin{pmatrix} -1 & 0 & 3 \\ 3 & 0 & -8 \\ 1 & \frac{1}{2} & -2\frac{1}{2} \end{pmatrix}$; $\begin{pmatrix} 4 & -1 & 1 \\ 16 & -4 & 5 \\ -7 & 2 & -2 \end{pmatrix}$.

15. $\mathbf{A}^{-1} = \begin{pmatrix} -6 & -2 & 2\frac{1}{2} \\ 9 & 3 & -3\frac{1}{2} \\ -4 & -1 & 1\frac{1}{2} \end{pmatrix}$;

$\mathbf{M} = \begin{pmatrix} -1 & 0 & \frac{1}{2} \\ -22 & -6 & 9 \\ -24 & -6 & 10 \end{pmatrix}$.

16. (i) $\mathbf{P(Q+R)}$; (ii) $\mathbf{(PQ+R)P}$; (iii) —;
(iv) $\mathbf{(A+B)(A-C)}$; (v) $\mathbf{(A-I)(A+I)}$ or
$\mathbf{(A+I)(A-I)}$; (vi) —.

17. (i) $\mathbf{B-A}$; (ii) $\mathbf{A}^{-1}\mathbf{B}$; (iii) $\mathbf{BA}^{-1}$;
(iv) $\mathbf{A}^{-1}\mathbf{BA}^{-1}$; (v) $\mathbf{(A+I)}^{-1}\mathbf{B}$; (vi) $\mathbf{A}^{-1}\mathbf{B}^{-1}$;
(vii) $\mathbf{B}^{-1}\mathbf{A}^{-1}$; (viii) $\mathbf{B}^{-1}\mathbf{A}^{-1}$.

18. $\pm\begin{pmatrix} 0 & 1/\sqrt{2} \\ \sqrt{2} & \sqrt{2} \end{pmatrix}$ or $\begin{pmatrix} \sqrt{6}/3 & \sqrt{6}/6 \\ \sqrt{6}/3 & 2\sqrt{6}/3 \end{pmatrix}$.

19. $\begin{pmatrix} 1 & 0 & 0 & -2 \\ 0 & 0 & -\frac{1}{3} & -\frac{4}{3} \\ 0 & 0 & 0 & -1 \\ -\frac{1}{2} & \frac{1}{2} & \frac{1}{6} & \frac{5}{3} \end{pmatrix}$.

Qu.27 (i) $\lambda^2\Delta$; (ii) $\lambda^3\Delta$; (iii) $\lambda^n\Delta$.

Qu.28 No change.

Qu.29 (i) 373 806; (ii) 0.

Qu.30 (i) Interchanges 2nd and 3rd columns;

(ii) $\begin{pmatrix} 0 & 1 & 0 \\ 1 & 0 & 0 \\ 0 & 0 & 1 \end{pmatrix}$.

Exercise 21b

1. (i) (a) Δ^2; (b) Δ^{n-1}; (ii) (a) $\Delta\mathbf{M}$; (b) $\Delta^{n-2}\mathbf{M}$.

6. Δ.

8. $(x-3)(x-4)$.

9. 0; $(x-2)(x-3)(x+5)$.

10. $70(x-2)(x-4)(x+3)^2$.

11. (i) $(x-y)(y-z)(z-x)$; (ii) $xyz(x-y)(y-z)(z-x)$.

12. $(x-y)(y-z)(z-x)(xy+yz+zx)$.

13. $(x-y)(y-z)(z-x)(xyz-x-y-z)$.

14. (i) and (ii) $(x+y+z)(x-y)(y-z)(z-x)$.

15. $(x-a)(x-b)(x-c)(a+b+c+x)$.

Miscellaneous exercise 21

1. (a) $\begin{pmatrix} 3 & 0 & 0 \\ 0 & 3 & 0 \\ 0 & 0 & 3 \end{pmatrix}$, $\frac{1}{3}\mathbf{B}$; 3, 1, 0; (b) -5, 7.

2. $\mathbf{B}^2$, (i) $\begin{pmatrix} -\frac{1}{2} & \frac{1}{2} \\ -1 & \frac{1}{2} \end{pmatrix}$, (ii) $k=2$.

3. $a=1$, $b=-4$; $\begin{pmatrix} \frac{1}{2} & -\frac{3}{2} \\ 1 & 0 \\ -\frac{1}{2} & \frac{1}{2} \end{pmatrix}$.

4. $\begin{pmatrix} 1 & 0 & 0 \\ -2 & 1 & 0 \\ 0 & 0 & 1 \end{pmatrix}$, $\begin{pmatrix} 1 & 0 & 0 \\ 0 & 1 & 0 \\ 0 & 1 & 1 \end{pmatrix}$, $\begin{pmatrix} 1 & 0 & -1 \\ 0 & 1 & 0 \\ 0 & 0 & 1 \end{pmatrix}$;

$\begin{pmatrix} 3 & -1 & -1 \\ -2 & 1 & 0 \\ -2 & 1 & 1 \end{pmatrix}$.

5. $\begin{pmatrix} 2 & -2 & 0 \\ 0 & 2 & 0 \\ 0 & 0 & 2 \end{pmatrix}$.

6. $-(a-b)(b-c)(c-a)(ab+bc+ca)$; $x = \dfrac{-abc}{ab+bc+ca}$.

7. (i) $\begin{pmatrix} a & b & c \\ 0 & a & b \\ 0 & 0 & a \end{pmatrix}$.

8. $a=2$, $b=\frac{1}{4}$, $c=-\frac{1}{64}$ or $a=-2$, $b=-\frac{1}{4}$, $c=\frac{1}{64}$.

Chapter 22

Qu.1 (i) $\frac{1}{6}$; (ii) $\frac{1}{4}$; (iii) $\frac{3}{4}$.

Qu.3 (i) $\frac{1}{6}$; (ii) $\frac{1}{3}$; (iii) $\frac{1}{2}$.

Qu.4 (i) $\frac{2}{9}$; (ii) $\frac{1}{3}$.

Qu.5 $\frac{3}{5}$.

Qu.6 A and C.

Qu.7 $\frac{5}{12}$.

Qu.9 $\frac{19}{64}$.

Qu.10 (i) $\frac{11}{120}$; (ii) $\frac{4}{25}$.

Qu.11 $\frac{19}{66}$.

Qu.12 $\frac{29}{72}$.

Exercise 22a

1. (i) $\frac{1}{4}$; (ii) $\frac{3}{4}$; (iii) $\frac{1}{13}$; (iv) $\frac{4}{13}$; (v) $\frac{9}{13}$; (vi) $\frac{51}{52}$.

2. (i) $\frac{1}{12}$; (ii) $\frac{5}{18}$; (iii) $\frac{1}{6}$; (iv) $\frac{11}{36}$; (v) $\frac{3}{4}$.

3. $\frac{1}{32}$, $\frac{5}{32}$, $\frac{10}{32}$, $\frac{10}{32}$, $\frac{5}{32}$, $\frac{1}{32}$.

4. (i) 90; (ii) $\frac{2}{15}$; (iii) $\frac{1}{2}$; (iv) no difference.

5. $\frac{1}{78}$.

6. (i) $\frac{1}{15}$; (ii) $\frac{2}{5}$.

7. (i) $^{26}C_{13}$; (ii) $^6C_3 \times {}^{20}C_{10}$.

8. 0·48.

9. $\dfrac{^kC_r \times {}^{n-k}C_{m-r}}{^nC_m} \left(= \dfrac{^{n-m}C_{k-r} \times {}^mC_r}{^nC_k} \right)$.

10. 0·84.

11. $\frac{15}{16}$.

12. (i) $\left(\frac{5}{6}\right)^2 \cdot \frac{1}{6}$; (ii) $\frac{1}{6} + \left(\frac{5}{6}\right)^2 \cdot \frac{1}{6} + \left(\frac{5}{6}\right)^4 \cdot \frac{1}{6} = \frac{2821}{7776}$; $\frac{6}{11}$.

14. $\frac{36}{91}$, $\frac{30}{91}$, $\frac{25}{91}$.

15. (i) Each party has at least one supporter among the four men.
(ii) 0·784; (iii) 0·216.

16. (i) $\frac{7}{32}$; (ii) $\frac{9}{16}$.

17. (i) $\frac{11}{20}$; (ii) $\frac{6}{11}$.

18. $\dfrac{1}{n+1}$.

19. 0·16, 0·44, 0·34, 0·06.

20. $\frac{1}{100}$.

21. (i) $\frac{511}{512}$; (ii) $\frac{7}{8}$.

22. (i) $\frac{1}{3}$; (ii) $\frac{1}{3}$, $\frac{2}{5}$, $\frac{1}{5}$, $\frac{1}{15}$.

23. (i) (a) $\frac{1}{2}$; (b) $\frac{1}{3}$; (c) $\frac{3}{8}$.

25. $\frac{4}{5}$.

Qu.13 (i) 0; (ii) 1.

Qu.14 (i) $\frac{3}{10}$; (ii) $\frac{1}{3}$; (iii) $\frac{4}{7}$.

Qu.15 (i) $\frac{1}{13}$; (ii) $\frac{1}{13}$.

Exercise 22b

1. (i) $\frac{5}{12}$; (ii) $\frac{1}{2}$; (iii) $\frac{1}{3}$; (iv) $\frac{1}{2}$; (v) $\frac{3}{5}$; (vi) $\frac{2}{5}$; (vii) $\frac{2}{3}$; (viii) $\frac{4}{7}$.

2. (i) $\frac{1}{4}$; (ii) $\frac{1}{3}$; (iii) $\frac{1}{2}$.

3. (i) $\frac{2}{5}$; (ii) $\frac{5}{11}$.

4. $\frac{6}{7}$.

6. $P(A).P(B) = 0$, i.e. at least one of A and B cannot happen.

7. False.

8. (i) False (consider $A \cap B = \emptyset$; (ii) true; (iii) true.

10. (i) $(1-p)^2$; (ii) $\dfrac{1-p}{2-p}$; tends to 0; tends to $\frac{1}{2}$; p as small as possible.

11. (i) $\frac{6}{11}$; (ii) $\frac{671}{1296}$; (iii) $\frac{125}{671}$.

Qu.16 (i) $\frac{2}{23}$; (ii) $\frac{4}{37}$.

Qu.17 $\frac{4}{7}$.

Qu.18 $\frac{4}{87}$.

Exercise 22c

1. (i) $\frac{7}{100}$; (ii) $\frac{3}{7}$; (iii) $\frac{19}{31}$.

2. (i) $\frac{128}{209}$; (ii) $\frac{16}{97}$.

3. (i) $\frac{2}{3}$; (ii) $\frac{1}{5}$.

4. (i) $\frac{3}{4}$; (ii) $\frac{27}{28}$.

5. $\frac{10}{43}$.

6. (i) $\frac{2}{5}$, $\frac{4}{15}$; (ii) $\frac{4}{37}$.

Miscellaneous exercise 22

1. (a) (i) $\frac{1}{30}$; (ii) $\frac{1}{5}$; (iii) $\frac{3}{10}$; (b) (i) $\frac{1}{8}$; (ii) $\frac{5}{16}$.

2. (a) $\frac{9}{20}$; (b) (i) $\frac{1}{100}$; (ii) $\frac{7}{400}$.

3. (i) $\frac{7}{216}$; (ii) $\frac{11}{36}$; $\frac{1}{8}$; $\frac{7}{8}$.

4. (a) $\frac{1}{6}$; (b) $\frac{11}{20}$; (c) $\frac{73}{400}$.

5. $\frac{5}{24}$, $\frac{1}{24}$, $\frac{5}{6}$, $\frac{1}{12}$; not independent and not mutually exclusive.

6. (i) $\beta + \frac{1}{5}(\alpha - \beta)(\alpha + 4\beta)$.

7. 0·8125, 0·25.

8. (a) $\frac{1}{6}$; (b) $\frac{7}{66}$; (c) $\frac{19}{33}$; (d) $\frac{2}{11}$; (e) $\frac{2}{11}$.

9. (a) $1 - \left(\frac{5}{6}\right)^7$; (b) (i) $\frac{5}{17}$; (ii) $\frac{4}{5}$; (iii) $\frac{1}{17}$; $\frac{23}{51}$.

10. (a) $\frac{1}{221}$; (b) $\frac{32}{221}$; (c) $\frac{13}{17}$; $\frac{25}{33}$.

11. Put $n=1$ in probabilities given.

12. (a) $\frac{105}{286}$; (b) $\frac{3}{5}$, $\frac{4}{5}$.

Index